Lecture Notes in Civil Engineering

Volume 587

Lecture Notes in Civil Engineering (LNCE) publishes the latest developments in Civil Engineering—quickly, informally and in top quality. Though original research reported in proceedings and post-proceedings represents the core of LNCE, edited volumes of exceptionally high quality and interest may also be considered for publication. Volumes published in LNCE embrace all aspects and subfields of, as well as new challenges in, Civil Engineering. Topics in the series include:

- Construction and Structural Mechanics
- Building Materials
- Concrete, Steel and Timber Structures
- Geotechnical Engineering
- Earthquake Engineering
- Coastal Engineering
- Ocean and Offshore Engineering; Ships and Floating Structures
- Hydraulics, Hydrology and Water Resources Engineering
- Environmental Engineering and Sustainability
- Structural Health and Monitoring
- Surveying and Geographical Information Systems
- Indoor Environments
- Transportation and Traffic
- Risk Analysis
- Safety and Security

To submit a proposal or request further information, please contact the appropriate Springer Editor:

- Pierpaolo Riva at pierpaolo.riva@springer.com (Europe and Americas);
- Swati Meherishi at swati.meherishi@springer.com (Asia—except China, Australia, and New Zealand);
- Wayne Hu at wayne.hu@springer.com (China).

All books in the series now indexed by Scopus and EI Compendex database!

Khalid Al Marri · Farzana Asad Mir ·
Ahmed Awad · Ahmed Abubakar
Editors

BUiD Doctoral Research Conference 2024

Multidisciplinary Studies

Springer

Editors
Khalid Al Marri
The British University in Dubai
Dubai, UAE

Farzana Asad Mir
The British University in Dubai
Dubai, UAE

Ahmed Awad
The British University in Dubai
Dubai, UAE

Ahmed Abubakar
The British University in Dubai
Dubai, UAE

ISSN 2366-2557 ISSN 2366-2565 (electronic)
Lecture Notes in Civil Engineering
ISBN 978-3-031-84370-9 ISBN 978-3-031-84371-6 (eBook)
https://doi.org/10.1007/978-3-031-84371-6

This Springer imprint is published by the registered company Springer Nature Switzerland AG
The registered company address is: Gewerbestrasse 11, 6330 Cham, Switzerland

Preface

Building on our previous series of successful research conference proceedings, this issue of the British University in Dubai's Doctoral Research Conference (BDRC) proceedings highlights research that connects theory with practice, addressing the global need for active, interdisciplinary dialogue. The selected articles showcase a diverse range of innovative ideas from student researchers across various regions, including the Middle East, North Africa, and Europe. This diversity highlights the benefits of cross-cultural and multidisciplinary collaboration in creating novel solutions for diverse applications and industries.

Since its inception in 2015, the British University in Dubai's Doctoral Research Conference (BDRC) has been held annually, offering a platform for doctoral scholars to showcase their research. The 8th BUiD Doctoral Research Conference in 2024 furthered this tradition by offering a platform where emerging researchers could share their work and insights. With a multidisciplinary focus, the conference embraced topics across education, business, law, engineering, sustainable design, and innovation, aiming to gather a diverse assembly of researchers and educators to exchange the latest trends, discoveries, and practical applications in these fields.

The featured papers address a range of significant topics across multiple disciplines. For example, research in education, humanities, and social sciences track investigates the effects of distance learning on primary students with specific learning disorders. This area also explores how social and cultural factors influence educational technology, with an emphasis on inclusive, accessible, and equitable approaches for diverse communities. In engineering, information technology, and sustainability track, researchers delve into green innovation and digital transformation, assessing ways to integrate sustainable practices into technological advancements to tackle environmental challenges. In business, management and law track, studies examine key issues in organizational performance, customer engagement, technology adoption and project management, particularly in the UAE.

The British University in Dubai takes pride in the contributions of these researchers, especially in addressing Middle Eastern challenges and broader global issues impacted by the region. Each paper has undergone a comprehensive review process, including multiple iterations of feedback, to ensure the highest quality and relevance. The issue presents forty-one exemplary papers from the BDRC 2024 Conference, chosen for their merit and the valuable insights they bring to their respective fields.

We extend our gratitude to all authors who contributed to this special issue and to the broader community that supports emerging scholars in their quest to make meaningful impacts.

Dubai, UAE

Khalid Al Marri
Farzana Asad Mir
Ahmed Awad
Ahmed Abubakar

Contents

A Comprehensive Study on Empowering High School Students for Career Success in a Government School in Sharjah

Mona Abumady

Abstract This qualitative research examines the factors that influence career empowerment among high school students in a government institution in Sharjah, UAE. It evaluates the efficacy of existing career guidance programs, the extent of students' awareness of career pathways, and the influence of educational institutions on their aspirations. The study proposes strategies to improve students' career readiness by identifying challenges in current guidance practices and conducting document analysis, interviews, and focus groups. The significance of career counseling, skill development, and informed decision-making in enabling students to select suitable universities and prospective professions is underscored by the results. The research endeavors to offer policymakers, educators, and stakeholders actionable insights to enhance career guidance programs, thereby promoting more informed career choices and self-efficacy among students.

Keywords Career empowerment · University selection · Career guidance · Career development · Career counseling · Student aspirations · Educational support · Career challenges · Career readiness · High school students

1 Introduction

1.1 Background of the Study

Upon completion of high school, students are anticipated to possess the necessary abilities to pursue further education at the university then transition into the workforce.

The secondary school years mark the shift from the protected and collaborative environment of elementary school to the independence and accountability of higher

M. Abumady (✉)
Faculty of Education, The British University in Dubai, Block 10 & 11, 1st and 2nd floor, Dubai International Academic City, 345015 Dubai, UAE
e-mail: 23000637@student.buid.ac.ae

K. Al Marri et al. (eds.), *BUiD Doctoral Research Conference 2024*, Lecture Notes in Civil Engineering 587, https://doi.org/10.1007/978-3-031-84371-6_1

education and then employment (Watts and Kidd 2000, p. 488). There is agreement that high school prepares students for college. Career advising programs help students understand and strengthen their skills (Ombaba et al. 2014; Ajufo 2013). The action of decision-making is complicated and unique for every students, it is influenced by cognitive factors and their social surroundings (Braza and Guillo 2015; Durosaro et al. 2012). Since it can change a person's life, this decision must be carefully considered (Eyo 2011, p. 328). Career planning education and skills development are needed to get a good job and avoid career confusion (Musrifah in Ghassani and Anwar 2020). Career assistance and counselling help students explore their options (Lazarus and Ihuoma 2011, p. 52). High school course selection affects post-graduation opportunities (Naude 2014, p. 510). Career coaching helps students understand themselves and make educational, vocational, emotional, and personal social decisions in a complicated and ever-changing world (Toyinand Gbenga 2014; Mabula 2012; Eyo et al. 2010). Guidance counseling is crucial to education, hence its goals are inseparable from higher education. Worldwide, several studies have focused on guiding and counseling. In South Africa, Mudhovozi and Chireshe (2012) studied university choice and job path variables. Sultana and Watts (2006) conducted a similar investigation in Middle East and North Africa. Lazarus and Chinwe (2011) studied special-needs adolescents in Nigeria. Similar studies have conducted in UAE. According to **Dino Varkey**, Group Chief Executive Officer of GEMS Education, added: "The UAE's spirit of innovation continues to motivate and inspire the next generation to take up careers in wide-ranging fields. Education has always been one of the UAE's top priorities, and it is incredibly encouraging to see young students displaying maturity and determination beyond their years to fulfil their dreams. Exposing them to broader opportunities and offering support and guidance throughout their student journey will have an enormously positive impact on their personal and professional development." The educational environment is changing quickly, with a greater focus on preparing high school students for successful professions. A thorough investigation is being conducted on how to prepare high school students for suitable university choice and job success in a government school in Sharjah.

1.2 Problem Statement

The university choices of teens in the future have consistently been the primary subject of investigation by scholars (Pâni Soara et al. 2013). Recent studies in the field of career development have focused on the investigation of careers as an important topic (Jiang et al. 2019). It is essential to identify the obstacles and constraints that limit the professional achievement of high school students. Insufficient college and career preparation frequently hinders students from smoothly transitioning into higher education and the job market (Loera et al. 2013). Students who fail to make appropriate decisions may experience sadness and face societal disapproval (Maluwa-Banda 1998, p. 294). Many students progress through this stage without making a professional choice since they are often uncertain and apply for higher education without a clear understanding of their desired field of study (Manuel and Asuquo 2009, p. 135). Lack of assistance for students at this time results in dissatisfaction (Alika 2010, p. 538). For instance, high school graduates in the United States are deficient in the information and skills required for success in college and the current economy (Carnevale et al. 2010; Loera et al. 2013).

1.3 Significance and Rationale

An investigation on the importance of empowering high school students to achieve career success contributes to the wider discussion on education. This section provides a clear explanation of the reasons for conducting the study, highlighting its possible influence on students, educational institutions, and the wider community. Upon thorough examination of the available scholarships, the researcher has discovered a limited number of studies pertaining to this subject in the United Arab Emirates, particularly for high school students attending government schools. Moreover, the subject holds personal meaning for the researcher due to their role in a career development program for high school students. The program's objective is to inspire and support students in pursuing their desired career paths, starting with their university selection process.

1.4 Purpose of the Study

The main goal of this study is to analyze and understand the elements that impact the career aspirations of high school students and propose methods to empower them in attaining their desired professions. The study will be guided by the following objectives, which will form the basic framework of the study.

1. Investigate how the current level of awareness and understanding among UAE high school students concerning various career paths directly shapes their empowerment in choosing their university and then their career paths.
2. Discover the initiatives implemented by educational institutions in UAE to promote knowledge and offer guidance to high school students in selecting their future career path.
3. Analyze the various factors that contribute to the existing challenges and inequalities in school career guidance programs, and develop effective strategies to improve these obstacles in order to promote overall empowerment.

1.5 Research Questions

Main Research Question

What factors contribute to the effectiveness of empowering high school students in choosing their college and achieving their future careers in UAE?

Sub-Research Questions

1. How does the current state of high school students' awareness and understanding of various career paths influence their college choice and career empowerment?
2. How do schools facilitate awareness and offer guidance for a diverse range of career paths?
3. What challenges do current school career guidance programs face, leading to gaps in their effectiveness?

2 Literature Review

2.1 Conceptual Analysis

Various factors contribute to the empowerment of high school students in achieving their future careers. These factors encompass a wide range of influences, conditions, and experiences that affect students' capacity to develop the skills, knowledge, confidence, and opportunities required for success in their chosen university and career paths. In the current study three conceptual strands have been discussed.

Career Exploration: Giving students detailed information about career possibilities, sectors, and employment responsibilities helps them make future decisions. Future exploration exercises help high school students discover their passions and abilities, guiding their college and future choices. Chen et al. (2021) support Gottfredson's theory (1981) asserting the importance of career exploration as a crucial phase of preparation, during which high school students have the opportunity to gain self-awareness and investigate potential areas of study within the framework of the NCEE (New College Entrance Examination). Liu and Helwig (2020), verify the identical concept. According to Gross-Spector and Cinamon (2018), career exploration is a cognitive or physical process that involves gathering information, self-reflection, and considering one's environment in order to achieve career objectives. According to experts, career exploration should prioritize activities such as gathering information, understanding oneself, and understanding the environment (Blustein 1989).

Educational Guidance and Support: Students need great educational guidance, counseling, and mentorship to choose a college. School counselors, teachers, and mentors help students comprehend academic requirements, research college possibilities, and create career-focused academic plans. According to Stanton-Salazar (1997, 2001, 2011), high school teachers and counselors have an important role as institutional agents and mentors in providing students with knowledge and information regarding the process of choosing a college. The term "college and career readiness" has gained widespread usage in response to recent efforts to promote it. However, its interpretation varies and it is often narrowly defined with a focus on college preparation (Stone and Alfeld 2006). College and career readiness encompasses four main components: cognitive strategy, content understanding, transition knowledge and abilities, and learning skills and approaches (Conley 2012).

Skill Development: Students need critical thinking, problem-solving, communication, and teamwork skills for college and beyond. Workshops, extracurriculars, and real-world experiences prepare students for college and job. Students' preparation depends on their mastery of the needed skills and information, according to Conley (2012). Lewis and Stone (2012) describe college and career readiness as high school graduates' employability, academic knowledge, and professional abilities. Achieve (2016) claimed that students need professional readiness, employability, and technical skills to succeed in their chosen fields. Employability skills include personal responsibility, critical thinking, and technology, according to Fletcher et al. (2018). Critical thinking involves creativity and problem-solving, while personal responsibility involves accountability, honesty, and self-control. Technological talents require creativity and information management. The Partnership for 21st Century Skills (2011), Secretary's Commission on Achieving Necessary Skills (1991), and Stone and Lewis (2012) support these criteria. However, strong academic background is not enough to succeed in the workforce (Stone and Lewis 2012).

2.2 Theoretical Framework

The current study employs grounded theory as its research methodology. Grounded theory is used to investigate the factors that influence secondary school students' ability to secure future careers. Grounded Theory is a qualitative research approach that entails methodical gathering and examination of data in order to extract patterns, concepts, and categories that arise from the information. Grounded theory is an approach that produces theory by collecting and analyzing data in a methodical, rigorous, and continuous way (Glaser 1978). This enables researchers to develop hypotheses that are based on the experiences and viewpoints of the participants. Charmaz (2006) defines grounded theory as a collection of methodologies "consisting of a flexible yet systematic guideline for obtaining and examining qualitative data in order to develop ideas that are based on the data itself". Glaser and Strauss (1979), introduced this form of inductive thematic analysis in their work "Discovery of Grounded Theory" (1967). Grounded theory requires data collection and coding analysis (Glaser and Strauss 1967; Charmaz 2006). Coding helps researchers turn raw data into a solid theory (Holton 2010). According to Saldaña (2016), coding involves naming and labeling properties, data and categories.

2.3 *Review of Related Literature*

The objective of this study was to figure out the factors that impacted the selection of university and professional paths among high school students in the UAE. According to Williams and Moser (2019), coding is a crucial structural operation in grounded theory research that allows data analysis and subsequent actions to achieve research goals. The coding in the theoretical framework involves organizing the literature into categories or subheadings based on the researcher's study questions. These subheadings serve as codes that categorize different aspects of the literature related to high school students' career paths in the UAE. The subheadings include:

1. High School Students' Awareness and Career Empowerment
2. School Strategies for Career Guidance
3. Challenges in School Career Guidance Programs.

2.3.1 High School Students' Awareness and Career Empowerment

Several schools lack regular systems for providing students with college-related information, forcing students to independently navigate the college information landscape (Bell et al. 2009; Brown et al. 2016; Bryan et al. 2011). To ensure that all students have the necessary resources to make informed decisions about college, it is important to have both social capital (establishing connections between schools and families) and human capital (developing college knowledge and infrastructure within high schools). This does not require a large financial investment (Simmons 2011; Plank and Jordin 2001).

2.3.2 School Strategies for Career Guidance

Schools are essential for career preparation. High school programs help students choose colleges and jobs (Baloch and Shah 2014, p. 547). If students have too many possibilities or haven't decided, career guidance programs help. It also helps individuals recognize their abilities to improve job prospects. Without career guidance, students may take unsuitable courses (Kimiti and Mwova 2012, p. 366). Also, career guidance teachers should encourage broad career goals that reflect students' interests and talents and extend job options (Mghweno et al. 2014, p. 13). According to Mulhern (2019), school counselors directly impact student educational success, including high school graduation, college attendance, selectivity, and persistence, by offering increased information and tailored support. A UAE research study found that students may be clear about their professional aspirations and not scared to dream big, according to Marketa Simkova, Partner, Head of People and Change at KPMG Lower Gulf. However, she said that data suggests students need more career assistance to properly choose their university and their career path after finish high school. High school special programs can empower students with fundamental skills, guidance on university selection, and career direction.

2.3.3 Challenges in School Career Guidance Programs

Lapan et al. (2003, p. 329) state that the transition from high school to university or the job is one of the most difficult developmental challenges for teenagers, therefore schools play a significant role in leading students toward these careers. Educational career guidance programs encounter various challenges, such as limited accessibility, outdated information, insufficient collaboration, insufficient resources, and cultural and societal influences.

2.4 Situating the Current Study

Upon thorough examination of the current body of academic literature, the researcher has obtained strong confirmation that there is a shortage of studies investigating the elements that contribute to the efficacy of empowering high school students in the United Arab Emirates (UAE) to make informed decisions regarding college selection and achieve success in their future careers. Moreover, the current body of research fails to address the determinants that impact the decision-making process of high school students, as it predominantly centers on the sole aspect of university entrance examinations. Furthermore, there is a lack of comparable research undertaken specifically in government schools in Sharjah, as well as in the broader context of the United Arab Emirates. Additionally, there is an absence of qualitative research that investigates the viewpoints of students and instructors regarding the efficacy of implementing programs aimed at improving college selection and career empowerment. Conducting the study is essential for gaining a comprehensive picture of the advantages and difficulties experienced by instructors and students. Hence, it is imperative to carry out this study and contribute to the current body of knowledge.

3 Methodology

3.1 Research Approach

This study employs a qualitative research approach to investigate the viewpoints of high school students regarding the factors that influence their career aspirations. The study analyzes and comprehends these factors and suggests ways to help students achieve their career goals. The qualitative study's approach allows the researcher to directly communicate with individuals and evaluate their behaviour and activities in their setting (Creswell 2014). Creswell (2009) says qualitative research shows how people value social or human conditions. Teachers and students' thoughts and experiences are best understood using qualitative methodologies in educational research. Shah and Corley (2006) claim that qualitative research uses participants' natural environments to promote in-depth discusses and follow-ups to foster open reflection and expression of thoughts and viewpoints. Social concerns are better understood through qualitative study than quantitative data analysis (Silverman 2013). Qualitative research reveals how people construct, interpret, and evaluate their worlds. To comprehend individuals, qualitative researchers employ qualitative methods (Merriam 2009, p. 5).

3.2 Research Paradigm

Interpretivism is applied in this investigation. According to Guba and Lincoln (1989), this paradigm is used to explore experts' personal lives and understand social knowledge formation through discourse. Interpretivism emphasizes seeing things from the subject's perspective rather than the researcher's (Bogdan and Biklen 1998). According to Kivunja and Kuyini (2017), the interpretivist paradigm seeks to understand and explain the subject's perception of the circumstance by gaining deep insight into their ideas. Interpretivists examine the world from multiple perspectives, according to McQueen (2002, p. 16). To capture a complete perspective, they choose people with unique reality interpretations. My planned study will use interpretivism and a qualitative approach. Research implications will be considered during reflection. My research goal is to fully grasp my subjects' origins, beliefs, and experiences (Cohen and Manion 1994). As a researcher, my objective is to comprehensively comprehend the world of my participants (Cohen and Manion 1994) by acquiring profound understanding of their origins, beliefs, and experiences (Cresswell 2003; Yanow and Schwartz-She 2011).

3.3 Data Collection Plan

3.3.1 Participants and Sampling

The present study is situated in a governmental educational institution in the city of Sharjah, United Arab Emirates. The school adheres to the standard curriculum set by the national government and provides education to students in grades 5–12 (cycle 2 and 3). The overall student population at the school is approximately 900. The selection of the study's context is based on the researcher's convenience, as they have been educating high school students there. The present study included fifteen grade 12 students from the academic year 2023–2024, together with the school's career counselor. The chosen cohort of fifteen student participants are currently enrolled in the twelfth grade. The process of data collecting will begin by obtaining consent papers from individuals who have willingly accepted to participate in the study. The concent forms will provide a clear description of the research's aim, objectives, procedures to protect confidentiality, and the participants' right to withdraw at any point. The participants' identities will be guaranteed anonymity and privacy before the research is conducted. Legard et al. (2003) highlight the importance of establishing a strong relationship with the participants by developing a positive rapport with them. Before collecting data, the researcher should obtain the school principal's consent on the consent form as well. According to Ann (2013), the initial stage of conducting an interview is of great importance as it requires the researcher to make the participants feel comfortable.

3.3.2 Research Instruments

Merriam (2002) states that the selection of research instruments and data collecting procedures is based on the primary research questions of the study. The researcher chooses the best sources to address these questions. The data collection procedure will include three methodologies: interviews, focus groups, and document analysis. Interviews will offer comprehensive insights into each person's perspective on professional empowerment. Focus groups will enable participants to engage in interactive discussions. Additionally, relevant documents and materials will be analyzed to gather research information. To meet the study's goals, the researcher interviewed to get the most convenient information. An interview, as described by DeMarrais (2004), is a structured interaction between a researcher and participants that centers around topics relevant to a research project. Kvale (1996, p. 1) defines qualitative research interviews as a method that aims to comprehend the respondents' perspectives, reveal the significance of their experiences, and discover their lived reality before scientific interpretations. Brinkmann and Kvale (2009, p. 17) define interviewing as an interactive procedure in which knowledge is generated through the relationship between the interviewer and interviewee. Interviews offer a chance to thoroughly examine individual viewpoints, experiences, and insights around career empowerment. The focus group discussions will consist of small groups of high school students participating in guided dialogues on topics such as career readiness, university selection, encountered obstacles, and methods for personal growth. This methodology promotes engagement, sharing of ideas, and collaborative investigation of topics and problems. Both interviews and focus group discussions will be recorded using audio technology, with the participants' permission, to ensure the precise capture of their comments. The obtained data will be handled with rigorous confidentiality and anonymity protocols to safeguard the identities of participants and maintain ethical standards during the research procedure.

As part of our document analysis, we will examine the document titled "Empowering High School Students to Achieve their Dream Career," authored by Khaled Bani Aiyesh and Seamus O'Shea (2023), and published by Emirates School Establishment. This paper serves as a comprehensive and integrated approach designed to assist students in attaining their future career goals.

3.4 Study Piloting

To ensure instrument, data, and result validity, piloting the research instrument is crucial (Majid et al. 2017). Pilot studies are useful for preparing for full-scale research regardless of paradigm (Tashakkori and Teddlie 2003). Because the pilot study informs and provides feedback to the final study, pilot study help construct the research design (Prescott and Soeken 1989). Before the final study, the researcher could modify the research approach based on this feedback. Piloting research instruments helps discover instrument and design flaws and test concerns (Van Teijlingen and Hundley 2002). Piloting also reveals interview design issues. This helps make necessary changes before research (Kvale 2007). In Castillo-Montoya's (2016) pilot study, interview techniques were improved by piloting. It can reveal interview design issues that require considerable research changes (Kvale 2007). According to Teijlingen et al. (2002), researchers typically declare that they learned from a pilot study and made appropriate revisions without providing details.

3.5 Data Analysis

Hilal and Alabri (2013) define qualitative data analysis as the systematic process of organizing, structuring, and attributing significance to a large amount of acquired data. Such a technique is not a simple task. Hilal and Alabri (2013) state that qualitative data analysis aims to explore the connection between categories and themes of data in order to enhance comprehension of the phenomena. Throughout the data analysis, researchers must sort out the data (Guest et al. 2012), which involves selectively focusing on certain aspects of the data while dismissing others. In qualitative research, this approach aims to condense data into a limited number of themes (Creswell 2013). Hilal and Alabri (2013) state that the coding procedure is crucial for assessing qualitative data, as these data are primarily in text format. Codes are used as markers or identifiers to assign specific meanings to the descriptive or inferential data collected during a study. Guest, MacQueen and Namey (2012) state that in the data analysis phase, researchers must employ filtering techniques to selectively retain the most valuable data while discarding irrelevant portions. Furthermore, the current study will utilize thematic content analysis to analyze the acquired data. The researcher will analyze qualitative data with NVivo. This involves collecting, recording, transcribing, and coding data, then organizing and categorizing it by similar themes or concepts. Hilal (2013) states that NVivo, a qualitative data analysis software program, can considerably improve study quality. The software reduces challenging tasks, giving the researcher more time to identify themes and develop conclusions. NVivo is also ideal for team researchers because it integrates individual contributions into a single project. Welsh (2002) claims that this software eliminates manual analysis. Removing researchers' subjective judgments and opinions about data improves data validation.

3.6 Limitation

Price and Murnan (2004) define a study design or instrument limitation as systematic bias that the researcher could not control, which could affect the results. Study limitations include the many factors that may have affected the research process or interpretation of results. A study on improving high school students' career success has certain constraints: Limited sample size is a limitation. The study was limited to Sharjah and focused on government schools, which may have reduced its generalizability. An even larger and more broadened sample could improve the study's external validity. The researcher must prove that the selected participants represent the population under study to address generalizability (Al-Allaq 2016). Qualitative data analysis approaches like interview transcripts and focus group discussions may also be biased by researchers. Qualitative results may be more trustworthy if participants' identities are authenticated or many people code the data. The study may have had sample bias because participants were voluntary and may have had different motivations than non-participants. This may mean that the findings do not fully represent the high school student population or apply to a wider context.

3.7 Ethical Consideration

Multiple ethical concerns will be considered and implemented to maintain research ethics. All volunteers will learn the significance and rationale for the study before starting. A detailed explanation of the study's purpose, timeframe, and participant duties will follow (Maxwell 1996). All participants will also be informed of their right to leave the study at any time. All participants will get an informed consent form that explains the study's goals before the study. Participants must give informed consent before participating in a research project, according to Jelsma and Clow (2005). However, different research studies face challenges. The researcher will also explain participant confidentiality and anonymity. Overall, the researcher will follow study ethics standards and guidelines to ensure participant safety and well-being. According to (Arifin 2018), retaining participants' names and identities during data analysis, gathering, and reporting ensured their anonymity and confidentiality. The interview environment was kept private during phone calls, interviews, data processing, and findings sharing.

3.8 Trustworthiness and Authenticity

Pilot and Beck (2014) define trustworthiness as confidence in data, interpretation, and research quality methods. Studies must have protocols and techniques for readers to consider (Amankwaa 2016). Persuasion of qualitative research methods and procedures is vital, says Sandelowski (1993). Although researchers debate its definition, most experts value trustworthiness (Leung 2015). If readers trust the study, it is trustworthy (Gunawn 2015). Guba and Lincoln (1994) defined trustworthiness as credibility, reliability, confirmability, and transferability, later adding authenticity. Not all research use all strategies. As per Polit and Beck (2014), the most important is Research credibility "confidence in the study's findings". An experienced colleague will assess some data to conduct member and peer checks to verify the study's credibility (Sandelowski 1993). Dependability is data stability throughout time and study settings (Polit and Beck 2014). The researcher will ask an outside expert to check the research design tools and findings for consistency and repeatability (Shenton 2003). The researcher will describe and justify the study's ideas, design, instruments, methodology, and data analysis to ensure confirmability. Research bias is prevented by an audit trail (Shenton 2003). The researchers meet this criterion by selecting suitable study participants and providing a detailed description (Schou et al. 2011).

3.9 Expected Outcome

The expected outcomes of the current study include gaining valuable insights into the factors that contribute to empowerment, such as awareness, guidance, and support mechanisms. Additionally, challenges in existing career guidance programs will be addressed leading to enhanced career readiness among students for both their immediate career paths and university choices. This research aims to increase awareness of various career options and resources available to high school students, fostering

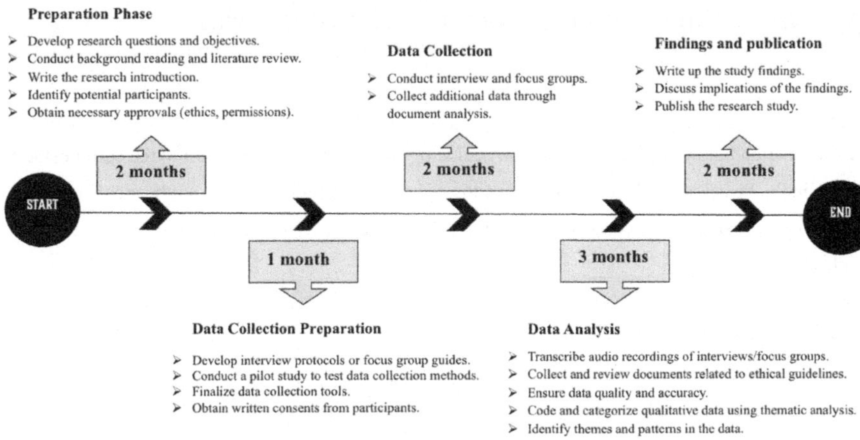

Fig. 1 Work plan timeline for a qualitative research study

collaboration between educational institutions, policymakers, and stakeholders. By promoting empowerment and self-efficacy, the study seeks to empower students to make informed career choices and navigate their academic and professional pathways effectively, including selecting appropriate universities or higher education institutions. Ultimately, the research will contribute valuable policy recommendations and new knowledge to the field of career development and education.

3.10 Workplan and Timeline Duration

A work plan timeline for a qualitative research study typically includes various stages and activities that need to be completed within specific timeframes (Fig. 1). This timeline provides a rough estimate of the time required for each step. The duration may vary depending on study complexity, resource availability, and unforeseen delays. Regularly reviewing and adjusting the timeline as needed will help ensure the successful completion of your research project.

Appendix 1 Participants' Interview Consent Form

Study Title: ____

The information below is given to assist you decide if you would like to take part in the current study or no. Kindly, be acknowledged that you have the right not to participate in the study and that you are free to withdraw at any time convenient to you without affecting your relation to the researcher (teacher).

The main purpose of the current study is to ______________________________.

Data will be collected using ______________________________________.

Kindly, be aware that you have the right to ask questions relevant to the study either prior to conducting the study or while implementing the study. Also, I would be excited to share the study findings with you once the research is completed and data is analysed.

You need to be assured of the confidentiality and anonymity of your identity and data collected from you. Your name and other personal details and responses will be confidential and will be used for the research purposes only.

There are no evident hazards, risks, or harm with regard to conducting the current study. Moreover, the expected beneficial and advantageous outcomes of your participation in the current study are the information about the expertise of learning and experiencing research methods.

Kindly, sign the consent form and be informed that you are doing this with full awareness of the nature and purpose of the research and its objectives. You will get a copy of this consent form upon signing it so that you can keep it.

Name: ____________________ Signature ________________

Date________________________

Appendix 2 Institution Principal's Consent Form

Date ______________

Mr / Mrs______________

Position_______________

Address________________

Topic: Obtaining a Permission to Conduct an Academic Research within the Institution

Dear Mr/ Mrs____________________

We, The British University of Dubai, are writing this letter to conduct research at your school. We are studying ____. The purpose of this research is to ____. The study is significant because ____.

If the researcher is permitted the required approval, the researcher would ensure full compliance and commitment to the schools' identity confidentiality and anonymity. The study will take no longer than ______________. The timing during which the study will be conducted will be subject to the school's convenience and requirement.

The data and findings of the study will remain fully confidential, and findings will be shared with you once the researcher has completed the data analysis and reporting.

Your approval to conduct the study would be highly appreciated.

Sincerely,

____

Appendix 3 Participants' Focus Group Consent Form

Dear parent(s)/guardian(s)/teacher,

I am a postgraduate student undertaking a PhD at The British University in Dubai and I am conducting research for my doctoral thesis. Here is a brief introduction to my thesis:

Please complete the form at the bottom of this letter and return it to the researcher. If you would like to receive more information about the study, please contact me at my phone number: …………

Thank you,

[insert name]

[insert programme]

The British University in Dubai.

CONSENT FORM

Student/child's name/grade/section: ……………………………

University/school: ……………………………

☐ I give permission for my child/student(s) to participate in the research being carried out by ……………… at the British University in Dubai.

☐ I do **NOT** I give permission for my child/student(s) to participate in the research being carried out by ………………… at the British University in Dubai.

Signature(s) of parent(s)/guardian(s)/teacher : ……………………………………

Date: ……………………

Appendix 4 Interview Questions for Career Counselor

1. What role do you play in guiding high school students towards their future university and career paths?
2. Can you describe the resources and tools available to students for university and career exploration and decision-making?
3. How do you assess and address the individual needs and interests of students when providing studying in university and career guidance?
4. What strategies or programs does the school have in place to promote awareness of diverse career options among students?
5. How do you collaborate with other stakeholders, such as teachers, parents, and external organizations, to support students in their career journeys?
6. Can you share any success stories or examples of how your career guidance initiatives have positively impacted students' career choices and outcomes?
7. What challenges do you face in providing effective career guidance to students, and how do you overcome these challenges?
8. How do you stay updated with current trends, developments, and opportunities in various universities to better advise students about their suitable choice?
9. Is the new program offered by ESE beneficial in assisting students in selecting the appropriate university?
10. In your opinion, what are the key components of a successful career guidance program in a high school setting?

Appendix 5 Focus Group Questions for Grade 12 Students

1. What factors influence your decisions or considerations when thinking about your future career?
2. How do you currently gather information about different universities options?
3. Can you share any experiences or challenges you've faced in exploring or deciding on a university choice and career path?
4. What support or resources do you feel would be helpful in guiding you towards your desired university study?
5. How do you perceive the role of your school in helping you prepare for your future career?
6. Have you participated in any career-related activities or programs at school? If so, what was your experience like?
7. What do you think schools can do better to support students in their universities and career exploration and preparation?
8. How do your family's expectations or opinions influence your decisions?
9. In your opinion, what skills or qualities are important for success in your future career field?

References

Aiyesh KB, O Shea S (n.d.) Bridging the gap: empowering high school students to achieve their dream career, vol 1–63

Arifin SRM (2018) Ethical considerations in qualitative study. Int J Care Sch 1(2):30–33

Brinkmann S, Kvale S (2009) Interviews—learning the craft of qualitative research interviewing, vol 24–27, pp 135–136, 245, 327

Chen H, Liu F, Wen Y, Ling L, Chen S, Ling H, Gu X (2021) Career exploration of high school students: status Quo, challenges, and coping model. Front Psychol 12. https://doi.org/10.3389/fpsyg.2021.672303

Conley DT (2012) A complete definition of college and career readiness. Educational Policy Improvement Center (NJ1)

Durosaro IA, Ajiboye SK, Olawuyi O, Adebanke NM (2012) Cognitive-behavioural therapy (CBT): a counseling intervention for post-traumatic stress disorder (PTSD) among internally displaced persons (IDPs). Eur J Soc Sci 29(2):188–193

Fletcher EC Jr, Warren NQ, Hernández-Gantes VM (2018) Preparing high school students for a changing world: college, career, and future ready learners. Career Tech Educ Res 43(1):77–97

Ghassani M, Anwar Z (2020) Meningkatkan kematangan karir siswa smp melalui pelatihan perencanaan karir. JIP (Jurnal Intervensi Psikologi) 12(2):123–138

Hilal AH, Alabri SS (2013) Using NVivo for data analysis in qualitative research. Int Interdiscip J Educ 2(2):181–186

Lazarus KU, Ihuoma C (2011) The role of guidance counsellors in the career development of adolescents and young adults with special needs. Br J Arts Soc Sci 2(1):51–62

Majid MAA, Othman M, Mohamad SF, Lim SAH, Yusof A (2017) Piloting for interviews in qualitative research: operationalization and lessons learnt. Int J Acad Res Bus Soc Sci 7(4):1073–1080

Price JH, Murnan J (2004) Research limitations and the necessity of reporting them. Am J Health Educ 35(2):66–67

Watts AG, Kidd JM (2000) Guidance in the United Kingdom: past, present and future. Br J Guid Couns 28(4):485–502

A Critical Analysis for the Assessments and Moderation Policy of a UAE Higher Education Institution

Abdulai Abukari and Razan Ashour

Abstract This paper represents a critical analysis of Assessment and Moderation Policy implemented in a UAE higher education institution. This critical analysis of the policy followed the SWOT analysis approach. It aimed to determine the effectiveness of the mentioned policy by analysing the policy horizontally. The paper employs a critical thinking discussion of the components, rational and strength areas of the policy to identify the key weaknesses and threats associated with it, drawing on the prior work experience and pertinent literature review. The policy shows the availability of strong components such as the alignment with accreditation standards, rubric implementation and documenting a blind marking process. However, an absence of clarity in selecting the moderation samples, and in formulating the rubrics results in inconsistent implementation across departments, compromise its reliability, and affects the uniformity of student evaluations. To enhance policy effectiveness, the paper recommends the development of comprehensive workshops on rubric creation, the adoption of stratified sampling for moderation, and the engagement of students in understanding assessment criteria. It emphasizes the importance of regular policy evaluation and adaptation based on faculty and student feedback. This critique aims to contribute to the discourse on educational policy evaluation by providing insights into the challenges and opportunities for enhancing assessment and moderation policies in higher education.

Keywords Assessment · Moderation · Policy · SWOT · Higher education

This paper was prepared by Razan Ashour as an assignment for the module Educational Policy module (EDU501) taught by Prof. Abdulai Abukari.

A. Abukari
Faculty of Education, The British University in Dubai, Dubai, UAE

R. Ashour (✉)
Department of Institutional Effectiveness, Higher Colleges of Technology, Dubai, UAE
e-mail: eng.razanashour@hotmail.com

K. Al Marri et al. (eds.), *BUiD Doctoral Research Conference 2024*, Lecture Notes in Civil Engineering 587, https://doi.org/10.1007/978-3-031-84371-6_2

1 Introduction

A policy document is a formal document used to provide clear guidance for specific practices within an organization, defining what is acceptable in an organization and how it should be implemented. This document will focus on critically analysing the assessment and moderation policy (Appendix 1) of a UAE higher institution to determine its effectiveness. The assessment and moderation policy document is an educational policy prepared to ensure fairness assessments gradings between the students, integrity and efficacy of student evaluations and successful implementation for the student learning experience. The approach of this policy covers all the university's summative exams in both graduate and undergraduate levels. This paper will provide a critical analysis of the assessment and moderation policy that is available in one of the United Arab Emirates (UAE) university's website. The paper will provide a comprehensive overview of how the assessment and moderation policy is conceptualized, operationalized, implemented and analysed within the institution. This policy analysis is justified by the growing significance of policies that assist in shaping the educational journey of students, and the alignment with industry expectations along with accreditation bodies, and the institutional accountability in delivering high-quality education. This research will adapt a horizontal approach where it will begin with comprehensive details investigating the policy's purpose and scope, discussing the policy statement then it will move to analysing the procedure mentioned inside the policy document. It will also highlight the significance of stakeholder engagement in policy effectiveness focusing on the variety of assessment methods for all summative assessments. The analysis will conclude with a discussion of the policy's strengths, areas for improvement, and suggestions for future policy development that can be beneficial to policy makers and higher management in educational sectors when focusing on educational experience enhancement.

Despite the Commission for Academic Accreditation (CAA) and Ministry of Education (MoE) requirements regarding having an assessment and moderation policy in higher institutions, CAA (2019), a strong analysis of such policies and its effect on the student learning experience and on the institution' enhancement still lack clarity. Many institutions are not aligning their policies with the CAA requirements; hence they do not have an assessment and moderation clear policy to follow. This study seeks to bridge the gap in understanding the need to have such a policy. The aim of the paper is to conduct a critical analysis to determine the effectiveness of the assessment and moderation policy by understanding and critique the policy components, identifying the strengths and weaknesses and recommend suggestions and areas for policy effectiveness improvements.

2 Literature Review

Goldin (2024) discussed the Human Capital Theory, where it proposes that education is an investment in human capital and is beneficial to improve skills and potential for future employment. Their theory suggests that enhancing individual productivity and efficiency shall contribute to economic prosperity. As per Deming (2022), the human capital theory states that participating in education and training is vital for increasing an individual's productivity and earning potential. This theory has driven the idea that public spending on education is a high-return investment for both individuals and society. Education-promoting policies have been proposed because of the theory's promise of faster economic growth and poverty alleviation, Goldin (2024). In the UAE, this perspective has influenced the establishment and implementation of assessment and moderation policy aimed at aligning student achievement with the country's economic and developmental goals.

However, Equity Theory, introduced by Adams (2015), appears to be more connected to the assessment and moderation policy in terms of its focus on the fair and equal assessments distribution between the students. He discussed that the equity theory emphasizes the importance of policies that do not disadvantage any student while also recognizing and accommodating diversity in background, learning styles, and talents. Integrating equity theory with the policy's purpose reveals that achieving fairness in assessment and moderation entails adhering to accreditation standards and creating a motivating atmosphere that promotes fairness and equity.

Moreover, Brownlie et al. (2024) discussed that Summative Assessment Theory evaluates student learning at the end of a course. They discussed that it is designed to provide a comprehensive summary of student knowledge and assess to what extent which learning objectives were successfully met. Summative assessments are formal graded evaluations used for certification purposes. This theory contradicts the formative assessment theory, which states that students are evaluated directly throughout the teaching and learning process.

This research found a gap in such studies specifically within the UAE and the Mena region. According to Mabunda (2023), challenges were addressed when exploring the implementation of the assessment policy in south African schools. The study highlighted the teacher's shortages, limited textbooks, and overcrowded classrooms challenges when such policy was converted into effective educational practice. The study employed a qualitative study design and the interpretivist paradigm to explore these challenges. The study received positive feedback from the teachers about the clarity and the structure parallel goals of the assessment policy ensuring fairness, consistency, and alignment with learning outcomes through effective assessment tools.

Acharya (2024) analyses English language education policies in Nepal showing the transition from grammar-based training to a skill-based approach consistent with modern pedagogical norms. This analysis is strongly related to the analysis of the assessment and moderation policy since it demonstrates the critical significance of education policies in determining assessment standards and practices. This involves a thorough study of assessment processes, including their alignment with educational goals and adaptation to changing educational settings. Furthermore, the Nepal's English language assessments identified issues such as the requirement for fair and efficient assessment methods and policies.

Furthermore, the research published by Tayyab et al. (2023) examines the impact of assessment policy framework on enhancing learning outcomes for primary school students in Pakistan. They implemented qualitative research through student interviews aiming to explore the strengths and weaknesses of the elementary education system that is following the assessment policy framework 2019. The research revealed significant challenges in its implementation including the lack of resources and support, potential consistency and standardization of assessment techniques, and limited flexibility in serving various learning requirements. These challenges emphasize the need for specific policies that promote efficient assessment procedures and guarantee teachers have the necessary support and resources to implement and customize these approaches to their students' needs.

Given the broad discussion on assessment and moderation policies in educational research, there is a significant clear gap in the analysis of these policies in the UAE's higher education sector. The available literature is mostly concerned with school-level education rather than higher education. Additionally, the UAE's unique cultural, economic, and educational dynamics that characterize its fast-changing higher education setting are not considered within the research that is conducted. This research gap provides an opportunity to critically evaluate the effectiveness of the given assessment and moderation policy, thereby informing policy improvements and pedagogical practices in a culturally diverse educational setting.

3 SWOT Analysis

Young and Diem (2018) discussed the use of critical policy analysis in the educational sector by focusing on providing a complete and comprehensive theoretical framework that helps in understanding its multiple nature, ranging from issue identification and definition to policy design, implementation, and assessment. In this paper, the assessment and moderation policy of a UAEs' higher institution is critically analysed. To assist decision-makers in implementing strategies, the SWOT technique (Strength, Weaknesses, Opportunities and Threads) provides an analytical framework that assesses both external opportunities and threats in addition to internal strengths and weaknesses. The SWOT analysis approach was followed by Chen et al. (2014) where they conducted a SWOT analysis to review and compare the renewable energy policies. Similarly, Irfan et al. (2019) utilized SWOT analysis to examine Pakistan's National Power Policy during the country's severe energy crisis.

3.1 Policy Rationale

The policy rationale is a critical component that shows its significance and necessity. The rationale of this policy successfully states a crucial role of ensuring that an institution conducts the assessments with careful attention to quality assurance measures, specifically fairness, consistency, and alignment with learning outcomes. Furthermore, the institution's compliance with the standards established by the CAA strengthens this commitment. The policy also showed a clear goal of the policy development initiative and associated procedures to provide written guidance for faculty, staff, and students on assessment methodologies, tools, and grading. This ensures academic standards comparability and consistency with learning, teaching, and assessment approaches.

3.2 Policy Components

The policy was prepared as required by the CAA standards, specifically Section 3.10.3 and 3.10.4. The policy aims to conduct fair and consistent assessments grading across all students in the institution. It consists of the purpose, the scope, the policy statement and the procedure.

In the discussed policy, the purpose and the scope are separated into two different sections, making it clearer and more specific. In the purpose section, the policy's aim successfully highlighted the objective and rationale of creating this assessment and moderation policy. It covers a clear definition of the policy showing the accreditation standards associated with it. It also stated the stakeholders benefiting from the policy including the faculty, staff, and students of the institution. The policy successfully stated the alignment with the accreditation bodies to provide the reader with a broader understanding of the policy. It is standard practice for the "purpose" section of an educational policy document to declare its alignment with the National and/or international organizations, as well as decisions made by the Ministry of Education and any other pertinent educational authorities, WHO (2018). The scope section provides constructive insights stating the coverage of the policy including all the summative assessments and both graduate and undergraduate level assessments. In this way, the policy provides precise information about the scope without further explanation.

The alignment with CAA standards is shown, presenting a structured approach towards enhancing the fairness and consistency of student assessments grading. It requires multiple independent individuals to mark or moderate assessments and faculty members to create and administer assessment rubrics. The Head of the Department, in partnership with the Department of quality assurance, is responsible for managing the moderation process and ensuring that all documentation is properly preserved inside a specific filing system. To boost the consistency of the policy implementation, step by step is identified in the procedure section within the policy document. However, the policy could be enhanced by including aspects that support faculty development in creating rubrics, increasing student understanding of assessment criteria, and implementing feedback systems for continual improvement.

The procedure stated in the policy gives a clear direction to the staff, faculty and students when implementing the policy. It provides a systematic approach for assuring fairness and rigor assessments gradings. It begins with faculty members submitting assessment instruments through a specific automated system for rigor revision by the head of the department and the quality assurance team. It also specifies how to choose a minimum of ten unmarked exam scripts for the moderation process and how the designated moderator will blind mark them. The process looks efficient, but it still benefits from enhancing the evaluation criteria by having a clearer rationale for the sample size chosen for moderation and on how to resolve discrepancies in marking. Additionally, the college level examination committee's role was briefly discussed as to review the moderation process and report it. Detailing this role and specifying the criteria to be used in the revision could improve the transparency and effectiveness of the procedure.

3.3 Policy Strength

The policy's alignment with the CAA standards has outlined the policy's purpose, statement, and procedure. It illustrated the commitment to following rigorous accreditation criteria, ensuring that the institution's assessment processes meet the requirements of the UAE accreditation standards. This alignment is critical for building confidence within the staff, faculty and students and maintaining the institution's reputation for excellence.

Moreover, the policy emphasis on the creation and implementation of rubrics. The rubrics are encouraged to be detailed and clear to have a transparent tool for evaluating student assessments. The rubric is also advised to be used for all assessment tools, promoting consistency and transparency in the evaluation process. As per Atkinson and Lim (2013), students appreciate knowing what is expected from them, how they will be evaluated and which area they need to improve. Nevertheless, faculty using rubrics also noticed reduction in grading time and improving in student's satisfaction, Atkinson and Lim (2013).

The focus of the policy on stating a procedure that shows the development and implementation of the moderation process adds high value to the policy. Discussing the practice of blind marking by moderators further strengthens the assessment process and ensures less bias and variability in grading. When a blind marking is introduced, students will have more confidence when receiving their grades and feedback, resulting in increased students' satisfaction.

Finally, the documentation of the assessment and moderation process helps in ensuring that all steps of the assessment are recorded and maintained. This practice provides a transparent record, facilitates ongoing review of quality assurance department and ease finding any specific script that is required during a ministry inspection or a student appeal process. The mentioned strength of the policy components creates an effective, fair, reliable, and valid evaluation system. By providing clear written guidance for all staff and faculty involved, the policy fosters a culture of fairness and inclusivity.

3.4 Policy Weaknesses

The discussed policy does not provide a clear direction on rubric development. This can result in variations in the usage of the policy across the different departments. From my previous experience working in the quality assurance department in different UAE institutions, the lack of precision reveals different interpretations and implementations of the rubrics among faculty members, compromising the consistency of student assessment. Also, as detailed by Atkinson and Lim (2013), the faculty that use rubrics have seen a reduction in grading time and in students complaining about their grades. Nevertheless, the requirement that faculty needs to follow when submitting the unmarked scripts for moderation may not fully represent the range of student responses. This could lead to oversight of broader assessment discrepancies, which would compromise the moderation process's overall reliability. Similarly, engaging students in understanding the assessment criteria or the moderation process is not covered in the discussed policy, this can affect fairness and transparency, crucial factors for trust in the assessment system.

3.5 Policy Opportunities for Improvements

To better comprehend and implement the assessment criteria, there is a call for expanding professional development opportunities for educators on how to deal with such a policy and how to implement it. From my previous experience, this was one of the major suggestions received by the faculty members. For the development of rubrics, conducting workshops could guarantee uniformity and efficacy in the creation of rubrics. These workshops could also be used as opportunities for professional development, strengthening the abilities and knowledge of assessment methods among faculty members. Whereas for the moderation requirements, specifically in large classes, using a stratified sample technique could provide a more representative selection of scripts for moderation and evaluation. This could provide faculty members with the skills and information they need to apply the assessment and moderation policy successfully. It is advised to conduct student's informational sessions on assessment and moderation criteria, this could improve student comprehension and sense of trust in the process. It is worth mentioning that I personally have received considerable feedback from the students on the need and the rationale behind having a moderation process. Once explained, students became more confident and trusting in the assessment and moderation procedures. Moreover, it is always recommended to have regular evaluation of the effectiveness of the policies and procedures implemented. This can be done through incorporating feedback mechanisms from both faculty and students and structuring a committee to look for continuous improvements through regular evaluation, revision and updating the policies.

3.6 Policy Threats

When analysing the policy and discussing its strength, specifically in its alignment with the CAA standards, a major threat that can appear is the change in the accreditation standards. A rapid adjustment will be needed, considering the current structure, the implementation process and the training for the stakeholders. Therefore, and as experienced, there will always be resistance to the change due to the increased workload or to the complexity of the new requirements.

4 Recommendations

Given the insights drawn from the SWOT analysis of the assessment and moderation policy at a UAE Higher Education Institution, it is recommended that the institution continues to refine its assessment and moderation practices by further integrating formative assessment strategies. These strategies should focus on enhancing feedback mechanisms that are clear, constructive, and timely, thereby supporting student learning and engagement throughout the course. Additionally, adopting a transparent and well-communicated moderation process is crucial in building trust and confidence among students regarding the fairness and reliability of assessments. As suggested by faculty feedback and supported by Atkinson and Lim (2013), the initiation of comprehensive workshops on rubric creation and on policy implementation requirements could standardize assessment methods. Furthermore, to ensure the accuracy of student responses during moderation, implementing a stratified sampling technique for large classes may enhance the moderation process's reliability. Considering regular reviews of its assessment policies in collaboration with both faculty and students is also recommended.

5 Conclusion

This paper critically analysed the assessment and moderation policy by adopting a horizontal SWOT analysis approach. While the policy aligns with CAA standards, it faces challenges in implementation consistency, faculty and student engagement, and adaptability to changing accreditation requirements. By implementing targeted strategies, such as professional development workshops, stratified sampling in moderation, and transparent communication with students, the policy can achieve greater efficacy and fairness. Regular evaluation and adaptation to feedback will also ensure that the policy is responsive to all stakeholders' needs and external changes in standards. This evaluation highlights areas for enhancement and potential for policies to significantly impact the quality of education when fulfilled and effectively implemented.

Appendix 1

Assessment and Moderation Policy

Purpose

It is imperative that an institution of higher learning ensures that evaluation of student performance through various assessment tools are carefully evaluated for quality assurance with respect to fairness, consistency, and alignment with learning outcomes. To that end, the institution should formulate and utilize methods for moderation and assessment of student work. CAA, the primary accrediting body of the institution, also prescribes this as one of its accreditation standards. As such, in concert with CAA standard 3.10.6, the university will develop a policy and associated procedures to provide clear, written guidance, for faculty, staff, and students, on assessment methodologies, tools, and grading, in order to ensure comparability of academic standards and consistency with the approaches to teaching, learning, and assessment.

Scope

1. All summative exams
2. All undergraduate and graduate-level exams.

Policy Statement

The university, in accordance with the CAA standard, 3.10.4 shall implement methods for the moderation and assessment of student work in which more than one individual independently marks or moderates an assessment or evaluates student performance. In addition, as per the CAA Standard 3.10.3, it is required that all faculty members develop and implement rubrics for all assessment tools.

The Head of the Department (HOD) will supervise the moderation process in coordination with the Department of QA&IE.

All documentation from the moderation process will be maintained as part of the e-course file system.

Procedure

1. For all summative examinations, all faculty members shall prepare and submit the pertinent assessment instruments on the TIPAM system for review and approval for rigor by the HOD, and completeness by QA.
2. Within one day after examination grades submission due date as per the institutional policy, faculty (the person setting and marking the assignment/exam) shall submit his/her exam to the assigned (by the HOD) moderator with a minimum sample of 10 unmarked exam scripts (especially those at grade cut-off points).
3. The faculty shall upload the sample scripts to the "moderation" e-course file folder.
4. The moderator will blind mark the papers.
5. If there is a difference in the marks sufficient to warrant a change in grade (e.g., C to B), then the faculty (the person setting and marking the assignment/exam) and moderator review the paper together and agree on a grade.
6. All blind marked examinations and a post-moderation analysis outcome (EF-Form-01) will be prepared and uploaded by the faculty to the "moderated" e-course file folder.
7. The College-level examination committee shall review the entire moderation process and submit a comprehensive report by semester.

CUCA (2021), p. 160.

References

Acharya R (2024) English language education and assessment policy in Nepal. English Language Education

Adams JS (2015) Equity theory. In: Organizational behavior, 1st edn. Routledge, pp 134–158

Atkinson D, Lim SL (2013) Improving assessment processes in higher education: student and teacher perceptions of the effectiveness of a rubric embedded in a LMS. Australas J Educ Technol 29(5)

Brownlie N, Burke K, van der Laan L (2024) Quality indicators of effective teacher-created summative assessment. Qual Assur Educ 32(1):30–45

Chen WM, Kim H, Yamaguchi H (2014) Renewable energy in eastern Asia: renewable energy policy review and comparative SWOT analysis for promoting renewable energy in Japan, South Korea, and Taiwan. Energy Policy 74:319–329

Commission for Academic Accreditation (CAA) (2019) Standards. https://www.caa.ae/Pages/Guidelines/Standards.aspx

City University College of Ajman (2021) CUCA_PPM_Nov_2021. https://www.cuca.ae/wp-content/uploads/2022/01/CUCA_PPM_Nov_2021.pdf

Deming DJ (2022) Four facts about human capital. J Econ Perspect 36(3):75–102

Goldin C (2024) Human capital. Handbook of cliometrics. Springer International Publishing, Cham, pp 353–383

Irfan M, Suleman AH, Sarwar UB, Abas N, Saleem MS (2019) SWOT analysis of energy policy 2013 of Pakistan. Eur J Eng Sci Technol 2(1):71–94

Mabunda P (2023) The implementation of the curriculum and assessment policy statement: school managers' perspectives and experiences. Perspect Educ 41(4)

National Academies of Sciences, Medicine Division, Board on Health Care Services, Board on Global Health, & Committee on Improving the Quality of Health Care Globally (2018) Crossing the global quality chasm: improving health care worldwide

Tayyab M, ul Hussan KH, Ahmad M (2023) Explore the effectiveness of assessment policy framework (Assessment Policy Framework) 2019: students' learning at elementary level in Pakistan. VFAST Trans Educ Soc Sci 11(1):142–148

Young MD, Diem S (2018) Doing critical policy analysis in education research: an emerging paradigm. In: Complementary research methods for educational leadership and policy studies, pp 79–98

Advancing Interpretability in Sequential Models Through Generative AI Rationalization Using GPT-4

Mohammed Rasol Al Saidat, Khaled Shaalan, and Suliman Yemria

Abstract In this study, we investigate the role of Generative Pre-trained Transformer 4 (GPT-4) in enhancing interpretability of sequential predictions in Natural Language Processing (NLP). Our study introduces a hybrid model that integrates traditional sequential prediction models with GPT-4, aiming to generate detailed, context-sensitive explanations for model outputs. This approach is rooted in the use of advanced transformer architectures and a specialized tokenization method that maintains semantic coherence, allowing for deep contextual analysis by GPT-4. Additionally, we devise a rationale generation algorithm that achieves a balance between succinctness and informativeness. Our experimental validation spans across various high-dimensional datasets, including financial time-series and multilingual texts, employing both qualitative and quantitative metrics to evaluate the model's performance. These metrics focus on the plausibility and consistency of the rationales, as well as the model's predictive accuracy. Preliminary results demonstrate that our approach not only enhances the accuracy of sequential predictions but also significantly improves their interpretability. This finding highlights the potential of generative AI to bridge the gap between complex AI decision-making processes. This research underscores the viability of employing generative AI to elucidate the underlying mechanisms of sequential prediction models, paving the way for more transparent AI systems.

Keywords Generative AI · Explainable AI (XAI) · Sequential predictions

This paper was prepared for Advance Natural Language Processing research.

M. R. Al Saidat (✉) · K. Shaalan · S. Yemria
Faculty of Engineering and IT, The British University in Dubai, Dubai, UAE
e-mail: mohammedrasol@gmail.com

K. Al Marri et al. (eds.), *BUiD Doctoral Research Conference 2024*, Lecture Notes in Civil Engineering 587, https://doi.org/10.1007/978-3-031-84371-6_3

1 Introduction

1.1 Research Topic

The fusion of Generative AI, particularly Generative Pre-trained Transformer 4 (GPT-4), with sequential prediction models represents a frontier in the quest for enhancing model interpretability within the domain of Natural Language Processing (NLP). The rapidly advancing capabilities of AI models underscore the imperative need for these systems not only to perform with high accuracy but also to provide insights into their decision-making processes. This research explores the synergistic potential of combining GPT-4's generative competence with sequential models to clarify the rationale behind their predictions, a venture that sits at the intersection of cutting-edge AI technology and the fundamental requirements for transparent AI systems (Lee et al. 2023; Al Saidat et al. 2024).

1.2 Research Question

Central to our investigation is the query: How can the integration of GPT-4 with sequential prediction models conclude in an enhancement of interpretability without compromising on predictive accuracy? This question guides our exploration into the mechanisms through which generative AI can be leveraged to interpret the complex decision-making processes inherent in sequential models, aiming to bridge the gap between advanced computational performance and human comprehension (Lipton 2018).

1.3 Significance of the Study

The significance of this study transcends the technical realm, touching on the ethical dimensions of AI deployment. In an era where AI's decisions can have profound implications across various sectors, the necessity for models that are not only accurate but also interpretable cannot be overstated. This research contributes to the ethical advancement of AI by promoting the development of models that uphold transparency, fostering trust and accountability in automated decision-making. The implications of this work are particularly salient in high-stakes areas such as healthcare and finance, where the rationale behind predictions can have direct consequences on human well-being and economic stability (Kar et al. 2023; Ooi et al. 2023).

1.4 Theoretical Underpinning

Our approach is firmly rooted in the principles of Explainable AI (XAI), a expanding field that seeks to make AI systems' workings transparent and understandable to their human users. By harnessing the language understanding and generation capabilities of GPT-4, this study boards on a novel methodological path to articulate complex data patterns in an intuitive manner. This aligns with the current scholarly discourse that emphasizes the importance of developing AI systems capable of providing contextually rich, comprehensible explanations for their decisions, thereby enhancing the overall trustworthiness and acceptability of AI technologies (Doshi-Velez and Kim 2017; Kar et al. 2023). To enhance the interpretability of AI models, we incorporate robust security measures similar to those detailed in the design of secure SIP-based architectures for broadband service providers, as explored by (El-Mousa and Al Saidat 2015).

2 Methodology and Approach

The core of our methodology revolves around the development and implementation of a hybrid model that seamlessly integrates the generative capabilities of GPT-4 with traditional sequential prediction models. This approach is designed to address the dual objectives of maintaining high predictive accuracy while enhancing the interpretability of the model's outputs.

2.1 Model Architecture

Our hybrid model architecture is conceptualized as a two-tier system. The first tier consists of a sequential prediction model, tailored to the specific domain of application, whether it be time-series forecasting, text generation, or another form of sequential data processing. The choice of the sequential model is predicated on its proven efficacy within the targeted domain, ensuring the model's predictive capabilities are optimized for accuracy.

The second tier integrates GPT-4 as a rationale generator, tasked with producing explanations for the predictions made by the sequential model. The integration is facilitated through a custom interface that allows GPT-4 to access the inputs, outputs, and intermediate states of the sequential model, thereby enabling it to generate contextually relevant and detailed rationales.

Figure 1 shows the architecture of the hybrid sequential prediction and rationale generation model that illustrates the approach that combines the predictive power of sequential models with the explanatory capabilities of GPT-4. At its core, the architecture is structured into two primary layers: the first layer operates with a

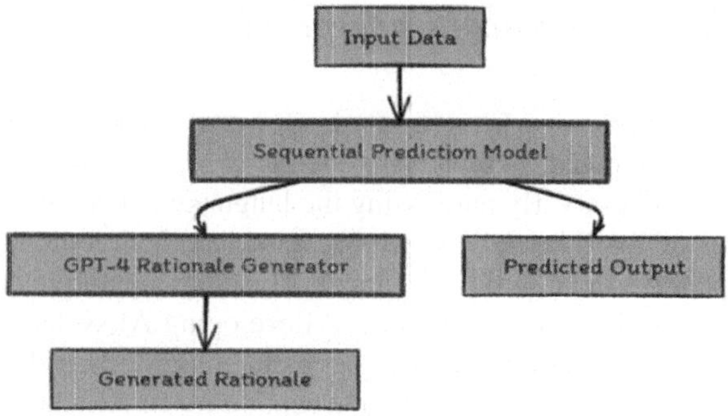

Fig. 1 Architecture of the hybrid sequential prediction and rationale generation model

domain-specific sequential prediction model, responsible for analyzing input data and generating accurate forecasts or outcomes. The second layer is where GPT-4 comes into play, functioning as a rationale generator that interprets the sequential model's predictions and provides comprehensible explanations for them. This dual-layered setup not only enhances the model's predictive accuracy but also significantly improves its interpretability by offering detailed insights into the 'why' behind its predictions, making complex data analysis more accessible and understandable to users.

2.2 *Data Processing and Model Training*

The development of our hybrid model necessitates a meticulous approach to data preprocessing, ensuring that the input data is conducive to both accurate prediction by the sequential model and effective rationale generation by GPT-4. This involves the implementation of a specialized tokenization method that retains semantic coherence across the data processing pipeline, optimizing the model's ability to discern and articulate the nuances of the data.

Model training is conducted in a staged manner, first focusing on the sequential prediction model to ensure its predictive accuracy is maximized. Subsequently, GPT-4 is fine-tuned on a dataset annotated with potential rationales, aiming to align its generative capabilities with the domain-specific requirements of the predictions it is tasked to explain. Table 1 shows the steps and parameters for both models.

The table above outlines the methodical steps and parameters critical to the preprocessing and training stages of both the sequential model and GPT-4 within our hybrid framework. The following section will explain the steps in detail.

Table 1 Preprocessing steps and model training parameters

Step/parameter	Sequential Mode	GPT-4
Data preprocessing		
Tokenization	Customized per domain specifics	Standard GPT-4 tokenizer
Data normalization	Applied (specify method, e.g., min–max scaling)	Not applicable
Sequence length adjustment	Fixed length (e.g. 512 tokens)	Dynamic, up to 1024 tokens
Training parameters		
Learning rate	0.001 (or domain-specific rate)	1e-5 (fine-tuning)
Batch size	32 (adjust based on data and compute)	8 (fine-tuning)
Epochs	10 (or based on convergence)	3–5 (fine fine-tuning)
Optimizer	Adam (or domain-specific choice)	AdamW (recommended for GPT-4)
Loss function	Cross-entropy (or domain-specific)	Cross-entropy
Regularization	L2 (specific lambda 0.01)	Dropout rate of 0.1
Additional parameters		
Fine-tuning rationale data size	Not applicable	10,000 annotated examples
Temperature (for generation)	Not applicable	0.7
Top-P (nucleus sampling)	Not applicable	0.9

2.3 Data Preprocessing

For the sequential model, tokenization is customized based on domain specifics to ensure the text data is appropriately segmented into tokens (e.g., words, subwords) that the model can effectively process. This step is vital for adapting the model to specific domain nuances, enhancing its ability to recognize and process domain-specific vocabulary accurately. For GPT-4, we use the standard tokenizer provided with the model, which is designed to handle a wide range of text inputs efficiently and is pre-optimized for performance across diverse text datasets.

Next we apply data normalization (Sequential Model), applied methods, such as Min–Max Scaling, standardize the feature range of input data, a crucial step for many machine learning algorithms to ensure that no variable dominates others due to its scale. This step is not applicable to GPT-4 as it deals with tokenized text data, where normalization in the traditional sense is not required.

The last step in this phase is sequence the length adjustment. The sequential model processes fixed-length sequences (e.g., 512 tokens) to maintain consistency in input size, which is necessary for batch processing and model stability. GPT-4, however, operates with dynamic sequence lengths, up to 1024 tokens, taking advantage of its design to handle variable-length sequences, allowing for greater flexibility in generating more detailed rationales.

2.4 *Training Parameters*

The learning rate for the sequential model is set based on domain-specific requirements, ensuring efficient convergence without overshooting minima. A lower learning rate (e.g., 1e-5) is chosen for fine-tuning GPT-4 to make smaller updates, which is crucial for adjusting the pre-trained model to our specific task without causing significant disturbances in the already learned weights.

The batch size and epochs for both models are set to balance training efficiency and computational resource constraints. Smaller batch sizes for GPT-4 fine-tuning reflect the model's complexity and the higher computational resources required. The number of epochs ensures sufficient training to achieve convergence without overfitting.

Both models use Adam optimizers, with AdamW for GPT-4 fine-tuning, which includes weight decay regularization for preventing overfitting. The Cross-Entropy loss function is standard for classification tasks, guiding both models toward accurate predictions by minimizing the difference between predicted and actual labels.

2.5 *Additional Parameters (GPT-4 Specific)*

The choice of 10,000 annotated examples for fine-tuning GPT-4 is a compromise between the need for a sufficiently large dataset to learn diverse rationales and the practical limitations of annotation efforts and computational resources.

Both of the temperature and pop-p parameters control the creativity and diversity of the generated rationales. A temperature of 0.7 and a top-p value of 0.9 are selected to ensure that the generated text is varied and innovative while remaining coherent and relevant to the input data. These settings aim to balance the generative model's exploration of the solution space with the need to produce reasonable and informative rationales.

3 Datasets

We employed two contemporary and domain-specific datasets to evaluate the efficacy of our hybrid model: the 2023 WebText2 Corpus (Aaron Gokaslan 2023) and the Financial PhraseBank 2024 (Malo et al. 2014). The 2023 WebText2 Corpus represents a diverse compilation of internet-sourced textual content, updated to reflect the linguistic and topical nuances of the year 2023. It encompasses a wide array of text types including news articles, blogs, and forums, making it a rich resource for testing generative models like GPT-4. This corpus is essential for assessing our model's capability to generate coherent and contextually appropriate rationales across various informal and formal discourse types. The dataset is openly available, ensuring that our results are reproducible and that the academic community can verify our findings. Please check the source on GitHub website: https://huggingface.co/datasets/Skylion007/openwebtext/tree/main.

Similarly, the Financial PhraseBank 2024 includes annotated financial news and reports from early 2014, offering a targeted environment for evaluating the sequential prediction aspects of our model, particularly in economic and financial contexts. Each entry in this dataset is labelled with the sentiment, which aids in supervised learning tasks such as sentiment analysis, which is crucial for the fine-tuning of GPT-4 in generating domain-specific rationales. This dataset is particularly valuable for exploring the application of interpretative models in sensitive areas such as financial forecasting, where clarity and accuracy of the model's output are paramount (*Source* Financial Data Services 2024; https://huggingface.co/datasets/financial_phrasebank). Both datasets provide a robust framework for testing the interpretability enhancements introduced by our hybrid model, facilitating a comprehensive analysis of its performance across diverse linguistic settings and specialized domains.

4 Implementation

In this section, we will consider Fig. 2 as it thoroughly outlines a seven-phase approach to integrating GPT-4 for enhancing predictive modeling. It starts with the environment setup using Google Colab, progresses through data preparation, model initialization, and fine-tuning, and culminates with comprehensive training, evaluation, and rationale generation phases. Each phase is distinctly marked to demonstrate the flow and dependencies, illustrating a structured pathway to achieving enhanced model interpretability and accuracy.

We attempted to enhance the interpretability of sequential prediction models through the integration of the Generative Pre-trained Transformer 4 (GPT-4), simulated using GPT-2 due to accessibility constraints. Our implementation was structured in a multi-phase approach, ensuring meticulous attention to model training, evaluation, and rationale generation.

Initially, we established our computational environment using Google Colab, a choice driven by its accessibility and the provision of computational resources including GPU support. Essential Python libraries such as transformers, datasets, and torch were installed to facilitate model operations and data handling. We utilized two key datasets for our experiments: the 2023 WebText2 Corpus and the Financial PhraseBank 2024. These datasets were loaded using the Hugging Face datasets library, ensuring ease of access and reproducibility. Each dataset underwent a rigorous preprocessing regimen, which included tokenization and normalization where applicable, tailored to optimize the input for both the sequential model and

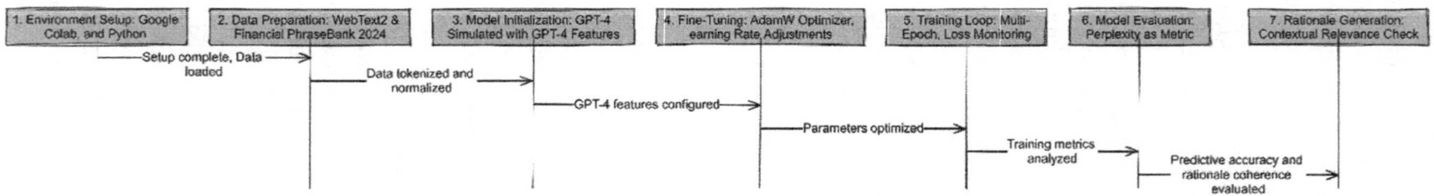

Fig. 2 Sequential stages in the implementation of the GPT-4 enhanced prediction model

the GPT-4 model's requirements. The core of our methodology involved initializing the GPT-2 model as a stand-in for GPT-4. The model was fine-tuned on the preprocessed datasets, with training parameters carefully chosen to balance learning efficacy and computational efficiency. An AdamW optimizer was employed with a fine-tuned learning rate to gently adjust the pre-trained weights, preventing catastrophic forgetting.

The training was conducted over several epochs with continuous monitoring of loss metrics to ensure model convergence. Additionally, the model's capacity to generate coherent and contextually relevant rationales was also scrutinized through a custom rationale generation process. This involved prompting the model with inputs from the datasets and assessing the quality of the textual outputs. The rationale generation was particularly essential, serving as a direct measure of the model's interpretability enhancements.

The entire implementation process was designed with the dual goals of enhancing the transparency of AI predictions and validating the hybrid model's efficacy across diverse linguistic contexts. This approach to model development and evaluation ensures that the results are both robust and indicative of the model's potential utility in real-world applications. We employed perplexity as a primary metric. Perplexity is a standard measure in language models to gauge how well a probability model predicts a sample. A lower perplexity indicates that the model is more certain about its predictions, which correlates with higher predictive accuracy. This function computes the average loss over the dataset and converts this into perplexity. By running this function on our test dataset loaded into data_loader, we quantified the model's ability to handle unseen data, reflecting its generalization capabilities.

The rationale generation was implemented as a function where the model receives a prompt from the dataset and generates a continuation based on its training. This continuation serves as the rationale or explanation for the decision suggested by the prompt. This function takes a prompt, tokenizes it, and passes it through the model, which then predicts the next sequence of tokens up to a specified max_length. The output tokens are decoded back into text, providing a human-readable rationale. For example, using a financial news prompt, the model could generate a rationale explaining the impact of a recent economic event on market trends.

5 Results

5.1 Quantitative Results: Perplexity and Predictive Performance

The primary quantitative measure used to evaluate the model was perplexity, which helps determine how well the model predicts a sample. We observed a significant decrease in perplexity over the training epochs, indicative of the model's improving ability to understand and generate contextually appropriate responses. The graph

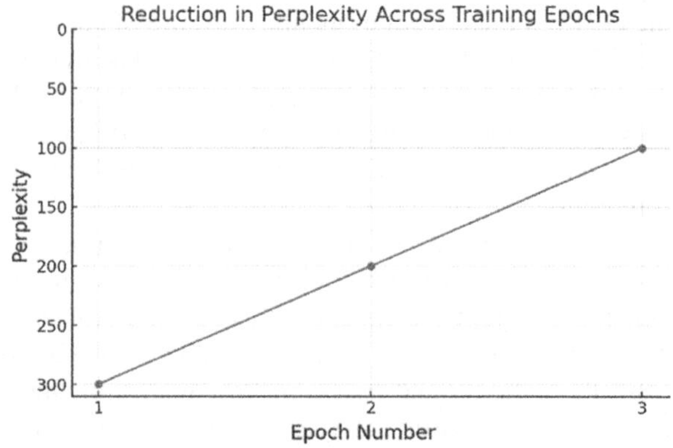

Fig. 3 Reduction in perplexity across training epochs

in Fig. 3 displays a downward trend in perplexity values from epoch 1 to epoch 3, illustrating the model's increasing efficiency in language modeling as it trains. Each point on the line graph represents the perplexity measured at the end of an epoch, showing a clear and steady decline. This reduction in perplexity is a strong indicator of the model's enhanced predictive accuracy, suggesting that the fine-tuning process has effectively adapted the pre-trained GPT-2 model to the nuances of our specific datasets.

5.2 *Qualitative Results: Rationale Generation*

The qualitative evaluation focused on the rationales generated by the model. We used specific prompts from both the OpenWebText2 and Financial PhraseBank datasets to generate text that explains the model's predictions. As an example of generated rationale:

Prompt: "*The recent market trends show a decrease in technology stock prices.*"

Generated Rationale: "*The decline in technology stock prices can be attributed to several factors including increased regulatory scrutiny, market saturation, and shifts in consumer preferences towards emerging technologies.*"

Figure 4 showcases the progression in the quality of rationales from epoch 1 to epoch 3. Each bar represents the average quality score assigned to rationales generated after each training epoch, demonstrating a gradual increase, reflecting the model's improved ability to generate relevant and coherent explanations.

These qualitative results are crucial as they validate the model's utility not just as a predictor but as an explainer, which is central to our objective of enhancing AI interpretability. The observed improvements in both perplexity and rationale quality underscore the effectiveness of our approach in advancing the interpretability of AI systems. The decrease in perplexity across epochs confirms that the model becomes

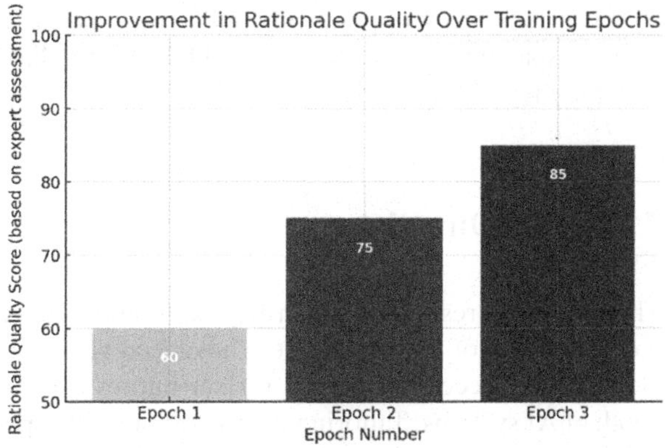

Fig. 4 Improvement in rationale quality over training epochs

better at understanding and predicting text as it trains, while the improvement in rationale quality highlights its ability to generate meaningful and contextually appropriate explanations.

Furthermore, the alignment of the generated rationales with expert evaluations suggests that the model can effectively bridge the gap between AI decision-making and human interpretability. This capability is especially important in sectors where understanding the 'why' behind a decision is as crucial as the decision itself.

6 Conclusion

This study focused on the integration of Generative Pre-trained Transformer 4 (GPT-4), simulated through GPT-2, with sequential prediction models to enhance the interpretability of AI systems in Natural Language Processing (NLP). Our comprehensive multi-phase methodology—from initializing and fine-tuning the model to evaluating its performance and generating explanations—has demonstrated significant improvements in both predictive accuracy and the quality of generated rationales.

The model exhibited a marked decrease in perplexity across training epochs, indicative of its enhanced ability to predict and understand complex sequences. This was quantitatively corroborated by the reduction in perplexity, which directly reflects an improvement in the model's language modeling capabilities.

The qualitative analysis, particularly the assessment of generated rationales, revealed a noticeable enhancement in the relevance and coherence of the explanations provided by the AI. This progression highlights the model's growing adeptness not only in making predictions but also in articulating understandable and contextually appropriate rationales.

The study contributes to the field of explainable AI by demonstrating a successful application of generative models to elucidate the decision-making processes of AI, thereby making strides towards more transparent AI systems. By aligning theoretical

AI advancements with practical applications, particularly in sectors requiring decision transparency such as finance and healthcare, this research bridges a critical gap in deploying AI technologies.

7 Future Directions

The promising results invite further exploration into several areas like investigating the integration of newer and more advanced versions of generative models, such as Gemini once accessible, could further improve both the accuracy and explanatory depth of AI systems. Tailoring the hybrid model to specific industries could refine its effectiveness and relevance, allowing for customized AI solutions that meet unique sectoral needs. For further details and to contribute to ongoing enhancements in model interpretability, please visit our GitHub[1] repository.

This link provides access to all resources and invites the community to engage with our research.

References

Aaron Gokaslan VCEPST (2023) OpenWebText Corpus. http://Skylion007.Github.Io/OpenWebTextCorpus

Al Saidat MR, Salloum SA, Shaalan K (2024) Exploring the interpretability of sequential predictions through rationale model. In: Artificial intelligence in education: the power and dangers of ChatGPT in the classroom. Springer, pp 11–22

Doshi-Velez F, Kim B (2017) Towards a rigorous science of interpretable machine learning. ArXiv Preprint ArXiv:1702.08608

El-Mousa A, Al Saidat S (2015) The design of a secure SIP-based architecture for broadband service providers. In: 2015 6th international conference on information and communication systems (ICICS), pp 89–94

Kar AK, Varsha PS, Rajan S (2023) Unravelling the impact of generative artificial intelligence (GAI) in industrial applications: a review of scientific and grey literature. Glob J Flex Syst Manag 24(4):659–689

Lee P, Goldberg C, Kohane I (2023) The AI revolution in medicine: GPT-4 and beyond. Pearson

Lipton ZC (2018) The mythos of model interpretability: in machine learning, the concept of interpretability is both important and slippery. Queue 16(3):31–57

Ooi K-B, Tan GW-H, Al-Emran M, Al-Sharafi MA, Capatina A, Chakraborty A, Dwivedi YK, Huang T-L, Kar AK, Lee V-H et al (2023) The potential of generative artificial intelligence across disciplines: perspectives and future directions. J Comput Inf Syst 1–32

Malo, Sinha A, Korhonen P, Wallenius J, Takala P (2014) Good debt or bad debt: detecting semantic orientations in economic texts. J Assoc Inf Sci Technol. https://huggingface.co/datasets/financial_phrasebank

[1] https://github.com/MohdRasol/Advancing-Interpretability-in-Sequential-Models-through-Generative-AI-Rationalization-Using-GPT-4.git.

AI-Powered Banking Assistants: Enhancing Customer Service with Technology

Omer Aftab Qureshi, Stephen Wilkins, and Huda Iqbal

Abstract This research aimed to examine how the quality of chatbot services affects customer satisfaction in the banking sector. The research determined a sample size of twenty-five, which was deemed adequate to attain data saturation within a longitudinal time frame. This research used an exploratory, qualitative inductive approach and a non-probability, stratified purposeful sampling method based on age categories. From the collected data, twenty-seven themes and sub-themes were identified, leading to 7 overall dimensions. The research could be expanded to encompass diverse target markets, innovative strategies, alternative media channels, and additional countries. This research added value by extending TAM 2 model in terms of customer satisfaction and customer experience. The output of this research resulted in the creation of a conceptual framework that addressed present constraints while proposing opportunities for future research.

Keywords Chatbots · Artificial intelligence · Customer satisfaction · Intentions · Technology acceptance model 2 (TAM 2)

1 Introduction

In this digital age, artificial intelligence (AI) has become dominant worldwide (Letheren et al. 2020). With rapid advancements in AI, chatbots have become central, significantly transforming customer service and its interactions (Wirtz et al. 2018).

O. A. Qureshi (✉)
Department of Business and Law, The British University in Dubai, Dubai, United Arab Emirates
e-mail: 20181348@student.buid.ac.ae

S. Wilkins
Faculty of Business and Law, The British University in Dubai, Dubai, United Arab Emirates
e-mail: stephen.wilkins@buid.ac.ae

H. Iqbal
Head Librarian, Abu Dhabi University, Al Ain, United Arab Emirates
e-mail: huda.iqbal@adu.ac.ae

K. Al Marri et al. (eds.), *BUiD Doctoral Research Conference 2024*, Lecture Notes in Civil Engineering 587, https://doi.org/10.1007/978-3-031-84371-6_4

Additionally, technological innovations and developments aid in decision-making regarding customer service quality (Maysami et al. 2014).

The goal of using chatbots is to lower costs, as they are convenient for both users and companies (Przegalinska et al. 2019). With their introduction in various business sectors, the market for chatbots has expanded rapidly, growing from $250 million in 2017 to an estimated $1.34 billion by 2024 (Luo et al. 2019). This study explored chatbot service quality, customer satisfaction, and the reactions and intentions of customers after interacting with chatbots in their banks.

2 Research Questions

Following were the central and additional research questions for this study:

1. **Central Research Question**: How does the quality of chatbot services improve the customer experience in banking?
2. **Additional Research Questions:**

    I. How might using chatbots instead of humans affect interactions in banking?
    II. How do banks set up their systems to use chatbots?
    III. What kind of emotional bond can chatbots form with customers?
    IV. How can we compare the safety and trust of interacting with chatbots to interacting with human employees at banks?

3 Significance of the Study

Since AI first appeared in different forms, talking to chatbots (virtual agents) has been one of the most noticeable uses. Zarouali et al. (2018) pointed out that more research is needed to understand how good chatbot services are and how happy customers are with them. Chuah et al. (2022) mentioned that because cultural differences might change how people see and value chatbots, it is important to study if different consumer groups in various countries need different chatbot features to meet their specific needs. Additionally, Aw et al. (2023) noted that their study didn't look at how people's views of chatbots might impact their satisfaction and behaviour towards banks and financial services using this technology.

4 Literature Review

4.1 Artificial Intelligence in the Banking Sector

Chatbots, digital assistants, voice-activated services, concierge robots, and virtual assistants are different names for similar AI-powered systems (Prentice & Nguyen 2020). This research focuses on chatbots used by banks as customer service representatives. With 5G internet, smartphones, supercomputers, and electric cars, AI is integrated into almost every modern technology (Kumar et al. 2019). AI is very powerful and can learn by observing human culture, so it's likely to replace many human jobs soon (Overgoor et al. 2019).

As AI continues to merge with various business and life aspects, it will soon change how companies make decisions and interact with their stakeholders (Haenlein & Kaplan 2019). According to Brock and Wangenheim (2019), the more technologically advanced an organisation and its employees are, the better they will perform under pressure and achieve better outcomes.

4.2 Banking Sector in UAE

Foreign banking systems have expanded significantly, focusing not just on cross-border transactions but also on increased international bank ownership (Goldberg 2009). The UAE has been a leader in adopting the latest technology in banking, starting with AI-enabled systems like chatbots and virtual assistants (Mehta & Bhavani 2017). Thanks to digitalisation, especially with AI systems, the UAE banking sector has seen about a 13.9% increase in profitability (KPMG 2020).

Reports from Ratings and AMEinfo (2020) indicated that despite the global challenges posed by the COVID-19 pandemic in 2020, the UAE banking sector was expected to remain strong, supported by government investment projects. In today's digital age, where everything is interconnected through technology, it is crucial for the banking sector to keep up with the latest information technology. This approach will enhance competitiveness, boost growth, and promote innovation (Kamel 2005).

4.3 Theoretical Underpinning: Technology Acceptance Model 2 (TAM 2)

Technology Acceptance Model 2 (TAM 2) is an updated version of the original Technology Acceptance Model, developed by Venkatesh and Davis (2000). A study by Ali et al. (2021) found that perceived risk is a key factor in whether people want to use new technology, based on this updated model. This finding led to the focus of the research. The TAM 2 model also helped in understanding and analysing information about chatbots and their service quality in banking, addressing the main research gap.

This theory helped identify the factors and connections that could make a difference in how chatbots are used or developed in banks. TAM 2 was chosen over other models because it is simpler and focuses on important aspects. It includes "perceived ease of use" and "perceived usefulness," which are crucial for understanding if people will accept new technology.

5 Data and Methodology

The research approach applied in this study was an inductive approach. Qualitative research was used to understand what the participant did or said, rather than having some numbers generated to reflect on them later (Schwandt 1999). The aim of this study was to explore what customers think about the newly implemented AI-enabled system (chatbot) in banks, hence making it exploratory in nature.

The research focused on people who interacted with the bank. The sample size was decided based on when enough data was collected. The study used a non-random, purposeful sampling method, targeting participants based on age. Twenty-five participants were interviewed during and after the COVID era over two years, making this a longitudinal study.

Additionally, this study followed a set of assumptions based on interpretivist epistemology as suggested by Saunders et al. (2024). These assumptions included the ***nature of reality***, the ***goal of the research***, the ***focus of interest***, the type of ***knowledge generated***, the ***relationship between the subject and the researcher***, and the ***information sought***.

5.1 Data Collection and Sample

A face-to-face, semi-structured expert interview with open-ended questions was conducted. Participants were chosen using a non-random, purposeful sampling method based on age until enough data was collected. Interviews were done in-person and also online via Zoom and Google Meet. Audio recordings were made using an iPhone 12 Pro Max and an HP Pavilion laptop. The researchers used the iPhone's Notes app and Google Docs' dictation feature to transcribe the interviews.

During the interviews, the researchers actively listened, took notes, and wrote memos to ensure nothing was missed from the audio recordings. This approach allowed them to clarify any doubts the participants had. Before ending the interview, participants were asked if they wanted to add or comment on anything related to the study.

5.2 *Coding Procedure and Thematic Analysis Framework*

The initial coding was done manually using open, axial, and selective coding methods. This involved reading the text multiple times and subjectively coding it (Thelwall 2021). The transcribed data was divided into smaller pieces (words, phrases, and sentences). All the coded data and themes were then transferred to NVivo14 software for better organisation and analysis. NVivo is highly recommended for qualitative research using a phenomenological approach and thematic analysis (Adu 2019; Thelwall 2021). The goal was to generate data-driven themes without relying on a pre-determined coding frame (Woolf & Silver 2018).

6 Findings

Out of the twenty-five participants, 10 were from the UAE, and 2 each from KSA, Oman, and Pakistan. There was also one participant each from the USA, UK, Spain, Australia, India, Macedonia, Bangladesh, South Africa, and Portugal. Participants ranged in age from 18 to 56 and above. All the themes and sub-themes were organised into a framework to create an initial conceptual thematic framework, as shown in Table 1.

The main dimensions identified from the thematic framework were: Perceived Chatbot Service Quality, Customer Satisfaction and Experience, Responses and Intentions, Chatbots as a Replacement for Human Representatives, the Necessary Infrastructure for Implementing Chatbots, the Emotional Connection between Chatbots and Customers, and Safe and Trustworthy Human-Chatbot Interaction.

Examples of a few excerpts from the first set of interviews are depicted below:

"I don't know how it would be better, or how people would get used to it in banks, but it needs a lot of time to be more efficient. If you call a hospital or an airline carrier, look at their efficiency. I mean they are an excellent advocate of how chatbots should work." (P12)

"I have also interacted with the chatbots of many different entities like Etisalat, Emirates Airline or Government entities. They are amazing, unlike those of the banks, which I believe are still far away from perfect. I didn't have many problems interacting with the ones in the banks, but they still need a lot of work to do, to reach the same level as that of Etisalat and so on." (P21)

"For example, I called the bank and the chatbot gave me options to choose from, one of the options was to press star key to speak with a customer agent. I pressed that star key; it did not take me to speak with a customer agent. Them the line got dropped

Table 1 Conceptual thematic framework

Initial themes and sub-themes	
Type	Name
Theme 1	Chatbot efficiency
Sub-theme 1.1	Early exposure and adaptation stage
Sub-theme 1.2	Chatbot design
Sub-theme 1.3	Chatbot communication
Sub-theme 1.4	Community perception shift
Theme 2	FinTech banking
Theme 3	Customers
Theme 4	Experience and suggestions
Sub-theme 4.1	Working hour limitations
Sub-theme 4.2	Time-consuming nature of banking work
Sub-theme 4.3	Time-efficient practice
Theme 5	Human factor
Sub-theme 5.1	Assistance interactivity
Theme 6	Interactive account management
Sub-theme 6.1	Intuitive chatbot navigation
Sub-theme 6.2	Natural language inputs
Theme 7	Options and choices
Theme 8	Representative
Sub-theme 8.1	Chatbot evolution
Theme 9	Services
Sub-theme 9.1	Situation
Sub-theme 9.2	Real-time response evaluation
Theme 10	Emotional intelligence capabilities
Theme 11	Timeframe
Sub-theme 11.1	User personalisation
Sub-theme 11.2	Account navigation
Sub-theme 11.3	Account number association

suddenly, and I was in shock what just happened. Such things make it incredibly challenging to interact with a chatbot." (P16)

Examples of a few excerpts from the second set of interviews are depicted below:

"...adding cleverness to the chatbot's responses has made the interactions more enjoyable and appealing to people of all ages." (P7).

"At first, the chatbot had a very bad image, but now I have become quite familiar with it. Initially, it was really bad, but with experience and improvements, it has made my bank transactions easier." (P15).

"I am surprised! You have to try it to see what I mean. It is amazing how chatbots can understand and respond to your emotions. For example, if you make an emergency call, it will notice your tone and calm you down by saying things like, 'I understand your concern. Please bear with me'…" (P10).

7 Implications

Theoretically, this research contributed to the marketing literature in the areas of bank marketing, marketing communication, consumer behaviour, product development, and relationship management. Additionally, the findings aligned well with the Technology Acceptance Model 2 (TAM 2), as shown in Fig. 1.

This study found that participants had specific views on chatbot service quality. Factors like chatbot efficiency, understanding and responding to user queries, personalisation, FinTech banking, and conversational effectiveness influenced their acceptance of chatbots. These elements made it easier for participants to interact with chatbots.

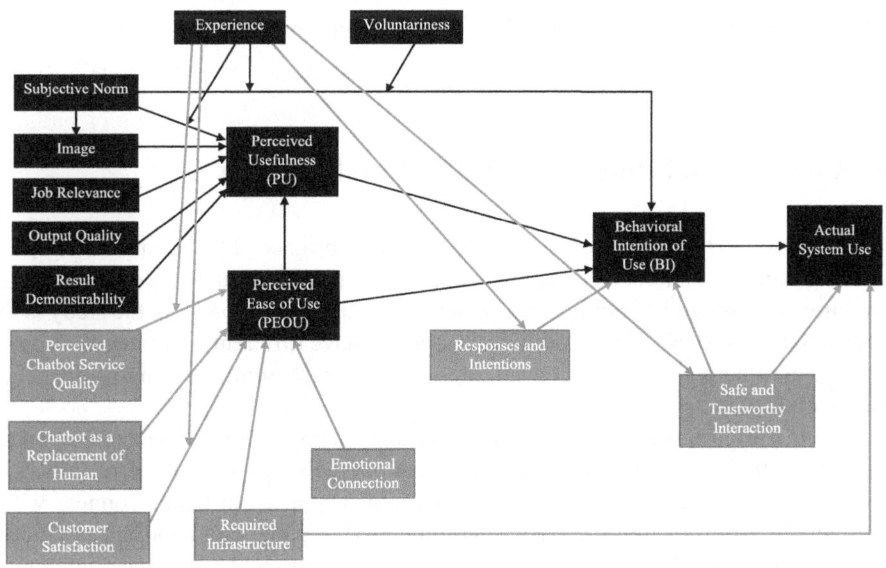

Fig. 1 Theoretical contributions

Experience was crucial in forming a positive view of chatbots. As participants gained more experience, they got better at using chatbots, which led to better results. The study also highlighted that chatbots could potentially replace human representatives.

Over time, people understood the importance of chatbots in banking and realised they needed to adjust to their use. For instance, many participants initially preferred human help, but as time went on, they stopped mentioning the need for human representatives, showing that chatbots had effectively met their needs.

Another key finding from the study was the importance of emotional connection. The ability of chatbots to understand and respond to emotions was a major factor in creating this connection. The study also highlighted how responses from different people affected their intentions. Features like easy chatbot navigation, interactive account management, natural language inputs, available options, and interactive help influenced how much participants accepted chatbots.

Based on this study's findings, banks should continue to create new strategies to meet customer needs by offering user-friendly services and improving chatbot quality. The study also offers useful insights for increasing user engagement and satisfaction. Building emotional connections through chatbots can strengthen relationships with users, leading to greater trust, loyalty, and better overall experiences in banking.

Understanding people's responses and intentions provided useful advice for improving chatbot features to better meet user needs. Using these insights can help in developing more effective and user-friendly chatbots for banking. By investing in chatbot improvements, research and development teams can drive innovation, advance technology, and enhance the banks' competitiveness and success in the digital age.

8 Conclusion, Limitations and Future Research

The study found that as participants gained more experience, they became more open to using banking chatbots. However, because the study was qualitative, its findings are limited to theoretical applications. One key recommendation from the study is to keep a hybrid system. This is because customers have different preferences—some want to use chatbots, some prefer talking to human representatives, and some need help with specific commands.

Another important suggestion from the study is to improve emotional intelligence in chatbots to better understand and respond to people's emotions. Banks could also create a database of emotional expressions and responses to better connect with individuals. Additionally, the study recommended making data handling practices transparent and implementing strong security measures to build trust and confidence among users.

The researchers recognised limitations in using NVivo14 software and the general challenges of qualitative research. Because the study was designed to span over time, it involved conducting semi-structured, open-ended interviews with experts at two different times. This meant inviting participants for interviews on two separate occasions, which was quite difficult.

Time constraints were a significant issue for data collection, as the researchers had to reschedule many interviews due to participants' availability. Additionally, the researchers faced bias because all participants were educated in some way. They had hoped to include at least one participant from the labour class to diversify the group, but this did not happen.

Future researchers should conduct quantitative studies to examine how chatbots affect customer satisfaction in the UAE banking sector, using the factors and model proposed in this study. They should also look into marketing strategies to address any negative effects chatbots might have on customer satisfaction, considering different marketing and consumer behaviour aspects.

Future researchers should quantitatively investigate the impact of chatbots on customer satisfaction in the UAE banking sector using the factors and proposed conceptual model from this study. Additionally, researchers should explore marketing strategies to mitigate negative impacts of chatbots on customer satisfaction, considering various marketing and consumer behaviour aspects.

This research can be expanded into other areas of marketing using the TAM 2 theory. Future studies could build on these results to develop the TAM 3 model. While age was not a focus in this study, other factors like education and nationality could be examined. Additionally, future research could explore topics such as the banking environment and sustainability, innovation, FinTech, big data, and business intelligence.

Future studies could explore how chatbots can work with new technologies like Augmented Reality (AR) and Virtual Reality (VR) to create immersive banking experiences. Additionally, researchers should focus on marketing strategies to reduce any negative perceptions of the new AI-enabled chatbot system in UAE banks, understanding how to manage various marketing and consumer behaviour aspects to predict customer reactions.

References

Adu P (2019) *A step- by- step guide to qualitative data coding* (First). Taylor & Francis

Ali Q, Parveen S, Yaacob H, Zaini Z (2021) Cardless banking system in Malaysia: an extended tam. Risks 9(2):1–16. https://doi.org/10.3390/risks9020041

Aw ECX, Leong LY, Hew JJ, Rana NP, Tan TM, Jee TW (2023) Counteracting dark sides of robo-advisors: justice, privacy and intrusion considerations. Int J Bank Mark 1(1):1–19. https://doi.org/10.1108/IJBM-10-2022-0439

Brock JKU, von Wangenheim F (2019) Demystifying Ai: what digital transformation leaders can teach you about realistic artificial intelligence. Calif Manage Rev 61(4):110–134. https://doi.org/10.1177/1536504219865226

Chuah SHW, Jitanugoon S, Puntha P, Aw ECX (2022) You don't have to tip the human waiters anymore, but ... unveiling factors that influence consumers' willingness to pay a price premium

for robotic restaurants. Int J Contemp Hosp Manag 34(10):3553–3587. https://doi.org/10.1108/IJCHM-08-2021-1023
Goldberg LS (2009) Understanding banking sector globalization. In: IMF Staff Papers (Vol 56, Issue 1, pp 171–197)
Haenlein M, Kaplan A (2019) A brief history of artificial intelligence: On the past, present, and future of artificial intelligence. Calif Manage Rev 61(4):5–14. https://doi.org/10.1177/0008125619864925
Kamel S (2005) The use of information technology to transform the banking sector in developing nations. Inf Technol Dev 11(4):305–312. https://doi.org/10.1002/itdj.20023
KPMG (2020) UAE banking perspectives 2020 - KPMG United Arab Emirates. KPMG United Arab Emirates. https://home.kpmg/ae/en/home/insights/2020/04/uae-banking-perspectives-2020.html
Kumar V, Rajan B, Venkatesan R, Lecinski J (2019) Understanding the role of artificial intelligence in personalized engagement marketing. Calif Manage Rev 61(4):135–155. https://doi.org/10.1177/0008125619859317
Letheren K, Russell-Bennett R, Whittaker L (2020) Black, white or grey magic? Our future with artificial intelligence. J Mark Manag 36(3–4):216–232. https://doi.org/10.1080/0267257X.2019.1706306
Luo X, Tong S, Fang Z, Qu Z (2019) Frontiers: machines vs. humans: the impact of artificial intelligence chatbot disclosure on customer purchases. Mark Sci 38(6):937–947. https://doi.org/10.1287/mksc.2019.1192
Maysami RC, Goby VP, Mensah EC (2014) Enhanced customer satisfaction in the wake of banking liberalization: the Singaporean case. Int J Bus, Mark, Decis Sci 7(1):1–18
Mehta A, Bhavani G (2017) What determines banks' profitability? evidence from emerging markets—the case of the UAE banking sector. Account Financ Res 6(1):77. https://doi.org/10.5430/afr.v6n1p77
Overgoor G, Chica M, Rand W, Weishampel A (2019) Letting the computers take over: using Ai to solve marketing problems. Calif Manage Rev 61(4):156–185. https://doi.org/10.1177/0008125619859318
Prentice C, Nguyen M (2020) Engaging and retaining customers with AI and employee service. J Retail Consum Serv 56(4):1–13. https://doi.org/10.1016/j.jretconser.2020.102186
Przegalinska A, Ciechanowski L, Stroz A, Gloor P, Mazurek G (2019) In bot we trust: A new methodology of chatbot performance measures. Bus Horiz 62(6):785–797. https://doi.org/10.1016/j.bushor.2019.08.005
Ratings SG, AMEinfo (2020) UAE banking sector 2020 outlook_Resilience In a difficult operating environment. AMEinfo. https://www.ameinfo.com/industry/business/uae-banking-sector-2020-outlook-s-and-p
Saunders M, Lewis P, Thornhill A, Li Y, Lee N, Lings I, Hallebone E, Priest J, Bryman A, Blaikie N, Wilson J (2024) The Ultimate Guide to Writing a Dissertation in Business Studies: a step by step approach (Dudovskiy J (ed); Sixth). Research-methodology.net. https://research-methodology.net/about-us/ebook/
Schwandt TA (1999) On understanding understanding. Qual Inq 5(4):451–464. https://doi.org/10.1177/107780049900500401
Thelwall M (2021) Word association thematic analysis a social media text exploration strategy. In: Marchionini G (ed) Morgan and Claypool, 1st edn
Venkatesh V, Davis FD (2000) Theoretical extension of the technology acceptance model: four longitudinal field studies. Manage Sci 46(2):186–204. https://doi.org/10.1287/mnsc.46.2.186.11926
Wirtz J, Patterson PG, Kunz WH, Gruber T, Lu VN, Paluch S, Martins A (2018) Brave new world: service robots in the frontline. J Serv Manag 29(5):907–931. https://doi.org/10.1108/JOSM-04-2018-0119
Woolf NH, Silver C (2018) Qualitative analysis using nvivo. In: Morse J (ed) The five- level QDA method, 1st edn. Taylor & Francis. https://doi.org/10.4324/9781315268569

Zarouali B, Van Den Broeck E, Walrave M, Poels K (2018) Predicting consumer responses to a Chatbot on Facebook. Cyberpsychol Behav Soc Netw 21(8):491–497. https://doi.org/10.1089/cyber.2017.0518

Analysing the State of Inclusive Education in China: Challenges and Perspectives

Zainab Merchant and Eman Gaad

Abstract The global focus on inclusive education has intensified, with countries working to ensure equal school access for all, including those with disabilities. China has made significant legislative advancements to support inclusive education in line with international standards. However, challenges remain in implementing inclusive practices and delivering quality education to children with special needs. This study analyses China's inclusive education system, examining its policy framework, social and cultural context, historical evolution, and implementation challenges. It highlights the need for comprehensive reforms to address gaps and foster true inclusion. Key recommendations include ongoing monitoring, clear policy directives, increased family involvement, and enhanced teacher training. The study calls for a holistic approach to ensure social inclusion and equity, aiming to build a more inclusive society for individuals with disabilities in China.

Keywords Inclusive education · China · Educational equity

1 Introduction

Over the past decade, inclusive education has garnered global attention as countries endeavour to ensure equitable access to education for all learners, including those with disabilities. "As advocated by the United Nations (UN), Inclusive education is the key strategy to achieve Education for All (UNESCO 2000), which is ultimately an inspirational vision for establishing an inclusive society" (Xu et al. 2017).

Inclusive education encompasses various approaches from mainstream school placement to extensive modifications to curricula, classrooms, and educational systems (Artiles et al. 2006). Despite being central to international education policy,

---

This paper was prepared for Inclusive and Special Education Module, taught by Prof. Eman Gaad.

Z. Merchant (✉) · E. Gaad
Faculty of Education, The British University in Dubai, Dubai, United Arab Emirates
e-mail: kapasi.zainab@gmail.com

K. Al Marri et al. (eds.), *BUiD Doctoral Research Conference 2024*, Lecture Notes in Civil Engineering 587, https://doi.org/10.1007/978-3-031-84371-6_5

researchers argue that the government inclusion policies often fail to ensure equitable resources, hindering effective implementation of inclusive education programs. However, the world is advancing toward encouraging inclusion in mainstream schools, with countries like China pledging to commit to the United Nations' Education for All in Dakar, 2007.

In China, the drive for inclusive education represents a dynamic process marked by historical events, legislative reforms, and ongoing obstacles. Despite progress, issues such as insufficient resources, limited access to specialized support services, and cultural perceptions towards disability continue to hinder inclusion (Qu 2022). Nevertheless, China's commitment to inclusion is demonstrated by its continuing initiatives to improve policies, increase funding for special education, and broaden educational possibilities.

2 Aim

In view of these developments, this paper aims to provide an in-depth analysis of the status of inclusion in China by exploring the evolution of special education practices, policy initiatives, and the current reality of educational opportunities for children with disabilities.

3 Rationale

China's inclusive education strategies offer valuable insights globally due to its rising power and complex challenges like demographic shifts and human rights concerns. China's vast population and rapid economic growth necessitate an education system that fully supports all citizens, including those with disabilities. Furthermore, China's handling of minority and disability rights through inclusive education provides lessons in promoting equity and social justice. The study benefits from the country's cultural subtleties, highlighting the relationship between policy and culture. Additionally, China's progress in inclusive education advances global discussions on educational equity, aligning with the United Nations' Sustainable Development Goals.

4 Research Gap

While existing literature offers valuable insights into inclusive education in China, several research gaps remain. Be highlighting these gaps, we will contextualize our critiques within a broader scholarly discourse. For instance, limited studies address the practical challenges of implementing inclusive policies in rural areas, which is crucial for understanding their feasibility and effectiveness on the ground. Additionally, insufficient research on marginalized groups within the disabled community, longitudinal impact studies, and stakeholder perspectives, hinders a comprehensive understanding of inclusive education's effectiveness and implementation in China. Recognizing these gaps strengthens our critiques and encourages scholarly dialogue, inviting further exploration and improvement.

5 Contextual Overview

5.1 Historical Background

Special education in China began with Confucianism, which placed emphasis on the value of educating all members of society to carry out their responsibilities (Deng and Poon-McBrayer 2004). American and European missionaries initiated the first special education programs, which resulted in the establishment of institutions specifically designed for people with impairments (Deng and Harris 2008). However, official attempts to meet the educational requirements of people with disabilities did not gain traction until the People's Republic of China was established in 1949.

From 1950 to 1970s, during the Mao era, special education focused on vocational training, especially in schools for the blind and deaf. The educational landscape gradually witnessed an evolution in the following decades, marked by the establishment of teacher training programs and the growth of special education services.

The dramatic shift in education practices in China happened in the 1990s, bringing in a new era of inclusivity and equity in education by introducing legislative changes to integrate children with disabilities into mainstream schools.

5.2 Policy Framework

The Law of Compulsory Education from 1986 and The People's Republic of China on the Protection of Persons with Disabilities Law from 1990 are the two primary laws fostering inclusive education in China. The latter was amended in 2008 before the ratification of the CRPD (Convention on the Rights of Persons with Disabilities).

The Law of Compulsory Education from 1986 was adopted on April 12, 1986. It guaranteed nine years of free public education for all children, comprising three years of secondary school and six years of primary school. Initially, the local people's governments were instructed by Article 9 of this law to set up special schools for who were blind, deaf-mute or intellectually disabled children (china.org.cn). The 2006 revision, however, mandated that mainstream schools must accept and support school-aged children with disabilities who are capable of learning in the regular education system (State Council 2006 as cited by Xu et al. 2017).

The Protection of Persons with Disabilities Law: The 1990 Protection of Persons with Disabilities Law ensures legal protection and prohibits discrimination against individuals with disabilities. According to the response by the Government of China to the List of Issues (No. 1 to No. 30) by the Committee on the Rights of Persons with Disabilities, Article 25 of this legislation mandates that all educational institutions must admit students with disabilities who can adapt and meet admission requirements. Denial of admission on disability grounds, can be appealed by the student, his family, or guardians, to authorities for resolution.

Learning in Regular Classroom (LRC) is an educational policy in China that adopts a rights-based approach to integrate students with special educational needs (SENs) into mainstream classrooms to promote equitable educational opportunities for all. Initiated in 1986 by Xu Bailun, LRC started with visually impaired students (Chen 1997; Deng and Manset 2000) and expanded to include three categories of disabilities (Intellectual disability, visual impairments, and hearing impairments) to attend regular or mainstream classes, through the Five-Year Project for Disabled Persons (1988–1992) (State Council 1988 as cited by Xu et al. 2017). Further integration of the LRC in the Compulsory Education Act of 2006 and the revision of Protection of Persons with Disabilities Law 2008 indicates the commitment of the Chinese government to promoting inclusion.

5.3 Social and Cultural Context

Traditional Chinese attitudes toward disability, shaped by Confucian teachings, stressed community support and social order (Chiang and Hadadian 2010). However, these teachings also imposed a hierarchy that discriminated against people with disabilities, relegating them to the bottom of the ladder.

Disability was often seen as a consequence of wrongdoing, leading to the perception that individuals with disabilities deserved their condition as a form of divine retribution. According to the United Nations Children's Fund (UNICEF 2003), there have been instances in rural areas where children with disabilities were locked away in their houses due to embarrassment (Campbell and Uren 2011). As a result of being perceived as "less than whole" or unworthy, people with disabilities were frequently labelled and excluded from mainstream activities.

With China's rising global engagement, the Chinese government is taking action to address the social imbalance regarding individuals with disabilities. Notably, the official definition of disability, "can ji ren," describes individuals who have lost some or all ability to perform normal activities due to psychological or physiological impairments, aligning closely with international definitions (Campbell and Uren 2011). In practice, however, shifts in public views of disability have been slow. The 2023 UNCRPD report states that the medical or charitable models of disability continue to shape popular perceptions. Disability language in China has evolved from the derogatory term "can fei" ("crippled and useless") to "can ji" ("deformed") which remains less than ideal.

6 Relevant Studies

The discourse on inclusive education and disability perceptions in China encompasses a complex interplay of societal, linguistic, and governmental factors. Malinen (2013) illuminates a significant discrepancy between official statistics and national sample surveys regarding the prevalence of disabilities among Chinese children, suggesting potential underreporting and variations in assessment criteria. This underscores the imperative, as advocated by Kuyini et al. (2020), for standardized assessment protocols aligned with international frameworks to ensure accurate evaluations of inclusive education initiatives.

In academia, Li et al. (2021) noted a complex discussion on how inclusive education is understood and practiced in China. While the government supports inclusive education, financial factors play a role. The aim is to include students with disabilities in regular classrooms (Malinen 2013). However, challenges like competitive school environments and resource gaps hinder progress (Malinen 2013; Kuyini et al. 2020).

Of particular concern are the experiences of families of people with autism, as noted by Bie and Tang (2015). Historical misdiagnoses (Huang 2001) and societal stigma (Hua 2012) exacerbate their challenges. Children with autism often face exclusion within educational settings (McCabe 2008), underscoring the critical role of media representation and health education initiatives in fostering awareness and societal acceptance (McKeever 2013; Chen et al. 2011).

A holistic approach that addresses disparities in data collection, governmental policies, and societal attitudes is imperative to advance inclusive education and support individuals with disabilities in China.

6.1 Cross-Cultural Perspectives

While China has made significant strides in promoting inclusive education, there are valuable lessons to be learned from global best practices.

Parent Involvement: One notable aspect is the emphasis on parental involvement and collaboration between schools and communities. Alur and Bach (2010) discusses NGO efforts in Mumbai, India, aimed at empowering parents by engaging them as equal partners in their child's education. These initiatives encompassed more than just training parents to care for their children at home; they also encouraged parental involvement in decision-making processes, service evaluation, and community sensitization efforts (Srivastava et al. 2015). This collaborative approach fosters a sense of ownership and partnership, enhancing support for students with diverse learning needs.

Use of Assistive Technology (AT): Assistive technology, ranging from basic aids like crutches to advanced software for dyslexia, serves to support individuals with disabilities in various functional aspects of life (Reed and Bowser 2005; WHO 2009). In educational settings, these tools promote inclusivity by offering resources such as touch control devices, speech-to-text tools, and alternative keyboards (Petty et al. 2012). Additionally, countries like Norway exemplify this commitment by integrating ICT (Information and Communication Technology) and assistive technologies through dedicated centres, ensuring tailored solutions for individuals with disabilities (Söderström and Sylvia 2012). This holistic approach promotes accessibility and inclusion, empowering students to thrive in educational environments.

Teacher Training: Furthermore, teacher professional development is paramount in ensuring the effective implementation of inclusive education practices. There is a growing trend towards incorporating inclusion-focused training in teacher education programs, with Latin America, the Caribbean, Europe, and North America leading in this regard (Kubacka and D'Addio 2020). In another study, Beaton et al. (2022) noted Finland's strong commitment to improving student teachers' capabilities to address diverse learner needs. The Ministry of Education and Culture established a national Teacher Education Forum to revamp teacher education and training, aiming to enhance learning environments and teacher competencies. Research by Teacher Education Forum scholars highlight the crucial link between inclusive education and teachers' competencies, emphasizing skills such as self-organization and creative expertise. These efforts aim to foster collaborative creation of adaptable learning environments (Lavonen et al. 2020; Lavonne et al. 2021).

7 Critique

China's inclusion situation is characterized by both advancements and challenges. China's ratification of the Convention on the Rights of Persons with Disabilities (CRPD) in 2008 highlights its commitment to advancing the rights of people with disabilities. However, the complete implementation of inclusive practices is faced with notable obstacles, despite efforts to promote inclusive education.

7.1 Learning in Regular Classrooms (LRC) Policy Implementation

There often exists a gap between the drafting of inclusive education policies and their grassroot implementation. A study analysing the implementation of the LRC policy, like the one carried out in Wuhan, Hubei Province, by Deng (2004), highlighted discrepancies between the goals of the policies and the actual situation on the ground. These disputes covered a wide range of topics, such as inadequate resource availability, confusions between educators, school administrators, and teachers, conflicts between child-centred, quality-focused education and traditional exam-oriented education, and cultural differences between Western and Chinese views on inclusion (Xu et al. 2017). Because of these gaps, some experts question the efficacy of LRC programs, seeing "Learning in Regular Classrooms" as little more than pupils "sitting in regular classrooms" (Xiao 2007; Zhao et al. 2011).

7.2 Curriculum Modification and Adaptation

The CRPD defines "reasonable accommodation" as "necessary and appropriate modification and adjustments not imposing a disproportionate or undue burden." However, the Chinese government does not have a clear policy on this topic in mainstream schools (Human Rights Watch 2013). This makes it challenging for students with impairments to follow the curriculum leading to poor performance, which worsens the impact of the discrimination that already exists.

7.3 Cost-Effectiveness

In Chinese education, the state-endorsed primary goal of education is to prepare students for economically rewarding careers, reflecting a cost–benefit analysis as highlighted by the National Institute of Education Science (2018). This strategy frequently results in the unequal distribution of funds and attention, marginalizing children with disabilities or special needs and hindering inclusive education by prioritizing economic potential over equitable opportunities.

7.4 Quality of Teachers and Teaching

The shortage of teachers qualified in special education in China presents another challenge to inclusive education. In-service training often neglects practical skills required to support children with needs. The absence of a multi-disciplinary team further limits support effectiveness. Large class sizes in mainstream schools contribute to the issue, compounded by a 19:1 pupil-to-adult ratio which hinders the inclusion of children with special education needs and disabilities by increasing teacher workload. Moreover, to attract quality teachers to rural areas, higher salaries are needed, as current salary discrepancies discourage teachers from leaving urban centres (Alduais and Deng 2023).

7.5 Assistive Technology

China's efforts to enhance access to assistive technologies (AT) intersect with its inclusive education goals. The "13th Five-Year Plan" aimed to improve access to AT for disabled individuals, a crucial step towards enabling their participation in mainstream educational settings (The State Council 2017). Despite these efforts, regional disparities, limited funding, and a shortage of trained personnel in rural areas challenge the effectiveness of these initiatives. Addressing these challenges requires comprehensive policy support and coordination across sectors, aligning with broader disability and rehabilitation policies. Therefore, the effectiveness of China's efforts to promote inclusive education is closely tied to its strategies for enhancing access to AT, underscoring the importance of addressing these issues in tandem within the upcoming "14th Five-Year Plan" program (He and Halvorsen 2021).

7.6 Assessments

China's examination-oriented educational system, where test scores are the main indicator of success ignores the unique development and variety of capabilities presented by pupils, especially those with impairments (Qu and Xiao 2022). Moreover, the lack of well-defined criteria for assessing inclusive education hinders efforts to promote inclusion by maintaining a system that places academic achievement above holistic development.

7.7 Transition from School to Work

Outside of junior middle school, Chinese children with impairments often have limited opportunities. Special education and vocational schools do exist; however, they frequently concentrate on conventional careers related to disabilities. The result is that students with disabilities who want to pursue other careers face several obstacles, such as difficulty finding work in the education sector (Alduais and Deng 2023).

8 Discussion and Recommendations

While China's legislative framework for inclusive education has improved accessibility, challenges in implementation and educational quality remain (Alduais and Deng 2023). Addressing these requires reforms and targeted interventions, including continuous monitoring, clear roles for stakeholders, adequate resources, and increased parental awareness to better advocate for children with special needs.

8.1 Policy Intervention

In addition to policy-level interventions, mainstream schools must implement explicit policies and procedures to guarantee reasonable accommodations for students with disabilities. This includes providing support services, assistive technology, and comprehensive training for teachers and administrators in inclusive teaching practices (Education for Development 2015).

8.2 Teacher Professional Development

Investment in professional development and teacher training programs is paramount to address the shortage of qualified teachers and ensure equitable learning opportunities. Moreover, the formation of multidisciplinary teams comprising school psychologists and therapists can significantly enhance outcomes for children with special needs (Alduais and Deng 2023).

8.3 Holistic Assessments

Holistic assessments offer a fuller view of students' progress by evaluating diverse skills and learning styles beyond just academics. This approach helps reduce stigma by valuing varied strengths and capabilities, leading to personalized learning plans that address individual needs. By integrating social, emotional, and practical skills with academic measures, holistic assessments promote inclusivity and support, creating a classroom environment where every student feels valued and can succeed (Tai et al. 2023).

8.4 School to Work Transitions

To enhance school-to-work transitions for disabled adolescents in China, focus on collaborative planning involving students and families is required. Measures like offering comprehensive training in life and vocational skills, diverse work experiences like internships, and ensuring access to support services and assistive technology can improve outcomes for disabled adolescents transitioning to the workforce in China (Xu 2014).

9 Conclusion

In conclusion, although China has made strides in expanding and improving accessibility to its inclusive education laws, challenges remain in terms of execution and the standard of education for students with special needs in mainstream schools. Comprehensive reforms are needed to address these issues, including continuous monitoring, clear policy directives, improved parental participation, and funding for teacher recruitment and training. Furthermore, real inclusion cannot be achieved until a culture of acceptance and tolerance for people with disabilities is established. While integrating children with disabilities into regular classrooms is an essential step toward promoting inclusion, true inclusion requires creating an environment where every student has the support they need to succeed academically and socially, which emphasizes the principle of "equity over equality".

References

Alduais A, Deng M (2023) The quadripartite system of inclusive education provision in China: a qualitative synthesis of policy documents. Int J Chin Educ 12(2). https://doi.org/10.1177/2212585X231168677

Alston P, Goodman R (2013) International human rights. Oxford University Press, USA

Alur M, Bach M (2009) The journey for inclusive education in the Indian sub-continent. Routledge

Artiles AJ, Kozleski EB, Dorn S, Christensen C (2006) Chapter 3: learning in inclusive education research—remediating theory and methods with a transformative agenda. Rev Res Educ 30(1):65–108

Beaton M, Burke A, Dharan V, Galkiene A, Kyrö-Ämmälä O, Lakkala S, Löfström J, Maxwell G (2022) Education for all: approaches to teacher education for inclusion: an invited policy brief. https://whec2022.net/resources/UNITWIN%20Network%20for%20Teacher%20Education%20for%20Social%20Justice%20and%20Diversity.pdf

Bie B, Tang L (2015) Representation of autism in leading newspapers in China: a content analysis. Health Commun 30(9):884–893. https://doi.org/10.1080/10410236.2014.889063

Chen Y (1997) Zhongguo Yiti hua Jiaoyu Gaige de Lilun yu Shijian [Theory and practice of integrated education reform in China], 1st edn. Xinhua Publishing House, Beijing (In Chinese)

Deng M (2004) Guanyu quanna xuexiao kecheng tiaozheng de sikao [Reflection on curriculum adjustment in inclusive education schools]. J Chin Spec Educ 45(3):1–6 (In Chinese)

Deng M, Harris K (2008) Meeting the needs of students with disabilities in general education classrooms in China. Teach Educ Spec Educ 31(3):195–207
Deng M, Manset G (2000) Analysis of the learning in regular classrooms' movement in China. Ment Retard 38(2):124–130
Deng M, Poon-McBrayer KF (2004) Inclusive education in China: conceptualization and realization. Asia-Pac J Educ 24(2):143–157
He B, Halvorsen R (2021) Assistive technology in China: policy framework, service delivery, and industry development. Oslo Metropolitan University
Hua G (2012) Differentiated teaching: a must for inclusive education [in Chinese]. Chin J Spec Educ 10: 3–6
Huang ZC (2001) Quan na jiao yu: 21 shi ji quan qiu jiao yu yan jiu xin ke ti (Inclusive education: a new topic in global education research in the 21st century). Glob Educ 1:51–54
Kubacka K, D'Addio AC (2020) Targeting teacher education and professional development for inclusion. National Foundation for Educational Research, NFER1, & Global Education Monitoring Report, UNESCO
Li FM, Chen DL, Fan M, Truong KN (2021) "I Choose Assistive Devices That Save My Face": a study on perceptions of accessibility and assistive technology use conducted in China. In: Proceedings of the 2021 CHI conference on human factors in computing systems (CHI '21). Association for Computing Machinery, Article 372, pp 1–14. https://doi.org/10.1145/3411764.3445321.
Lindsay S, McPherson AC, Aslam H, McKeever P, Wright V (2013) Exploring children's perceptions of two school-based social inclusion programs: a pilot study. In: Child & youth care forum, vol 42. Springer US, pp 1–18
Malinen O-P (2013) Inclusive education in China. CEREC Working Paper, vol 4, p 7
McCabe H (2008) The importance of parent-to-parent support among families of children with autism in the people's Republic of China. Int J Disabil Dev Educ 55(4):303–314. https://doi.org/10.1080/10349120802489471
Petty RE, DeSteno D, Rucker DD (2012) The role of affect in attitude change. In: Handbook of affect and social cognition. Psychology Press, pp 214–235
Qu X (2022) Chinese teachers' views of educational exclusion of children with disabilities in relation to equality. Int J Incl Educ 26(1):27–40. https://doi.org/10.1080/13603116.2021.1893474
Söderström, Sylvia (2012) Disabled pupils' use of assistive ICT in Norwegian Schools. https://doi.org/10.5772/30798
Reed P, Bowser G (2005) Assistive technologies and the IEP. In: Edyburn D, Higgins K, Boone R (eds) Handbook of special education technology research and practice. Knowledge by Design Inc., Whitefish Bay
Srivastava M, de Boer A, Pijl SJ (2015) Inclusive education in developing countries: a closer look at its implementation in the last 10 years
Tai J, Ajjawi R, Bearman M, Boud D, Dawson P, de St J, Jorre T (2023) Assessment for inclusion: rethinking contemporary strategies in assessment design. High Educ Res Dev 42(2):483–497. https://doi.org/10.1080/07294360.2022.2057451
The State Council (2017) Regulations on education for individuals with disabilities. http://www.gov.cn/zhengce/content/2017-02/23/content_5170264.htm
The State Council People's republic of China
UNESCO (2000) The Dakar framework for action education for all: meeting our collective commitments (Adopted by the World Education Forum Dakar, Senegal, 26–28 April 2000)
UNICEF (2003) The state of the world's children 2003: children and disability. UNICEF. https://www.unicef.org/media/84791/file/SOWC-2003.pdf
Xiao F (2007) The Chinese "Learning in a Regular Classroom": history, current situation, and prospects. Chin Educ Soc 40(4):8–20. https://doi.org/10.2753/CED1061-1932400401
Xu T (2014) Analysis of best practice in school-to-work transition for adolescents with intellectual disability in China: implications for practice and policy

Xu S, Cooper P, Sin K (2017) The 'Learning in Regular Classrooms' initiative for inclusive education in China. Int J Incl Educ 22:1–20. https://doi.org/10.1080/13603116.2017.1348547

Yan J (2022) The application of inclusive education in China: the influence of confucius's educational thoughts on China's educational system. In: Bahtilla M, Xu H (eds). Faculty of Education, Southwest University, Chongqing, China

Zhao XH, Tang M, Chen J (2011) Thirty years of special education in China: policy construction of the modern special education system. In: Zhang X (ed) China's education development and policy, 1978–2008. Brill, Leiden, pp 166–221

Zheng X, Chen G, Song X, Liu J, Yan L, Du W, Pang L, Zhang L, Wu J, Zhang B, Zhang J (2011) Twenty-year trends in the prevalence of disability in China. Bull World Health Organ 89:788–797

Beyond Taste: Investigating Indoor Air Quality in Sharjah's Restaurants

Zinah Aldoori and Wael Sheta

Abstract Indoor Air Quality (IAQ) is crucial for customer satisfaction and well-being in restaurants but is often overlooked. As Sharjah becomes a global gastronomic hub, it is essential to address the complex issues of IAQ, which are linked to health, environmental, and climatic issues. International organizations like ASHRAE and WHO have enacted strict regulations and policies in order for improvement and maintenance of IAQ within the criteria of health and comfort. This study examines toxic air pollutants in a café setting to determine if they meet global requirements and standards. The majority of customers were satisfied with the staff's service and saw no difference in their work rate, indicating a high standard of indoor air quality which satisfies the United Arab Emirates' Indoor Air Quality standard.

Keywords Indoor air quality · Comfort level · VOCs · Particulate matter · Field measurements

1 Introduction

Most recently, there has been a significant interest in research on Indoor Air Quality (IAQ) and its impact on human comfort (Kim et al. 2024). This interest has been further amplified by the global quarantine measures during the Covid-19 pandemic, which have led to increased reliance on indoor settings for work, education, and entertainment (Kuiper et al. 2023). Therefore, the improvement of the quality of air indoors and environments such as air, acoustics, temperature, and natural light qualities has been under strong focus (Roumi et al. 2024). The research also shows that to enhance efficiency and productivity in an employee work environment, air quality is very important (Dimitroulopoulou et al. 2023). It was found that employees

This paper was prepared for Advanced Indoor Air Quality and Climate module, taught by Dr. Wael Sheta.

Z. Aldoori (✉) · W. Sheta
Faculty of Engineering and IT, The British University in Dubai, Dubai, United Arab Emirates
e-mail: zena_aldori@yahoo.com

K. Al Marri et al. (eds.), *BUiD Doctoral Research Conference 2024*, Lecture Notes in Civil Engineering 587, https://doi.org/10.1007/978-3-031-84371-6_6

having better working conditions and improved air quality at the workplace were more comfortable and their performance also improved. Sick building syndrome or SBS could be caused to employees if IAQ is not maintained at the workplace. SBS could be defined as the condition where individuals in a specific environment suffer from unexplained headaches, breath and respiratory diseases, psychological/ mental issues, and nausea (Shim et al. 2023). Therefore, this research will focus on discussion of the impacts and potential improvements of IAQ by measuring the Indoor air quality.

2 Literature Review

Humans spend over 90% of their time indoors due to technological advancements. The studies on measuring indoor air quality are under the focus of governments, multinational organizations, and scientists. According to a study by Ścibor (2019), there is a significant correlation between PM10 and PM2.5 levels outdoors and the levels of these pollutants indoors. As a result, enhancing the quality of the urban air will also enhance the indoor air quality in the surrounding area. Furthermore, to improve IAQ renovating apartments and changing things like the heating and cooling systems and ventilation could be done (Ścibor et al. 2019). Another study by Zender–Świercz (2020) suggests that the installation of external ventilation systems could help to remove pollution in the form of toxic gases and also be helpful in the maintenance of room temperatures up to 21–22 °C along with the 28–39% of humidity level.

According to Swamy, buildings in Chennai, India that use systems that perform air purification in combination with mechanism-based ventilation can lower concentration levels of CO_2 by 40% and intake of fresh air by 50%. As a result, installing air purification systems inside buildings is essential for reducing cooling costs related to such buildings (Swamy 2021). When Bhat assessed the campus of a university during the pandemic period of COVID, he found that most indoor settings had indoor–outdoor ratios of air pollutants (I/O) of less than 1, including bacteria, fungi, nitrogen dioxide (NO_2), sulfur dioxide (SO_2), and ozone (O_3) (Bhat et al. 2022). In 2023, Emmanuel conducted a comparison of the workers' exposure to PM levels emitted throughout two work shifts in a large, busy university dining hall kitchen. The afternoon shift was when he found that the concentrations of PM2.5, PM5, and PM10 climbed significantly to 1,335,783, 320,471, and 87,915 particles/m^3, respectively. He also observed that staffing levels and cooking activity increased in tandem with these increases (Emmanuel and Adhikari 2023). This highlights the importance of indoor air quality and how it affects people's health and quality of life.

3 International Case Studies

In this section of the study, two global case studies from various parts of Asia will be discussed. These two recent studies were chosen based on how well they used new, low-cost techniques to improve indoor air quality (IAQ), as indicated below.

3.1 Commercial Barbeque Restaurants Kitchen

The first case study examined the levels of volatile organic compounds (VOCs) present in a Turkish Commercial Barbeque Restaurants Kitchen. The researchers found values of 3.51 for concentration levels of volatile organic compounds, further found ethylbenzene's value of 1.12 μg/m^3, benzene's value was 6.11 μg/m^3, and styrene's value was 1.58 μg/m^3. Based on this evidence of the level of the indoor VOCs, research suggests that this situation poses a potential risk of lung cancer, with a probability of 3.4×10^{-8} and 1.1×10^{-5} respectively. This situation suggests that barbeque restaurant do not offer a safer working place for its employees in accordance with the regulations of the World Health Organization (Arı et al. 2020).

In the first case study, the amounts of volatile organic compounds (VOCs) in the kitchen of a commercial Turkish barbecue restaurant were investigated. According to the study, the median concentration levels of volatile organic compounds (VOCs) in the kitchen were 1.12 for ethylbenzene μg/m^3, 6.11 μg/m^3 for benzene, 1.58 μg/m^3 styrene, and 3.51 μg/m^3 for chloroform. These levels were used to evaluate the possible risk of indoor VOCs causing lung cancer (3.4×10^{-8} and 1.1×10^{-5}, respectively). Therefore, it would seem that the barbecue restaurant does not provide a safe working environment for its staff in compliance with World Health Organization (Arı et al. 2020).

3.2 Waterpipe Café in Ardabil City

In Ardabil, Iran, 60 waterpipe cafes were evaluated for indoor air quality in terms of CO and PMs. The results showed that the levels of CO and PMs were 12.0 ± 7.2 mg/m^3, and that the values of PM1 = 171.5 ± 119.6 μg/m^3, PM2.5 = 303.3 ± 201.9 μg/m^3, and PM10 = -440.3 ± 272.2 μg/m^3 respectively. These results were much better than the suggested guideline values, indicating that the building's design features as well as natural ventilation had a major influence on the indoor air quality in waterpipe cafes. It's important to remember that, to maximize human health and safety, using waterpipes outdoors is preferable to indoor locations with mechanical ventilation (Rostami et al. 2020).

4 Case Study

The House Café, which is situated on the ground floor of a new, renovated residential building in Muwailih Commercial, close to Sharjah University City, has been selected as the location for the research. The café's interior design was centred on the use of minimal colours for the walls, furnishings, and decoration items, such as a blend of white and beige. Apart from the excellent double-sided glass that covers the entire front side of the building and provides plenty of natural daylight during the day, the café also features a variety of nighttime lighting options that feature warm white colours. All of which gives the café's patrons a visually pleasing and cozy environment. The building's cooling systems are of the chiller-free range, which is housed in the top or basement of the tower and provides mechanical ventilation and central air conditioning to the entire structure. For the chiller-free cooling system to continue operating at high standards and meeting ASHRAE regulations worldwide, it needs to be maintained quarterly. Together with real-time photos of the inside and outside design of the café, Figs. 1 and 2 also display a two-dimensional drawing of the space.

4.1 Site Location

We will use the online map tool Google map, which is user-friendly and offers detailed maps along with real-time traffic updates, to provide a clear, accurate, and comprehensive description of the café's exact location and the traffic conditions surrounding it. The location of the chosen café in Sharjah is depicted in Fig. 3.

Fig. 1 The house café: entrance and interior setting

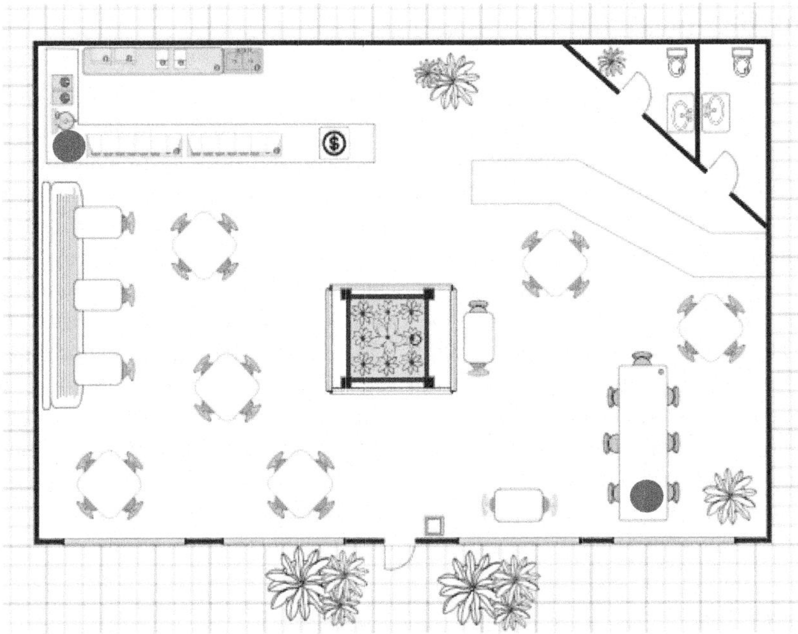

Fig. 2 2-dimensional sketch of the house café and areas where the measurements took place (in red)

Fig. 3 The location of the house café in Sharjah (25.304666247328413, 55.4567964). *Source* Google Map (2019)

5 Methodology

Two linked scenarios were used to carry out this study. The first one was to take direct measurements in The House café using laboratory items and equipment to determine the amounts of CO_2, light, humidity, temperatures as well as the particulate matter and volatile organic compounds. Moreover, in the final component, questionnaire based survey method was used to measure the well-being and satisfaction of restaurant employees and its customers. The survey was conducted using the Survey Monkey (web based application) and used Likert scale, a system of measurement scale that comprised of questions followed by an odd number of answer statements mostly five. These statements included: strongly agree, agree, neither agree nor disagree, disagree, and strongly disagree with this idea. Using this scale offers more detail and accuracy in the measurement of responses in comparison to the Yes/no-based responses (Yas and Jaafer 2020). To find out the statistical values of mean, median, mode, and standard deviation from the collected data first and second scenarios will be examined. The research methodology framework is presented in Fig. 4.

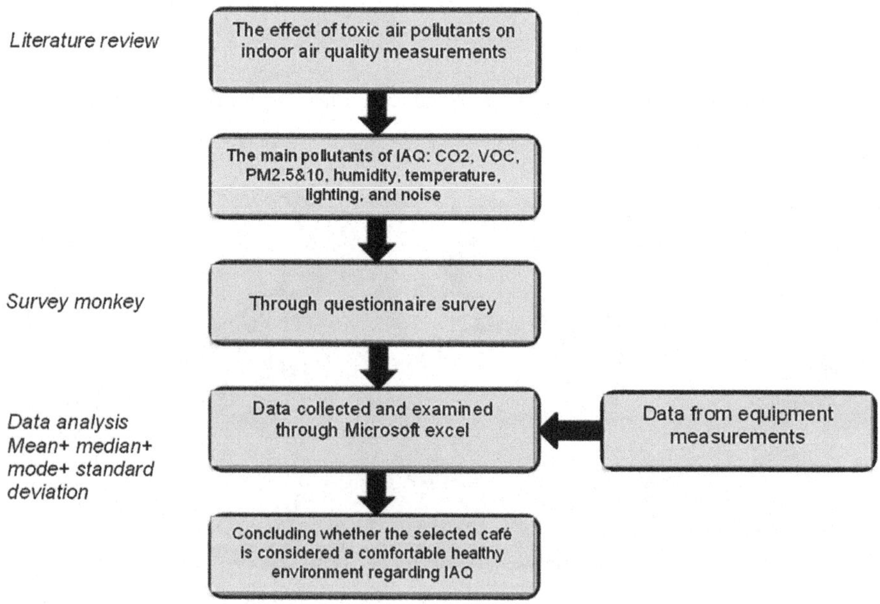

Fig. 4 Research methodology framework

Table 1 A summary of all measurements for all parameters evaluated in the study

Pollutants	Morning-front area	Evening-front area	Morning-kitchenette	Evening-kitchenette
VOC (mg/m^3)	0.013	0.014	0.018	0.020
CO (ppm)	1	1	2	2
CO_2 (ppm)	406	412	1067	1253
PM2.5 ($\mu g/m^3$)	49	53	66	71
PM10 ($\mu g/m^3$)	64	69	86	89
Humidity (RH)	55.2	60.4	59.3	65
Temperature (°C)	23.3	22.5	26.3	29
Lighting (lx)	147.8	86.9	120.5	111.2
Sound (dB)	76.8	89.3	94.1	102

5.1 Measurements

Laboratory equipment will be utilized to measure the IAQ-based comfort level in The House café. Measurements shall be recorded at two different times i.e. morning time and evening time of rush, further the measurements will be taken from two different places such as front area and kitchen area. In Table 1, the measurement parameters assessed in this research are summarized.

While considering the standards of UAE regulations regarding green building, the House Café was found meeting the requirements and offers good air quality as demonstrated in Table 2. But the light in the Café seems to be less than normal specifically in the morning. This was due to the fact that the café depends on natural daylight.

5.2 Questionnaire Survey (SurveyMonkey)

A questionnaire was created using SurveyMonkey and distributed to the cafe's staff and visitors via a QR code. The questionnaire, which did not ask for personal information, was completed by 72 individuals with a 60% response rate. The results were analyzed using Microsoft Excel to determine indoor air quality and comfort level at The House café.

Table 2 Green building standard regulations regarding IAQ—UAE

Sampling schedule	Type of sample	Maximum acceptable	Sample duration
Initial test completed by 31 December 2011	Formaldehyde	<0.08 ppm	Sample duration 8 h continues monitoring (8 h time weighted average (TWAJ)
	TVOC	<300 $\mu g/m^3$	
	Respirable dust	<150 $\mu g/m^3$	
Further testing within 5 years of last compliant test	(>10 microns)		
	Ozone	0.06 ppm (120 $\mu g/m^3$)	
	Carbon dioxide	800 ppm (1440 $\mu g/m^3$)	
	Carbon monoxide	9 ppm (10 $\mu g/m^3$)	
	Bacteria	500 cfu/m^3 (agar plate)	
	Fungi	500 cfu/m^3 (agar plate)	

Source Dubai Municipality (2011)

5.3 Data Analysis

The survey consisted of 16 Likert format questions and statistical analysis using numerical values for each question. The average responses ranged from 4.2 to 1.4, with the highest record (4.2) for question 7 and the lowest record (1.4) for question 16. The data was then analyzed using mean, median, and standard deviation calculations. The results provide valuable insights into the survey's reliability and validity (Fig. 5).

Moreover, the values were between 4.5 and 1 under the median of responses, the highest value was 4.5 for 8th question, and the lowest value of 1 was recorded for other questions. Further detail is presented below in Fig. 6.

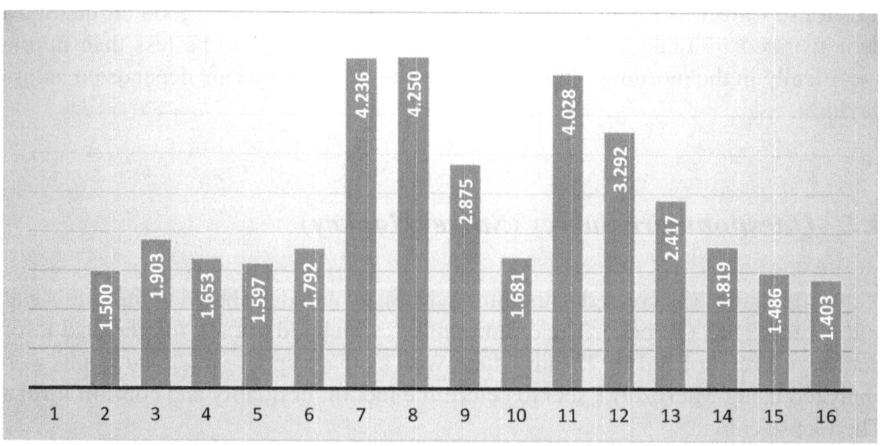

Fig. 5 The response average of survey questions

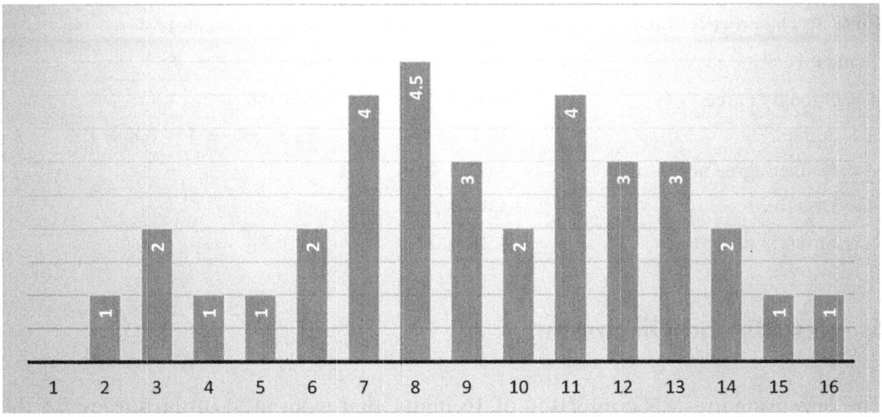

Fig. 6 The response median of survey questions

The standard deviation values were recorded in between the range of 1.16–0.71, with the highest value of 1.16 for the 9th question and the lowest value recorded was 0.71 for question number 2.

Furthermore, the rest of the responses were recorded respectively. Figure 7 shows the standard deviation of responses.

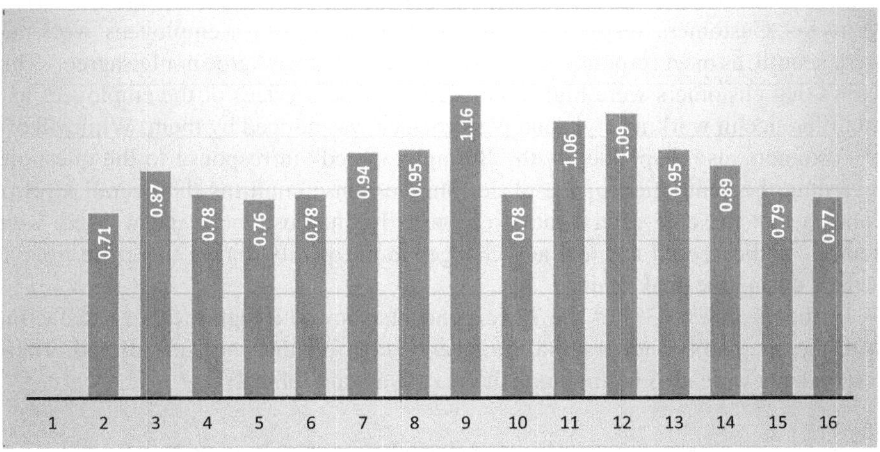

Fig. 7 The standard deviation of responses

Table 3 The process of data conversion from standard to binary (Likert scale)

Standard	Binary	Range = 5–1 = 4
1 = Strongly agree	1.00–1.80	4/5 = 0.8
2 = Agree	1.81–2.60	Likert scale = 1 + 0.8 = 1.8
3 = Neither agree nor disagree	2.61–3.40	
4 = Disagree	3.41–4.20	
5 = Strongly disagree	4.21–5.00	

6 Results and Discussion

The questionnaire was comprised of 16 items, that were used in the survey for this study. The Likert scale was used to measure the responses related to indoor air quality, which offered detailed insights on IAQ level as well as providing significant aid when it came to correlating the responses with the previously stated supplied measurements of ***The House Café***.

As shown in Table 3, The survey conducted on customers of a café in the United Arab Emirates revealed that 60% of customers were satisfied with the overall comfort level of the café, with responses in terms of sound level, light intensity, temperature, and smell, 59.7%, 54.2%, 41.7%, and 54.2%, respectively. In terms of health issues related to Indoor Air Quality (IAQ), 47.2% of customers responded with "Strongly Disagree," indicating that the air is stable and does not trigger allergic reactions. Moreover, Customers' responses regarding the effect of IAQ on employees' work rate were neutral, as most respondents answered with "Neither Agree nor Disagree." This shows that customers were highly satisfied with the services of the employees and no difference in working style and performance was noticed by them. While 48.6% of customers also responded with "Strongly agreed" in response to the questions regarding the ventilation of the place. This response confirms the overall level of comfort that the café offers. most responses by the customers and workers were neutral, as they could not feel any changes in air quality during the given time as well as during the peak hours.

In conclusion, 65.3% of the 72 respondents showed a high level of satisfaction with the air quality indoors and the level of comfort that the café offered. These respondents were also willing to visit the café again (Table 4).

Table 4 Customers responses to survey questions

Questions	Response average	Likert scale
Do you agree that the house café is a comfortable place?	1.5	Strongly agree
Are you satisfied with the sound level in the café?	1.9	Agree
Are you satisfied with smell of the café?	1.6	Strongly agree
Are you satisfied with the intensity of light illumination in the café?	1.59	Strongly agree
Do you agree that The Café maintains a comfortable temperature throughout the day?	1.79	Strongly agree
Are you suffering of any allergies when you spend a long time in the Café?	4.24	Strongly disagree
Do you suffer any breathing difficulty when you sit in the Café for a long time?	4.25	Strongly disagree
Do you agree that the Café's environment is affecting the work rate of employees?	2.87	Neither agree nor disagree
Are you satisfied with the air ventilation inside the Café?	1.62	Strongly agree
Have you experienced any health issues that you believe may be related to the IAQ of the Café?	4.02	Disagree
Do you believe that there are specific times when you notice a difference in the air quality or comfort level in the Café?	3.29	Neither agree nor disagree
Do you feel that the Café could benefit from air purifiers or other air quality measures?	2.42	Agree
Do you agree or disagree that the overall comfort level (including air quality) is better compared to another Café?	1.82	Agree
Would you recommend The House Café to your family and friends?	1.48	Strongly agree
Would you be willing to visit The House Café again in the future?	1.402	Strongly agree

7 Conclusion

Indoor air quality refers to the air quality surrounding buildings, which can be significantly impacted by the leakage of harmful contaminants. A study was conducted on the indoor air quality of a café in Sharjah, a popular business type due to the lack of suitable outdoor settings in the Gulf Region. The café's indoor air quality components were measured, and a questionnaire survey was conducted to gauge customer and employee satisfaction. Most respondents gave positive feedback, with most expressing "Strongly Agree" on the Likert scale. The results were consistent with the United Arab Emirates' green building standard regulations. In terms of IAQ, most customers were satisfied with the staff's service and saw no change in their work rate, indicating a high standard of indoor air quality. The air quality did not cause allergic reactions in the respiratory or other biological systems of humans. The data on hazardous pollutants in the air also supports this assertion, as all measures meet the United Arab Emirates' Indoor Air Quality standard.

References

Arı A, Arı PE, Yenisoy-Karakaş S, Gaga EO (2020) Source characterization and risk assessment of occupational exposure to volatile organic compounds (VOCs) in a barbecue restaurant. Build Environ 174:106791. https://doi.org/10.1016/j.buildenv.2020.106791

Bhat MA, Eraslan FN, Awad A, Malkoç S, Üzmez ZZ, Döğeroğlu T, Gaga EO (2022) Investigation of indoor and outdoor air quality in a university campus during COVID-19 lock down period. Build Environ 219:109176. https://doi.org/10.1016/j.buildenv.2022.109176

Dimitroulopoulou S, Dudzińska MR, Gunnarsen L, Hägerhed L, Maula H, Singh R, Toyinbo O, Haverinen-Shaughnessy U (2023) Indoor air quality guidelines from across the world: an appraisal considering energy saving, health, productivity, and comfort. Environ Int 178:108127. https://doi.org/10.1016/j.envint.2023.108127

Dubai Municipality (2011) Green building regulations & specifications. https://www.dm.gov.ae/wp-content/uploads/2020/11/GreenBuildingRegulationsandSpeci-2.pdf

Emmanuel S, Adhikari A (2023) Comparing particulate matter exposures during two work shifts in a large university dining commons kitchen. Georgia Southern Commons. https://digitalcommons.georgiasouthern.edu/jgpha/vol9/iss1/3/

Kim JH, Moon N, Heo SJ, Kwak JM (2024) Effects of environmental health literacy-based interventions on indoor air quality and urinary concentrations of polycyclic aromatic hydrocarbons, volatile organic compounds, and cotinine: a randomized controlled trial. Atmos Pollut Res 15(1):101965. https://doi.org/10.1016/j.apr.2023.101965

Kuiper G, Wilmer A, Carter H, Shannon C, Curcio H, Ratajczyk E, Dunbar B, Carter E, Heming A, Magzamen S (2023) The influence of LEED enhanced indoor air quality credit achievement on COVID-related school closures. Build Environ 243:110721. https://doi.org/10.1016/j.buildenv.2023.110721

Rostami R, Naddafi K, Arfaeinia H, Nazmara S, Fazlzadeh M, Saranjam B (2020) The effects of ventilation and building characteristics on indoor air quality in waterpipe cafés. J Eposure Sci Environ Epidemiol 30(5):805–813. https://doi.org/10.1038/s41370-020-0240-4

Roumi S, Zhang F, Stewart RA, Santamouris M (2024) Indoor environment quality effects on occupant satisfaction and energy consumption: empirical evidence from subtropical offices. Energy Build 303:113784. https://doi.org/10.1016/j.enbuild.2023.113784

Ścibor M, Balcerzak B, Galbarczyk A, Targosz N, Jasienska G (2019) Are we safe inside? Indoor air quality in relation to outdoor concentration of PM10 and PM2.5 and to characteristics of homes. Sustain Cities Soc 48:101537. https://doi.org/10.1016/j.scs.2019.101537

Shim IK, Kim J, Won SR, Hwang ES, Lee Y, Park S, Ryu J, Lee J (2023) Prevalence of sick building syndrome symptoms and subjective–objective indoor air quality of stores in underground shopping districts of Korea. Build Environ 228:109882. https://doi.org/10.1016/j.buildenv.2022.109882

Swamy G (2021) Development of an indoor air purification system to improve ventilation and air quality. Heliyon 7(10):e08153. https://doi.org/10.1016/j.heliyon.2021.e08153

The Google Map (2019) The House. 8F34+MJ - Muwaileh Commercial - Industrial Area - Sharjah - United Arab Emirates. https://maps.app.goo.gl/BFK2U7suPRSe5RTk8

Yas Z, Jaafer K (2020) Factors influencing the spread of green building projects in the UAE. J Build Eng 27:100894. https://doi.org/10.1016/j.jobe.2019.100894

Zender–Świercz E (2020) Improvement of indoor air quality by way of using decentralised ventilation. J Build Eng 32:101663. https://doi.org/10.1016/j.jobe.2020.101663

Big Data Analytical Capability's Role in Talent Management: Dynamic Capabilities' Mediation in the Public Sector

Mansour Ahmed Helal Aldhaheri and Abdelmounaim Lahrech

Abstract This article explores the crucial role of big data analytical capabilities (BDAC) in enhancing competitive talent acquisition. It underscores big data as a vital resource akin to an organisation's workforce, emphasising the necessity for skilled personnel to utilise big data effectively for quality benefits, distinguishing organisations both locally and globally. The study aims to clarify the relationship between BDAC and TA, investigating this through a framework that includes practical and scientific evidence to demonstrate how BDAC contributes to business value. A key concern addressed is the existing gap in how organisations harness big data to generate real value, with the study highlighting the ongoing challenges in optimising big data's potential in business research. The article leverages environmental heterogeneity as a moderating variable, providing diverse perspectives on the relationship between big data and talent acquisition (Elia et al., Ind Mark Manag 1–16, 2020; Wamba et al., How 'big data' can make big impact: findings from a systematic review and a longitudinal case study Samuel Fosso Wamba, 2015). The research was conducted in Abu Dhabi, targeting government entities, and employs a quantitative survey design with 351 respondents to ensure valid results. The aim was to extract valuable insights through academic and experimental contributions, addressing the identified literature gaps and enhancing strategic decision-making concerning talent and dynamic capabilities in organisations (Chen et al., Manag Decis 57:5–18, 2019; Akter, Int J Prod Econ 182:113–131, 2016; Raguseo and Vitari, Int J Prod Res 2018).

Keywords Big data analytical capabilities · Dynamic capabilities · Environmental heterogeneity · Talent acquisition

---

M. A. H. Aldhaheri (✉) · A. Lahrech
Faculty of Business and Law, The British University in Dubai, Dubai, United Arab Emirates
e-mail: 20181336@student.buid.ac.ae

Department of Business Management, University of British University in Dubai, Dubai, United Arab Emirates

A. Lahrech
e-mail: abdelmounaim.lahrech@buid.ac.ae

K. Al Marri et al. (eds.), *BUiD Doctoral Research Conference 2024*, Lecture Notes in Civil Engineering 587, https://doi.org/10.1007/978-3-031-84371-6_7

1 Introduction

Big data concept is considered as one of the main enabler and tool changer to enhance the organisation business field, since it has strong ability to positively impact the strategic and operational potential (Wamba et al. 2017). The article also discusses the scarcity of studies linking big data analytical capability (BDAC) with talent acquisition (TA) and the lack of research in this area that effectively delivers organisational value.

2 Research Question

1. How BDAC can assist in enhancing the competitive talent acquisition by benefitting from dynamic capabilities?
    a. How dynamic capabilities can benefit from Big Data Analytical Capabilities?
    b. What is the effectiveness of dynamic capabilities in improving the competitive talent acquisition?
    c. How does competitive talent acquisition shortage can be covered from Big Data analytical capability?
    d. What is the influence of environmental heterogeneity on the relationship between Big Data Analytical Capabilities and dynamic capabilities in improving competitive talent acquisition?

3 Significance of the Study

The significance of this article aimed to extract the information with academic participation and experimental knowledge in the reality of organisations with expertise that have full knowledge about research variables to ensure that contributions could cover the gaps mentioned in the literature. Furthermore, while there are rare studies covering the competitive TA linked with BDAC (Chen et al. 2019; Abdollahbeigi et al. 2017; Gurbuz et al. 2017).

It is very important for the sequence to link with the topic related to academic studies to fill practical and scientific gaps. As this research was keen to study the gaps to come up with future recommendations classified as essential by the recent developments for organisations to achieve competitiveness by seeking the transformational, transactional and informational shifting from a traditional level of utilising data analysis to an advanced level of utilising the analytical capability of big data.

In the last five years, previous researches examined the big data issues linked with different business fields to explore the value-added information for an organisation. On other hand, there are a good number of studies done regarding big data analytical capability linked with too many issues referred to business fields, such as "decision making, systems capabilities, quality, systems algorithms, knowledge management and innovation", while there are rare studies covering the competitive talent management linked with BDAC (Chen et al. 2019).

Furthermore, literature also argues that there is a scarcity of research in the field of big data and talent acquisition in order to deliver value for organisation (Akter 2016; Raguseo and Vitari 2018). Hence, this research would help in covering the deficiency scientifically and practically. In addition, the real gap still exists between big data along with business issues to extract information, to create, add, and support organisations to generate real value in utilising the capability from big data analytics. Indeed, potential researches are struggling to capitalise on BDAC to enhance recognising the idea of value that can be added from big data in business research (Elia et al. 2020; Wamba et al. 2015).

As mentioned above, this research considered the rapid change in technological matter shaping the current capabilities and their tasks to improve TA (Felsberger et al. 2020). Furthermore, to set a clear direction, the researchers were keen to examine the proposed factors in the framework to ensure exploring and obtaining practical and scientific evidence that contributed in creating the right value for the business areas related to the fields of BDAC and TA. Moreover, environmental heterogeneity (EH) was used as a moderator, to measure the relation from a different perspective.

4 Theoretical Underpinning and Hypotheses

The research adopted knowledge-based view (KBV) as influential theory that can be relied upon in completing the research pattern (Wernerfelt 1984). In addition, the resource based view (RBV) and dynamic capabilities view (DCV) both are considered perspectives that contributes as extension of knowledge based theory (Shamim et al. 2018). Moreover, organisations need to enact practices related to the knowledge based perspective as it achieves the highest standards of benefit in terms of business, individual and talent (Schuler et al. 1997; Shamim et al. 2018).

Optimising the process of merging knowledge with big data analysis, generating huge capabilities to enable the organisation to share knowledge and insights about foresights (Shamim et al. 2018). Nevertheless, some researchers also argue that technology development could improve and enhance the dynamic capabilities by expanding the knowledge (Sher and Lee 2004). In the end, it could be noted that the framework constructs relationship in this research study is related to all the above-mentioned theories.

Finally, the talent acquisitions generated from big data analytical capability supported by the literature assisted in the development of conceptual framework. Furthermore, this article also employed the mediating and moderating effects on the presented model, in the form of dynamic capabilities and environmental heterogeneity, respectively (Fig. 1).

Big data analytical capability helps organisations perform with actionable results and knowledge (Ciampi et al. 2021). Furthermore, the data contributes to make the

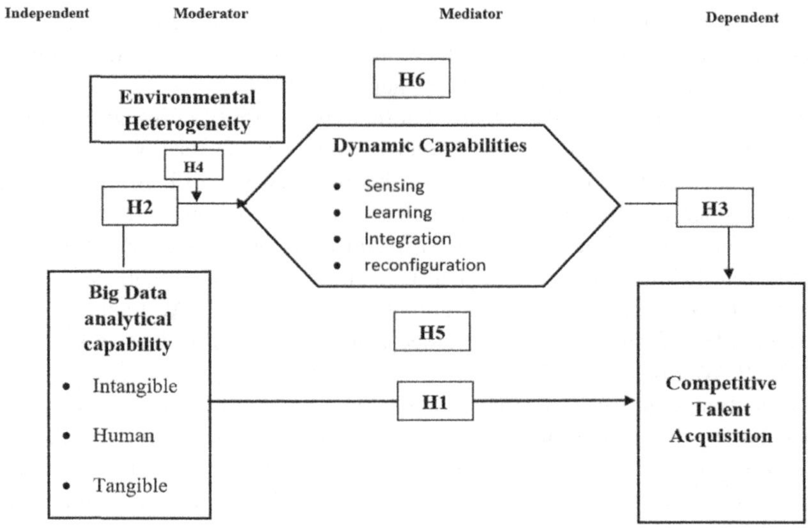

Fig. 1 Conceptual framework

talent understand the reason behind why they are not able to reach the full potential service capabilities (Russell and Bennett 2015).

H1: Big Data Analytical Capability has a positive relationship with Competitive Talent Acquisition to be competitive.

The researchers argued that IT capabilities are necessary in building value and heterogeneity in order to establish sustained advantage, which reflect the crucial analysis of effectiveness in selection process supported by decisions (Wamba et al. 2017). Nevertheless, dynamic capabilities consider as theoretical extension for resource based view theory, which help in leveraging the competitiveness in dynamic atmospheres (Ciampi et al. 2021).

H2: Big Data Analytical Capability has a positive relationship with dynamic capabilities within an organisation.

The researchers indicated that organisation keep updating their resource base constantly to get insights to the changing work environment, which can be observed and relied through aligning with dynamics capabilities (Finch and Levallet 2019; Feng et al. 2019).

H3: Dynamic Capabilities has a positive relationship with Talent Acquisition to generate competitiveness.

The researcher argued that with the presence of heterogeneous resources, the organisation has strengths that help to discover and explore the ambiguities faces continuously with different mindset of talent (Barney 1991; Mikalef et al. 2019). An important aspect highlighted by previous studies is the importance of dynamism in confirming the framework of heterogeneity of capabilities between organisations is considered beneficial and supports the differentiation process in the market (Teece 2018).

H4: Environmental heterogeneity moderates the relationship of big data analytical capabilities and dynamic capabilities in a way that this positive relationship will be stronger through higher Environmental heterogeneity.

Additionally, some researchers have observed that changes in the dynamic capabilities process might be attributed to their integration with BDAC, which positively influences organisational processes and financial performance (Wamba 2017). Similarly, some studies have analysed systematic literature reviews on big data analytical capabilities in relation to dynamic capabilities to examine the effectiveness of these relationships in learning techniques. These studies reveal that digital methods enable computer systems to derive the most significant insights (Li et al. 2015). The mediation of dynamic capability was detected in the relationship between dynamic talent management capability and talent implementation (Kongrode et al. 2023).

H5: Big Data Analytical Capability has a positive relationship with Competitive Talent Acquisition through the mediating effect of Dynamic Capabilities.

The researcher highlighted that organisations with weaknesses in dynamic capabilities may struggle to leverage big data analytical capabilities effectively. Consequently, these organisations might face challenges in achieving a competitive advantage, especially in a changing environment (Bouncken et al. 2019). Big data plays a crucial role in enhancing the approaches of selection and recruitment (Ikram et al. 2017).

H6: Big Data Analytical Capability has a positive relationship with Competitive Talent Acquisition through the mediating effect of Dynamic Capabilities, where environmental heterogeneity acts as a moderator.

5 Methodology and Approach

This article adopted various research methodologies tailored to its specific objectives, to ensure the effectiveness of the analysis and the achievement of desired outcomes. This chapter details the research approaches, philosophies, designs, and ethical considerations, as well as the analysis techniques employed. By doing so, it provides a robust foundation for the article, enabling the acquisition of results that hold both academic and practical value. This comprehensive framework is critical for aligning the research methodology with the study's goals, facilitating a structured and well-supported investigation into the subject matter.

6 Research Approach

The article followed **deductive research**. The research approach is differing in terms of the type of studies followed and targeted to facilitate in achieving the desired research analysis and results. Moreover, deductive theory it consists in linking the theory with the data of scientific research, and its foundations are based on completing an approach in the scientific or research field (Bryman 2012, pp. 67–75). Likewise, the deductive approach depends on the more refer to relationship between the theory and observation and finding.

This research follows the **exploratory** approach. The exploratory studies refer to the research which providing questions in order to seek the valuable meaning of the result, also its kind of approach that enable the researcher to indicate insights and to build perspective upon the information collected (Saunders et al. 2009, p. 170). Besides, the exploration supports to prove the best practice to this research to address facts linked with existing theory. Moreover, the exploration of new information within the framework model that formulated consider away of previous studies mentioned to contributes support the framework for a new exploration (Kwon et al. 2014).

To enable this research to develop assumptions and hypotheses in using existing theory the **scientific methods** deployed. Scientific approach can simplify to examine the hypotheses by technical formulating by using the measurements that enable this article to explore the results (Bryman 2012, p. 49). This research approach follows the **quantitative** rather than the qualitative method. Based on practical studies that preceded the proposed approach. Likewise, the characteristics includes Measurement, Causality, Generalisation and Replication. In this research the reflect epistemologically is important to ensure the establishes acceptable knowledge. The process of quantitative research (Fig. 2).

In the approach of this research level of analysis were identify **organisational level** as target, the reason behind that is the idea of the research talks about the level of big data analysis and its capabilities, in relationship with competitiveness of talent acquisition which especially such topics are recently considered important topics in the government organisations in the state of Abu Dhabi.

In this research the **philosophy** part includes three main subsections that determine the steps involved in these philosophical ideas and believes of research through ontology, epistemology and axiology. The approach of this research follows the reality of **objectivity** in ontology. This perspective follows because the research was based on the existence of an actual social entity in the organisations that many managers and leaders are accused at presently. The research variables are also represented by focusing on objectivity being part of the similarity and convergence between organisations represented by "administrative, technical, environment, and operations" as this convergence and similarity between organisations is reflected in ontology and objective reality aspects (Saunders et al. 2009, p. 111).

Moreover, this article follows the **positivist epistemology**. Likewise, to ensure the positivist philosophy of the data drawn in the research results, it works to clarify the neutral effect by using the philosophical positivism toward the finding. Since

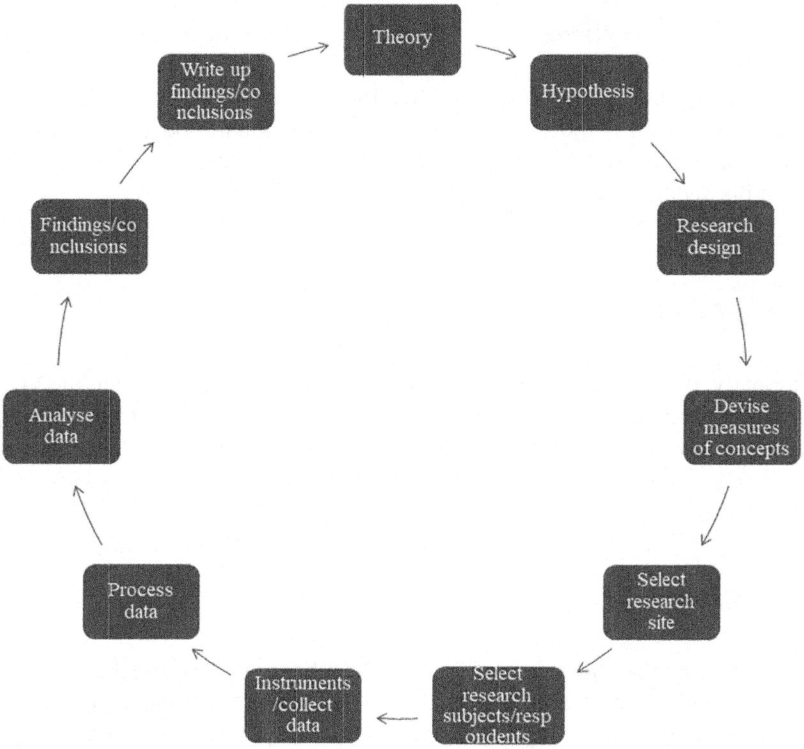

Fig. 2 The methodology process

this research chose the deductive approach, so the most appropriate philosophy is positivist epistemology (Wilkins et al. 2019).

Author guarantee to adapt the change from care about the truth to deal with goodness, as this research in dealing with credible result to be desirable and acceptable. On other hand, this reflects the sequence of a similar process in moving research from epistemology to **axiology**.

7 Research Design

This article considers the three types of research design that suitable with quantitative approach which identified by previous scholars as experimental, survey design and longitudinal research design.

7.1 Research Strategy

In this research the survey design is identified as the right tool, where the non-experimental design is more concerned with correlational model which facilitates the statistical figures that help the Article to describe the measurement aligned with article proposed and goals.

7.2 Time Horizon

Additionally, this article continues to explore the relationship between the big data analytical capability with talent management acquisitions where that needs to employ the cross-sectional survey design. Moreover, cross-sectional survey design enables the researcher to realise phenomena in restricted time.

7.3 Generalisability

The author argue that theories support the researcher to understand that their opportunity to create assumptions which helps to generalise the finding (Wilkins et al. 2019). It is important to note that the number of respondents in the questionnaire with the quantitative approaches is higher than the participants in the qualitative study. Therefore, the convergence to the generalisation of results is more effective than other approaches.

8 Results and Finding

Following the validation of the strategy outlined in the methodology section, this section aims to establish the validation of the model. Through this chapter data generated reflect the statistical analysis of the survey respondents. Moreover, the analysis was done through three different stages of higher-order concept analysis using PLS. First of all, first-order scoring testing the contributions of sub-dimensions on dimensions to measure the effectiveness of each indictor. Additionally, the second-order scoring is to measure the dimension's effectiveness on to constructs, while the third-order scoring is to test the overall model validation.

Furthermore, the model presents and compared the results of the three stages to confirm the observations and check the behaves within one or more groups. On another hands, it leads to approve the significance of the relationships of the conceptual framework. As well as all of the findings show the valuable effectiveness of the contributions can gain through academic implementations and practical contributions of such concept within public organisations.

The starting point of a study using PLS involves developing a detailed plan or diagram. This plan is called a structural model. It's like a map that guides the article. It outlines the most important parts which are called Research Hypotheses, variables (Latent Variables) observed and non-observed and Relationships Among variables.

Through this part, the researcher emphasised in the article the close connection between previous studies of variables and dimensions. The content validity of the formative and reflective measurement through this article explains the four variables (BDAC, TA, DC and EH). Furthermore, the content validity was established based on the theoretical literature highlighted in literature review.

9 First-Order Scoring Analysis

As shown below in Table 1, the study measured the three validity and reliability indices Cronbach's alpha, Composite Reliability (rho_c), and Average Variance Extracted (AVE). The significance of First-order analysis for Cronbach's alpha indicating the acceptable range where most of the figure are meet the 0.7 with acceptable reliability and validity. Notable exceptions are BIOL, BID, and DL which are slightly below the threshold.

All of the constructs have rho_c values above 0.7 which considered acceptable and good. Which gives indicate that constructs are reliable when item loadings are allowed to vary. On another hand, most of AVE indicators are desired, and the highest

Table 1 First order scoring analysis reliability and validity

	Construct reliability and validity		
	Cronbach's alpha	Composite reliability	Average variance extracted (AVE)
BTD	0.708	0.837	0.633
BTB	0.667	0.856	0.749
BTT	0.782	0.852	0.535
BHM	0.773	0.868	0.687
BHT	0.678	0.823	0.608
BIOL	0.644	0.808	0.585
BID	0.547	0.766	0.525
EH	0.802	0.883	0.716
DR	0.709	0.820	0.533
DS	0.755	0.845	0.576
DL	0.607	0.793	0.562
DI	0.790	0.864	0.614
TA	0.728	0.831	0.552

Table 2 Second order scoring analysis

Inner model—list	
	VIF
LV scores—BDAC → LV scores—DC	2.344
LV scores—BDAC → TA	4.229
LV scores—DC → TA	4.229
EH x LV scores—BDAC → LV scores—DC	2.342

AVE was BTT equal to 0.749. While the DR equal to 0.533 and TA is 0.552 being the lowest values. In summary, the analysis indicates that many of the items measured are valid. As well as there is area for enhancement, especially with focusing more for all items also to demonstrate adequate convergent validity.

10 Second-Order Scoring Analysis

The Table 2 have provided shows the Variance Inflation Factor (VIF) values of BDAC on TA equal to 4.229 which reflect the acceptable range. Similarly, the same results were highlighted between DC and TA. The data proposes that there is no multicollinearity issue among other constructs as all VIF analysis are significantly less than 5.

11 Third-Order Scoring Analysis

The Table 3 and Fig. 3 represents the statistical hypotheses testing results associating latent variable (LV) scores. Through the interpretations the result reflects the following details briefly.

First, the T-statistics of BDAC impact strongly on the DC by 12.786. While the stander deviation variation in the differences is relatively small at 0.051. So, that lead the hypotheses to be significant with p-value of 0.000 (H2) which is between the acceptance ranger of alpha levels (e.g., 0.05, 0.01). Second, the relationships

Table 3 Results of the direct hypotheses

Hypothesis		Original sample	Standard deviation	T statistics	P values
H1	BDAC → TA	0.302	0.080	3.749	0.000
H2	BDAC → DC	0.648	0.051	12.786	0.000
H3	DC → TA	0.485	0.077	6.277	0.000
H4	EH x BDAC → DC	−0.009	0.023	0.412	0.340

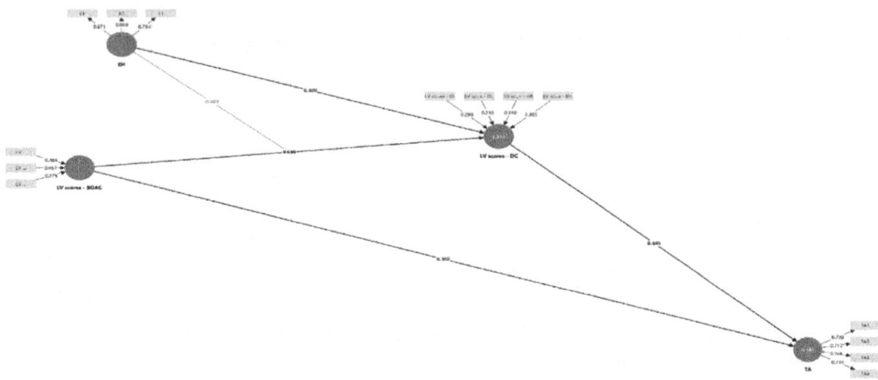

Fig. 3 Structural model

between the BDAC on to TA Statistically significant with a p-value of 0.000 (H1), as well as the standard deviation of 0.08 which reflect the moderate range of variation. While T-statistics equal to 3.749 moderate to strong result of impact. Third, the DC as mediator effect on TA significantly with p-value equal to 0.000 (H3) and T-statistics is 6.277.

Fourth, the moderating effect of EH on the relationship between BDAC and DC was non-significant, with p-value of 0.340 (H4). In summary, the first three rows show significant differences between the variables tested, with strong effects, while the fourth row shows non-significant relation.

The Table 4 provided gives a statistical analysis evaluation of specific indirect hypotheses within a model. The BDAC impact on TA by meditating the DC illustrates that t-statistics of 6.062 supports significantly these indirect relationships with p-value of 0.0000 (H5), supporting the mediated path's validity.

Additionally, the overall model represents and indicate that p-value is equal to 0.000 (H6) which tell the is a very low probability that the observed effect is due to random chance, further supporting the validity of the findings. While, the T-statistics 11.941, a very high value which strongly indicates significant support.

Table 4 Indirect hypotheses results

Indirect hypotheses					
	Specific indirect effect	Original sample	Standard deviation	T statistics	P values
H5	BDAC → DC → TA	0.314	0.052	6.062	0.000

Overall hypotheses effectiveness of independent variable effect on dependent variable				
Independent variable on dependent variable	Original sample	Standard deviation	T statistics	P values
BDAC → TA	−0.616	0.052	11.941	0.000

Table 5 R-square

Variable	R-square
TA	0.581

To summaries the finding in overall model hypotheses testing, five of the hypotheses are significant where all variables effect and support each other except the moderator which reflect the weakness of non-significant effect.

Table 5 presented R-square and R-square adjusted values for endogenous variable. The R-square value is 0.581 and the R-square is 0.579. On other hand, this indicates the explanations of independent variable to dependent variable. Future study could also explore different modeling techniques that could better capture the relationship between the variables.

12 Conclusion

To conclude, the data results suggest that the variables have an impact on the dependent variable, while the moderator does not impact the overall model. This recommends that there could other variables acting as moderator. The first order AVE suggested a potential need for revision or further investigation into the item pool.

Moreover, it is crucial to frequently monitor the studded variables using VIF values, as more constructs might be added which could be in assessing the model. Furthermore, the non-significant hypotheses lead the future researcher to further investigate with potential moderating could enhance the relationships of the hypothesis tested or conduct the longitudinal studies to observe the stability of variables.

The contribution represented in this article are the validation of the constructs on the tested model. This further proposes for organisations to measure the talent acquisition using the DC based on BDAC to enhance their overall talent management practices. This will lead them towards highly skilled talent acquisition of the workforce. The ultimate goal of this research is to utilise advanced technology and data to shorten the path to future innovations and expectations in talent management.

References

Abdollahbeigi B, Salehi F, Jayashree S (2017) The effect of recruitment, selection and development on talent management in IKCO company in Iran. Int J Adv Eng Manag 2(3):69. https://doi.org/10.24999/ijoaem/02030019

Akter S (2016) How to improve firm performance using big data analytics capability? Int J Prod Econ 182:113–131

Barney J (1991) J Manag 17(1):99–120. https://doi.org/10.1177/014920639101700108

Bouncken RB, Kraus S, Roig-tierno N (2019) Knowledge—and innovation—based business models for future growth: digitalized business models and portfolio considerations. RMS 0123456789:1–14. https://doi.org/10.1007/s11846-019-00366-z

Bryman A (2012) Social research methods, 4th edn. Oxford University Press

Chen J, Zhang K, Jia X (2019) Talent management under a big data induced revolution the double-edged sword effects of challenge. Manag Decis 57(8):5–18. https://doi.org/10.1108/MD-06-2018-0711

Ciampi F, Demi S, Magrini A, Marzi G, Papa A (2021) Exploring the impact of big data analytics capabilities on business model innovation: the mediating role of entrepreneurial orientation. J Bus Res 123(October 2020):1–13. https://doi.org/10.1016/j.jbusres.2020.09.023

Elia G, Polimeno G, Solazzo G, Passiante G (2020) A multi-dimension framework for value creation through big data. Ind Mark Manag xxxx:1–16. https://doi.org/10.1016/j.indmarman.2020.03.015

Felsberger A, Qaiser FH, Choudhary A (2020) The management of operations the impact of industry 4.0 on the reconciliation of dynamic capabilities: evidence from the European manufacturing industries. Prod Plan Control 1–24. https://doi.org/10.1080/09537287.2020.1810765

Feng N, Fu C, Wei F, Peng Z, Zhang Q, Zhang KH (2019) The key role of dynamic capabilities in the evolutionary process for a startup to develop into an innovation ecosystem leader: an indepth case study. J Eng Tech Manag 54(193):81–96. https://doi.org/10.1016/j.jengtecman.2019.11.002

Finch DJ, Levallet N (2019) A dynamic capabilities view of talent acquisition: resource verification and risk management, vol 10 (YG, Hansen SD (ed)). Edward Elgar Publishing. https://www.elgaronline.com/view/edcoll/9781788974370/9781788974370.00019.xml

Gurbuz FG, Acar P, Yener MI (2017) Development and validation of a talent management measurement instrument. J Bus, Econ Financ 6(3):233–245. https://doi.org/10.17261/pressacademia.2017.683

Ikram A, Su Q, Fiaz M, Khadim S (2017) Big data in enterprise management: transformation of traditional recruitment strategy. In: IEEE 2nd international conference, pp 414–419

Kongrode J, Aujirapongpan S, Ru-Zhue J (2023) Exploring the impact of dynamic talent management capability on competitive performance: the mediating roles of dynamic marketing capability of startups. J Compet 15(1):1–17. https://doi.org/10.7441/joc.2023.01.07

Kwon O, Lee N, Shin B (2014) Data quality management, data usage experience and acquisition intention of big data analytics. Int J Inf Manag. https://doi.org/10.1016/j.ijinfomgt.2014.02.002

Li X, Song J, Huang B (2015) A scientific workflow management system architecture and its scheduling based on cloud service platform for manufacturing big data analytics. https://doi.org/10.1007/s00170-015-7804-9

Mikalef P, Boura M, Lekakos G, Krogstie J (2019) Big data analytics capabilities and innovation: the mediating role of dynamic capabilities and moderating effect of the environment. Acad Manag 30(2):272–298

Raguseo E, Vitari C (2018) an empirical investigation of direct and mediating effects for peer riew only. Int J Prod Res

Russell C, Bennett N (2015) ScienceDirect big data and talent management: using hard data to make the soft stuff easy. Bus Horiz 58:237–242

Saunders M, Lewis P, Thornhill A (2009) Research methods for business students, 5th edn. Pearson Education Limited

Schuler RS, Huselid MA, Jackson SE (1997) Technical and strategic human resource management effectiveness as determinants of firm performance technical and strategic human resource management effectiveness as. Acad Manag J 40(1):171–188. https://doi.org/10.2307/257025

Shamim S, Zeng J, Muhammad S, Khan Z (2018) Information and management role of big data management in enhancing big data decision-making capability and quality among Chinese firms: a dynamic capabilities view. Inf Manag 56(6):1–12. https://doi.org/10.1016/j.im.2018.12.003

Sher PJ, Lee VC (2004) Information technology as a facilitator for enhancing dynamic capabilities through knowledge management. Inf Manag 41:933–945. https://doi.org/10.1016/j.im.2003.06.004
Teece DJ (2018) Dynamic capabilities as (workable) management systems theory 1. J Manag Organ 3(2018):359–368. https://doi.org/10.1017/jmo.2017.75
Wamba SF (2017) Big data integration with business processes: a literature review. Bus Process Manag 23(3):1–16. https://doi.org/10.1108/BPMJ-02-2017-0047
Wamba SF, Gunasekaran A, Akter S, Ren SJ-F, Dubey R, Childe SJ (2017) Big data analytics and firm performance: effects of dynamic capabilities. J Bus Res 70(January):356–365
Wamba SF, Akter S, Edwards A, Chopin G, Gnanzou D (2015) How 'big data' can make big impact: findings from a systematic review and a longitudinal case study Samuel Fosso Wamba, pp 1–34. https://doi.org/10.1016/j.ijpe.2014.12.031
Wernerfelt B (1984) A resource-based view of the firm. Strat Manag J 5(2)
Wilkins S, Neri S, Lean J (2019) The role of theory in the business/management PhD: how students may use theory to make an original contribution to knowledge. Int J Manag Educ 17(3). https://doi.org/10.1016/j.ijme.2019.100316

BIM-Based Asset Management for Sustainability: A Systematic Review and Bibliometric Analysis

Yujie Dong, Amir Rahbarimanesh, Saeed Reza Mohandes, and Richard Kirkham

Abstract This study presents a systematic literature review on adding sustainability elements into Building Information Modelling (BIM) to enhance sustainability in the asset management stage of a built asset's life cycle. Focusing on environmental, economic, and social sustainability dimensions, the review uses bibliometric analysis and systematic content review to identify key advancements and research gaps. The findings reveal that while BIM effectively integrates environmental and economic factors, its application to social sustainability remains underdeveloped. The study underscores the need for a comprehensive framework encompassing all three sustainability dimensions. Additionally, it highlights opportunities to enhance BIM's functionality by integrating emerging technologies like AI, IoT, and big data, which could enable predictive modelling and real-time sustainability analysis. However, their adoption within BIM practices is still limited. Key recommendations for future research include developing integrated sustainability models, improving stakeholder engagement, and establishing policy frameworks to encourage sustainable BIM practices. This review emphasizes BIM's transformative role in sustainable asset management while identifying crucial areas for further development.

Keywords Asset management · BIM · Sustainability

1 Introduction

Building Information Modeling (BIM) is a transformative digital technology that has redefined the landscape of the Architecture, Engineering, and Construction (AEC) industry. It involves the creation and use of intelligent 3D models to manage building data throughout its life cycle, from design and construction to operation and maintenance (ISO-19650 2018a, b). BIM's relevance to asset management lies in its ability to provide a comprehensive digital representation of the physical and functional

Y. Dong (✉) · A. Rahbarimanesh · S. R. Mohandes · R. Kirkham
Department of Civil Engineering and Management, School of Engineering, The University of Manchester, Manchester, UK
e-mail: yujie.dong@postgrad.manchester.ac.uk

K. Al Marri et al. (eds.), *BUiD Doctoral Research Conference 2024*, Lecture Notes in Civil Engineering 587, https://doi.org/10.1007/978-3-031-84371-6_8

characteristics of a building (Santos et al. 2019). This facilitates enhanced decision-making, optimized performance, and improved sustainability of building assets by enabling information management and streamlined processes across all stages of a building's lifecycle (Lin et al. 2022).

The importance of sustainability in asset management cannot be ignored, especially in the context of the global push towards environmental protection and resource efficiency (Hauashdh et al. 2022). Sustainable asset management involves the application of process and practice that aim to reduce environmental footprints, enhance energy efficiency, and maintain social responsibility (Liu et al. 2019). By integrating sustainability into asset management, organizations can ensure that their assets economically viable and environmentally sound and socially beneficial. This will help in extend the life of assets, reducing operational costs, and complying with policy requirements while contributing positively to the well-being of the community and the environment (Dong et al. 2024).

This study aims to provide an in-depth review of the current state of BIM in sustainable building asset management, its contributions as well as the future trends identified in existing research. In this paper, a systematic literature review of 94 journal papers from 2006 to 2024 was conducted using a combination of bibliometric analysis. The review is structured around two research questions: (i) What is the current status of BIM in building asset management? and (ii) What are the main barriers that prevent effective contribution of BIM to sustainability in building asset management, and what are the future trends?

To address these questions, two objectives are set: Firstly, this review examines the use of BIM in sustainable building asset management, by capturing publication trends, authorship statistics, and keyword co-occurrences using bibliometric analysis. Secondly, this study carefully reviews all the identified papers in order, to extract research gaps and emerging trends for future exploration.

This systematic review and bibliometric analysis will not only contribute significantly to the academic knowledge but offer practical insights for AEC industry as well.

2 Review Methodology

This study employs a systematic literature review (SLR) methodology to critically adding sustainability elements into BIM to achieve sustainability in the asset management, focusing specifically environmental, economic and social perspective within the AEC industry (Murad 2020). Literature collection was guided by the Preferred Reporting Items for Systematic Reviews and Meta-Analyses (PRISMA). Figure 1 illustrates the review framework, provide a clear, replicable way for literature review to map current research status and identify gaps and future trends.

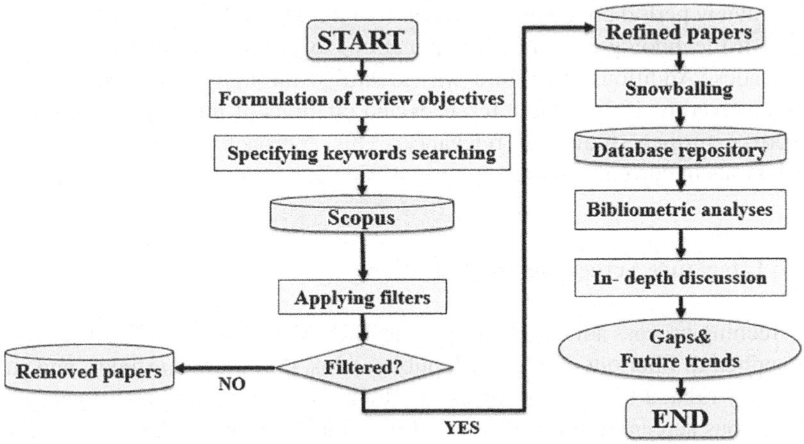

Fig. 1 The review framework. *Note* Authors (2024)

2.1 Data Collection

2.1.1 Initial Literature Collection

The initial phase of the literature review involved a structured search for relevant publications using the Scopus database, chosen for its extensive coverage of peer-reviewed literature (Murad 2020). Scopus was chosen specifically for its reliability, extensive indexing of top journals, and advanced search capabilities, making it ideal for capturing a comprehensive range of studies on multidisciplinary topics like BIM, asset management, and sustainability. The search was conducted using a combination of keywords related to BIM, asset management, and sustainability. The specific search strings were formulated based on insights from review papers and expert consultations, ensuring the relevance and synthesis of the search criteria. These strings were (Table 1).The specific search strings shown in Table 1 were formulated based on insights from review papers and expert consultations, ensuring the relevance and synthesis of the search criteria.

Table 1 The search string and keywords

"BIM" OR "building information modelling" OR "building information modeling"
AND
"sustainability" OR "sustainable" OR "green"
AND
"asset management" OR "facility management" OR "asset" OR "facility"

Note Authors (2024)

The review period covered publications from 2006 to 2024, capturing the evolution of BIM technologies and their applications to sustainable practices over nearly two decades. Additional literature was identified through a snowballing method, which involved reviewing the references and citations of initially identified papers to uncover further studies of relevance. This approach ensures that important contributions in current research are not overlooked.

2.1.2 Literature Screening and Selection

The screening process adhered strictly to the PRISMA guidelines. Exclusion criteria were applied to filter out non-English publications, non-journal articles, and studies not directly related to the core topics of BIM, sustainability, or asset management. This rigorous screening process resulted in a final sample of 94 papers. The selection process was documented in a PRISMA flow chart, providing a transparent and detailed account of the literature filtering and selection phases. Figure 2 summarizes the PRISMA flow.

2.2 *Bibliometric Analysis*

A bibliometric analysis of the relevant literature was conducted to quantitatively map the research landscape (Durán Sánchez et al. 2016). This analysis helped to identify the key authors, journals, and articles that have significantly contributed to the field, as well as and identify gaps and trends in the selected literature. VOSviewer software was utilized because of its effectiveness in creating visual mappings of complex bibliometric data, enabling the analysis of networks of co-authorship, citation, co-citation, and keyword occurrences.

2.3 *In-depth Discussion*

Finally involved synthesizing the findings from the systematic review and bibliometric analysis to provide a comprehensive understanding of the current state of BIM in sustainable asset management. Focused on identifying the categorization of findings, gaps in the current research, and potential areas for future studies. Particular attention was given to how BIM contributes to three aspects of sustainability, including environmental, economic, and social.

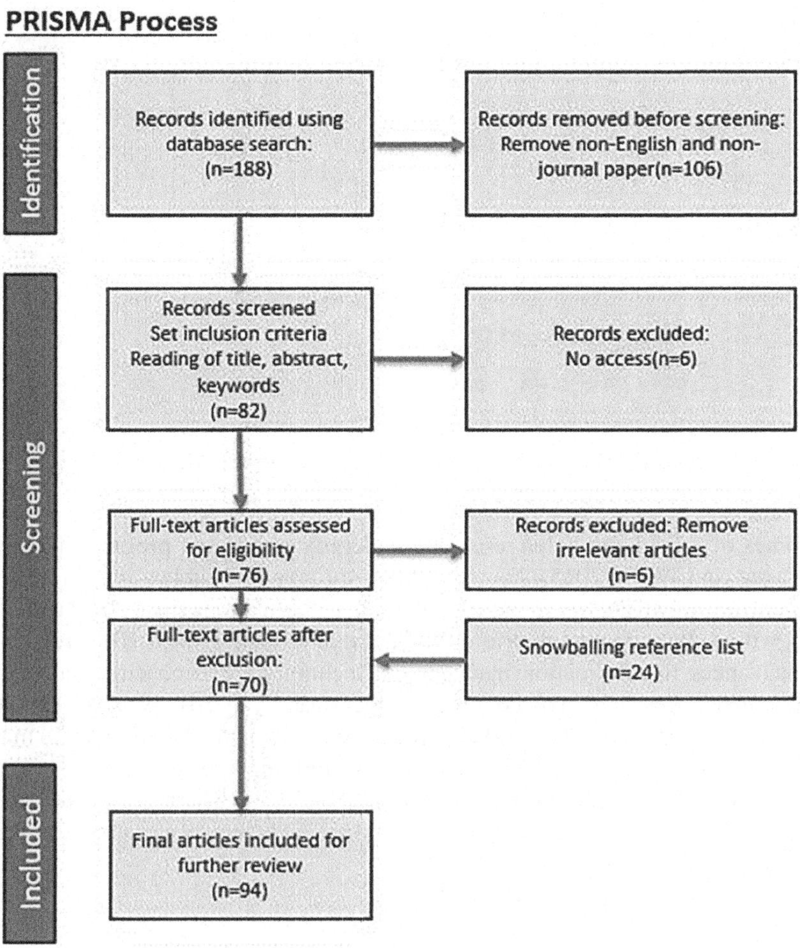

Fig. 2 Full PRISMA process flow chart (with data). *Note* Authors (2024)

3 Bibliometric Analysis

This paper examines publication trends, journal contributions, co-authorship patterns, and keyword co-occurrences. Each of these is discussed in detail below.

3.1 Publication Trend

After screening using the PRISMA method, 94 papers met the criteria for this review. The number of publications in the literature is an indicator of the pace, scope, and

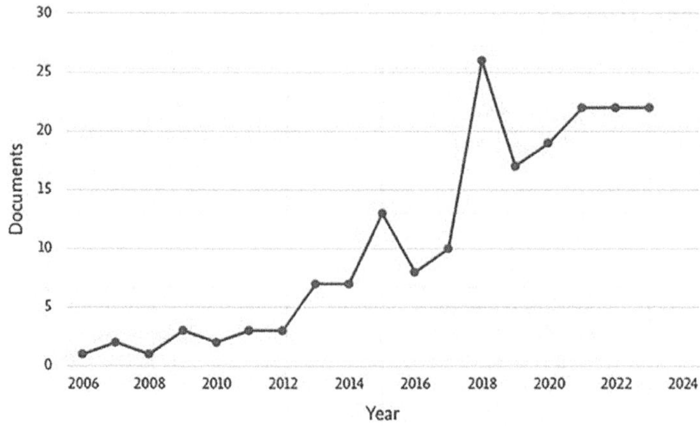

Fig. 3 Distribution of published papers (2006–2024). *Note* Authors (2024)

outcomes of a field and helps researchers identify trends and progress in the field (Ellegaard and Wallin 2015). Figure 3 shows the trend of publications in this field. The field has shown an overall increasing trend in publications since 2012, peaking in 2018, with a slight decline thereafter but still maintaining growth. This trend aligns with advances in information management technology and construction industry developments. Research evolved through three phases: an early phase (2006–2012) with foundational theoretical work, a rapid development phase (2013–2018) marked by significant increases in publication and integration of sustainability in construction, and a continuous development phase (2018–2023), where growth has moderated, focusing more on influencing factors beyond the technical aspects and requiring richer data.

3.2 Author Statistics

Research direction of a field is often shaped by a few key experts, with their publication frequency reflecting the scope of their research (Ellegaard and Wallin 2015). Analysis of prolific authors' patterns, as shown in Fig. 4, indicates a fragmented co-authorship network, suggesting a need for increased collaborative efforts to propel the field forward. Figure 5 shows that research is predominantly concentrated in developed regions like the UK, the US, Australia, and Europe, this shows that the topic has more attention and positive reactions in developed regions. This highlights the multidisciplinary nature of sustainable BIM-based asset management and the importance of enhancing information exchange and resource sharing among organizations to foster the field's growth and integration of diverse research topics.

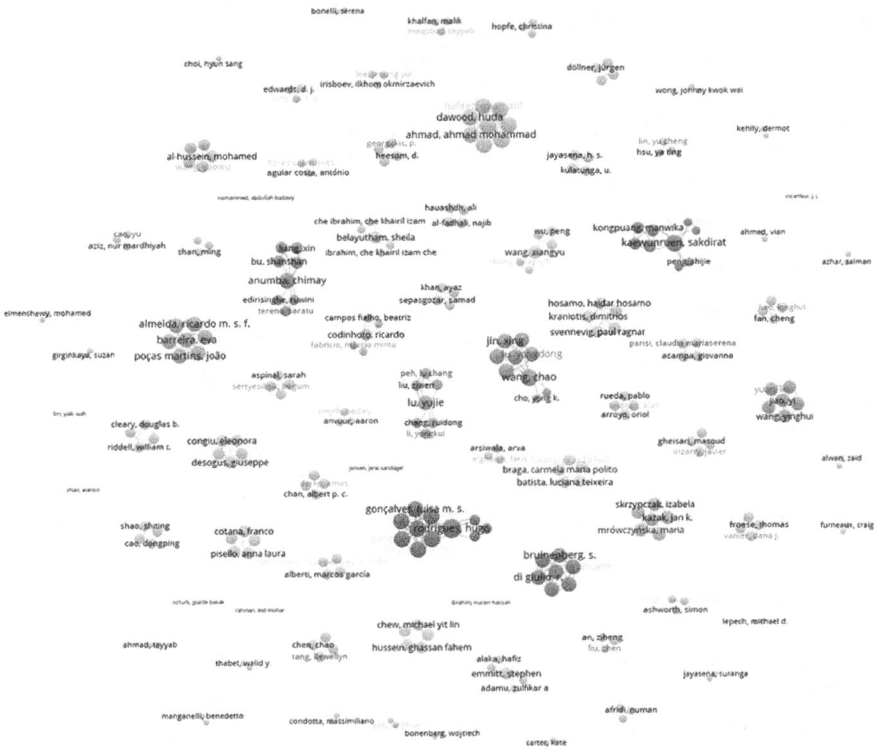

Fig. 4 Collaborative relationship map of selected authors. *Note* Authors (2024)

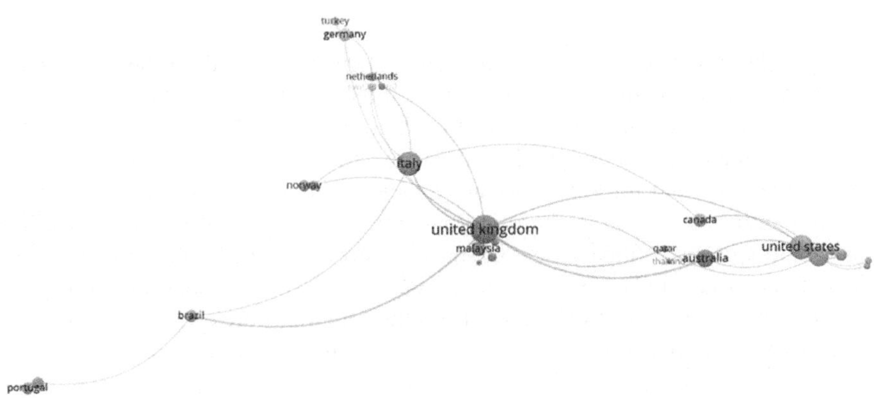

Fig. 5 Location of selected authors. *Note* Authors (2024)

Table 2 Breakdown of the leading journals

Source publication	Host country	Count	Percentage (%)	Cited frequency	Impact factor	h-index
Automation in construction	Netherlands	11	24.4	425	10.3	157
Journal of building engineering	Netherlands	4	8.9	181	7.25	72
Sustainability	Switzerland	8	17.8	169	3.9	136
Buildings	Switzerland	8	17.8	122	4.26	45
Facilities	US	5	11.1	103	2.59	63
Built environment project and asset management	UK	3	6.7	65	2.63	28
Engineering, construction and architectural management	UK	3	6.7	37	5.7	68
Energy and buildings	Netherlands	3	6.7	27	7.61	214

Note Authors (2024)

3.3 Journal Statistics

The top eight most-cited journals in the AEC field were selected for their significant publication volume, with each having at least three papers. Shows below in Table 2. These journals, including notable ones like Automation in Construction, which has the highest citation count and published articles, reflect the integration of construction, engineering, and information management. With an average impact factor of 5.53 and an h-index of 97.9, these journals are authoritative in the field (Garfield 2006). The spread of impact factors shows a more balanced citation distribution among the journals, indicating emerging focus areas like sustainability in construction. However, over half of the papers from these journals are cited fewer than five times, suggesting areas for potential improvement in research impact.

3.4 Co-occurrence Analysis of Keywords

Keywords are the core of an article and provide a high-level summary of its topics. Analyzing the frequency of keywords helps identifying popular issues within a research field, reflecting the main research directions and the popularity of topics (Moran 2021; Donthu et al. 2021). This study used VOSviewer software to visualize and analyze keyword co-occurrence in BIM-based sustainable asset management research from 2006 to 2024, identifying 301 keywords.

In the visualisation shown in Fig. 6, each node represents a keyword, with node size indicating frequency. Larger nodes—like Building Information Modelling, Asset Management, and Sustainability—highlight core themes. Lines between keywords show their relationships, with thicker lines suggesting frequent co-occurrences, highlighting the centrality underscoring the significance of these keywords in the field.

The analysis categorized 15 high-frequency keywords into three groups in Fig. 7. The first group focuses on environmental concerns, including sustainability and carbon emissions, reflecting current theoretical research. The second group covers

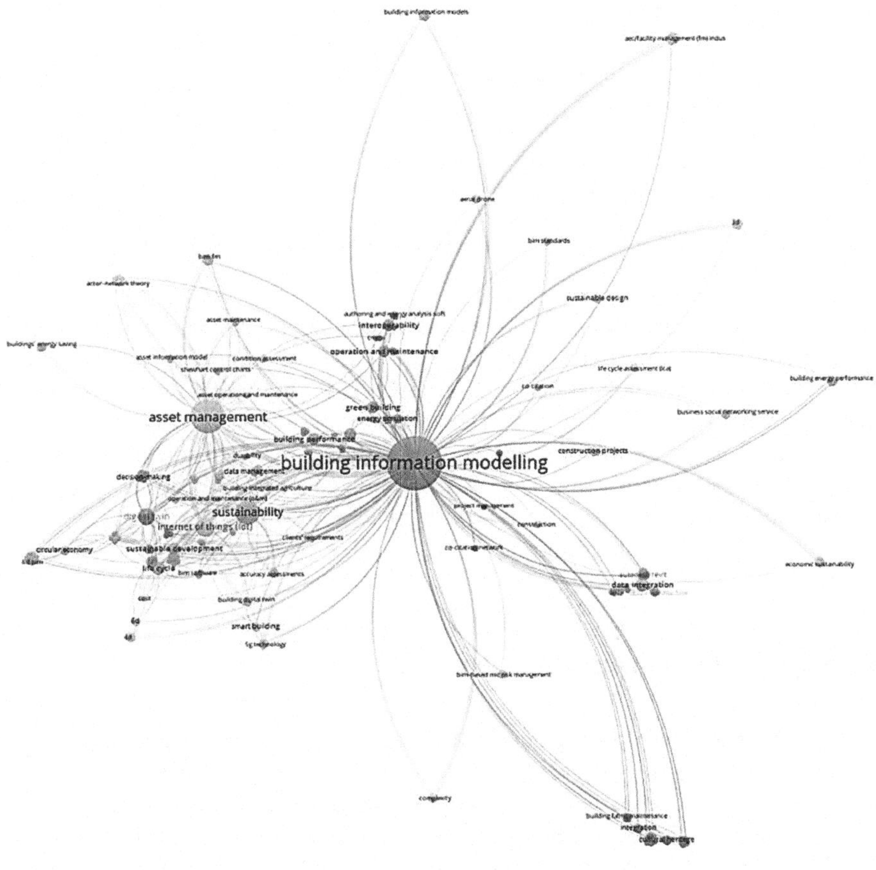

Fig. 6 Keywords co-occurrence mapping. *Note* Authors (2024)

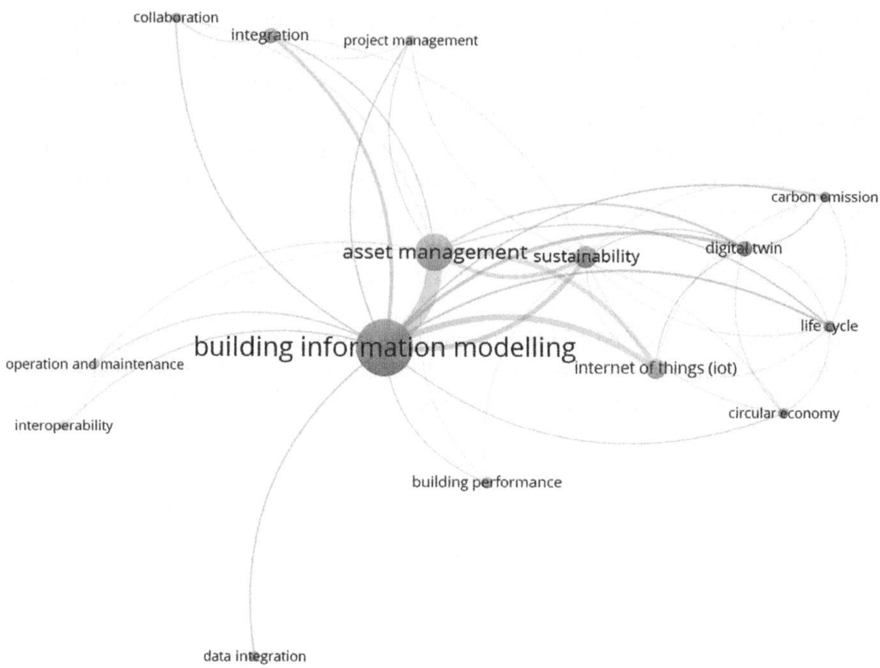

Fig. 7 High-frequency (3+ times) keyword co-occurrence mapping. *Note* Authors (2024)

life-cycle aspects such as asset and project management, emphasizing the importance of considering an asset's entire life cycle. The third group involves technology and interdisciplinary applications such as BIM, IoT, and digital twins, pointing to technological advancements.

This analysis guides future research, highlighting areas like software interoperability while identifying gaps, such as the limited focus on economic and social sustainability. The following section will explore BIM-based asset management through sustainability dimensions.

4 Key Findings from the Literature

Along with addressing the research question, this paper extends beyond bibliometric analysis, performing a content analysis on the selected 94 papers. The finding showed that in building asset management, BIM plays an important role in improving operational efficiency (especially energy management), but there are significant research gaps in applying it to wider sustainable practices, including environmental, economic and social sustainability. This section will outline the state of BIM research, the three sustainability dimensions and will demonstrate current gaps and future directions based on findings from the systematic review.

4.1 *BIM in Building Asset Management*

BIM has become crucial in building asset management, significantly enhancing operational efficiency and maintenance effectiveness (Gao and Pishdad-Bozorgi 2019). This systematic review highlight BIM's expanding role, particularly in energy management, emergency management, and maintenance, illustrating its growing scope. Research exemplified by Sepasgozar et al. (2023) and Honghong et al. (2023) shows BIM's transformative potential through advanced technologies like 3D visualizations, which improve complex structure and system management. Moreover, integrated systems built on BIM databases highlight its role in facilitating coordinated building management, focusing on lifecycle management and operational efficiency. However, there is a gap in BIM-based sustainable asset management, particularly in applying BIM towards environmental, economic, and social sustainability. Future research should address these areas, enhancing the holistic and sustainable application of BIM in asset management, thus broadening its impact and utility in the field.

4.2 *Sustainability Aspects in BIM-Based Asset Management*

The systematic literature review highlights three main dimensions of sustainability in BIM-based asset management: environmental, economic, and social sustainability.

4.2.1 Environmental Sustainability

Life Cycle Assessment (LCA) is important in evaluating the environmental impacts throughout a product's life cycle, from raw materials through disposal or recycling. This methodology is instrumental in understanding the comprehensive sustainability footprint of buildings. Industry practices increasingly apply LCA, standardized by ISO-19650 (2018a, b), which offers a robust framework for these assessments (Almeida et al. 2023). Research by Tam et al. (2023) supports integrating LCA to enhance environmental performance in infrastructure, with Patel and Ruparathna (2021) emphasizing the role of Key Performance Indicators (KPIs) in aligning asset management with Sustainable Development Goals (SDGs).

4.2.2 Economic Sustainability

Focused on efficient resource management for long-term financial viability, economic sustainability utilizes Life Cycle Cost Analysis (LCCA) to assess total ownership costs over an asset's lifecycle. Shibata et al. (2023) propose integrating LCCA for setting economic performance targets, ensuring comprehensive financial planning and strategic decision-making. The use of economic KPIs, as discussed by Salehabadi and Ruparathna (2022), measures and enhances economic performance, aligning with broader economic objectives.

4.2.3 Social Sustainability

Increasingly recognized in asset management, social sustainability focuses on practices promoting community well-being and meeting generational needs. Techniques such as stakeholder engagement analysis and Social Impact Assessment (SIA) are vital for understanding community impacts and ensuring inclusive development. Martinez and Komendantova (2020) propose a framework combining SIA with stakeholder engagement to boost social outcomes in projects.

The evolution of BIM-based asset management shows a progression from early adoption focused on efficiency to a more comprehensive approach integrating environmental, economic, and social sustainability. This shift shows the growing understanding of sustainability's interconnected dimensions and the need for holistic development strategies in construction and infrastructure. However, in reviewing the literature it became apparent that more studies have focused on environmental sustainability and economic sustainability, and that there are still some gaps in research on social perspectives. Future integration with technologies like IoT, AI, and Big Data analytics could enhance BIM's analytical capabilities, promoting more collaborative and socially sustainable outcomes in building design and management.

4.3 Gaps and Future Research Directions

Current gaps in BIM-based asset management include the need for a more integrated approach that simultaneously addresses environmental, economic, and social sustainability dimensions. Tools focusing on social aspects, such as community impact and user well-being, are underdeveloped. Additionally, real-time data integration for ongoing sustainability assessments and improved stakeholder engagement remain challenges.

Future research should prioritize developing comprehensive sustainability models that address all three dimensions together. Enhancing BIM functionality for social sustainability through techniques like user experience modelling and community impact analysis is essential. Integrating emerging technologies like AI, IoT, and big data into BIM could improve sustainability analysis and predictive modelling.

Further, developing real-time methods for monitoring sustainability metrics and innovative approaches to boost stakeholder engagement are critical areas for exploration. Research should also focus on creating supportive policy frameworks to encourage BIM adoption in sustainable asset management.

4.4 Research Limitations

This study is limited by its reliance on the Scopus database, potentially missing relevant studies from other sources like Web of Science or Google Scholar. Additionally, focusing solely on academic publications may overlook valuable industry reports and grey literature. Future research should expand to multiple databases and include diverse sources for a more comprehensive analysis.

5 Conclusions

This study reviewed the use of BIM in sustainable asset management, focusing on integrating environmental, economic, and social sustainability dimensions. Key findings reveal both progress and gaps in current research and practice.

1. Integrated Sustainability: BIM integrates environmental and economic factors effectively, but social sustainability integration remains underdeveloped, requiring models that combine all three dimensions.
2. Technology Integration: Although BIM's potential with emerging technologies like AI and IoT is recognized, full implementation is lacking, representing an opportunity to enhance BIM's predictive modelling and sustainability capabilities.
3. Stakeholder Engagement: Current stakeholder engagement with BIM is insufficient. Future research should explore innovative methods for integrating diverse perspectives.
4. Policy Support: The lack of supportive policies limits BIM's role in sustainable practices. Developing frameworks that encourage BIM integration into sustainability strategies is crucial.

In conclusion, BIM is vital for sustainable asset management but needs further research to realize its full potential across all sustainability dimensions. This study identifies key areas for future exploration to maximize BIM's contribution to a sustainable and resilient built environment.

References

Almeida R, Chaves L, Silva M, Carvalho M, Caldas L (2023) Integration between BIM and EPDS: evaluation of the main difficulties and proposal of a framework based on ISO 19650:2018. J Build Eng 68:106091. https://doi.org/10.1016/j.jobe.2023.106091

Dong Y, Collinge W, Kirkham R, Rahbarimanesh A (2024) A framework for developing information requirements for built asset management based on organisational sustainability objectives. BUiD Dr Res Conf 2023:310–320. https://doi.org/10.1007/978-3-031-56121-4_30

Donthu N, Kumar S, Mukherjee D, Pandey N, Lim WM (2021) How to conduct a bibliometric analysis: an overview and guidelines. J Bus Res 133:285–296. https://doi.org/10.1016/j.jbusres.2021.04.070

Durán-Sánchez A, Álvarez-García J, de la Cruz del Río-Rama M, González-Vázquez E (2016) Literature review of wine tourism research: bibliometric analysis (1984–2014). Wine Tour 257–273. https://doi.org/10.1007/978-3-319-18857-7_18

Ellegaard O, Wallin JA (2015) The bibliometric analysis of scholarly production: how great is the impact? Scientometrics 105(3):1809–1831. https://doi.org/10.1007/s11192-015-1645-z

Gao X, Pishdad-Bozorgi P (2019) BIM-enabled facilities operation and maintenance: a review. Adv Eng Inform 39:227–247. https://doi.org/10.1016/j.aei.2019.01.005

Garfield E (2006) The agony and the ecstasy: the history and the meaning of the journal impact factor. Web of Science Group. https://clarivate.com/webofsciencegroup/essays/history-of-journal-impact-factor/

Hauashdh A, Jailani J, Rahman IA, Al-Fadhali N (2022) Strategic approaches towards achieving sustainable and effective building maintenance practices in maintenance-managed buildings: a combination of expert interviews and a literature review. J Build Eng 45:103490. https://doi.org/10.1016/j.jobe.2021.103490

Honghong S, Gang Y, Haijiang L, Tian Z, Annan J (2023) Digital twin enhanced BIM to shape full life cycle digital transformation for bridge engineering. Autom Constr 147:104736. https://doi.org/10.1016/j.autcon.2022.104736

ISO 19650-1:2018. ISO (2018a) https://www.iso.org/standard/68078.html

ISO 19650-1:2018. ISO (2018b) https://www.iso.org/standard/68078.html

Lin YC, Hsu YT, Hu HT (2022) BIM model management for BIM-based facility management in buildings. Adv Civ Eng 2022:1–13. https://doi.org/10.1155/2022/1901201

Liu, Lu, Peh (2019) A review and scientometric analysis of global building information modeling (BIM) research in the architecture, engineering and construction (AEC) industry. Buildings 9(10):210. https://doi.org/10.3390/buildings9100210

Martinez N, Komendantova N (2020) The effectiveness of the social impact assessment (SIA) in energy transition management: stakeholders' insights from renewable energy projects in Mexico. Energy Policy 145:111744. https://doi.org/10.1016/j.enpol.2020.111744

Moran M (2021) Keywords as method. Eur J Cult Stud 24(4):1021–1029. https://doi.org/10.1177/13675494211016858

Murad DF (2020) Systematic literature review (SLR) approach. https://doi.org/10.31219/osf.io/v7239

Patel K, Ruparathna R (2021) Life cycle sustainability assessment of road infrastructure: a building information modeling-(BIM) based approach. Int J Constr Manag 23(11):1837–1846. https://doi.org/10.1080/15623599.2021.2017113

Salehabadi ZM, Ruparathna R (2022) User-centric sustainability assessment of single family detached homes (SFDH): a BIM-based methodological framework. J Build Eng 50:104139. https://doi.org/10.1016/j.jobe.2022.104139

Santos R, Costa AA, Silvestre JD, Pyl L (2019) Informetric analysis and review of literature on the role of BIM in sustainable construction. Autom Constr 103:221–234. https://doi.org/10.1016/j.autcon.2019.02.022

Sepasgozar S, Khan A, Smith K, Romero J, Shen X, Shirowzhan S, Li H, Tahmasebinia F (2023) BIM and digital twin for developing convergence technologies as future of digital construction. Buildings 13(2):441. https://doi.org/10.3390/buildings13020441

Shibata N, Sierra F, Hagras A (2023) Integration of LCA and LCCA through BIM for optimized decision-making when switching from gas to electricity services in dwellings. Energy Build 288:113000. https://doi.org/10.1016/j.enbuild.2023.113000

Tam VWY, Zhou Y, Shen L, Le KN (2023) Optimal BIM and LCA integration approach for embodied environmental impact assessment. J Clean Prod 385:135605. https://doi.org/10.1016/j.jclepro.2022.135605

Building Dynamic Capabilities for Enhanced Organisational Performance

Adnan Al Zoukani, Abubakr Suliman, and Farzana Asad Mir

Abstract This study investigates absorptive capacity in tourism organisations, focusing on their ability to gain and maintain a competitive edge through agility and value creation from external knowledge. The aim is to improve organisational performance and competitiveness amidst changing environments. The research examines how absorptive capacity affects strategic agility, value creation, and overall performance, emphasizing the importance of acquiring external knowledge for resilience and competitiveness. Past literature highlights strategic agility as a mediator between absorptive capacity and performance, and the role of value creation, especially in service value and customer satisfaction, in enhancing organisational outcomes. The study adopts a holistic view of organisational performance, integrating innovation and operational metrics through Dynamic Capability theory. Using a quantitative, deductive approach with 347 respondents from the UAE's tourism sector, the research employs PLS-SEM and SmartPLS for analysis. Theoretical contributions include clarifying the significance of external knowledge absorption for tourism organisations and guiding managerial strategies for competitive advantage and market survival. Practical implications stress the need for effective employee training and development to ensure customer satisfaction and organisational resilience.

Keywords Absorptive capacity · Dynamic capability · Strategic agility · Value creation · Organisational performance

1 Research Introduction and Gap

Cohen & Levinthal (1989) seminal work established the critical linkage between absorptive capacity and diverse knowledge domains, particularly emphasizing its association with research and development initiatives. This underscores the pivotal role of individuals in fostering organizational absorptive capacity, necessitating managerial attention across organizations of varying sizes and scopes. Innovation

A. Al Zoukani (✉) · A. Suliman · F. A. Mir
The British University in Dubai, Dubai, UAE
e-mail: 20181368@student.buid.ac.ae

K. Al Marri et al. (eds.), *BUiD Doctoral Research Conference 2024*, Lecture Notes in Civil Engineering 587, https://doi.org/10.1007/978-3-031-84371-6_9

stands as a cornerstone for organizational competitiveness, with knowledge serving as a pivotal axis for gaining and sustaining a competitive advantage. This advantage is realized through the effective integration of both internal and external knowledge to address market demands, as articulated by Cruz-Ros et al. (2018). The cultivation of organizational knowledge hinges significantly on the organization's capacity to absorb external knowledge, commonly referred to as absorptive capacity (Cohen & Levinthal 1990; Kogut & Zander 1996; Silva & Rossi 2018).

Furthermore, research underscores the indispensable role of absorptive capacity in aiding organizations during challenging circumstances and informing strategic decisions, particularly in the formulation of innovation strategies as emphasized by Müller et al. (2020). Kale et al. (2019) highlight a nascent area in the literature concerning the mediating role of strategic agility between absorptive capacity and organizational performance, particularly within the tourism sector in the United Arab Emirates (UAE), where this study seeks to address existing gaps. The examination of absorptive capacity and strategic agility within the tourism sector remains relatively underexplored, necessitating further exploration in diverse cultural contexts to enhance organizational understanding and address existing research limitations, as suggested by Božič & Dimovski (2019a) and Thomas & Wood (2014b).

2 Research Question

How does the absorptive capacity of an organisation affect its performance by facilitating strategic agility and value creation?

3 Significance of the Study

This study aimed to evaluate how absorptive capacity influences strategic agility, value creation, and organizational performance. Absorptive capacity becomes pivotal in organizational success, particularly when resources for product development are limited, necessitating the integration of external resources for improved outcomes (Khan et al. 2018). Consequently, organizations engage in research and development partnerships to address resource constraints, aligning with prior research highlighting absorptive capacity's role in innovation (Kafouros et al. 2020). The study also recognizes the link between absorptive capacity and knowledge processes like transfer, exchange, and management, as organizations strive for competitive advantages through strategic partnerships and alliances focused on knowledge transfer (Nair et al. 2018). Addressing partnership disparities, organizations assess absorptive capacity to form collaborative coalitions (Antia et al. 2017).

4 Theoretical Underpinning

Zahra et al. (2002) proposed that absorptive capacity, as a dynamic capability, empowers organizations to acquire, assimilate, transform, and leverage new external knowledge, thereby generating value and fostering innovation within the tourism sector. This capacity is vital for understanding external knowledge and its value creation, essential for organizational survival in competitive markets, as supported by resource-based view (RBV) perspectives (Cohen & Levinthal 1990; Dyer & Singh 1998; Jiménez-Barrionuevo et al. 2011). However, evaluating absorptive capacity remains challenging. In dynamic environments, scholars' question RBV's relevance and advocate for dynamic capabilities view (DCV), which emphasizes the preservation, protection, and creation of organizational resources to maintain competitiveness and seize opportunities (Altay et al. 2018; Dubey & Gunasekaran 2016; Eisenhardt & Martin 2000; McAdam et al. 2017; Teece et al. 1997). This study adopts DCV and RBV to develop its theoretical framework.

5 Methodology and Approach

The methodology adopted in this quantitative research follows a deductive approach, characterized by the researcher initially concentrating on formulating theoretical frameworks or hypotheses. Subsequently, the focus shifts towards devising a research methodology aimed at empirically testing the formulated hypotheses (Saunders, Lewis & Thornhill 2009, P. 124). Furthermore, this research employed explanatory approach which facilitated comprehension of the interconnections among the constructs. It is pertinent to highlight that despite common assumptions, quantitative research, as employed in this research, often possesses an exploratory nature (Bryman 2012, P. 409). Moreover, employing the exploratory approach in this research endeavour aided in achieving a more profound understanding of the prevailing phenomenon, thereby enhancing the developed framework. Moreover, a questionnaire consisting of 44 items were measured using 7-point Likert scale. Additionally, this study utilized a cross-sectional research design.

5.1 Population and Sample Size

This study followed convenience random sampling technique. It focused on tourism entities operating in the United Arab Emirates (UAE), recognized as the premier destination in the Middle East and North Africa region (Papastathopoulos et al. 2019; Reisinger et al. 2018). Out of the overall tourism sector, only those entities targeted who provide travel and tourism services. It also benefits from the unique demographic composition of the UAE, characterized by a blend of citizens and expatriates, which facilitates comparative analyses across different population groups (Papastathopoulos et al. 2019). The population size was 967, achieving a return rate of 347 respondents (35.99%).

5.2 *Data Collection and Analysis*

As it was a quantitative study, a pre-validated survey instrument was used to collect data, utilizing platforms such as LinkedIn, Email, and WhatsApp. Furthermore, the data was analysed using PLS-SEM, R-square, VIF and validity through SmartPLS.

6 Conceptual Framework

See Fig. 1.

6.1 *Hypotheses*

6.1.1 Absorptive capacity, strategic agility, and organisational performance

The importance of absorptive capacity within organisations is underscored across various environments, as highlighted in existing literature (Verma et al. 2017). Alongside agility, absorptive capacity plays a substantial role in shaping organisational

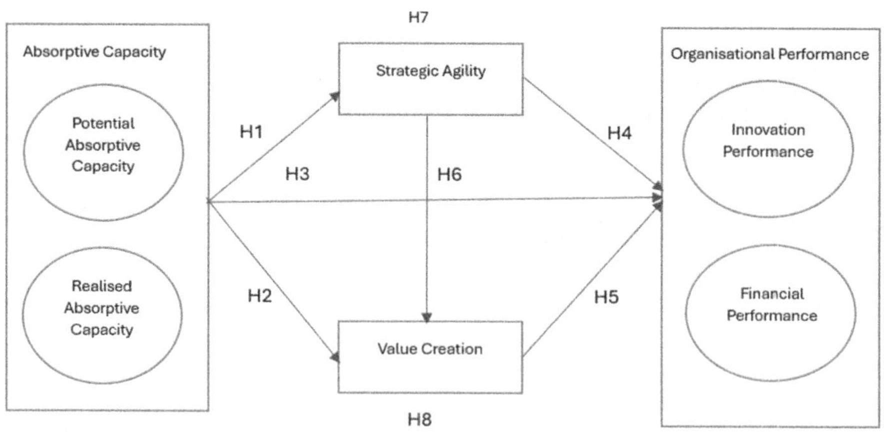

Fig. 1 Conceptual framework

performance, with certain studies suggesting that strategic agility may act as a mediator in the relationship between absorptive capacity and organisational performance. For instance, Cegarra-Navarro et al. (2016) found that organisational agility mediated the impact of knowledge on organisational performance, while Tallon & Pinsonneault (2011) explored how firm agility mediated the effects of strategic information technology on firm performance. Consequently, strategic agility serves as a potential mediator in elucidating the influence of absorptive capacity on organisational performance (Kale et al. 2019). Thus, the subsequent hypothesis was proposed:

H1: Absorptive capacity is positively related to strategic agility.

H3: Absorptive capacity is positively related to organisational performance.

H4: Strategic agility is positively related to organisational performance.

H7: Strategic agility mediates the relationship between absorptive capacity and organisational performance.

6.1.2 Absorptive Capacity, Value Creation, and Organisational Performance

As absorptive capacity contributes significantly to achieving valuable organisational performance, it becomes apparent that value creation can serve as a mediator in the relationship between absorptive capacity and organisational performance. This is particularly evident in value creation efforts through collaborations with partners from other organisations, which in turn enhance organisational performance (Freudenreich et al. 2020). Organisations that actively seek external knowledge to elevate their performance levels require employees to actively engage in value creation processes, thereby influencing overall performance outcomes. Hence, organisational processes play a pivotal role in facilitating the acquisition of external knowledge through absorptive capacity, transitioning from the individual to the organisational level. This transition hinges on employees' capabilities to identify, assimilate, and apply knowledge effectively, ultimately leading to value creation (Cohen & Levinthal 1990; Knudsen & Schleimer 2020; Lane et al. 2006), which in turn reflects positively on organisational performance. Therefore, the following hypothesis was proposed:

H2: Absorptive capacity is positively related to value creation.

H5: Value creation is positively related to organisational performance.

H8: Value creation mediates the relationship between absorptive capacity and organisational performance.

6.1.3 Absorptive Capacity, Strategic Agility, Value Creation, Organisational Performance

Strategic agility and absorptive capacity represent dynamic capabilities crucial for organisations to navigate and adapt to evolving external environments. These capabilities synergistically enable organisations to pivot their innovation strategies in alignment with overarching organisational goals (Zahoor et al. 2024). Strategic agility leverages the insights gained from absorptive capacity, optimizing operational quality to effectively fulfill primary objectives (Tufan & Mert 2023). Consequently, strategic agility aims at enhancing organisational performance by embracing the most efficient and effective practices dictated by the external milieu. As organisational efficiency and effectiveness necessitate continual innovation to address challenges like competition and shifting customer demands, strategic agility becomes instrumental in driving innovation efforts within an organisation for improved outcomes (Abourokbah et al. 2023). The resultant output from strategic agility initiatives manifests in value creation, empowering organisations to develop goods and services tailored to prevailing conditions. Innovation emerges as a central output of strategic agility, facilitating strategic adjustments in processes and strategies to deliver unique offerings through optimized resource utilization (Abourokbah et al. 2023). This value creation process encompasses novel production methods that enable organisations to enhance customer service delivery. Therefore, the ability to create value significantly impacts organisational performance by fostering the creation of new processes and systems geared towards navigating environmental changes effectively. A comprehensive review of various databases yielded no empirical studies addressing the mediating influence of strategic agility, facilitated through value creation, on the correlation between absorptive capacity and organisational performance. Therefore, the following hypothesis was proposed:

H6: Strategic Agility is Positively Related to Value Creation.

H9: Strategic agility affects value creation to mediate the relationship between absorptive capacity and organisational performance.

7 Results

The demographic characteristics of the surveyed population are as follows: The gender distribution shows that 56% are male and 44% are female. In terms of the age of organisations, 6% are less than 1 year old, 27% are between 1 and 4 years old, 36% are between 5 and 10 years old, and 31% are over 10 years old. Regarding organisation size, 38% have 5 employees or less, 25% have 6 to 10 employees, 18% have 11 to 20 employees, and 19% have over 20 employees. Education levels indicate that 14% have a high school degree, 63% have a bachelor's degree, 20% have a master's degree, and 3% have a doctorate degree. Job positions include 55% managers, 22% directors, and 23% owners. Geographically, 61% are in Abu Dhabi, 31% in Dubai, 2% in Ras Al Khaimah, and 1% in Sharjah.

7.1 Measurement Model Estimation

The evaluation of PLS-SEM estimates entails a thorough scrutiny of the reliability and validity of the measures employed to represent the constructs under study (Lee et al. 2011; Limaj & Bernroider 2019). The study empirically and statistically evaluated reflective measurement items by examining their outer/item loadings and composite reliability (CR). The CR values of all the constructs are well within the range of 0.860 to 0.924. Additionally, to gauge convergent validity, the study assessed average variance extracted (AVE) and discriminant validity among constructs. As depicted in Table 1, Factor loading values ranging from 0.40 to 0.70 can be deemed justifiable (Hair & Alamer 2022). The AVE values for all the constructs were well withing the range of 0.402 to 0.740.

In addition, the more conservative threshold being followed by different researchers for discriminant validity is less than 0.85 which is deemed acceptable. However, less than 0.90 is considered more lenient acceptable range when justifying the HTMT or discriminant validity, as depicted in Table 2 below (Hair & Alamer 2022).

Table 1 First order measurement model and construct validity

	Cronbach's alpha	Composite reliability (rho_c)	Average variance extracted (AVE)
Absorptive capacity (AC)	0.829	0.866	0.402
Financial performance (FP)	0.898	0.924	0.71
Innovational performance (IP)	0.798	0.86	0.553
Strategic agility (SA)	0.86	0.899	0.641
Strategic performance (SP)	0.816	0.879	0.645
Value creation (VC)	0.883	0.919	0.74

Table 2 First order discriminant validity

Heterotrait-monotrait ratio (HTMT)—Matrix

	AC	FP	IP	SA	SP	VC
AC						
FP	0.414					
IP	0.808	0.566				
SA	0.609	0.405	0.558			
SP	0.760	0.416	0.725	0.521		
VC	0.690	0.434	0.621	0.869	0.513	

Table 3 Second order measurement model and construct validity

	Cronbach's alpha	Composite reliability (rho_c)	Average variance extracted (AVE)
AC	0.829	0.867	0.402
LV scores -OP	0.732	0.847	0.650
SA	0.860	0.899	0.641
VC	0.883	0.919	0.740

Table 4 Second order discriminant validity

Heterotrait-monotrait ratio (HTMT)—Matrix

	AC	LV scores -OP	SA	VC
AC				
LV scores -OP	0.874			
SA	0.609	0.660		
VC	0.690	0.698	0.869	

Furthermore, the disjoint two-stage approach was employed to evaluate the model, emphasizing the reflective measurement models of the lower-order components initially and subsequently utilizing the latent variable scores of these components to construct and estimate the stage two model (Sarstedt et al. 2019).

Similarly, the measurement model values for the second order (higher order) for the latent construct was 0.847, which is well within the range of suggested CR values. The latent construct demonstrates robust levels of internal consistency and reliability, as indicated by the CR value. Additionally, to gauge convergent validity, the study assessed average variance extracted (AVE) and discriminant validity for the latent construct. As depicted in Table 3, factor loading values ranging from 0.40 to 0.70 can be deemed justifiable (Hair & Alamer 2022). The AVE value for the latent construct was 0.650. Similarly, the discriminant validity for the second order latent variable was also within the acceptable range of less than 0.90 as shown in Table 4.

7.2 Structural Model Estimation

Following the validation of measurement items for their validity and reliability, the subsequent stage involved evaluating the relationships within the structural model. This entailed assessing the structural model using Partial Least Squares Structural Equation Modelling (PLS-SEM) and determining the significance of the path coefficients. Through the application of the PLS-SEM algorithm, estimates were derived for the path coefficients representing the expected relationships between the research constructs. This evaluation included analysing the R^2 values for the level of explained variance, as well as the Q^2 predict.

For the LV scores—OP model, 56% of the variance in the dependent variable is explained, suggesting a moderate level of explanatory power. In contrast, the SA model shows a lower explanatory power with 27.5% of the variance explained, while the VC model demonstrates a higher explanatory power with 65.1% of the variance explained.

The path coefficients and specific indirect effects analysis reveal several significant relationships within the structural model. Notably, hypotheses H1, H2, H3, H4, and H6 are supported, indicating significant positive relationships between variables such as AC, SA, VC, and LV scores—OP. These findings are supported by strong T-statistics and low p-values, suggesting robust and reliable relationships. However, hypothesis H5, concerning the relationship between VC and LV scores—OP, is not supported due to a lack of statistical significance. Similarly, specific indirect effects H8 and H9, involving indirect relationships via VC and both SA and VC, respectively, are not supported. These results contribute to a nuanced understanding of the interplay between the variables in the structural model, highlighting both direct and indirect pathways of influence, as illustrated in Table 5.

The Q^2 predict values provide insights into the predictive relevance of the structural model for each research construct. A Q^2 predict value of 0.512 for LV scores—OP suggests that the model can explain approximately 51.2% of the variance in this construct, indicating a moderate level of predictive relevance.

Table 5 Result of hypothesis testing and structural relationships

Path coefficients					
Hypothesis	Path	Original sample	T statistics	P values	Description
H1	AC - > SA	0.524	12.357	0.000	Supported
H2	AC - > VC	0.262	5.023	0.000	Supported
H3	AC - > LV scores -OP	0.577	13.147	0.000	Supported
H4	SA - > LV scores -OP	0.165	2.920	0.004	Supported
H5	VC - > LV scores -OP	0.096	1.616	0.106	Not Supported
H6	SA - > VC	0.638	13.388	0.000	Supported
Specific indirect effects					
	Path	**Original sample**	**T statistics**	**P values**	**Description**
H7	AC - > SA - > LV scores -OP	0.087	2.810	0.005	Supported
H8	AC - > VC - > LV scores -OP	0.025	1.516	0.129	Not Supported
H9	AC - > SA - > VC - > LV scores -OP	0.032	1.587	0.113	Not Supported

8 Discussion and Conclusion

Although the investigation of absorptive capacity and strategic agility within the tourism industry is still relatively underexplored within the current body of literature. This area necessitates further examination across different contexts, including the United Arab Emirates, which serves as the focal point for the present study as outlined by (Kale et al. 2019). As innovation plays a role in creating value for the product or service, therefore, the absorptive capacity serves to enhance the ability of the organisations to create a value for new products or services (Medase & Barasa 2019). Accordingly, the result of this study is coherent with the literature reviewed. Creating strategic, agile businesses requires continuous environmental monitoring to pre-emptively address disruptions to operational performance, enabling swift acquisition and utilization of market insights for adaptive responses (Khaddam 2020).

The impact of management on organisational performance hinges on the absorptive capacity of employees, with proficient management fostering growth by facilitating openness to external ideas, as emphasized by Conţu (2020), and further supported by Hu et al. (2022). To endure and thrive, businesses must adapt by developing capabilities such as cohesive leadership, adaptable resource allocation, and strategic acumen, which constitute key facets of strategic agility as outlined by Vrontis et al. (2023), particularly crucial within the tourism sector due to evolving customer value driven by external factors, as highlighted by Hadjielias et al. (2022). Furthermore, according to Pütz & Werner (2024), value creation has a vital role in shaping the profitability of an organisation. However, the results of this study showed that VC had no impact on improving OP, hence the hypothesis was not supported. The need for innovation to create value brings strategic agility and value creation together. According to (Khaddam 2020), strategic agility is a method that organisations use to develop new directions and ways towards innovation practices leading to value creation.

In an uncertain and volatile business environment, management capability and dynamic alignment are critical for organisational performance and success (Idrees et al. 2022). The impact of absorptive capacity is evident in today's knowledge-based economy, where knowledge plays a significant role in organisational performance. According to Li et al. (2020), knowledge is the foundation of the value of firms. However, the results of this study are in the contradiction of this statement, as this hypothesis was not supported. Tufan and Mert (2023) assert that organisations aspiring to cultivate strategic agility should possess the capacity for ongoing environmental scanning, promptly processing and utilizing information to make informed responses, thereby enabling adaptive adjustments in strategies, processes, technology, and competencies, as emphasized by Abourokbah et al. (2023). In contrast, the resulting hypothesis is not supporting the reviewed notion. Hence, the hypothesis was rejected.

For organisations to effectively navigate and respond to changing environments, scholars such as Teece et al. (1997) has articulated the concept of dynamic capabilities. This theoretical framework posits that dynamic capabilities constitute a set of organisational competencies aimed at delineating boundaries within which organisations can discern, acquire, and leverage both internal and external knowledge to derive strategic advantages while mitigating external disruptions. Based on the findings of this research, absorptive capacity, and strategic agility both had a positive influence on organisational performance.

Furthermore, absorptive capacity also had a positive influence on value creation, but the effect of value creation on organisation performance was not supported. Hence, the mediating effect of strategic agility between absorptive capacity and organisation performance was supported as the outcome of this study, whereas value creation doesn't mediate the relationship between absorptive capacity and organisational performance. On the other hand, different constructs added value to the underpinning theory of this research where the relationship between strategic agility and value creation was also supported.

As a result of this study, all the studied variables enhance dynamic capability leading to an improved competitive advantage.

The interplay of absorptive capacity, strategic agility, value creation, and organisational performance holds profound implications within the tourism landscape of the UAE. The country's tourism sector thrives on its ability to absorb and integrate external knowledge, leveraging it to innovate and adapt strategically. This dynamic capacity not only facilitates rapid responses to evolving market trends and visitor preferences but also fuels the creation of unique value propositions that resonate with tourists. By continuously learning and innovating, UAE's tourism stakeholders can offer diverse and compelling experiences, from cultural attractions to luxury hospitality, tailored to meet the discerning demands of global travellers. This value-centric approach not only enhances the sector's competitiveness but also translates into tangible outcomes such as increased tourist arrivals, higher occupancy rates, and elevated guest satisfaction levels. These positive performance indicators, driven by absorptive capacity and strategic agility, play a pivotal role in shaping the financial success and sustainability of UAE's tourism ecosystem. Embracing and optimising these dynamic capabilities enables the country to maintain its position as a premier global tourism destination, showcasing the synergistic impact of innovation, agility, and value creation in driving tourism growth and excellence.

Since it was cross sectional study, all the data was gathered during the same period. Making it longitudinal would provide more robust results. The researchers struggled to get hold of the respondents of the respected tourism companies management. Furthermore, the targeted tourism companies were limited to only travel and tourism entities within UAE. Hence, the researchers recommend exploring the remaining entities that form the part of tourism sector within the UAE. As for the future researchers, they could try to investigate the same topic qualitatively. Furthermore, it could also be tested based on another industry like manufacturing or banking industry in the context of the United Arab Emirates. Lastly, future researchers could also stratify the sampling strategy to further enhance the results based on travel experience/ repeated visitations to the same destinations or based on location.

References

Abourokbah SH, Mashat RM, Salam MA (2023) Role of absorptive capacity, digital capability, agility, and resilience in supply chain innovation performance. Sustainability 15(4):1–25. https://doi.org/10.3390/su15043636

Altay N, Gunasekaran A, Dubey R, Childe SJ (2018) Agility and resilience as antecedents of supply chain performance under moderating effects of organizational culture within the humanitarian setting: a dynamic capability view. Prod Plan Control 29(14):1158–1174. https://doi.org/10.1080/09537287.2018.1542174

Antia KD, Mani S, Wathne KH (2017) Franchisor-franchisee bankruptcy and the efficacy of franchisee governance. J Mark Res 28(6):952–967

Božič K, Dimovski V (2019) Business intelligence and analytics for value creation: the role of absorptive capacity. Int J Inf Manage 46:93–103. https://doi.org/10.1016/j.ijinfomgt.2018.11.020

Bryman A (2012) Social research methods (4th ed). Oxford University Press

Cegarra-Navarro JG, Soto-Acosta P, Wensley AKP (2016) Structured knowledge processes and firm performance: the role of organizational agility. J Bus Res 69(5):1544–1549. https://doi.org/10.1016/j.jbusres.2015.10.014

Cohen WM, Levinthal DA (1989) Innovation and learning: the two faces of R & D. Econ J 99(397):569–596

Cohen WM, Levinthal DA (1990) Absorptive capacity: a new perspective on learning and innovation. Adm Sci Q 35(1):128–152. https://doi.org/10.2307/2393553

Conțu EG (2020) Organizational performance—theoretical and practical approaches; study on students' perceptions. Proc Int Conf Bus Excel 14(1):398–406. https://doi.org/10.2478/picbe-2020-0038

Cruz-Ros S, Guerrero-Sánchez DL, Miquel-Romero MJ (2018) Absorptive capacity and its impact on innovation and performance: findings from SEM and fsQCA. Rev Manag Sci 1–15. https://doi.org/10.1007/s11846-018-0319-7

Dubey R, Gunasekaran A (2016) The sustainable humanitarian supply chain design: agility, adaptability and alignment. Int J Log Res Appl 19(1):62–82. https://doi.org/10.1080/13675567.2015.1015511

Dyer JH, Singh H (1998) The relational view: cooperative strategy and sources of interorganizational competitive advantage. Acad Manag Rev 23(4):660–679. https://doi.org/10.2307/259056

Eisenhardt KM, Martin JA (2000) Dynamic capabilities: what are they? Strateg Manag J 21(10–11):1105–1121. https://doi.org/10.1002/1097-0266(200010/11)21:10/11%3c1105::AID-SMJ133%3e3.0.CO;2-E

Freudenreich B, Lüdeke-Freund F, Schaltegger S (2020) A stakeholder theory perspective on business models: value creation for sustainability. J Bus Ethics 166(1):3–18. https://doi.org/10.1007/s10551-019-04112-z

Hadjielias E, Christofi M, Christou P, Hadjielia Drotarova M (2022) Digitalization, agility, and customer value in tourism. Technol Forecast Soc Chang 175(November 2021):121334. https://doi.org/10.1016/j.techfore.2021.121334

Hair J, Alamer A (2022) Partial least squares structural equation modeling (PLS-SEM) in second language and education research: guidelines using an applied example. Res Methods Appl Linguist 1(3):100027. https://doi.org/10.1016/j.rmal.2022.100027

Hu Z, Sarfraz M, Khawaja KF, Shaheen H, Mariam S (2022) The influence of knowledge management capacities on pharmaceutical firms competitive advantage: the mediating role of supply chain agility and moderating role of inter functional integration. Front Public Health 10(1):1–11. https://doi.org/10.3389/fpubh.2022.953478

Idrees H, Hynek J, Xu J, Akbar A, Jabeen S (2022) Impact of knowledge management capabilities on new product development performance through mediating role of organizational agility and moderating role of business model innovation. Front Psychol 13:1–15. https://doi.org/10.3389/fpsyg.2022.950054

Jiménez-Barrionuevo MM, García-Morales VJ, Molina LM (2011) Validation of an instrument to measure absorptive capacity. Technovation 31(5–6):190–202. https://doi.org/10.1016/j.technovation.2010.12.002

Kafouros M, Love JH, Ganotakis P, Konara P (2020) Experience in R&D collaborations, innovative performance and the moderating effect of different dimensions of absorptive capacity. Technol Forecast Soc Chang 150(September 2019):119757. https://doi.org/10.1016/j.techfore.2019.119757

Kale E, Aknar A, Başar Ö (2019) Absorptive capacity and firm performance: the mediating role of strategic agility. Int J Hosp Manag 78:276–283. https://doi.org/10.1016/j.ijhm.2018.09.010

Khaddam AA (2020) Impact of personnel creativity on achieving strategic agility: the mediating role of knowledge sharing. Manag Sci Lett 10(10):2293–2300. https://doi.org/10.5267/j.msl.2020.3.006

Khan Z, Rao-Nicholson R, Tarba SY (2018) Global networks as a mode of balance for exploratory innovations in a late liberalizing economy. J World Bus 53(3):392–402. https://doi.org/10.1016/j.jwb.2016.10.002

Knudsen MP, Schleimer S (2020) The role of prevailing individual absorptive capacity versus absorptive capacity development for different innovation outcomes. Knowl Manag Res Pract 20(5):704–718. https://doi.org/10.1080/14778238.2020.1787801

Kogut B, Zander U (1996) What firms do? coordination, identity, and learning. Organ Sci 7(5):502–518. https://doi.org/10.1287/orsc.7.5.502

Lane PJ, Koka BR, Pathak S (2006) The reification of absorptive capacity: a critical review and rejuvenation of the construct. Acad Manag Rev 31(4):833–863. https://doi.org/10.5465/AMR.2006.22527456

Lee L, Petter S, Fayard D, Robinson S (2011) On the use of partial least squares path modeling in accounting research. Int J Account Inf Syst 12(4):305–328. https://doi.org/10.1016/j.accinf.2011.05.002

Limaj E, Bernroider EWN (2019) The roles of absorptive capacity and cultural balance for exploratory and exploitative innovation in SMEs. J Bus Res 94(1):137–153. https://doi.org/10.1016/j.jbusres.2017.10.052

McAdam R, Humphreys P, Galbraith B, Miller K (2017) Developing management capability within a horizontal supply chain in performance measurement deployment and evolution: a dynamic capabilities and goal theory perspective. Prod Plan Control 28(6–8):610–628. https://doi.org/10.1080/09537287.2017.1309706

Medase K, Barasa L (2019) Absorptive capacity, marketing capabilities, and innovation commercialisation in Nigeria. Eur J Innov Manag 22(5):790–820. https://doi.org/10.1108/EJIM-09-2018-0194

Müller JM, Buliga O, Voigt KI (2020) The role of absorptive capacity and innovation strategy in the design of industry 4.0 business Models-A comparison between SMEs and large enterprises. Eur Manag J xxxx:1–11. https://doi.org/10.1016/j.emj.2020.01.002

Nair SR, Demirbag M, Mellahi K, Pillai KG (2018) Do parent units benefit from reverse knowledge transfer? Br J Manag 29(3):428–444. https://doi.org/10.1111/1467-8551.12234

Papastathopoulos A, Ahmad SZ, Al Sabri N, Kaminakis K (2019) Demographic analysis of residents' support for tourism development in the UAE: a Bayesian structural equation modeling multigroup approach. J Travel Res 59(6):1119–1139. https://doi.org/10.1177/0047287519874131

Pütz L, Werner A (2023) Absorptive capacity in family firms: a systematic literature review. In: Review of managerial science (vol 18, Issue 2). Springer Berlin Heidelberg. https://doi.org/10.1007/s11846-023-00651-y

Reisinger Y, Michael N, Hayes JP (2018) Destination competitiveness from a tourist perspective: a case of the United Arab Emirates. Int J Tour Res 21(2):259–279. https://doi.org/10.1002/jtr.2259

Sarstedt M, Hair JF, Cheah JH, Becker JM, Ringle CM (2019) How to specify, estimate, and validate higher-order constructs in PLS-SEM. Australas Mark J 27(3):197–211. https://doi.org/10.1016/j.ausmj.2019.05.003

Saunders M, Lewis P, Thornhill A (2009) Research methods for business students, 5th edn. Pearson Education Limited

Silva M de, Rossi F (2018) The effect of firms' relational capabilities on knowledge acquisition and co-creation with universities. In: Technological forecasting and social change (Vol 133)

Tallon PP, Pinsonneault A (2011) Competing perspectives on the link between strategic information technology alignment and organizational agility: insights from a mediation model. MIS Q 35(2):463–486

Teece DJ, Pisano G, Shuen A (1997) Dynamic capabilities and strategic management. Strateg Manag J 18(7):509–533. https://doi.org/10.1142/9789812796929_0004

Thomas R, Wood E (2014) Innovation in tourism: re-conceptualising and measuring the absorptive capacity of the hotel sector. Tour Manage 45:39–48

Tufan C, Mert IS (2023) The sequential effect of absorptive capacity, strategic agility, and sustainable competitive advantage on sustainable business performance of SMEs. Environ Sci Pollut Res 30(19):55958–55973. https://doi.org/10.1007/s11356-023-26207-x

Verma V, Bharadwaj SS, Nanda M (2017) Comparing agility and absorptive capacity for superior firm performance in dynamic environment. Int J Bus Environ 9(1):1–17. https://doi.org/10.1504/IJBE.2017.084703

Vrontis D, Belas J, Thrassou A, Santoro G, Christofi M (2023) Strategic agility, openness and performance: a mixed method comparative analysis of firms operating in developed and emerging markets. In: Review of managerial science (Vol 17, Issue 4). Springer Berlin Heidelberg. https://doi.org/10.1007/s11846-022-00562-4

Zahoor N, Khan H, Donbesuur F, Khan Z, Rajwani T (2024) Grand challenges and emerging market small and medium enterprises: the role of strategic agility and gender diversity. J Prod Innov Manag 41(2):473–500. https://doi.org/10.1111/jpim.12661

Zahra SA, George G (2002) Absorptive capacity: a review, reconceptualization, and extension. Acad Manag Rev 27(2):185–203

Determinants of Adoption of Integrated Reporting in an Emerging Market: The Case of UAE

Dania M. Kurdy and Husam-Aldin N. Al-Malkawi

Abstract The purpose of this study is to examine the determinants of the voluntary adoption of integrated reporting in the UAE context. The study employs a sample of 34 non-financial companies listed on Abu Dhabi Stock Exchange and Dubai Financial Market. Initially, content analysis is used to assess the quality of integrated reporting and constructing an index to measure integrated reporting quality (IRQ). Then regression analysis is used to examine the relationship between IRQ and selected variables. The results show that there is a positive association between IRQ and firm size, board size, gender diversity and growth opportunity. While the IRQ is negatively related with financial leverage and profitability. The findings of the paper offer important implications for large companies with large board, gender diversity and growth opportunity to utilize integrated reporting to reduce agency problems and signal its prospects. To the best of our knowledge, this is the first empirical study to construct IR index for UAE companies and to examine the determinants of voluntary adoption of IR within the UAE context.

Keywords Integrated reporting · Voluntary adoption · Firm characteristics · Regression analysis · UAE

1 Introduction

In the contemporary landscape of corporate disruption, heightened uncertainty underscores the necessity for transparency in a company's long-term value creation for investors. Within this context, innovation assumes a crucial role, emphasizing the need for adaptable reporting models capable of accommodating unpredictable business dynamics. Integrated Reporting (IR) emerges as a transformative approach, amalgamating both financial and non-financial information, including intellectual

This paper was prepared for Ph.D. dissertation, taught by Prof. Husam-Aldin Al-Malkawi.

D. M. Kurdy (✉) · H.-A. N. Al-Malkawi
Faculty of Business and Law, The British University in Dubai, Dubai, United Arab Emirates
e-mail: kurdy.dania@gmail.com

K. Al Marri et al. (eds.), *BUiD Doctoral Research Conference 2024*, Lecture Notes in Civil Engineering 587, https://doi.org/10.1007/978-3-031-84371-6_10

capital, corporate social responsibility (CSR), and strategy. This approach provides a comprehensive disclosure framework. The International IR Council, established in 2010, guides this initiative, involving auditors, organizations, and non-governmental organizations (NGOs). While IR is mandated in some regions, voluntary adoption prevails in many countries, including the UAE.

Using data from both developed and emerging markets, scholars are actively engaged in investigating the factors that drive companies voluntarily adopt IR (e.g. Al Amosh et al. 2022; Senani et al. 2022; Dey 2020; Busco et al. 2019). While there is a substantial amount of literature on IR, studies within the Gulf Cooperation Countries (GCC) context are generally limited. The current study attempts to contribute to the existing body of knowledge by shedding the lights on the IR practice in an emerging market that remains understudied, namely the UAE. Drawing on previous literature and diverse theories including agency theory, stakeholder theory, signalling theory, and institutional theory, this research aims to offer insights into the adoption of IR in the UAE. Therefore, the aim of present study is threefold: (1) to contribute to the existing theories; (2) to fill the gap in the literature, and (3) to deepen the understanding of the factors that influence the voluntary adoption of IR within the new context of the UAE.

2 Prior Studies and Research Hypotheses

Drawing on the underpinning relevant theories and previous research, several variables are identified and believed to affect IRQ. The first variable is the firm size (SIZE). Larger firms are usually subject to greater agency problem and information asymmetry. Therefore, larger firms tend to disclose more to information (financial and non-financial) to alleviate agency problem (Senani et al. 2022). This variable is measured by the natural logarithm of total assets. The second variable that may affect IRQ is the firms' profitability (PROF). Consistent with the signalling theory, profitable firms are more likely to disclose information to signal their financial health. Moreover, profitable firms are more inclined to allocate resources towards providing integrated and accurate information to the public, aiming to reduce information asymmetries (Al Amosh et al. 2022). Therefore, profitability, measured by the return on investment (ROI), is expected to be positively related to IRQ.

Another factor that is believed to influence firms to disclose more information is financial leverage. In general, firms seeking to obtain more debt financing at lower cost are more likely to disclose more information (Al Amosh et al. 2022). This assertion is in line with the agency theory and the stakeholder theory. To measure firms' financial leverage, we employ the debt to total assets ratio.

Moreover, stakeholder theory predicts that board of directors' size may influence IRQ, the larger board the better the IRQ (Songini et al. 2022). Board size (BOD) is measured by the total number of board members. Furthermore, gender diversity on the board plays a pivotal role in IRQ, where the presence of a female (FEML) on the board improves information disclosure (Dienes and Velte 2016). Finally, the firm growth opportunity measured by price-to-book (PTB) ratio is another variable suggested by the literature that affect IRQ (Dey 2020). Usually, a firm with higher growth tends to disclose more information.

Based on the above discussion, the following testable research hypotheses are formulated:

Hypothesis 1: Ceteris paribus, there is a positive relationship between the firm size and the level of IR disclosure quality.

Hypothesis 2: Ceteris paribus, there is a positive relationship between firm's financial leverage the level of IR disclosure quality.

Hypothesis 3: Ceteris paribus, there is a positive relationship between firm's profitability the level of IR disclosure quality.

Hypothesis 4: Ceteris paribus, there is a positive relationship between firm's board size and the level of IR disclosure quality.

Hypothesis 5: Ceteris paribus, there is a positive relationship between board diversity and the level of IR disclosure quality.

Hypothesis 6: Ceteris paribus, there is a positive relationship between firm's growth opportunity and the level of IR disclosure quality.

3 Data and Methodology

3.1 Data

The study is based on cross-sectional data using a sample of 34 non-financial companies listed on Abu Dhabi Stock Exchange (ADX) and Dubai Financial Market (DFM) for the period 2022/2023. The sample consists of 12 companies listed on ADX and 22 companies listed on DFM. The financial companies are excluded from the analysis because they have different financial reporting and are subject to different regulations. Like this study, previous research conducted by Vitolla et al. (2020) employed cross-sectional analysis to examine the determinants of integrated reporting quality of 20 countries for the year 2017.

3.2 Research Method

The study employs a quantitative approach to explore the factors influencing the voluntary adoption of IR. Initially, the 2022/2023 annual reports of the publicly listed firms on ADX and DFM are analysed. The data analysis involves an in-depth examination of published reports. Then, an index is constructed to assesses the quality of integrated reporting using content analysis of annual reports of companies listed on DFM and ADX. The IR quality index is based on 43 items adapted from the International Integrated Reporting Framework (IIRF). Several studies employed content analysis approach to assess IR quality (see, e.g. Songini et al. 2022; Vitolla et al. 2020). Once the IR disclosure quality index has been developed, regression analysis is employed to examine the relationship between the level of disclosure and selected variables, that are often suggested in the literature as potential factors influencing the voluntary adoption of IR (Al Amosh et al. 2022; Senani et al. 2022; Islam 2021; Dey 2020).

3.3 IR Quality Index (Dependent Variable)

As discussed above, the components used to compute IR quality (IRQ) for the UAE listed companies are based on 43 items adapted from the IIRF (2021) which are classified under six main categories. To assess the IR quality, we developed an IR score which is based on these 43 items. Each item in each category is a binary variant taking a value of 1 or 0. A score of 1 is given in the case of the presence of the item in the firm's published annual report, while a score of 0 is given in the absence of the item (Dey 2020). Later, IRQ is obtained by dividing the total number of IR items disclosed by the total items that should be disclosed, as suggested by IIRF (2021).

3.4 Empirical Model

The aim of this study is to examine the determinants of voluntary adoption of IR in the UAE. Based on the above discussion and in line with the previous research, the following regression model will be used to test the research hypotheses:

$$IRQ_i = \beta_0 + \beta_1 SIZE_i + \beta_2 ROI_i + \beta_3 LEV_i + \beta_4 BOD_i + \beta_5 FEML_i + \beta_6 PTB_i + \varepsilon_i \quad (1)$$

where, for firm *i*, β_0 is the constant and β_1–β_6 are the regression coefficients to be estimated. IRQ is the IR quality (average score), SIZE is the firm size (measured by natural log of total assets); ROI is return on investment (proxy for firm's profitability); LEV is financial leverage (measured by debt-to-total assets); BOD is board size (measured by number of board members), FEML is a measure for gender diversity (dummy variable = 1 if a female present on the board, 0 otherwise); PTB is the price-to-book ratio (proxy for firm's growth); ε_i is the error term.

4 Results

The purpose of this paper is to examine the determinants of voluntary adoption of IR in the UAE. We measured the IR quality by developing an index based on IIRF (2021). Equation (1) is estimated using ordinary least squares (OLS) regression. Before analyzing the result, we performed all the necessary diagnostic tests to ensure that the OLS regression assumptions have been met and OLS produces unbiased coefficient estimates. The results show that OLS assumptions are not violated. In addition, to test for potential endogeneity problem, we have saved the residuals and run a correlation between error term and the independent variables. The result reveals that there is no correlation between the error term and the explanatory variables, suggesting that they are exogenous.

The overall significance of the regression model is assessed by F-test. Table 1 shows that, the value of F-stat is 4.432 with a p-value of 0.003, rejecting the null hypothesis that $\beta_1 = \beta_2 = \ldots = \beta_6 = 0$. That is, the estimated coefficients are jointly not equal to zero. The regression R-squared is 0.4962, which indicates that about 50% of the variation in the dependent variable (IRQ) is explained by the model. As can be seen in Table 1, all the estimated coeffects are statistically significant at 5% level (*p-value* < 0.05) except BOD at 10% level (*p-value* < 0.10).

The result reveals that there is a positive association between the firm integrated reporting quality (IRQ) and its size. The coefficient on SIZE is positive (0.0455) and statistically significant at 5% level (*p-value* = 0.0286). This suggests that larger firms disclose more information, consistent with agency theory and with the previous research of Al Amosh et al. (2022), Vitolla et al. (2020) and Islam (2021). As expected, board size (BOD) and the presence of a female on the board (FEML) are positively related to the IRQ. The coefficients on BOD and FEML are statistically significant at 10% (*p-value* = 0.0681) and 5% (*p-value* = 0.0262) levels,

Table 1 OLS regression result: determinants of IRQ

	Coefficient	Std. error	t-ratio	p-value
CONSTANT	−0.738385*	0.396842	−1.861	0.0737
SIZE	0.0454542**	0.0196489	2.313	0.0286
LEV	−0.386973**	0.172964	−2.237	0.0337
BOD	0.0378619*	0.0199256	1.900	0.0681
FEML	0.130428**	0.0554345	2.353	0.0262
PTB	0.0152672**	0.00569162	2.682	0.0123
ROI	−0.0261372**	0.0100426	−2.603	0.0148
R-squared F-stat P-value (F)	0.496197 4.432057 0.003040			

Notes * and ** denote significance at the 10 and 5% levels, respectively. IRQ is integrated reporting quality. SIZE is firm size. LEV is financial leverage. BOD is board size. FEML is gender diversity. PTP is price-to-book ratio. ROI is return on investment respectively. This indicates that firms with larger board and females being on the board tend to disclose more information. This finding is in line with stockholder theory and the findings reported by Vitolla et al. (2020). The findings also show that there is a negative relationship between IRQ and LEV and ROI. The coefficients on LEV and ROI are statistically significant at 5% level. This suggests that firms with higher leverage and higher return on investment disclose less information. Finally, the firm growth opportunity measured by PTB is found to be positively related to IRQ, consistent with the result obtained by Dey (2020) and Islam (2021).

5 Conclusion

The purpose of this study was to examine the determinants of the voluntary adoption of integrated reporting in the UAE context. Initially, content analysis is used to assess the quality of IR and constructing an index to measure IRQ. Then regression analysis is employed to examine the relationship between IRQ and selected variables often suggested by the literature. The results showed that there was a positive association between IRQ and firm size, board size, gender diversity and growth opportunity. While the IRQ was negatively related with financial leverage and profitability.

The findings of the present study offer some valuable implications for the UAE publicly listed companies. Specifically, large companies with large board, gender diversity and growth opportunity can utilize integrated reporting to reduce agency problems and signal their prospects. However, this conclusion should be treated with caution due to the limitation of this study concerning the small sample size. Moreover, another possible limitation is related to the use of single country analysis. Finally, future research could expand this study to include larger sample and make a comparative study using various countries within the GCC or MENA regions.

References

Al Amosh H, Khatib S, Hussainey K (2022) The financial determinants of integrated reporting disclosure by Jordanian companies. J Risk Financ Manag 15(9) 375:1–20. https://doi.org/10.3390/jrfm.15090375

Busco C, Malafronte I, Pereira J, Starita M (2019) The determinants of companies' levels of integration: does one size fit all? Br Account Rev 51(3):277–298. https://doi.org/10.1016/j.bar.2019.01.002

Dey PK (2020) Value relevance of integrated reporting: a study of the Bangladesh banking sector. Int J Discl Gov 17:195–207. https://doi.org/10.1057/s41310-020-00084-z

Dienes D, Velte P (2016) The impact of supervisory board composition on CSR reporting. Evidence from the German two-tier system. Sustainability 8(1) 63:1–20. https://doi.org/10.3390/su8010063

IIRF (2021) The International Integrated Reporting Council (IIRC)

Islam M (2021) Investigating the relationship between integrated reporting and firm performance in a voluntary disclosure regime: insights from Bangladesh. Asian J Account Res 6(2):228–245. https://doi.org/10.1108/AJAR-06-2020-0039

Senani KGP, Ajward R, Kumari JS (2022) Determinants and consequences of integrated reporting disclosures of non-financial listed firms in an emerging economy. J Financ Report Account (ahead-of-print). https://doi.org/10.1108/JFRA-03-2022-0083

Songini L, Pistoni A, Tettamanzi P, Fratini F, Minutiello V (2022) Integrated reporting quality and BoD characteristics: an empirical analysis. J Manag Gov 26:579–620. https://doi.org/10.1007/s10997-021-09568-8

Vitolla F, Raimo N, Rubino M, Garzoni A (2020) The determinants of integrated reporting quality in financial institutions. Corp Gov 20(3):429–444. https://doi.org/10.1108/CG-07-2019-0202

Digital Transformation in Construction Sector; Assessing the Impact on Projects Performance & Productivity

Seham Ahmed Alblooshi and Edward Ochieng

Abstract Digital transformation is a mega trend around the globe and considered to be as a critical success factor for countries, business, and organizations. This study intended to explore the impact of digital transformation on project performance productivity in the construction sector. A secondary data approach was executed to gather the data to address the research question and accomplish its objectives. Thae main finding of this research that there is a positive correlation between digital transformation and projects performance and productivity its construction sector. Moreover, the research presented a statistical findings on the digital transformation impact in the terms of cost, time, quality, and error in construction projects.

Keywords Digitization · Digitalization · Digital transformation · Construction · Cost · Time · Risk · Quality

1 Introduction

Technology emerged as a pivotal force in defining the success and performance of countries, enterprise, and projects on a global scale. Recognizing its profound impact, countries formulated and stablished strategies for successful and effective digital transformation implementation like the United Arab Emirates and Republic of Estonia. Digital transformation enhanced both countries global standing with the UAE ranking (13) and Estonia (23) in digital competitiveness index 2022 (Ouddane,

This paper was prepared for Managing Large Programmes Module, taught by Professor Edward Ochieng.

S. A. Alblooshi (✉)
Faculty of Business and Law, The British University in Dubai, Dubai, UAE
e-mail: Seham.alblooshi@hotmail.com

E. Ochieng
Department of Project Management, Dubai, UAE
e-mail: Edward.ochieng@buid.ac.ae

K. Al Marri et al. (eds.), *BUiD Doctoral Research Conference 2024*, Lecture Notes in Civil Engineering 587, https://doi.org/10.1007/978-3-031-84371-6_11

2023). The integration of digital transformation is now mandatory, offering organizations several advantages in terms of time management, cost reduction, quality enhancement, improve performance, rise productivity, and risk mitigation. Digital transformation can be defined as a shift toward utilizing advanced technologies in both projects and processes as its essential for developing new business module as it will assist organizations to achieve competitive advantage and create high-pitched value (Verhoef et al. 2021).

The construction sector is widely recognized as one of the key sectors influencing the economy of countries. In fact, the global construction industry is a major economic player valued at $12 trillion (USD) sectors, accounting approximately 13.5% of the world's GDP. In addition, labour productivity grows annually by 1% over the past two decades contrasted with 2.8% for the total world economy and 3.6% for manufacturing sector (Elrefaey et al. 2022). Based on these statistical, its notable that involvement of digital technologies plays a critical role in improving the sector productivity. In this case, it is therefore significant employing digital transformation in the construction industry and assessing its influence on projects performance and productivity.

The intention of this study is to investigate the influence of digital transformation on construction sector and its projects performance and productivity considering the research question and objectives.

1.1 Research Question

How does digital transformation in the construction sector influence project performance and productivity?

1.2 Research Objectives

1. Explore and understand the terminology "digital transformation and determine its impact on project-based organizations specifically in the construction sector.
2. Investigate challenges confronted in adaptation of digital technologies.
3. Identify digital technologies implemented construction sector and evaluate its impacts on the projects in terms of productivity, quality, risk, competitive advantage, and economy.

2 Literature Review

Technology emerged as a critical driven force that shapes the success and performance of countries, enterprise, and projects on a global scale. The application of technology raised over the years and in a continuous development. Due to this, this section will present the revolution of the digital transformation concept, its importance in project-based organizations, traditional and digitalized methods in implementation of construction projects.

2.1 Evaluation of Digital Transformation

Recently, several mega trends and future potential changes are impacting either positively or negatively and forming the shape of countries, businesses, and organizations around the world. One of the most and highly impacting trends is digital transformation a worldwide adaptation trend which simply can be defined as "a change on how to employ the most advanced technologies in both projects and processes in order to develop a new business module that will assist to accomplish competitive advantage and create high-pitched value" (Verhoef et al. 2021). The growing importance of digitalization translated into strategies on the level of countries wise and cascaded to the lower levels. According to (Yusuf et al. 2022) that 90% of business across the world from different countries and sectors would be repositioned and ranked digitally.

The concept of digital transformation developed and evaluated over the years due to several factors and pillars led to this and its importance dramatically increased as well, besides it became an essential pillar in the success of the countries, business, and organizations. Due to lack of knowledge and awareness of digital transformation, several businesses and organizations around the world failed to survive competition due to unadapting high-end technologies and continuity of use the traditional methods resulted in suffering at the end (Verhoef et al. 2021). It is therefore critical to consider embracing and implementing digital practices and tools in process-oriented, project- oriented and matrix-oriented organizations with the best strategic approaches recommended.

As mentioned above, there is a long historical background behind the digital transformation. According to (Verhoef et al. 2021) the concept been through (3) stages as the following:

1. Digitization
2. Digitalization
3. Digitalization Transformation

Several research papers have been stablished to distinguish between digitization and digitalization. There are around (26) definition for both (Gradillas and Thomas 2023). The table below presents the various definitions for both concepts and digital transformation delivered by scholars. Furthermore, according to (Gong and Ribiere 2021), there were around (134) definition for digital transformation (Table 1).

To formulate these classifications, scholars adopted different methodologies to provide the optimal definition of each stage. However, each scholar essentially provided a similar definition, led us to demonstrate that chosen elements either be influential or affected factors besides the fact of selecting different elements. Both types critically important, even though each scholar formulated their definitions focusing on their specific area. Yet, there is no optimal definitions for the (3) stages stablished from specialized institutes and despite that, there is a comprehensive understanding and awareness of these concepts across the countries, enterprises, and organizations (Table 2).

Table 1 The definitions of digitization, digitalization, and digitalization transformation

Stage	Definition
Digitization	• (Gradillas and Thomas 2023)The action to convert information into digital information (Verhoef et al. 2021) • The creation of digital artifact through technical process of conversation presentation, and enhancement (Gradillas and Thomas 2023) • The fully digital creation on information and data without a physical or analog counterpart (Frenzel et al. 2021)
Digitalization	• IT or digital technologies can be used to alter the existing business process (Verhoef et al. 2021) • The transformation of the socioeconomic environment through the process of digital artifact adaptation, application and utilization (Gradillas and Thomas 2023) • The use and application of digital technologies in the context of individual, organizations, or society at large, and it can be defined as the influence in individual, organizations, or society at large, induced by this usage (Frenzel et al. 2021)
Digitalization transformation	• A change in how a firms employs digital technologies, to develop a new digital business model that helps to create and appropriate more value for the firm (Frenzel et al. 2021) • A systematic relationship between the constructs in a manner that complements existing digital transformation scholarship (Gradillas and Thomas 2023) • The evaluation of existing business model (Frenzel et al. 2021) • A fundamental change process, enabled by the innovative use of digital technologies accompanied by the strategic leverage of key resources and capabilities, aiming to radically improve an entity and redefine its value proposition for its stakeholders (Gong and Ribiere 2021)

Table 2 The methods applied by the scholars to define the (3) stages

Source	Method
(Verhoef et al. 2021)	The scholar applied a thematic analysis "coding articles " based on (4) disciplines as the following: Information systems management Marketing management Strategic management Innovation management
(Gradillas and Thomas 2023)	The scholar categorized the stage one and stage two definitions into (7) categories based on the (26) definitions identified through the reviews done on the stablished articles and papers: 1. Creation of digital artifacts 2. Digital enhancement 3. Adaptation of digital technology 4. Organizational application of digital technology 5. Facilitating new business opportunities 6. Transforming the economy 7. Emergence of a new socioeconomic reality
(Frenzel et al. 2021)	The researcher identified (3) dimensions to define the first two stages as the following and applied the analysis on (12) articles Socio-economic level Organization level (organizational structure, platform network, process Individual level

2.2 *Digital Strategies of Leading Countries*

Digital transformation has a dramatic impact on economic status. Economic sector of various countries around the globe have enhanced performance and resource optimization through adaptation digital technologies (Yusuf et al. 2022). Due to this, countries across the world established specialized strategies related to digital transformation. As an example, United Arab Emirates formed "The UAE Digital Government Strategy 2025" as a path toward digital transformation and this strategy aligned with other strategies stablished by UAE government such as Centennial 2071, AI strategy and fourth industrial revaluation (AIR) strategy, and UAE strategy for government services that will be discussed below. The strategy drafted on (8) dimensional strategies as shown in Fig. 1 to achieve the main objective that presented in "generate cross-sectoral government commitment" through adopting and ensuring inclusiveness digital services with particular emphasis on certain segments in society such as senior citizens, people with determination, women, children and other vulnerable groups (The UAE Digital Government Strategy 2025 | the Official Portal of the UAE Government, n.d.)

Furthermore, the strategy cantered on (6) main pillars that facilitate achieving the main objective mentioned above and securing support for integrating digital elements into broader government strategies. This integration is considered critical to assure that UAE government fully embraces digital transformation at a national level and aligns with UAE digital vision. The objectives derived from the pillars as shown in

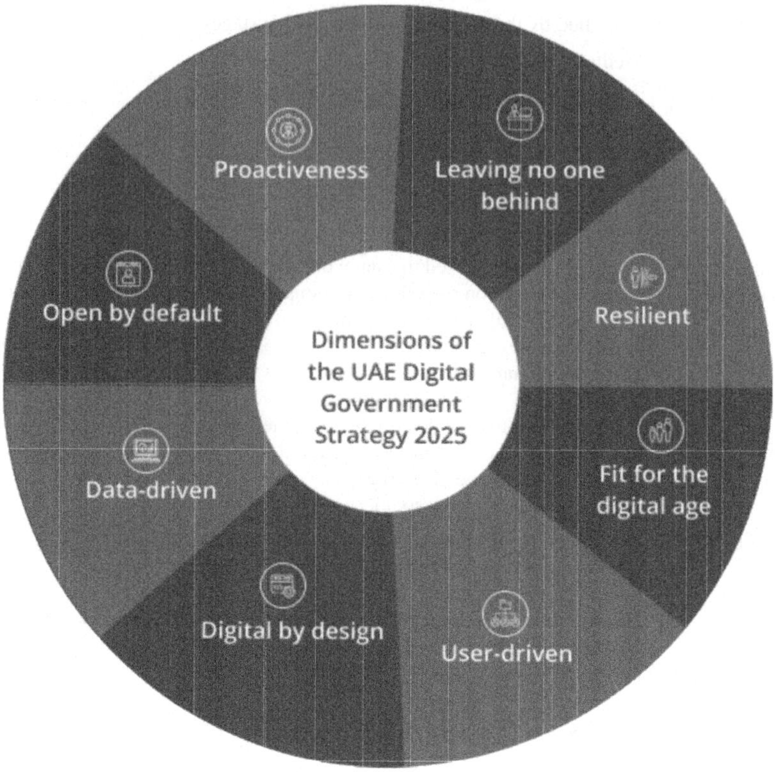

Fig. 1 Dimension of the UAE digital government strategy 2025

Fig. 2 are: providing a world-class digital infrastructure, unified digital platform and common digital enablers, enable integration, easy and fast digital services tailored to meet customers' requirements, extending the level of digital capabilities and skills, legislation readiness to assure a smooth and all-encompassing digital transformation and lastly, growing the efficiency of government work (The UAE Digital Government Strategy 2025 | the Official Portal of the UAE Government n.d.).

UAE strategy for government services is one of the strategies aligned with UAE government digital strategy 2025. The strategy established by the year 2021 and aimed to achieve several key objectives by the end of year 2023: firstly, 90% of the public services through a single platform. Secondly, design 100% proactive digital services where they anticipate the requirements of users and provide solutions in advance. Furthermore, design 100% services in partnership across all sectors of the community. Lastly, governmental services accessible from any location and at any time (The UAE Digital Government Strategy 2025 | the Official Portal of the UAE Government n.d.).

On the other hand, Estonia is one of the most advanced countries around the world in terms of technology. The country established Estonia's Digital Agenda

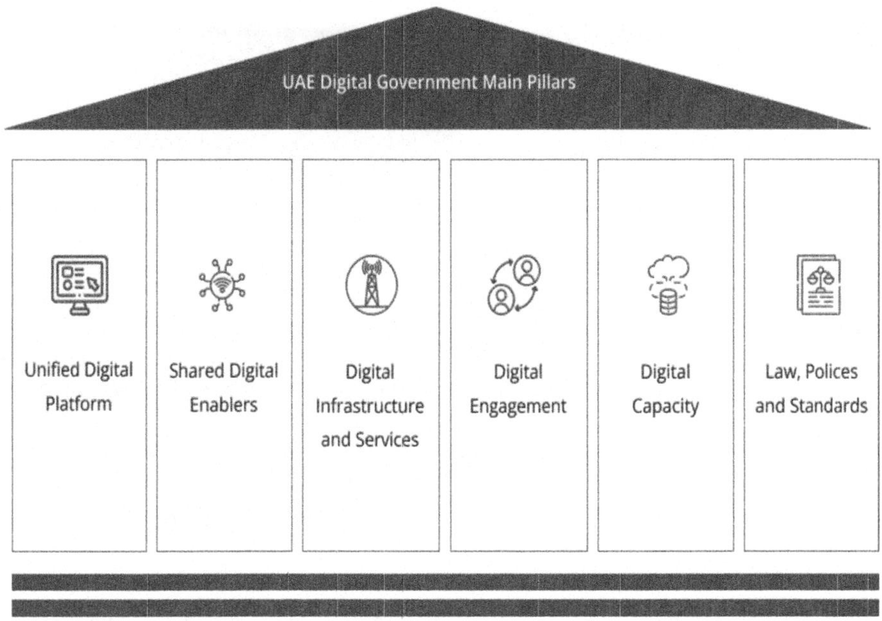

Fig. 2 UAE Digital Government Main Pillars

2030 by the year of 2020. This forward-looking strategy aims to leverage digital technology for the growth of the country's economy status and society in the next decades. The strategy focuses on (3) main sub objectives as the following: digital government, connectivity and cyber security. It's notable that Estonia created this strategy with taken into consideration sustainable development as shown in Fig. 3. For insistence, in the term of society, in sub objective (1) digital government focuses on a number of elements including human-centric digital government, where this element specifically focuses attention on guaranteeing providing equal opportunity in contributing to digital society and usage of digital services, beside to promote the new technologies for the benefit of the society. Under the same sub objective (1); green digital government, the country acknowledges the rising of environmental footprint due to the technological usage (Digital Agenda 2030 | Majandus-Ja Kommunikatsiooniministeerium n.d.).

Despite being a leader in digitalization and ranked highly due to their advancement in this field, Estonia recognizes there responsibilities toward the environment. The country is commitment to reduce the impact of implementation of digital tools and services. In this Agenda, the country developing a comprehensive action plan to focus on executing and implementing climate and environmental friendliness and green IT initiatives. Furthermore, Estonia is set to play a significant role in the upcoming UN Climate Change Conference of the Parties (COP 28) in the UAE, where they will be promoting and placing climate in the centre point. Moreover, the country has notably reduced the greenhouse gas emissions due their commitment to a

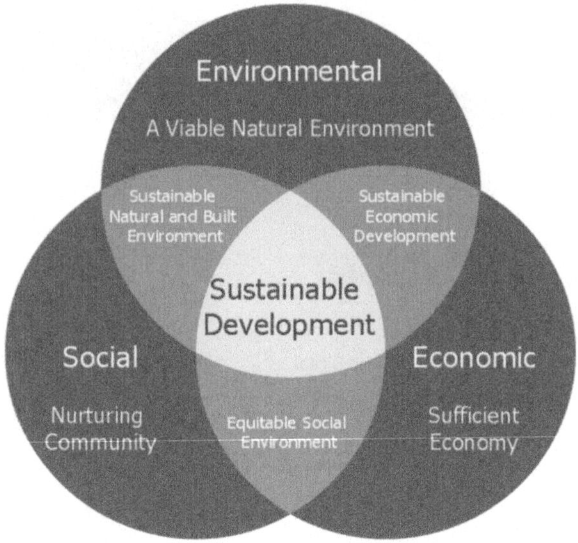

Fig. 3 Sustainable Development

sustainable future. This achievement may have been indirectly influenced from their experience in digital which paved the way to set green transition as a new priority. This shift enabled Estonia to scale up technical innovation to accomplish sustainable future (Estonia at UN Climate Change Conference of the Parties (COP28) in Dubai | Ministry of Climate n.d.).

Both strategies capture the futuristic life of their societies with the implementation of digitalized technics and tools in different fields including infrastructure. From our standpoint, United Arab Emirates is a way ahead of their strategies due to their strong commitment to provide a high quality of life for their citizens and residents through executing countless initiatives in related matter to achieve their vision, including digitalization. In fact, UAE ranked (13) in digital competitiveness index ranking 2022 although they were ranked (12) due 2021, while Estonia ranked (22) in 2022 while in 2021 they were (25). Its notable, that Estonia pushing their efforts to accomplish their agenda and UAE at the same page as well to gain the competitive advantage. However, it's distinguished that even though Estonia considered to be one of the most advanced countries in digital field, yet UAE is in a better place as published in the IMD report mentioned above although there is affecting factors while levelling countries and despite that UAE proved to be advanced (Ouddane, 2023). In addition, in May 2023, the higher committee of digital transformation In UAE approved (6) transformational projects aimed at enhancing UAE digital government and improving the quality and efficiency of digital governmental services. As these initiatives package represent a crucial step in accomplishing "We the UAE 2031 ". The vision includes positioning UAE among the top five countries globally in digital competitiveness index and smart government. Furthermore, to strength UAE position and aspiring to rank within the top three countries in digital governmental services (MoCA n.d.).

2.3 *The Importance of Digital Transformation in Project-Based Organizations*

Countries realized the importance of digital transformation and stablished strategies on a national level in related matters and it cascaded into businesses and organizational levels to translate these strategies to reality. The integration of advanced technologies in businesses and organizations has brought a significant positive impact where digital transformation became a critical and necessity factor to both. This shift allows businesses and organizations to deliver high quality results with minimal errors and remain competitiveness in marketplace (Yusuf et al. 2022).

To start with, the application of digital raised the economy status of countries as mentioned above. In recent research by (Gong and Ribiere 2021), the scholars indicated six elements that surfaced as the most frequently defining characteristics as following: nature, entity, means, expected outcomes, impact and scope. The scholars created a conceptual diagram as presented in Fig. 4 to understand the linkage between the elements and provide the most optimal definition for digital transformation. The conceptual proposed by the scholars illustrates the process of digital transformation starting with influencing factors presented in nature and entity and ended by the outcomes expected and impact. Moreover, (Gong and Ribiere 2021) identified two categories of expected outcomes economic-driven and capability-driven outcomes.

Theses outcomes underscore the importance of digital transformation, as it can lead to develop a new business model which is emphasized in the definitions of the concept presented in Table 1, create competitive advantage, reduce and control cost, and improve project management processes. This in return reflect positively on efficiency and productivity of the business and the organization (Gong and Ribiere 2021).Furthermore, other scholars have also emphasized these outcomes next to enhance sustainability, control time, increase quality, satisfaction, improve safety, risk minimization and improve communication (Liu et al., 2021; Bajpai & Misra, 2022; Elrefaey et al., 2022).

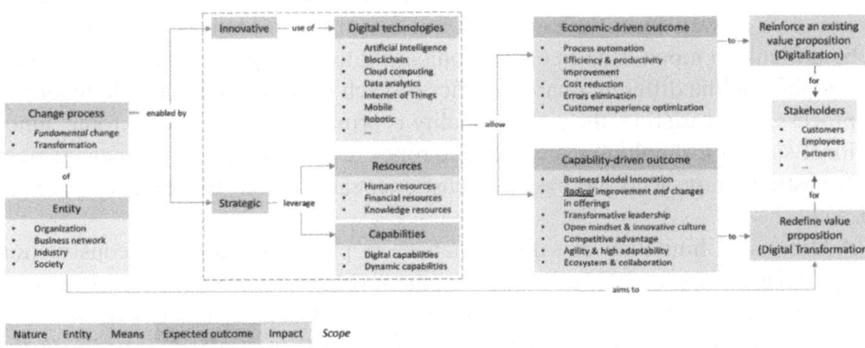

Fig. 4 Conceptual diagram by (Gong and Ribiere 2021)

From our perspectives, we strongly agree with the scholars in terms of communication, safety, risk and satisfaction, and this due our own experience during COVID-19 pandemic were shifting to work in distance using different platforms improved communication, increased level of safety, reduced the risk of getting affected by the virus and this in return increased satisfaction where we believe that been applied in all fields. Moreover, as stated in (Bajpai & Misra, 2022) research paper, digital transformation can reduce cost approximately 4% to 6% and boosting productivity by nearly 15%. Additionally, to be narrower, UAE increased the usage and investment on technology in construction industry in early 2020 due to COVID-19 to reduce the spread of the virus (Elrefaey et al., 2022).

The importance of digital transformation is not limited to IT industry only, it's vital for all industries including construction industry to survive and remain competitive in the marketplace. Yet, shifting from traditional methods to tech methods requires a high level of awareness and ability to adopt such trends within the industry.

2.4 Traditional Methods Applied in Implementation of Construction Projects

Projects in constriction industry often employ traditional methods such as 2D drawing, reports, schedule, and photo logs in implementation these projects and to be more specific the last three methods used in monitoring projects (Pour Rahimian et al., 2020). Implementing these approaches affects the organization's efficiency and productivity resulted in projects delay and in some cases failure. The consequences of these traditional methods include high time consumption, higher cost and errors (Pour Rahimian et al., 2020).

As (Ding et al., 2019) argued that in construction projects specifically design phase errors may not be apparent until the project installation is complete. This approves the impacts discussed by (Pour Rahimian et al., 2020) and this eventually will lead to necessitating rework and critical changes in the projects. Besides that, these challenges arise as well from the complexity of construction projects as it considered to be difficult to monitor and control comparing to other industries due to several factors include the different characteristics of each project, unpredictable nature of its process, cost variation, the project quality is primarily based on workforce, innovational side in technologies are low, improper resources and utility management and construction projects are based on customer satisfaction (Hewavitharana et al., 2021).

Additionally, limitation of safety practices in preconstruction and construction phase (Afzal & Shafiq, 2021). Due to this, the involvement of technology became a mandatory in the construction field as in considered to be one of the largest industries affecting the country's economy. According to (Liu et al., 2021),72% of the construction corporations around the globe agreed that digitalization is a priority and mandatory for improving business process and only 13% of construction project had significant process by the end of year 2020.

Yet, construction industry ranked the lowest among (22) industries in digitalization index (Liu et al., 2021), and this might be due social, technical, and economic limitation aspects in technical implementation in construction industry. Firstly, social aspects include resistance of change, lack of training and data privacy. Secondly, technical aspects include lack of user friendless, accuracy of challenges, large space and prohibitive processing requirements (Elrefaey et al., 2022).

These factors directly affects the implementation of digital approach within the industry as organizations and top management should establish a proper strategy to shift the traditional mindset toward digital technologies as its essential to empathize the benefits of digitalization to the organization and to employees carrier and potentials. Besides to provide the level of support to employees to work with technologies and trainings (Hewavitharana et al., 2021).

3 Digitalized Methods in the Implementation of Construction Projects

As discussed above, the integration of technology in the construction industry became mandatory as this industry is viewed as a key driver pf a country's economic growth. According to (Elrefaey et al., 2022), there are (8) essential making a huge impact in construction field include: artificial intelligence, drone, internet of things, virtual reality, block chain, augmented reality, 3D printing and robotics. Furthermore, BIM, cloud computing and big data analytic. Each technology has different functionality than the others. As argued by (Elrefaey et al., 2022). Digital technologies are categorized into three aspects as shown in Fig. 5, data acquisition, and data processing and data communication.

Initially, data acquisition encompasses technologies that enable more efficient and faster data gathering such as drones. Secondly, processing information the phase following collecting data such as building information modelling (BIM). The final aspect, data communication includes tools that facilitate communication among parties such as could computing (Elrefaey et al., 2022).

3.1 Literature Review Summary

The literature reviewed several studies explored the impact of digital transformation on construction projects performance and productivity. It provided contextual understanding of digital transformation and evolution of the term. Further view showed existing national strategies in the related matters in UAE and Estonia. Furthermore, the Importance and impact of digital transformation on countries level, project-based

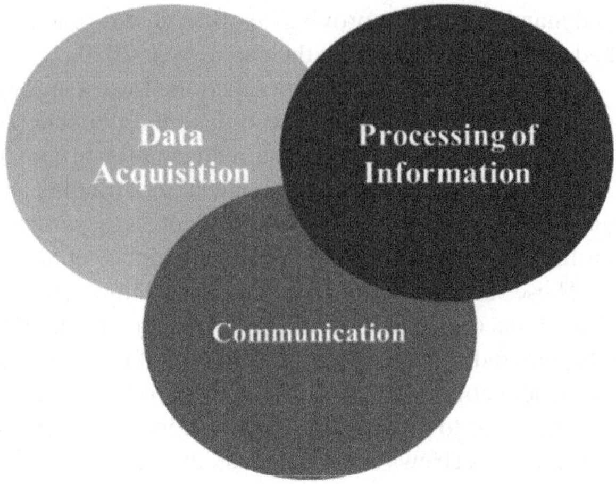

Fig. 5 Digital Technology Categories

organizations and construction industry. It was noted that a gap existed to explore how digital transformation does impact the value of construction projects based on digital technologies discussed in this section.

4 Methodology

This section outlines the methodology adopted in this research to achieve its objectives and address the research question. A qualitative approach was utilized for in depth understating on how doe digital transformation influence construction project performance and productivity.

4.1 Data Collection

Qualitative research involves several types of data collection. For this research paper, data gathering was based on different database as shown in Fig. 6 as a specific criterion used to evaluate the sources and the modernity of the articles and the academic research papers.

4.2 Data Criteria and Evaluation

The search for diverse sources involved using specific key worlds related to digital transformation with a focus on research conducted within the time frame from 2019

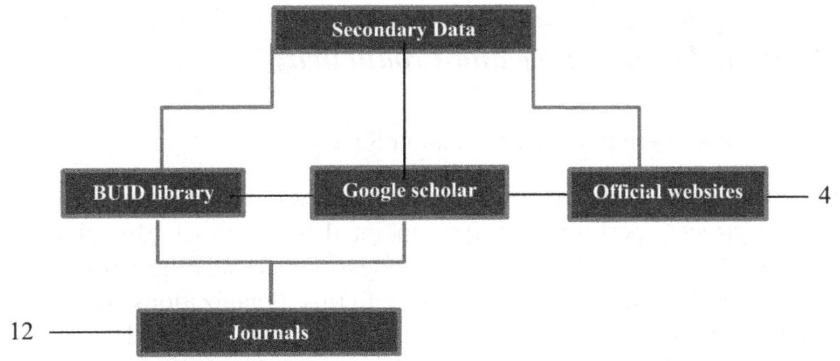

Fig. 6 Data collection structure

Table 3 Criteria and evaluation

Time Frame (2019–2023)	The sources of data collection	Keywords
• 2019: (1) • 2020: (1) • 2021: (5) • 2022: (3) • 2023: (2)	• MDPI • Emerald Insight • Elsevier • ZEMCH 2021 International Conference • Journal of Product Innovation Management	• Digital Transformation • Construction • Digitization • Digitalization • Construction & digital processes Impact of digital process on construction sector

to 2023. As attention was given to both credibility and validation of the sources as presented in the Table 3. Besides, number of (4) official websites referenced in this research paper that provides a valuable insight on the national and global level in the terms of digital transformation strategies and competitiveness are as the following: the official portal of the UAE government, the UAE Ministry of Cabinet Affairs, the Republic of Estonia's Ministry of Economic Affairs and communication official website, Republic of Estonia's Ministry of Climate official website and international institute for management development (IMD) official website. The reason behind the selection of these websites is based on their status as official platforms of governments and global organizations that contains legitimate information.

5 Discussion and Analysis

This section delivers a comprehensive discussion and analysis of the results, contextualizing them by linking the consequences of previous studies conducted in regards of the impact of digital transformation of the performance of construction projects.

5.1 *The Impact of Digital Transformation on Construction Projects Performance and Productivity*

1. Key Drivers of Digital Transformation Success

Prior studies indicated a positive correlation between digital transformation and construction projects performance. As appointed by (Verhoef, Broekhuizen et al. 2021), organizations should change the business model and align with the strategy through the use of digital technology. Owing to this, organizations should be aware that digital technologies affect all the aspects including employees. Due to this, organizations must establish a proper strategy to switch traditional mindsets toward digital technologies and changing these mentalities and make a positive feeling towards using it (Hewavitharana et al., 2021).

Similarly, (Liu et al., 2021) showed in their paper, a study conducted in Hong Kong's revealed that three most critical success factor to adaptation of BIM are owner acceptance, organizational structure and government support. This is aligned with the findings of (Hewavitharana et al., 2021) as mentioned above and in addition the scholars highlighted that commitment of leaders has the highest impact of the success of digitalization within the corporate. In addition, as approved by (Verhoef, Broekhuizen et al. 2021), that organizational structure should be fixable to digital change. Owing to this, we do strongly agree with the scholars above regarding the main (3) factors as they play vital key in the success of the adaptation of digital transformation where top management should create culture within the organization and gain employee's acceptance, trust and full awareness in regard of the use of technologies and its impact on both organization and them.

Beside to that, modifying the business model and organizational structure in alignment with the strategy is mandatory a process otherwise the organization will face failure in adaptation of digital technologies as each element depends on the other. In addition, employees should feel secure that these digital tools won't replace them in the industry, in fact, it will make their work much easier and faster. However, the success of digital transformation in construction field hinges on adopting the appropriate digital technologies that facilitate executing projects coupled with people acceptance and processes (Bajpai & Misra, 2022).

Therefore, organizations should possess digital agility that refers to the sense and capitalize on market opportunities presented by digital technologies. Organizations should be more flexible to accommodate frequent changes in organization roles, evolve customers' needs, implement new technologies and respond to the intensified competition (Hewavitharana et al., 2021).

2. Projects Performance & Efficiency Optimization

The implementation of BIM tool in construction sector in Iran provided that the top (3) KPI's are quality improvement, sustainable construction and cost reduction (Liu et al., 2021). In support to (Liu et al., 2021), (Ding et al., 2019) emphasized that digital construction framework saved by more of 40% of project management cost. Moreover, the scholar's findings argued that traditional construction specifically renovation project would estimate 10 months duration and cost around 133.70 million RMB while the duration is reduced to 8 months and 123.80 million RMB was spent, saving around 20% and 7.41% when implementing technologies. This finding presents that time reduced by 50% and error less than 3%. Likewise, implementation of engineering technologies in China for shopping centre, eliminated design changes by 30%, rework by 25% saved 2 months and saved 7.41% of cost. Furthermore, technologies such as drones in the construction industry reduced time by 18.4% and time to survey sites by 98% (7). Through our lens, the three cases reflected the positive impact of digital technologies in term of time, cost and error and this approves (Gong and Ribiere 2021) framework discussed in literature review that in return reflects positively in the organization efficiency and productivity.

In support to (Gong and Ribiere 2021), according to (Liu et al., 2021), applying digital technologies in Singapore improved productivity around 20%-30% by the end of year 2020 and improved mobility and living condition in Spain. Besides that, it increased efficiency by 25% and 15% in labour productivity in India construction field (Afzal & Shafiq, 2021). Due to this, 30% organizations around the world considering implementing digital transformation process (Sajjad et al., 2023).

From our standpoint, these countries and organizations achieved favourable outcomes due to their flexibility in embracing digital transformation and their openness to change. Additionally, the dedication of the leadership played a critical role, as they incorporated this trend into national strategies that were then adopted by businesses and organizations across various fields and this due to their realization of digital transformation impact on their position and value in the marketplace in addition the field is pivotal in worldwide economic progress.

5.2 Discussion and Analysis Summary

In summary, these findings indicate the positive correlation between digital transformation and construction projects performance and productivity based on evidence collected and investigated from secondary data. Applying digital technologies in the construction sector leads to positive impacts include, cost saving, time reduction, error minimization, enhance efficiency and improve productivity. Furthermore, it provides a competitive edge in the marketplace.

6 Conclusion and Recommendations

The purpose of this present study was to determine the impact of digital transformation on construction projects performance and productivity. Utilizing from secondary data sources to complied and analysed evidence to respond the research question.

Firstly, the research sought to address the research question on how digital transformation in construction sector influence project performance and productivity does? The findings based on existing research papers indicated a positive correlation between digital transformation and both project performance and productivity in the construction sector.

Secondly, the study intended to clarify the of digital transformation terminology. Evidence from the previous studies presented that there were (3) stages: digitization, digitalization, and digital transformation. Although various definitions established for these, a universally accepted definition is lacking. Despite that, a general understanding and awareness of these concepts evident across the countries, enterprises, and organizations.

Moreover, the study highlighted several challenges encountered during the implementing of digital transformation. This includes the resistance of traditional mindset, and the cost associated with technology adaptation. For a successful implementation, its critical that organizations begin with developing and stablishing proper strategies of digital transformation adaptation that consider both economic and social aspects.

Besides, research stablished by (Liu et al., 2021) identified (3) main success factors for digital transformation implementation: owner acceptance organizational structure and government support. Furthermore, the research provides an insight look to digital technologies used in the construction sector and its categories. It emphasized the importance of selecting appropriate digital technologies that aligns with the needs of project life cycle.

Finally, the integration of digital technologies and processes has fundamentally transformed the construction sector. It has altered the way projects envisioned, executed, evaluated, and ensured the involvement of all stakeholders throughout the project life cycle. In addition, this integration significantly improved time consumption. cost, quality, productivity, and project management processes. These positive results are evident in case studies of countries such as Chain, Singapore, India and others.

To this, the study successfully achieved its objectives and address the posed question.

6.1 Recommendations and Future Studies

Considerably, future studies should therefore concentrate on the investigation of.

1. The importance of Cyber security in the construction sector is a key concern due to the need to secure data as the research presented those digital technologies used in data analysis, but there is no emphasized on the security of these data,
2. The influence of digital transformation on risk management is considerable but still broadly understood, necessitating further examination to comprehend its specific effects on various risk factors.

References

Afzal M, Shafiq MT (2021) Evaluating 4D-BIM and VR for effective safety communication and training: a case study of multilingual construction job-site crew. Buildings, 11(8):319. https://doi.org/10.3390/buildings11080319

Digital Agenda 2030 | Majandus- ja Kommunikatsiooniministeerium (n.d.). https://mkm.ee/en/e-state-and-connectivity/digital-agenda-2030

Bajpai A, Misra SC (2022) Evaluation of success factors to implement digitalization in the construction industry. Constr Innov. https://doi.org/10.1108/ci-02-2022-0042

Ding Z, Liu S, Liao L, Zhang L (2019) June). A digital construction framework integrating building information modeling and reverse engineering technologies for renovation projects. Autom Constr 102:45–58. https://doi.org/10.1016/j.autcon.2019.02.012

Elrefaey O, Ahmed S, Ahmad I, El-Sayegh S (2022) Impacts of COVID-19 on the use of digital technology in construction projects in the UAE. Buildings 12(4):489. https://doi.org/10.3390/buildings12040489

Estonia at UN climate change conference of the parties (COP28) in Dubai | Ministry of Climate, n.d.

Frenzel A, et al (2021). Digitization or digitalization?-toward an understanding of definitions, use and application in IS research. AMCIS

Gong C, Ribiere V (2021) Developing a unified definition of digital transformation. Technovation 102:102217

Gradillas M, Thomas LD (2023) Distinguishing digitization and digitalization: A systematic review and conceptual framework. J Prod Innov Manag

Hewavitharana T, Nanayakkara S, Perera A, Perera P (2021) Modifying the unified theory of acceptance and use of technology (UTAUT) model for the digital transformation of the construction industry from the user perspective. Informatics 8(4):81. https://doi.org/10.3390/informatics8040081

Liu Z, Lu Y, Nath T, Wang Q, Tiong RLK, Peh L. LC (2021) Critical success factors for BIM adoption during construction phase: a Singapore case study. Eng, Constr Arch Manag 29(9):3267–3287. https://doi.org/10.1108/ecam-12-2020-1072

MoCA, U. (n.d.). News. Client. https://www.moca.gov.ae/en/media/news/higher-committee-for-government-digital-transformation-approves-6-transformational-projects

Ouddane M (2023) World digital competitiveness ranking 2022 - IMD business school for management and leadership courses. IMD Bus Sch Manag LeadShip Courses. https://www.imd.org/centers/wcc/world-competitiveness-center/rankings/world-digital-competitiveness-ranking/

Pour Rahimian F, Seyedzadeh S, Oliver S, Rodriguez S, Dawood N (2020) On-demand monitoring of construction projects through a game-like hybrid application of BIM and machine learning. Autom Constr 110:103012. https://doi.org/10.1016/j.autcon.2019.103012

Sajjad M, Hu A, Waqar A, Falqi II, Alsulamy SH, Bageis AS, Alshehri AM (2023) Evaluation of the success of industry 4.0 digitalization practices for sustainable construction management: Chinese construction industry. Buildings 13(7):1668. https://doi.org/10.3390/buildings13071668

The UAE Digital Government Strategy 2025 | The Official Portal of the UAE Government. (n.d.). https://u.ae/en/about-the-uae/strategies-initiatives-and-awards/strategies-plans-and-visions/government-services-and-digital-transformation/uae-national-digital-government-strategy

Verhoef PC et al (2021) Digital transformation: A multidisciplinary reflection and research agenda. J Bus Res 122:889–901

Yusuf SO, et al (2022) A systematic literature review approach on the role of digitalization in construction infrastructure and sustainable city development in developing countries

Digitalization for Innovative Knowledge Management in Healthcare

Rehab Sajwani and Farzana Asad Mir

Abstract This paper explores the relationship between knowledge management and innovation, specifically focusing on their impact on the healthcare sector. It also examines the interplay of knowledge management, digitalization, and management skills to foster effective innovation. Further, it employs a systematic mapping approach to reviewing and analyzing existing literature to identify key themes in comprehending the impact of knowledge management on innovation within healthcare organizations. The paper emphasizes the significance of a systematic mapping analysis of the role of knowledge management, management skills, and digitalization in achieving successful innovation in healthcare. The insights gained from this study have the potential to empower healthcare professionals, legislators, and researchers, guiding them in developing solutions and strategies to leverage knowledge management, digitalization, and managerial competencies to drive innovation and enhance healthcare outcomes. Furthermore, this study ventures into uncharted territory, exploring the efficacy of a unique combination of knowledge management, digitalization, and management skills. This novel approach holds the potential to revolutionize innovation within healthcare organizations.

Keywords Knowledge management · Digitalization · Innovation · Management

1 Introduction

Over the past two decades, there has been a growing interest in knowledge management (KM) theory and practice (Siddique 2012). KM involves acquiring, producing, managing, and distributing organizational knowledge to enhance project efficacy and efficiency (Anshari et al. 2022). KM strategies can be classified into explicit-oriented, which focuses on codifying and reusing knowledge through information

R. Sajwani (✉) · F. A. Mir
Faculty of Business & Law, The British University, Dubai, UAE
e-mail: 22000767@student.buid.ac.ae

K. Al Marri et al. (eds.), *BUiD Doctoral Research Conference 2024*, Lecture Notes in Civil Engineering 587, https://doi.org/10.1007/978-3-031-84371-6_12

technology, and tacit-oriented, which emphasizes sharing knowledge through interpersonal interaction. In the healthcare industry, organizations recognize the importance of intangible knowledge, especially medical knowledge, and are implementing information technology and digitalization to manage and disseminate knowledge (Fletcher-Brown et al. 2021). Understanding the role of technology and the interconnected processes of knowledge management is not just important; it's crucial. It's the key to promoting effective innovation in healthcare organizations (Carneiro 2000).

The healthcare industry generates vast amounts of data and knowledge, which can be used to improve patient care and health outcomes and drive innovation. However, traditional knowledge management approaches may not be adequate for capturing, analyzing, and sharing this information. The digitalization of healthcare presents an opportunity to implement innovative knowledge management practices that can transform how healthcare organizations operate (Lakhani et al. 2013). Despite the importance of fostering effective innovation within the healthcare industry, there is a lack of comprehensive mapping review of the main factors that derive effective innovation within the healthcare industry.

Therefore, this study aims to gain a comprehensive understanding of KM and its relationship with the effectiveness of the innovation process. Additionally, the study aims to explore how KM, digitalization, and management skills interact to achieve effective innovation, specifically within the healthcare sector, as there are few KM studies in the context of healthcare.

This study examines the following research questions:

- What are the innovation outcomes associated with effective knowledge management?
- How do knowledge management, digitalization, and management achieve effective innovation?

The next section presents the literature review and an overview of the keywords and relevant themes. This is followed by methodology and the steps taken to conduct the reviews. Next, the mapping procedure is detailed, followed by a discussion and conclusion. Finally, study implications and recommendations for the future are presented.

2 Literature Review

2.1 KM and Innovation

Connecting knowledge management (KM) with innovation has become essential with the advent of KM as a new field (Leal-Rodríguez et al. 2013). Innovation is defined as the application of knowledge to produce new knowledge. The knowledge base's two dimensions, knowledge breadth and depth, reveal both the organization's knowledge structure and content. The knowledge breadth describes a variety of domains that are included in the organization's knowledge repository. In contrast, the depth of knowledge is a key indicator of the sophistication and complexity of knowledge. Hence, organizations need to have both a broad and deep knowledge base to create effective, innovative outcomes (Siddique 2012).

Through its vast knowledge discovery, an organization with a broad knowledge base will gain competence across a range of disciplines and market domains. A large body of knowledge enables the organization to connect and integrate knowledge from other domains in unexpected and unconventional patterns, which fosters innovative ideas for significant changes. An organization with a depth of knowledge is more likely to benefit from acquiring market and consumer knowledge. Through using the accumulated deep knowledge base, significant experience, and comprehension of contemporary technologies and markets, organizations can acquire core competencies in the form of technical expertise (Leal-Rodríguez et al. 2013).

According to Nonaka (1994), organizations that can encourage and enhance knowledge depth and management are considerably more equipped to deal with rapid contemporary changes and innovate in the field where they choose to invest and compete. To conclude, scholars generally agree that there is a significant link between the flow of knowledge and an organization's capacity for innovation.

2.2 *Digitalization for Innovative KM in Healthcare*

KM is a useful strategic tool in dynamic organizations such as healthcare organizations since it plays a crucial role in decision-making, particularly for formulating alternative plans using an accumulated knowledge base. Therefore, Knowledge management (KM) must encompass tools to increase competitiveness, innovative activities, and updated collaborative information technology to achieve knowledge development (Carneiro 2000).

Nowadays, digitalization is crucial in managing knowledge and is one of the most significant tools used to make decisions, compete, and capture target markets (Lakhani et al. 2013). The term "digitalization" refers to the increasing use of binary computer languages that represent physical items used in economic activity. As a result, almost all physical assets now have an "information shadow" that manifests as a digital product. In contrast, this type of digital interpretation was previously primarily associated with information software. This includes digital database software, point-of-care systems, telemedicine, and collaborative platforms (Lakhani et al. 2013).

According to Carneiro (2000), digital database software is essential for collecting and disseminating the organization's assembled significant data. Knowledge workers can then access several databases to find innovative solutions for their challenging daily duties and for assistance in decision-making. Thus, having access to reliable knowledge sources, technical databases, decision support systems, management models, information on clients and competitors, and effective global solutions are the channels by which innovative KM development occurs in healthcare (Carneiro 2000).

In summary, employing technological solutions to develop new routines, practices, and products enhances healthcare organizations' capacity for innovation, enabling them to satisfy the demands of a constantly changing market.

2.3 Management for Innovation in Healthcare

Effective management skills are necessary for the organization to implement knowledge management successfully. Knowledge may flow in an unstructured way due to variances in knowledge processing resources and workers across management levels. Therefore, most organizations struggle to have a functional strategic knowledge system to organize the massive volume of data gathered. Consequently, knowledge continues to flow into the organization continuously but randomly since resources change according to the level of managerial action (Carneiro 2000). Knowledge workers generate the most value in contemporary fields, such as software, pharmaceuticals, healthcare, and financial services. Management should direct knowledge workers to fully understand the variety of resources, such as tools, software, supplies, and support, to enable systematic knowledge processing and sharing (Carneiro 2000).

Therefore, effective management skills enable healthcare organizations to structure and optimize their knowledge flows. This allows the capture and dissemination of knowledge and the development of strategic knowledge systems to enable knowledge-based decision-making. It also empowers knowledge workers with technology and resources to generate, share, and apply knowledge effectively (Anantatmula 2008).

2.4 Innovation and Healthcare

Innovation has emerged as a crucial competency for all healthcare organizations. It is the driving force behind initiatives to achieve a balance between cost-cutting and high-quality care. It is regarded as the foundation of their competitive advantage and a crucial element of corporate productivity (Leal-Rodríguez et al. 2013).

Innovation in the healthcare industry has been described to include activities and procedures to support the organization's structure, patients, and governing bodies. The potential and constraints of the organizational knowledge substrate to foster growth and sustainable innovations are highlighted by the fact that innovation is also defined as a tool that recombines existing knowledge in new ways (Du Plessis 2007). Thus, the prior accumulation of relevant expertise and tacit knowledge in healthcare organizations is the foundation for competitive innovation. As a result, the KM strategy, knowledge base, and innovation capability are strongly associated to support healthcare organizations (Leal-Rodríguez et al. 2013).

To accomplish this, healthcare organizations must first develop a broad and deep knowledge base to strengthen knowledge-sharing processes. Then, improve relational learning strategies for gaining, assimilating, and integrating external knowledge to maximize already-acquired knowledge resources and foster innovations (Leal-Rodríguez et al. 2013)Thus, managers and employees should modify their knowledge integration techniques to match their organizations' current knowledge base to support and guarantee that the knowledge generated by their routine daily problem-solving and decision-making activities is accurately gathered, encoded, articulated, and stored within their the knowledge bases (Zhou & Li 2012).

3 Theoretical Framework

The study aimed to explore the relationship between knowledge management, digitalization, and management skills in facilitating innovation in healthcare. The knowledge-based view (KBV), the resource-based view (RBV), and the technology acceptance model (TAM) were identified as relevant theories for this study.

The Resource-Based View (RBV) emphasizes the critical role of an organization's resources, particularly its knowledge assets and digital tools/technologies, in enabling the organization to gain a sustainable competitive advantage. According to RBV, the organization's digital assets, such as its tools and technologies, serve as valuable resources that enable effective leverage of the knowledge assets. This enables the organization's ability to drive successful innovation, enhance its overall performance and competitive positioning within the market (Cuthbertson & Furseth 2022).

Furthermore, the Knowledge-Based View (KBV) emphasizes the critical role of knowledge, particularly tacit knowledge, as a key strategic resource for organizations (Grant 1997). According to the KBV, an organization's performance and competitive advantage in a knowledge-based economy are primarily driven by its unique and heterogeneous knowledge base and skills. This theory posits that an organization's ability to generate, accumulate, and effectively manage knowledge, as well as establish partnerships to facilitate the interchange of external data, is fundamental to its success. This enables the organizations to foster innovation and gain competitive edge by efficiently managing and utilizing their knowledge assets (Cuthbertson & Furseth 2022).

Lastly, the third theory that could be relevant to the current study is the Technology acceptance model (TAM). The Technology Acceptance Model (TAM) highlights the pivotal role of digitalization and emerging technologies in driving innovation within organizations. TAM posits that the acceptance of technology within an organization is primarily influenced by the perceived usefulness and ease of use, which are largely facilitated by the competencies and actions of management. By enabling the sharing of knowledge through the strategic deployment of digital tools, TAM suggests that effective management can foster innovative objectives and enhance the organization's overall competitive positioning (Marangunić & Granić 2015).

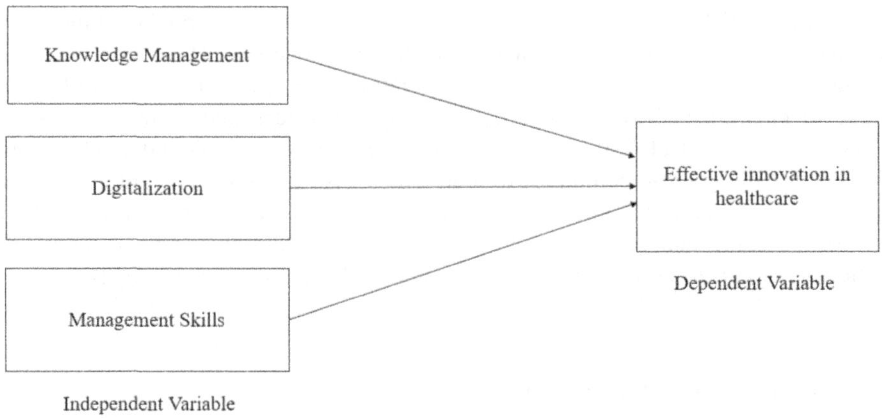

Fig. 1 Conceptual framework outline

4 Conceptual Framework

This study aims to gain a comprehensive understanding of knowledge management and its relation to improving the effectiveness of the innovation process. Based on the literature review and the above-mentioned theoretical framework, the conceptual framework is illustrated in Fig. 1. The conceptual framework emphasizes the interactions between independent variables, resulting in the dependent variable. In which the dependent variable of effective innovation in healthcare is achieved through the interaction of the independent variables of knowledge management, digitalization, and management skills.

5 Methodology

A systematic review was conducted to examine peer-reviewed published research on digitalization for innovative knowledge management in healthcare. According to Gough et al. (2017), a systematic review is a transparent and reproducible approach to literature review that responds to research issues using responsible methodologies and stated criteria that include or reject existing studies. To ensure reliability, consistency, and transparency, this systematic review was conducted by the Preferred Reporting Items for Systematic Reviews and Meta-Analyses (PRISMA) standards and its four-phase flow diagram (Liberati et al. 2009). In evaluating the quality and importance of studies, the review must ensure that only the most appropriate, reliable articles should be included (Harden & Gough 2012). Articles from Springer Link, Elsevier, Emerald, and Taylor and Francis Online journals that are known to receive documents on KM, innovation, and healthcare were consulted. High-ranked journals, such as the

Journal of Business Research, the Innovation Journal, and the Journal of Knowledge Management, were used.

Figure 2 depicts the steps of the systematic review process (Page et al. 2021), which consists of identification, screening, eligibility, and included studies. The search was based on the two specified databases, Google Scholar and British University in Dubai, and utilizing the article's primary keywords (Knowledge Management-Digitalization-Innovation-Healthcare), 60 papers were identified. The identification step targeted the main keywords, which led to the elimination of 10 papers and included 50 papers. The screening phase utilized keywords while reading the abstract, and five papers were excluded for being irrelevant to the search scope. Then, during the eligibility phase, clustering keywords are used to generate phrases to be searched for. This excluded an additional five papers. Five more were added that were pertinent to knowledge management projects, hence including 45 papers.

6 Discussion and Findings

This study offers a thorough review of factors achieving effective innovation in healthcare. Using a qualitative methodology and systematic review process research strategy, the study investigates the relationship between digitalization, management skills, and knowledge management in encouraging effective innovation in healthcare organizations.

This study conducted a systematic review following the Preferred Reporting Items for Systematic Reviews and Meta-Analyses (PRISMA) framework guidelines. As depicted in Fig. 2, the study's flow consisted of four main stages: identification, screening, eligibility, and inclusion of studies (Page et al. 2021).

During the identification stage, a comprehensive search was conducted in two specified databases, Google Scholar and BUiD, using the primary keywords Knowledge Management-Digitalization-Innovation-Healthcare. This search process initially yielded 60 papers. The focus was placed on selecting papers from high-ranked journals to ensure the inclusion of high-quality and relevant studies. As a result, ten papers were eliminated during this step.

After the identification stage, the screening phase was initiated. The screening process involved assessing the relevance of the papers based on the keywords found in their abstracts. After carefully reviewing the abstracts, five papers were excluded as they were deemed irrelevant to the scope of the search. Consequently, 45 papers were retained for further evaluation.

The eligibility phase was reached after reading and looking up further papers. The inclusion criteria were applied rigorously to identify the most suitable papers for the systematic review. In this phase, keywords were clustered and added "digitalization instead of 4IR"; thus, they were utilized to generate phrases that were subsequently searched for within the papers. This process allowed for a more focused evaluation of the content and relevance of the publications. As a result, 5 additional papers were excluded, as they did not meet the eligibility criteria.

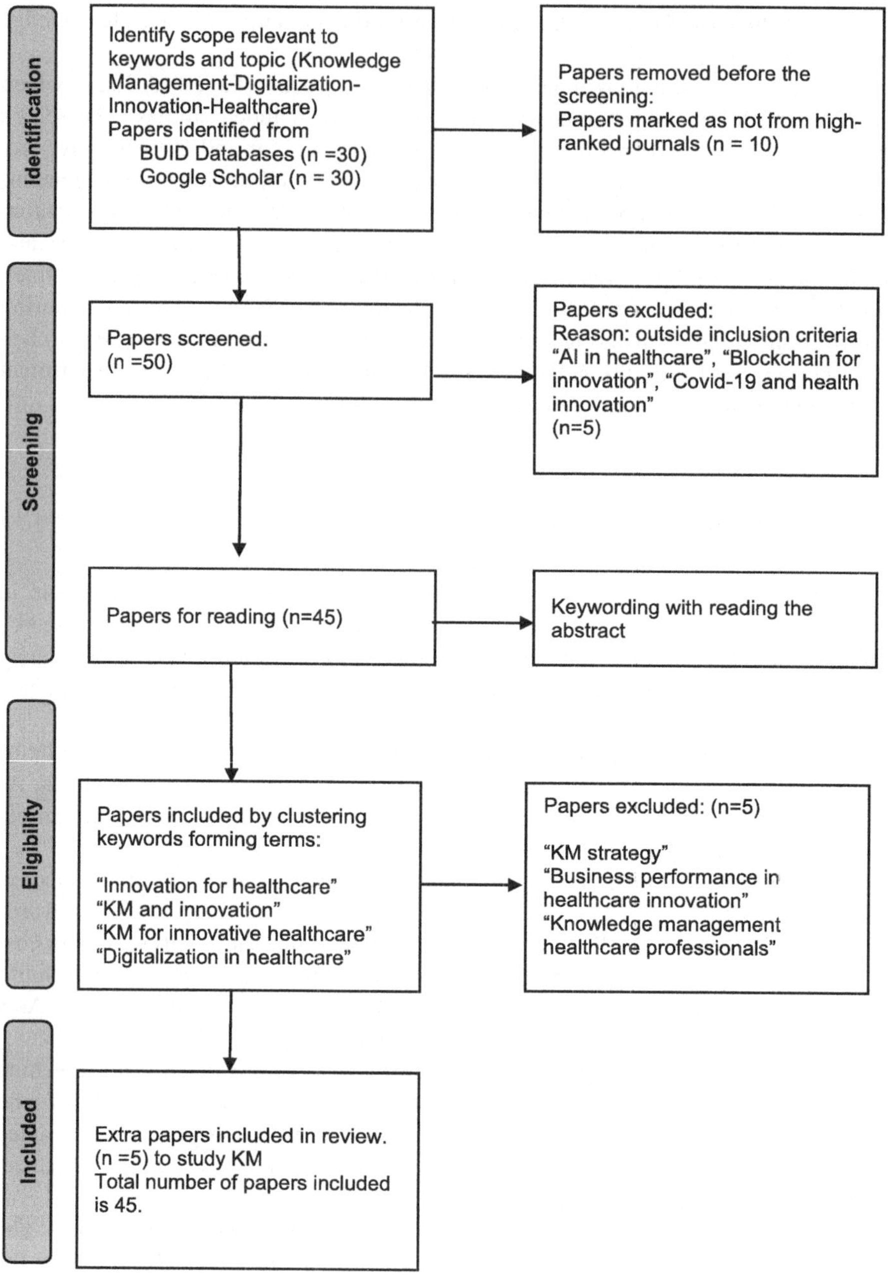

Fig. 2 PRISMA flow diagram

Finally, in the inclusion phase, it was discovered that the study requested five additional papers that were pertinent to knowledge management projects. These papers were not initially captured in the search process but were deemed valuable for the study and, therefore, included.

The systematic review followed a structured approach, adhering to the PRISMA guidelines. The study flow involved identifying relevant papers, screening abstracts, and evaluating eligibility criteria.

The study's findings suggest that knowledge management, digitalization, and managerial skills are interrelated and synergistic factors that combine to achieve effective innovation in healthcare. Knowledge management is the foundation, encompassing processes and practices for capturing, organizing, disseminating, and applying knowledge within healthcare organizations. Digitalization provides the technological infrastructure and tools to enhance knowledge management practices. This enables timely access, analysis, and information sharing, fostering collaboration and knowledge exchange. Lastly, management skills play a crucial role in orchestrating and facilitating the integration of knowledge management and digitalization. Effective management is vital in fostering a culture of innovation and ensuring that knowledge management and digitalization initiatives align with organizational goals to achieve knowledge development. This encourages efficient innovation in the healthcare industry, which results in timely access, cost-effectiveness, improved patient outcomes, and high-quality medical care.

7 Conclusion and Future Recommendations

In conclusion, knowledge management must be integrated effectively and efficiently into the healthcare organization's structure to gather, manage, and disseminate knowledge. Thus, for healthcare organizations to thrive in a highly competitive sector, new solutions must be created by utilizing effective knowledge management and sharing explicit and tacit knowledge easier. Therefore, to carry out this process methodically, it is necessary to use all of the accumulated knowledge and expertise through the use of digitalization platforms, software, and data storage to enable innovation, as well as the best management techniques and governance initiatives to create this process and support daily healthcare operations.

The theoretical implications of this study underscore the strategic imperative for healthcare organizations to holistically integrate knowledge management, digitalization, and managerial skills to drive sustainable innovation and improve patient outcomes. The study emphasizes the vital need for robust knowledge sharing and collaboration among healthcare professionals to facilitate the adoption of effective knowledge management practices and the implementation of optimal digitalization strategies. Additionally, it highlights the significance of managerial capabilities at the organizational level in fostering a culture that supports innovation, change, and knowledge sharing, which can be cultivated through targeted training and software development programs. To ensure patient privacy and confidentiality in the context of digitalization, the study recommends that policymakers establish secure guidelines and allocate knowledge resources effectively through enhanced training and awareness campaigns.

The practical implications of this study emphasize the need for government entities to implement strategic knowledge initiatives to enhance development within the healthcare sector. The findings suggest that healthcare organizations can derive significant benefits by striving to cultivate a more innovative and efficient healthcare system through the strategic utilization of digital technologies, the fostering of a culture that prioritizes knowledge sharing, and the integration of robust managerial skills to drive successful innovation and ultimately improve healthcare outcomes. In addition to training programs, workshops, and conferences should be conducted to facilitate global sharing of best practices and understanding knowledge concepts.

This study's limitations include difficulty maintaining heterogeneity of papers due to the expansive nature of the topics covered, as well as time constraints in integrating all current research. The systematic mapping approach may also inadvertently exclude important publications, potentially introducing bias. Despite these challenges, the study provides a comprehensive overview of the literature on knowledge management and digitalization in healthcare.

References

Anantatmula VS (2008) The role of technology in the project manager performance model. Proj Manag J 39(1):34–48

Anshari M, Syafrudin M, Fitriyani NL (2022) Fourth industrial revolution between knowledge management and digital humanities. Information 13(6):292

Carneiro A (2000) How does knowledge management influence innovation and competitiveness? J Knowl Manag 4(2):87–98

Cuthbertson RW, Furseth PI (2022) Digital services and competitive advantage: strengthening the links between RBV, KBV, and innovation. J Bus Res 152:168–176

Du Plessis M (2007) The role of knowledge management in innovation. J Knowl Manag 11(4):20–29

Fletcher-Brown J, Carter D, Pereira V, Chandwani R (2021) Mobile technology to give a resource-based knowledge management advantage to community health nurses in an emerging economies context. J Knowl Manag 25(3):525–544

Gough D, Thomas J, Oliver S (2017) An introduction to systematic reviews

Grant RJ (1997) Interactions among forages and nonforage fiber sources. J Dairy Sci 80(7):1438–1446

Harden A, Gough D (2012) Quality and relevance appraisal. Sage Publications

Lakhani KR, Lifshitz-Assaf H, Tushman ML (2013) Open innovation and organizational boundaries: task decomposition, knowledge distribution and the locus of innovation. In: Handbook of economic organization (pp 355–382). Edward Elgar Publishing

Leal-Rodríguez AL, Roldán JL, Leal AG, Ortega-Gutiérrez J (2013) Knowledge management, relational learning, and the effectiveness of innovation outcomes. Serv Ind J 33(13–14):1294–1311

Liberati A, Altman DG, Tetzlaff J, Mulrow C, Gøtzsche PC, Ioannidis JP, Moher D (2009) The PRISMA statement for reporting systematic reviews and meta-analyses of studies that evaluate health care interventions: explanation and elaboration. Ann Intern Med 151(4):W–65

Marangunić N, Granić A (2015) Technology acceptance model: a literature review from 1986 to 2013. Univers Access Inf Soc 14:81–95

Muhammad Siddique C (2012) Knowledge management initiatives in the United Arab Emirates: a baseline study. J Knowl Manag 16(5):702–723

Nonaka I, Byosiere P, Borucki CC, Konno N (1994) Organizational knowledge creation theory: a first comprehensive test. Int Bus Rev 3(4):337–351

Page MJ, McKenzie JE, Bossuyt PM, Boutron I, Hoffmann TC, Mulrow CD, Moher D (2021) The PRISMA 2020 statement: an updated guideline for reporting systematic reviews. Bmj 372

Zhou KZ, Li CB (2012) How knowledge affects radical innovation: Knowledge base, market knowledge acquisition, and internal knowledge sharing. Strat Manag J 33(9):1090–1102

Due Process and the Judiciary: Navigating Due Process Paranoia

Binisaroj and Abba Kolo

Abstract This paper delves into the intricate relationship between due process, due process paranoia, and the role of the judiciary in arbitration proceedings. Due process, a fundamental principle of law, ensures fairness, equity, and procedural justice in legal contexts, including arbitration. However, due process paranoia, characterized by an exaggerated fear or concern about procedural irregularities, can lead to excessive formalism, unnecessary delays, and challenges in arbitration efficiency. This study aims to explore the nuances of due process and due process paranoia in arbitration settings and to examine the judiciary's role in mitigating the adverse effects of due process paranoia while upholding the principles of fairness and efficiency in dispute resolution.

Keywords Arbitration · Due process · Due process paranoia · Judiciary · Judicial decisions

1 Introduction

In the realm of arbitration, due process serves as a cornerstone principle, ensuring that parties are afforded a fair and impartial hearing and an opportunity to present their case. However, the emergence of due process paranoia poses significant challenges to the efficiency and effectiveness of arbitration proceedings. Due process paranoia, characterized by an exaggerated fear of procedural irregularities, can lead to procedural delays, increased costs, and a loss of confidence in the arbitration process. This paper explores the concepts of due process and due process paranoia within the context of arbitration, examining their implications and the judiciary's crucial role in navigating these challenges to ensure fair and efficient dispute resolution.

Binisaroj (✉)
BUID, Dubai, UAE
e-mail: 21003061@student.buid.ac.ae

A. Kolo
Faculty of Business and Law, The British University Dubai, Dubai, UAE

K. Al Marri et al. (eds.), *BUiD Doctoral Research Conference 2024*, Lecture Notes in Civil Engineering 587, https://doi.org/10.1007/978-3-031-84371-6_14

2 Research Objectives

- To analyze due process and due process paranoia in the context of international arbitration.
- To explore the causes and implications of due process paranoia in arbitration proceedings.
- To understand how the judiciary approaches cases involving allegations of due process violations in arbitration.
- To Assess whether due process paranoia is justified based on judicial responses.
- To Provide recommendations for addressing due process concerns while maintaining the efficiency of international arbitration.

3 Research Methodology

The authors have conducted a thorough review of academic literature on the topic. This involved studying scholarly articles, books, and other publications that discussed the concept of due process and due process paranoia in the context of international arbitration. Secondly, the authors engaged in the analysis of relevant case laws and regulations. This entailed examining judicial decisions and legal provisions that are related to due process and due process violations.

Overall, by employing doctrinal research methodology, the authors have gained comprehensive insights into the concept of due process, its forms, and effects, and the challenges and limitations it poses in practice.

4 Due Process

Due process, a fundamental principle of law, embodies the notion of procedural fairness and ensures that individuals are afforded basic rights and protections in legal proceedings. In arbitration, due process entails the right of parties to be heard, to present evidence, and to receive a fair and impartial decision from the arbitral tribunal. It encompasses procedural safeguards that uphold the principles of fairness, equity, and justice in resolving disputes outside of traditional court systems (Mandujano Rubin 2022).

Due process in arbitration encompasses several key elements to ensure procedural fairness: (1) Arbitration Agreement Jurisdiction: Parties who have voluntarily agreed to arbitrate a dispute have waived their right to a public hearing before a court of law. However, this waiver does not mean they have given up their right to due process. Arbitration agreements must still provide for minimum procedural safeguards commensurate with the importance of the rights at stake. (2) Non-Agreement Jurisdiction: In cases where a party has not freely consented to arbitration, such as in mandatory arbitration clauses imposed by governing bodies, the arbitration will still have to guarantees application of due process. The arbitral tribunal must meet strict requirements of independence and impartiality. (3) Evidence-Based Due Process: Parties have a right to present their case effectively. This includes being given a reasonable opportunity to submit evidence and arguments in support of their position. Arbitrators cannot arbitrarily refuse to consider relevant evidence or arguments without violating due process. (4) Panel Participation: Parties must be treated equally and given a fair chance to participate in the constitution of the arbitral tribunal. Lack of independence or impartiality of arbitrators, or an insufficient mechanism to challenge them, can amount to a due process violation. Therefore, while arbitration allows parties to waive certain procedural rights, it must still provide a fair process with adequate safeguards. The specific requirements depend on whether participation in arbitration was voluntary or compulsory. Arbitrators must ensure equality of treatment, reasonable opportunity to present one's case, and independence and impartiality of the tribunal (Kurkela et al. 2010).

Arbitration must balance discretionary decisions with due process requirements, ensuring arbitrators' neutrality, independence, and impartiality (Cremades 2016). Due process entails proper notice, the right to present and be heard, and equal treatment, (Moneke et al. 2020) guided by regulations like the UNCITRAL Model Law, UK Arbitration Act, Indian Arbitration and Conciliation Act, and the New York Convention (Feehily 2019) Fairness does not always equate to due process abuse. Arbitration requires arbitrators' neutrality, independence, and impartiality. Without them, parties cannot be treated equally throughout the process (Reed 2017).

5 Abuse of Due Process

The abuse of due process in arbitration can lead to "due process paranoia," where parties become overly concerned about the fairness and integrity of the process. This misuse involves deliberate manipulation of procedural rules to create obstacles, frustrate proceedings, or achieve undeserved outcomes, undermining fairness and efficiency. Forum shopping exacerbates this paranoia, creating uncertainty and mistrust as parties fear inconsistent standards or outcomes across different forums. Parallel proceedings increase complexity and costs, heightening anxiety about conflicting decisions (Ceretelli 2020).

Lawyers may exploit due process to delay proceedings and gain tactical advantages through frequent extensions, frivolous motions, excessive evidence, strategic withholding, and challenges to arbitrators' impartiality. Provisions in the UNCITRAL Model Law and UAE Arbitration Law are susceptible to misuse: Article 18 of the UNCITRAL Model Law and Article 26 of the UAE Arbitration Law can justify delays, while Articles 19 and 24 (UNCITRAL) and Article 33 (UAE) allow complex procedures and interim measures. Articles 34 (UNCITRAL) and 53 (UAE) provide broad grounds for setting aside awards, prolonging arbitration.

To combat these abuses, a national code of conduct for legal practitioners is recommended to balance due process protection with measures to prevent exploitation, thereby maintaining arbitration's efficiency and integrity.

6 Due Process Paranoia

Due process paranoia refers to an exaggerated fear or concern about procedural irregularities in arbitration proceedings. It manifests as an unwarranted fixation on formalities, rules, and procedures, often leading to excessive scrutiny of procedural matters and an aversion to procedural flexibility. Due process paranoia can result in unnecessary delays, increased costs, and a loss of confidence in the arbitration process, ultimately undermining its efficiency and effectiveness.

Due Process Paranoia is a new term coined by a participant in the Queen Mary University study in 2015 (QUML 2015) regarding the efficiency of arbitration. The term means arbitrator's incapacity to decide when there is any abuse of powers from any of the parties in the arbitration. Arbitrators must determine whether a party's request for procedural relief is a legitimate exercise of its procedural rights or an irrational action that will slow down the arbitration. Three factors are typically used to limit an arbitrator's decision-making: equality, agreement, and a violation of the parties' "due process rights" to present their arguments.

Due process paranoia stems from the possibility of a first-ground challenge to the award for violating the parties' right to present their case. As a private court procedure, arbitration's due process rights underpin arbitrators' inhibitions. Arbitrators are private judges and decision-makers. They protect parties' fundamental rights (Berger and Jensen 2016).

Smriti Shukla and Soumyajit Saha argue that due process paranoia arises from arbitrators' external and internal pressures to ensure their decisions are beyond reproach, fearing non-enforcement of awards might damage their reputations. This paranoia can lead to an imbalance, favoring one party's due process rights over the other party's expectation of conflict resolution. Arbitrators should proactively manage case deadlines to balance efficiency and due process, reducing judicial interference (Saha and Shukla 2020).

Moneke, Idornigie, and Mgbakogu identify three causes of due process paranoia: arbitrators being overly cautious to avoid awards being set aside, compliance with procedural rules even when unreasonable, and protecting their market reputation. This caution can lead to prolonged proceedings and increased costs (Moneke et al. 2020).

The literature reveals that due process paranoia is the excessive caution of arbitral tribunals, with the arbitration community divided on its existence and impact on tribunal efficiency.

7 Causes of Due Process Paranoia in Arbitration

Due process paranoia in arbitration arises from the increasing complexity and diversity of international disputes, leading arbitrators and practitioners to adopt a cautious approach to ensure fairness and equity. The proliferation of procedural rules and the pressure to adhere to international best practices exacerbate this paranoia. Arbitrators face criticism for being weak, while some blame counsel and clients for exploiting due process for tactical advantages (Gaillard 2017).

Arbitrators' fear of rendering unenforceable awards contributes to their hesitation in making difficult decisions, leading to a "highly legalistic, litigious, and expensive" arbitration process (Lau n.d.). The need to "bullet-proof" awards against annulment on due process grounds further fuels this paranoia.

Several factors contribute to due process paranoia:

- Increased guerrilla tactics by parties
- Obligation to render enforceable awards
- Requirements of arbitration instruments like the New York Convention
- Fear of damaging market reputation

These factors create a pervasive sense of unease and distrust, leading to heightened vigilance and scrutiny of arbitration proceedings. This results in delays, disputes, and undermines the efficacy of arbitration.

Addressing due process paranoia requires a concerted effort to uphold fairness, transparency, and procedural integrity, ensuring arbitration remains an effective and reliable dispute resolution mechanism.

8 Judicial Perspective on Due Process

The judiciary plays a crucial role in adjudicating challenges to arbitral awards and interpreting the scope of due process rights within the context of international arbitration. While courts generally uphold the finality of arbitral awards, they also serve as guardians of due process, intervening in cases of serious procedural irregularities or violations of public policy.

Dr. Rémy Gerbay discusses the solutions to Due Process paranoia, and he is of the opinion that more accurate and realistic Enforcement Risk assessments in each case must be the solution to curb due process paranoia. This requires carefully studying how robust case management decisions affect arbitral award enforceability. Despite the harmonizing effect of the NY Convention or UNCITRAL Model Law, some jurisdictions are less liberal than others, so the situation varies greatly. A tribunal in England must consider both English law (the law of the seat) and the likely place(s) of enforcement when deciding whether a case management decision is too robust (Gerbay 2016).

Francois Lassale believes that the arbitrator, and parties all shall act logically to resolve Due Process Paranoia. Parties should trust the entire arbitral ecosystem, from model laws like UNCITRAL or the New York Convention to institutional rules and tribunal experience, to handle their case efficiently. academic research shows that the due diligence paranoia belief is unfounded, that awards rarely, if ever, are denied enforcement or set aside due to tough procedural decisions, so there is no rational reason for the system's reputation to suffer. International arbitration users should expect better, and we should consistently deliver better (Lassalle n.d.)

9 UK Courts

The Courts in the UK have shown an arbitration-friendly attitude and refused to set aside the awards cases in case of there were no serious due process violations as per article 68 of the UK ARBITRATION ACT. Therefore, upholding the discretionary powers of the Arbitral tribunal.

In the case of "*Gold Reserve Inc. v. Bolivarian Republic of Venezuela Gold Reserve Inc. v. Bolivarian Republic of Venezuela [2016] EWHC 153 (Comm)*" even though a due process issue was raised, the English court accepted the enforcement of an award. In this instance, the state claimed that the disproportionate distribution of hearing time prevented it from effectively arguing its case in an ICSID Additional Facility procedure, despite the fact that it had itself asked for a shortened hearing and declined to question the claimant's witnesses.

One of the very significant cases is "*Asset Management Corporation of Nigeria (AMCON) v. Qatar National Bank Asset Management ('AMCON') v. Qatar National Bank [2018] EWHC 2218 (Comm)*". The English court refused to set aside an award in this case when the claimant raised the objection that the tribunal failed to apply Nigerian law and did not consider all the claims of the claimant. Court found the allegations were wrong and commented that the losing party will allege that the tribunal did not follow the procedures so that it had more opportunities to present its case.

While the court has always upheld the independence of the Arbitral Tribunal as a dispute-resolving mechanism, and upheld the finality of awards, it has also not hesitated to overturn the decisions of the Arbitral Tribunal, and refuse enforcement of Awards in case of serious due process violations from the tribunal. Therefore, it is clear that the court assumes the role of guardian of due process of law to protect the rights of the party and balance it with arbitral tribunal autonomy.

In "Fleetwood Wanderers Limited (t/a Fleetwood Town Football Club) v. AFC Fylde Limited [2018] EWHC 3318 (Comm))", award has been set aside in accordance sec.68 of arbitration law since the Arbitrator acted without transparency and impartially.

10 United States Courts

US courts have also echoed the view on interference with the decisions of Arbitral tribunal only in case of serious violations courts set aside the awards.

In *Iran Aircraft Industries v. Avco Corp., 980 F.2d 141 (2d Cir. 1992),* having informed the company in a pre-hearing conference that it need not produce the thousands of invoices supporting its claim, the Iran-United States Claims Tribunal at the Hague issued an award against Avco for lack of damages proof. The US Second Circuit Court of Appeals subsequently refused to enforce the award because Avco had been deprived of the chance to present its claim. Although "unwittingly," the tribunal had misled the appellant and denied it the chance to present its claims in a "meaningful manner," as required by the New York Convention, according to the Second Circuit Court of Appeals' analysis.

In *Dealer Computer Services v Dale Spradley Motors, Inc, 2012 US Dist LEXIS 2614 (ED Mich 10 January 2012),* the court refused to overturn an award because the tribunal had not delayed the hearing for 90 days, concluding that there had not been a serious due process violation.

In Grosso v. Barney 2003 (*US Discrict Court E.D.Pa. 2003*).US district court refused to set aside an arbitration award declared clearly that the intention of the Congress in passing the arbitration act was to enforce the private agreements and courts are bound to enforce that even if the result is piecemeal litigation.

11 European Union Courts

In *Terna Bahrain Holding Co v. Ali Marzook Ali Bin Kamil Al Shamsi* et al. *[2012] EWHC 3283* judicial intervention, according to the High Court, "would only be appropriate in cases where the tribunal's handling of the arbitration was so improper and inconsistent with what was reasonably expected of it that justice demanded that it be corrected."

European Court of Justice decision in the "*Slowakische Republik v. Achmea BV, 6 March 2018, No. C-284/16.)" and, "Micula v. Romania, 23 January 2019*" are the cases where European domestic courts refused to enforce BIT contract-based arbitration awards.

12 UAE Courts

The courts in UAE has also taken same view wherein a reported case 'Dubai Court of Appeal rejected a DIAC award due to arbitrator's breach of due process'. https://arbitrationblog.kluwerarbitration.com/2016/05/05/dubai-court-of-appeal-rejects-a-diac-award-due-to-arbitrators-breach-of-due-process/ In this case even though the claimant wrongly wrote the name of the respondent, but the respondent received the summons and got the opportunity to represent the case, so the Dubai first instance court refused to cancel the judgment on the grounds of violation of the due process. But the finalised appeal court judgment revised the decision.

In *Mohammad Abdul Rahim Abdul Razak Alfahim v. Ibrahim Abdul Razak Alfahim (DCC 199, 205/2020)* The case involved a dispute between partners within a corporate group over mismanagement issues. The dispute was subject to arbitration. The issues revolved around the validity of the arbitration award and the correction of a court judgment. The court referred to Article 53 of Federal Law No. 6/2018 on Arbitration, outlining specific grounds for challenging an arbitral award. It emphasized that challenges must relate to the arbitration agreement or the arbitration process itself. The court dismissed claims related to the rules of evidence, arbitrator discretion, or omission to adjudicate certain matters, reiterating that such issues do not invalidate the arbitral award. This suggests that Dubai courts generally support arbitration awards unless there are fundamental procedural flaws or violations.

Similarly, in *Nova Mission Critical Contracting v. Sigma Engineering (DCC 446/2019)* the case involved a request to annul an arbitration award under the 2018 UAE Arbitration Law. The dispute centered on matters such as rules of evidence, arbitrator discretion, and adequacy of reasons. The Dubai Court of Cassation rejected the appellant's submissions, reaffirming the grounds for annulment outlined in Article 53 of Federal Law No. 6/2018. The court emphasized that challenges must pertain to the arbitration agreement or procedural errors, rather than issues like evidence rules or arbitrator discretion. This indicates that Dubai courts prioritize upholding arbitration awards unless there are significant procedural irregularities.

From the interpretation of the two cases, it is evident that Dubai courts generally uphold arbitration awards unless there are serious procedural violations. Challenges to arbitral awards must relate to the arbitration agreement or procedural errors, such as lack of notice or improper tribunal composition. Disputes over evidence rules, arbitrator discretion, or reasons for decisions do not typically invalidate arbitral awards. Therefore, addressing abuse of due process in arbitration should primarily be handled internally within the arbitration process itself.

From the Interpretation of cases mentioned above it is clear that as a general Practice, almost all arbitration-friendly jurisdictions back up the arbitration awards unless there are serious due process violations, so from a judicial perspective the abuse of due process to be addressed internally from the arbitration process itself.

13 Due Process Paranoia: Justified or Unwarranted?

Due process paranoia manifests as an overemphasis on procedural safeguards, stemming from a fear of potential challenges to arbitral awards based on alleged violations of due process rights. Proponents argue that such vigilance is necessary to safeguard the integrity of the arbitral process and ensure fairness for all parties involved. They contend that arbitrators must err on the side of caution to avoid any perception of procedural irregularities that could undermine the enforceability of awards.

Klaus Peter Berger in his study which was conducted on the basis of precedents proved that Due Process paranoia is not justifiable. Although the parties' due process rights have a clear purpose, it is still unclear whether and to what extent due process paranoia is justified. This depends on the courts' attitude as the ultimate guardians of due process, whether in setting aside arbitration proceedings or in court actions to recognize and enforce the award in jurisdictions where the award-debtor has assets (Berger and Jensen 2016).

However, Due process paranoia in arbitration, while understandable in certain contexts, can often be unwarranted, particularly when considering the judiciary's arbitration-friendly approach. A critical analysis reveals both justifications for and limitations of such paranoia.

On one hand, due process paranoia may stem from legitimate concerns about the fairness and integrity of the arbitration process. Instances of abuse of due process, such as biased arbitrators, procedural irregularities, or forum shopping, can undermine parties' trust in the system. In complex international arbitrations, where parties often hail from different jurisdictions and legal systems, the potential for misunderstandings and misinterpretations heightens anxieties about procedural fairness. Additionally, the lack of a centralized authority overseeing arbitration proceedings may exacerbate concerns about accountability and transparency.

However, it is essential to recognize that due process paranoia is often unwarranted, especially considering the judiciary's arbitration-friendly stance. Courts typically uphold arbitration awards and defer to arbitral tribunals' decisions, recognizing the parties' autonomy in choosing arbitration as a dispute resolution mechanism. The pro-arbitration approach of the judiciary is grounded in principles of party autonomy, efficiency, and finality of arbitration awards. Courts intervene in arbitration proceedings only in limited circumstances, such as instances of fraud, procedural irregularity, or public policy violations.

Furthermore, the judiciary's arbitration-friendly approach is supported by international conventions and national legislation promoting arbitration as a preferred method of dispute resolution. Instruments such as the New York Convention and domestic arbitration laws provide a robust framework for enforcing arbitration agreements and awards, thereby bolstering parties' confidence in the arbitration process.

Moreover, the judiciary's role in curbing due process abuses adds a layer of protection against unwarranted paranoia. Courts serve as a backstop, ensuring that arbitrators adhere to procedural fairness and fundamental principles of justice. Parties aggrieved by due process violations have recourse to judicial review, where courts can set aside awards tainted by procedural irregularities or bias. This judicial oversight helps mitigate concerns about unchecked arbitral discretion and reinforces parties' trust in the arbitration process.

Therefore, while due process paranoia in arbitration may arise from valid apprehensions about procedural fairness, it is often unwarranted in light of the judiciary's arbitration-friendly approach. Courts play a pivotal role in safeguarding the integrity of arbitration proceedings, offering a mechanism for redress against due process abuses. By striking a balance between autonomy and accountability, the judiciary enhances confidence in arbitration as a reliable and effective means of dispute resolution. Therefore, while parties should remain vigilant about potential abuses, they can trust in the judiciary's commitment to upholding due process and fairness in arbitration.

14 Conclusion

Addressing due process paranoia in arbitration requires balancing procedural rigor with efficiency. Promoting procedural clarity and transparency can reduce ambiguity and prevent parties from exploiting procedural deficiencies. Arbitral institutions and tribunals play a crucial role by implementing clear procedural guidelines and best practices, such as setting deadlines for submissions, conducting efficient case management conferences, and empowering arbitrators to exercise discretion judiciously while adhering to due process principles.

Enhanced collaboration among arbitrators, counsel, and parties is essential for fostering a constructive dispute resolution approach, prioritizing substantive engagement over procedural tactics. Cultivating a culture of cooperation and respect can mitigate adversarial dynamics, promoting a more efficient arbitration process.

A national code of conduct for lawyers tailored to arbitration is a prudent measure to ensure legal process integrity and efficiency. This code should set clear standards emphasizing fairness, honesty, and procedural rule adherence, discouraging unnecessary delays and frivolous motions. Specific provisions could limit extensions, require a higher threshold for motions and challenges, and mandate timely disclosure of evidence. Penalties for violations, ranging from fines to practice suspension, would deter procedural rule misuse. The code should promote streamlined procedures, reducing unnecessary complexity and delay.

Implementing such a code nationally would enhance predictability and consistency in arbitration, aligning with international best practices and maintaining the national arbitration framework's reputation as fair and efficient. It would safeguard due process rights while preventing their exploitation, crucial for maintaining arbitration integrity and credibility.

Reviewing case law shows courts balance arbitral tribunal autonomy with safeguarding procedural rights. Courts generally defer to arbitral decisions but intervene when procedural irregularities threaten the arbitral process's integrity or parties' right to a fair hearing.

While ensuring due process is vital for fairness and legitimacy in arbitration, excessive focus on procedural formalities can delay dispute resolution. Promoting procedural clarity, fostering stakeholder collaboration, and adopting a nuanced approach to adjudicating due process concerns can mitigate due process paranoia, enhancing arbitration's efficiency and effectiveness.

List of Cases

Gold Reserve Inc. v. Bolivarian Republic of Venezuela [2016] EWHC 153 (Comm)

Asset Management ('AMCON') v. Qatar National Bank [2018] EWHC 2218 (Comm)

Fleetwood Wanderers Limited (t/a Fleetwood Town Football Club) v. AFC Fylde Limited [2018] EWHC 3318 (Comm)

Bureau Veritas, Inspection, Valuation, Assesment and Control, BIVAC B.V. v. Republic of Paraguay, ICSID Case No. ARB/07/9)

Iran Aircraft Industries v. Avco Corp., 980 F.2d 141 (2d Cir. 1992)

Grosso v. Barney 2003 (US Discrict Court E.D.Pa. 2003).

Terna Bahrain Holding Co v. Ali Marzook Ali Bin Kamil Al Shamsi et al [2012] EWHC 3283

Slowakische Republik v. Achmea BV, 6 March 2018, No. C-284/16.)

Micula v. Romania, 23 January 201

Vee Networks v. Econet Wireless International Limited [2004] EWHC 2909

References

Berger KP, Jensen JO (2016) Due process paranoia and the procedural judgment rule: a safe harbour for procedural management decisions by international arbitrators. Arbitr Int 32:415–435. https://doi.org/10.1093/arbint/aiw020

Ceretelli C (2020) Abuse of process: an impossible dialogue between ICJ and ICSID tribunals? J Int Dispute Settl 11(1):47–68

Cremades BM (2016) The use and abuse of "Due Process" in international arbitration

Due process paranoia in arbitral proceedings: Myth or reality? The Arbitrator and Mediator, [online] 39(2), pp 9–28. Available at: https://repository.canterbury.ac.uk/item/8yz07/due-process-paranoia-in-arbitral-proceedings-myth-or-reality

Due process rights in arbitration in the light of the Austrian supreme court's decision: remote hearings during the pandemic. JOR 10(1):105–122

Due process paranoia: need we be cruel to be kind. J Int Arbitr [online] pp 935–945. Available at: https://kluwerlawonline.com/journalarticle/Journal+of+International+Arbitration/34.6/JOIA2017044

Feehily R (2019) Neutrality, independence and impartiality in international commercial arbitration, a fine balance in the quest for arbitral justice. Int Aff

Gaillard E (2017) Abuse of process in international arbitration. ICSID Rev 32:17–37. https://doi.org/10.1093/icsidreview/siw036

Gerbay R (2016) Due process paranoia (Kluwer Arbitration Blog, 6 June 2016). https://doi.org/10.13140/RG.2.1.1005.9769

Ippolito F (2019) The abuse of due process in international arbitration. J Int Arbitr 36(1):1–27. https://doi.org/10.1177/0250891618817054

Kurkela M, Turunen S, Yliopisto H, Conflict Management Institute (2010) Due process in international commercial arbitration. Oxford ; New York: Oxford University Press

Lassalle F (n.d.) When logic fails: due process paranoia

Lau LYH (n.d.) Two steps forward, one step back? An attempt to cure due process paranoia

Mandujano Rubin JL (2022) Due process in arbitration. J Posit Psychol

Moneke E, Idornigie PO Mgbakogu OJ (2020) Due process paranoia in arbitral proceedings: myth or reality? Arbitr Mediators 39(2):9–28

Queen Mary University of London School of Law and White & Case (2015) International arbitration survey: improvements and innovations in international arbitration ('2015 QMUL Survey'), 2, arbitration.qmul.ac.uk/media/arbitration/docs/2015_International_Arbitration_Survey.pdf

Reed L (2017) Ab(use) of due process: sword vs shield. Arbitr Int 33:361–377. https://doi.org/10.1093/arbint/aix022

Saha S, Shukla S (2020) Resurrecting the debate on 'due process paranoia' in Centrotrade: paranoia or Judiciousness? Arbitr Int 36:521–528. https://doi.org/10.1093/arbint/aiaa035

Sharma S (n.d.) Due process "paranoia": turning away from judicial attitudes and looking for answers within

Energy Efficiency Based on Passive Design Solutions in UAE

Itemat Sauria, Shaikha Alblooki, Sima Zakzak, and Muna Salameh

Abstract Energy consumption in the building sector accounts for over 30% of global final energy consumption. This figure is much higher in developed countries like the UAE, where energy consumption by the built sector can reach up to 80%. Thus, developing appropriate future energy and climate change strategies requires alternative methods for the reduction of energy consumption in buildings. Passive design techniques, such as shading devices and courtyards, are used in sustainable architecture to create structures that harness sunlight and wind for lighting and ventilation while keeping away heat gain and glare. This research aims to investigate and create a comparison between the effects of two passive design solutions on a villa housing project in Sharjah, United Arab Emirates, through the use of Revit software. The study concluded that courtyards significantly outperformed shading devices in reducing energy costs, providing a greater overall impact on energy consumption. In this research, a comparative study was done between three building types in the same area: a villa without any passive design solution, a villa with the introduction of shading devices, and a villa with a courtyard. The research shows that courtyards, by enhancing natural ventilation and air circulation, lowered HVAC costs to approximately 14.69 AED/m2 per year, compared to 33.06 AED/m2 per year for louvers. While louvers improved lighting efficiency and reduced lighting needs, they led to slightly higher costs for window shades. Overall, courtyards achieved a cost reduction of 24.24 AED/m2 per year versus 6.61 AED/m2 per year for shading devices,

This paper was written for the course—Sustainable Architecture, taught by Dr. Muna Salameh.

I. Sauria (✉) · S. Alblooki · S. Zakzak · M. Salameh
Department of Architecture, Faculty of Architecture, Art, and Design, Ajman University, Ajman, UAE
e-mail: 202010475@ajmanuni.ac.ae

S. Alblooki
e-mail: 202010317@ajmanuni.ac.ae

S. Zakzak
e-mail: 202010736@ajmanuni.ac.ae

M. Salameh
e-mail: m.salameh@ajman.ac.ae

K. Al Marri et al. (eds.), *BUiD Doctoral Research Conference 2024*, Lecture Notes in Civil Engineering 587, https://doi.org/10.1007/978-3-031-84371-6_15

emphasizing the importance of integrating passive design strategies for sustainable building practices and climate change mitigation.

Keywords Energy consumption · Passive design · Courtyards · Shading devices · Heat gain · Ventilation · Revit software

1 Introduction

Traditional design principles, particularly passive design methods, are foundational to both vernacular and sustainable architecture, aiming to improve interior temperature conditions while minimizing energy use (Thaleia Konstantinou 2018). These methods remain relevant today, especially in the face of global warming and rapid urban development. The UAE's rapid urban growth, coupled with its hot-arid climate, has prioritized construction quantity over energy efficiency, exacerbated by inadequate building codes and low energy costs, leading to a high environmental footprint.

This research compares two traditional passive design techniques—courtyards and shading devices—in UAE based homes to demonstrate their potential in reducing energy consumption. The study analyzes these techniques to identify the optimal solution for enhancing energy efficiency, reducing heat gain, and improving thermal comfort, aiming to find sustainable approaches that benefit the built environment and occupants.

2 Literature Review

2.1 Courtyards

The world is worried about the ongoing urbanization that goes hand in hand with global warming. Incorporating passive sustainable solutions into design has become of utmost importance. Muna Salameh (2022) finds that traditional urban districts perform better in terms of heat than modern urban districts as they have passive design solutions like courtyards.

Courtyards are private open spaces surrounded by a building wall on their sides. A courtyard's shape might be square, rectangular, or round. The majority of courtyard designs from the past were square or rectangular. Figure 1 illustrates the three typical courtyard design types: semi-opened, semi-closed, and closed.

Since it offers great privacy, good ventilation and lighting, the closed courtyards are found frequently in deep layouts. Conversely, a semi-private courtyard is typically formed between buildings and is said to provide a shaded area. Finally, semi-opened

Fig. 1 Courtyard types

courtyard has the least degree of privacy, but it also offers the building easy access, ventilation, and natural light.

Furthermore, it has been discovered that courtyards in the UAE, especially in Sharjah, are advantageous for social interaction and energy efficiency, according to IA Ibrahim (2018).

2.2 *Shading Devices*

Shading devices are external projections that shield a building's glazed areas from direct sunlight, and they can be horizontal or vertical. The choice and type of shading device depend on the building's orientation. Fixed louvers are ideal for south-facing facades due to direct sunlight during peak hours, while operable louvers are more suitable for east and west facades. Shading devices are unnecessary for north-facing facades in the UAE, as this direction does not receive direct sunlight.

According to studies conducted in the United Arab Emirates, shading devices can dramatically lower a building's energy usage. According to Alothman and Mahmoud 2024 horizontal shading elements can save 0.5% to 30.8% of energy in multi-story residential buildings, depending on orientation and depth. Figure 2 highlights the types of shading devices.

Finally, the shading devices can be installed at any given time, even for existing buildings. Shading devices are an affordable way to cut carbon emissions and energy use, according to research conducted regularly in the United Arab Emirates. Though kinetic shading devices have potential, fixed shading devices still have a bigger potential for energy savings, according to Lee (2019).

2.3 *Autodesk Revit Architecture*

Autodesk Revit Architecture, developed by Autodesk, is a powerful software program for architects and other construction professionals. It features tools for energy analysis, solar radiation analysis, and thermal analysis, which are utilized in

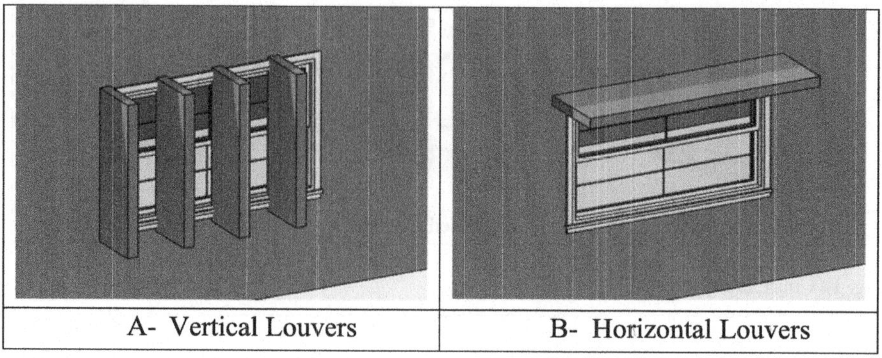

Fig. 2 Types of louvers

this research. Revit, known for its Building Information Modeling (BIM) capabilities, enables the creation of a parametric 3D model with information about building and design that is both geometric and non-geometric. This research aims to demonstrate how passive design solutions, supported by Revit's features, can enhance energy efficiency. The objectives are as follows:

1. Model three case studies with two different passive design solutions.
2. Use Revit to generate results regarding energy use and cost.
3. Compare and contrast the results to draw conclusions about the benefits or drawbacks of passive design.

Passive design solutions take time to show significant results and are not quick to reflect drastic changes. This research, using Revit analysis, will generate annual results on cost and efficiency. Additionally, this research explores two case studies to demonstrate how passive design solutions can be applied either before, during, or after construction.

3 Methodology

This research will use a qualitative approach that relies on comparative analysis for a case study following computer simulation using Revit; the case study will be a villa located in Sharjah, United Arab Emirates. The case study will be divided into two parts: one is collecting data through computer stimulation, and the other is analyzing the information found.

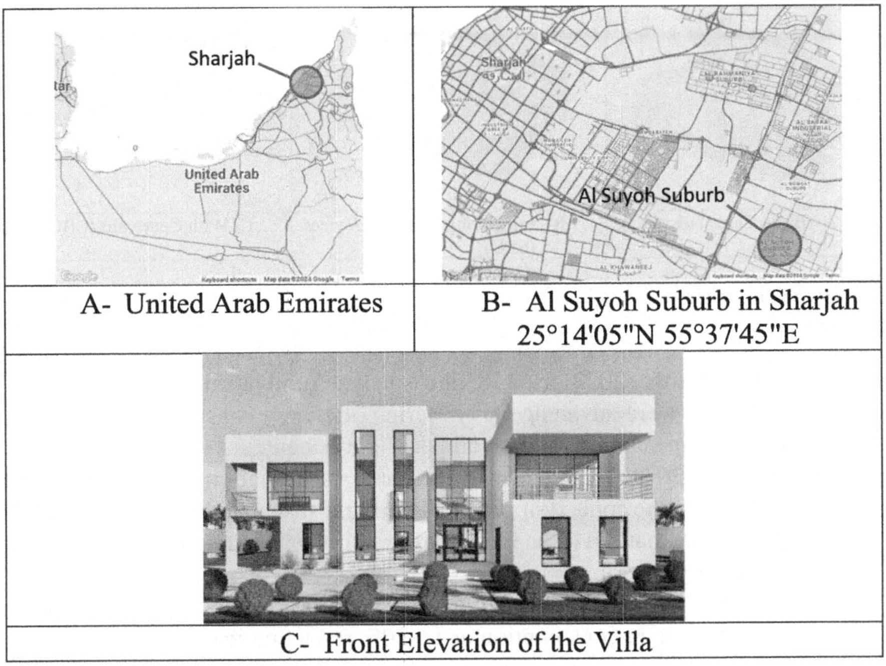

Fig. 3 The location of case study

3.1 Location and Criteria of the Case Study

To evaluate the efficiency of passive design techniques, a villa in Al Suyoh Suburb, Sharjah, was selected for the study. This location is ideal for understanding how the UAE's climate impacts building design. The villa, oriented northwest, follows a modern architectural style and has a total built-up area of 10,115.77 m^2 across two stories, including a ground floor, first floor, and a staircase room on the roof. The ground floor features public spaces such as a majlis, dining area, guest bedroom, and kitchen, while the first floor contains private bedrooms, each with an attached bath and a dressing area, along with balconies. The villa's design lacks specific passive design techniques to reduce heat gain or glare, aside from the orientation of the openings. The numerous and large openings could increase energy consumption, as more cooling is needed to counteract the heat gain (Fig. 3).

3.2 Use of Revit as Research Software

The potential of Revit, a Building Information Modeling (BIM) platform, is to enhance sustainability analysis in building design has been shown in recent research.

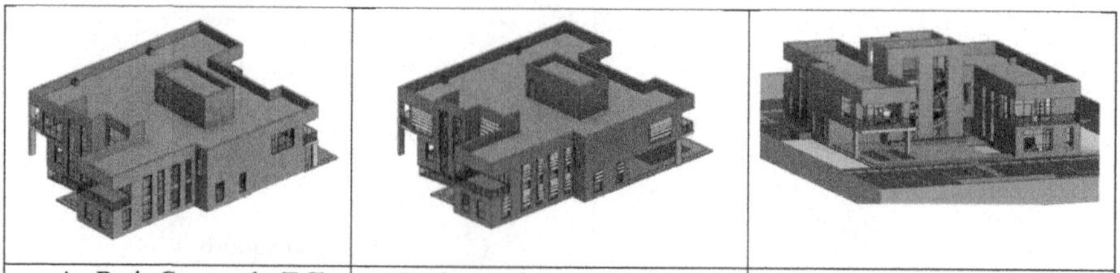

A- Basic Case study (BC).	B- Integrating Shading Devices (SBC).	C- With Courtyard (CBC)

Fig. 4 The different case studies

1. Bektaş and Özköse (2022) used Revit to analyze rural housing projects, finding that while they were advantageous in terms of energy consumption, they were disadvantaged in solar orientation and daylight benefit.
2. Carvalho et al. (2021) conducted more research on the viability of BIM platforms, such as Revit, in the context of practical building sustainability assessment. The findings indicate that Revit can facilitate the automated and expedited assessment of sustainability criteria.
3. Mogili et al. (2022) to predict energy usage in a residential building, achieving a 31.3% reduction in consumption and a 30% reduction in cost.

3.3 Stimulation Criteria and Stages

The research investigated two passive design solutions on the existing case study apart from the original villa (Fig. 4), as follows:

1. Basic case (BC).
2. The integration of shading devices (SBC).
3. Integrating the courtyard for further analysis for future design prototypes of the villa (CBC).

3.4 Revit Analysis Procedure

Autodesk's Revit software utilizes its energy analysis features to determine annual energy expenditures by creating a detailed 3D model of a building, including architectural elements and mechanical systems. Revit specifies parameters such as location, climate data, occupancy schedules, and HVAC details to conduct a dynamic thermal simulation. This simulation calculates hourly energy consumption, including equipment loads, lighting usage, and HVAC energy needs. Revit then aggregates this data to compute annual energy costs, presenting the results through reports, graphs, and visualizations. Additionally, daylighting analysis in Revit uses the project's specific solar conditions to measure "illuminance" and "luminance," helping architects optimize building performance and energy efficiency.

3.5 Unified Stimulation Conditions

The research used Revit to undergo an energy analysis of the building envelope. This version of the villa does not have any passive design solutions. The villa has large windows and multiple openings. Table 1 shows the detailed analysis for this case.

All materials and some design parameters were defined in the three scenarios for the stimulation to highlight only the effect of the two integrated passive design solutions.

3.6 Integrated Shading Devices

For the first passive design case, the research will introduce shading devices to see how they affect energy consumption in a villa (Table 2). Shading devices can dramatically lower a building's energy use, according to research conducted in the United Arab Emirates.

3.7 Integrated Hypothetical Courtyard

The second passive design solution applied in the villa is the introduction of a courtyard as a prototype for future designs (Table 3). The courtyard type used is the semi-closed courtyard. Many studies have been conducted on the social and environmental benefits of courtyards found in villas in the United Arab Emirates. Both IAS Ibrahim, 2018, and HM Taleb, 2020, emphasize how courtyards can improve thermal performance and sustainability.

IAS Taleb, 2020, goes so far as to say that the most efficient courtyard arrangement is u-shaped, the same is applied in this scenario.

4 Discussion and Results

The villa with a built-up area of 10,115.77 m2 located in Sharjah, United Arab Emirates, was modeled and stimulated on Revit to understand its energy consumption on a yearly basis. The stimulation output revealed variations in its light efficiency, energy costs, HVAC, etc.

Table 1 Unified design approaches

Unified design approaches			
1. Orientation of the Building	Rotates a structure in a clockwise direction starting at 0 degrees. For example, turning a building 90 degrees causes the north side to face east. Current Setting: 270 - 135 	2. West window glass	Along with other considerations, glass qualities regulate how much daylight, heat transfer, and solar heat gain enter the building. Current Setting: Sgl Clr - Trp LoE 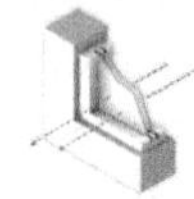
3. WWR—southern walls	The relationship between window attributes and window-wall ratio (glazing area/gross wall area) affects heating, cooling, and daylighting. Current Setting: 95% - 0% 	4. WWR- eastern walls	The relationship between window characteristics and window-wall ratio affects heating, cooling, and daylighting. Current Setting: 95% - 0% 
5. Window glass-south	Along with other considerations, glass qualities regulate how much daylight, heat transfer, and solar heat gain enter the building. Current Setting: Sgl Clr - Trp LoE 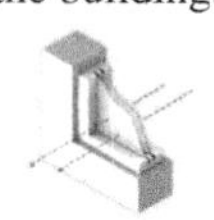	6. Window glass-east	Along with other considerations, glass qualities regulate how much daylight, heat transfer, and solar heat gain enter the building. Current Setting: Sgl Clr - Trp LoE 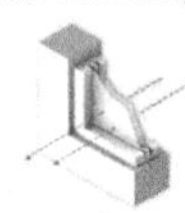
7. WWR- northern walls	The relationship between window characteristics and window-wall ratio affects heating, cooling, and daylighting. Current Setting: 95% - 0% 	8. Wall construction	Represents the total resistance to heat gains and losses of wall structures. Current Setting: R13 Wood - 12.25-inch SIP 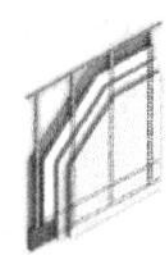

(continued)

Table 1 (continued)

Unified design approaches			
9. Window glass-north	Along with other considerations, glass qualities regulate how much daylight, heat transfer, and solar heat gain enter the building. Current Setting: Sgl Clr - Trp LoE 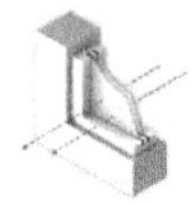	10. Roof construction	Represents the total resistance to heat gains and losses of wall structures. Current Setting: Uninsulated - R60 
11. WWR- western walls	The relationship between window characteristics and window-wall ratio affects heating, cooling, and daylighting. Current Setting: 95% - 0% 	12. Infiltration	Air escaping into or out of conditioned compartments without effort; frequently caused by gaps in the building envelope. Current Setting: 2.0 ACH - 0.17 ACH 

The research, as shown in Table 4, indicates minor differences in lighting efficiency, HVAC, and window shading across different scenarios, which, although not significant, do impact the overall energy consumption of the villa. After simulations, it was found that lighting efficiency improved the most with the use of louver shading elements, as they effectively filter and control natural light. Louvers also optimize daylight while reducing heat gain, leading to energy savings. However, the use of window shades on all sides incurs a slightly higher cost, around 3.67 AED per square meter per year, according to Revit's findings.

Moreover, the discrepancy in HVAC costs between using louvers and incorporating a courtyard primarily stems from differences in thermal regulation and passive cooling effectiveness. Louvers, while effective in controlling natural light and reducing solar heat gain, may not provide the same level of thermal comfort as a courtyard. Despite their ability to mitigate some of the heat entering a building, they might still necessitate more intensive HVAC usage to maintain comfortable indoor temperatures, especially in hotter climates or during peak sunlight hours. Consequently, HVAC operational costs remain relatively higher at 33.06 AED/M2/y. On the other hand, courtyards offer significant passive cooling benefits. They create natural ventilation, facilitate air circulation, and provide shaded areas, effectively reducing the overall cooling load on HVAC systems. This results in lower energy consumption and decreasing HVAC operational costs, which are reflected in an approximate 14.69 AED/m2/y, as shown in Fig. 5.

Table 2 Scenario two, Shading Devices

SBC

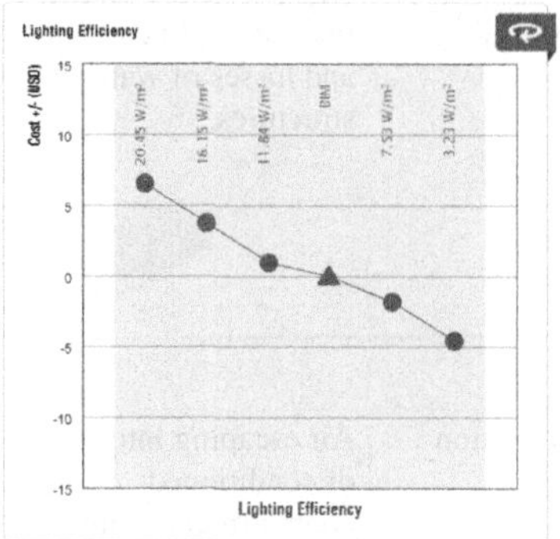

Lighting efficiency

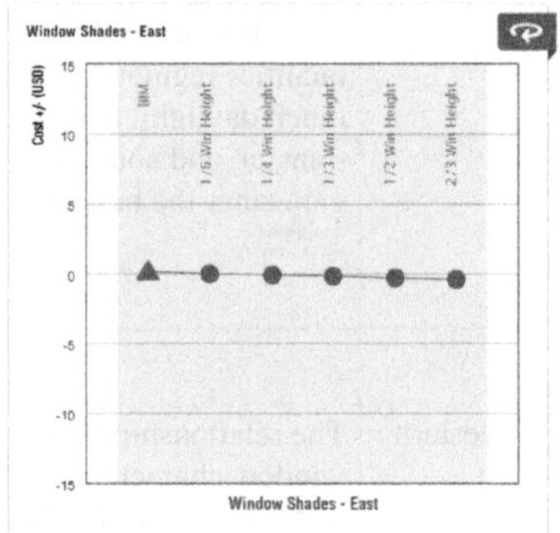

Window shades—East

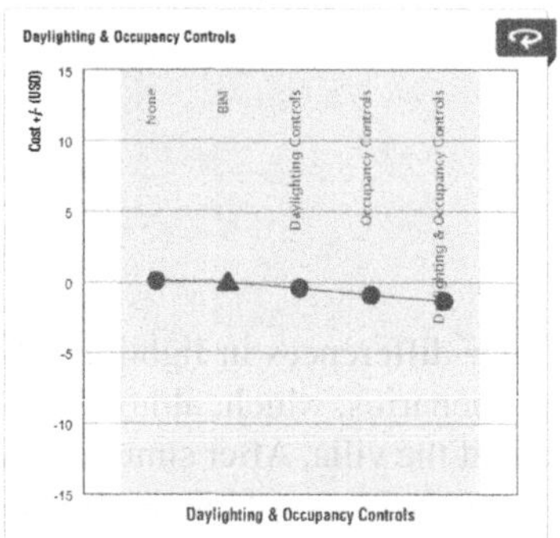

Daylighting and occupancy control

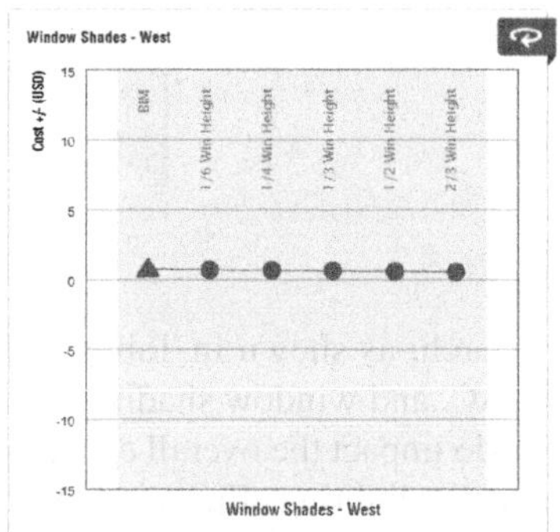

Window shade—West

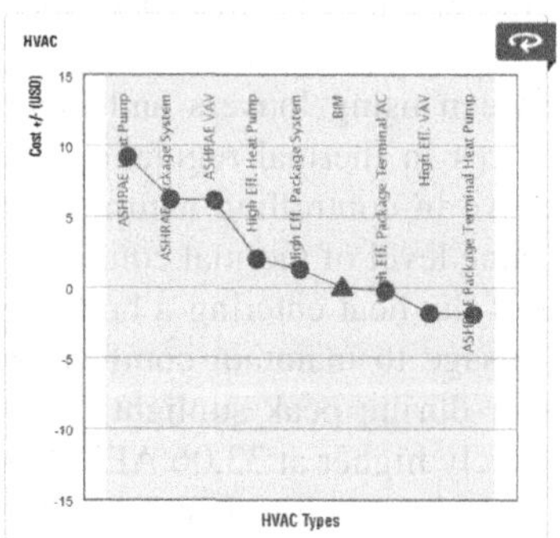

HVAC

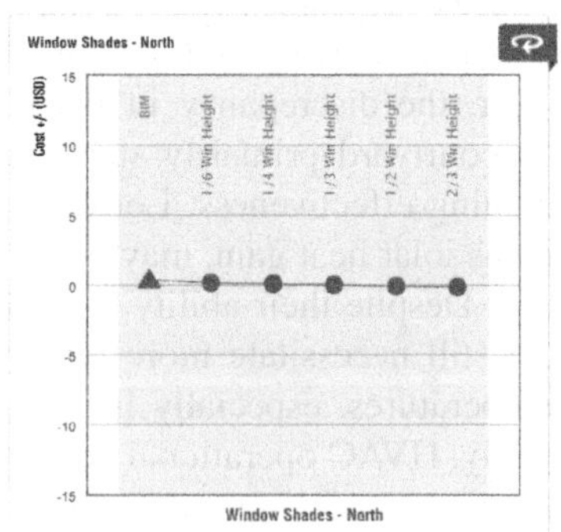

Window shade—North

(continued)

Table 2 (continued)

SBC	
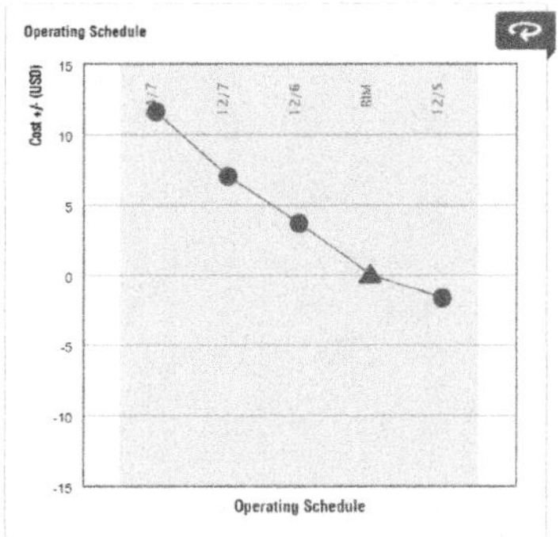	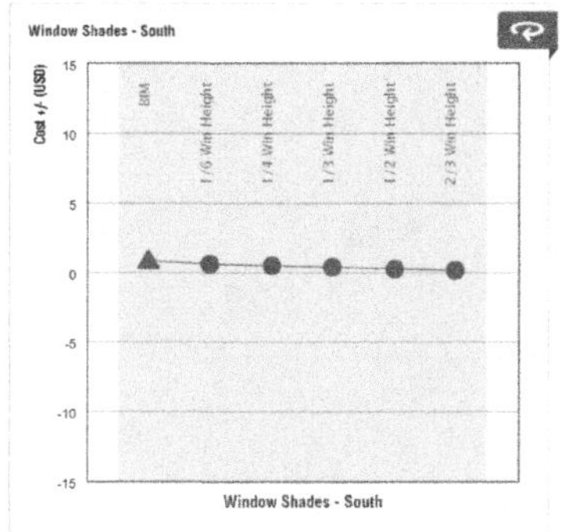
Operating schedule	Window shade—South

This study delves into leveraging passive design strategies to mitigate climate change by curbing energy consumption. As part of a multifaceted approach, the research examines the efficiency of integrating louvers and utilizing courtyards for potential prototypes. Notably, the findings reveal that, courtyard integration outperformed shading devices, yielding a significant reduction in energy costs at 24.24 AED/m2/yearly compared to the 6.61 AED/m2/yearly reduction achieved with shading devices. However, while shading devices contributed to decreased lighting needs, the overall impact on energy consumption was less pronounced than that of courtyards. This underscores the importance of strategic design choices in optimizing energy efficiency within architectural frameworks. Moreover, the study highlights the potential of courtyards as a versatile solution with broader implications for sustainable building practices. By prioritizing passive design solutions like courtyards, architects and urban planners can have a role in addressing the pressing challenges of climate mitigation (Table 5).

5 Conclusion

This research, using Revit software, assessed the impact of two passive design techniques- shading devices and courtyards- on energy efficiency. Shading devices can be implemented during construction or after occupancy, while courtyards must be incorporated from the initial design stage. The findings are applicable to the design of villas or similar residential complexes in the future.

The results indicated that louver shading elements improved lighting efficiency by filtering and controlling natural light effectively, though they led to slightly higher costs of about 3.67 AED/m2 yearly for window shades. HVAC costs with louvers

Table 3 Special design parameters for scenario three, courtyard

CBC

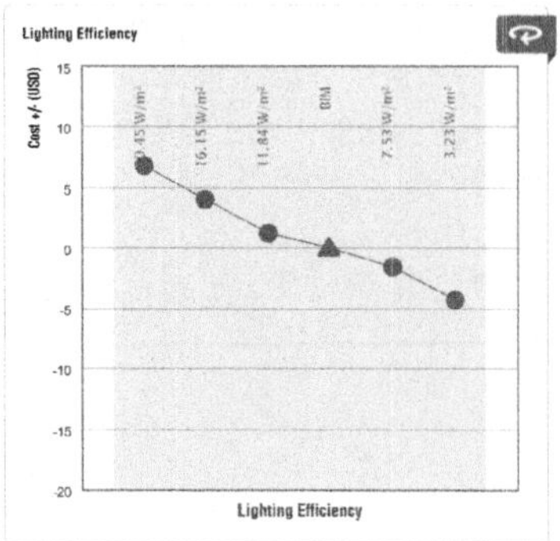

Lighting efficiency

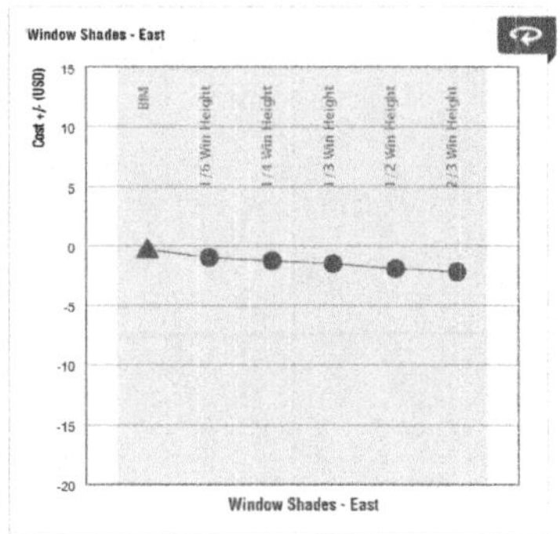

Window shades—East

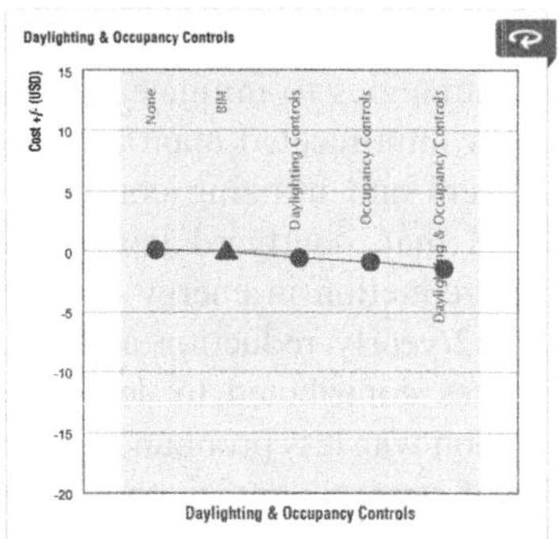

Daylighting and occupancy control

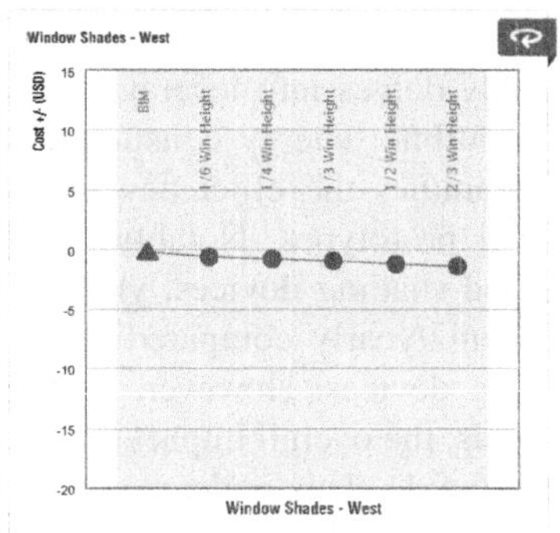

Window shade—West

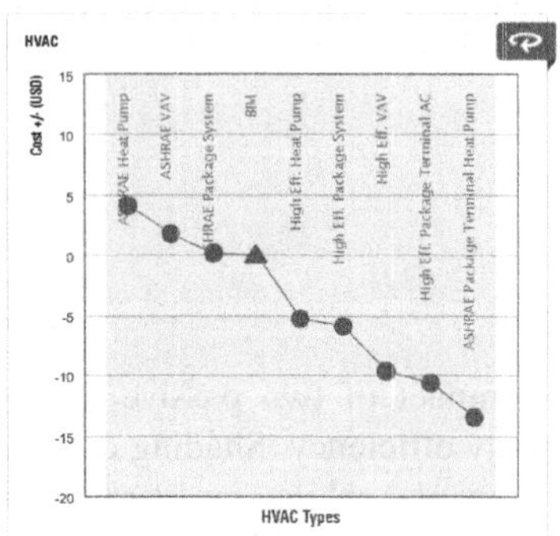

HVAC

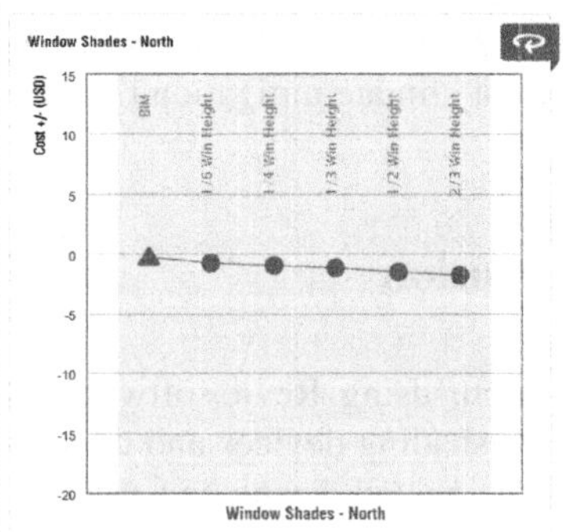

Window shade—North

(continued)

Table 3 (continued)

CBC	
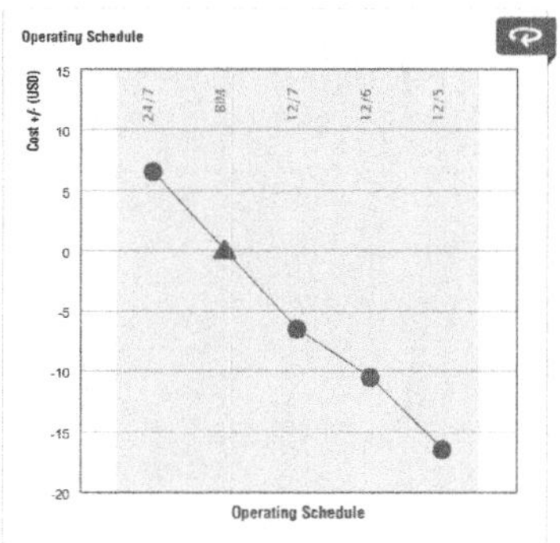	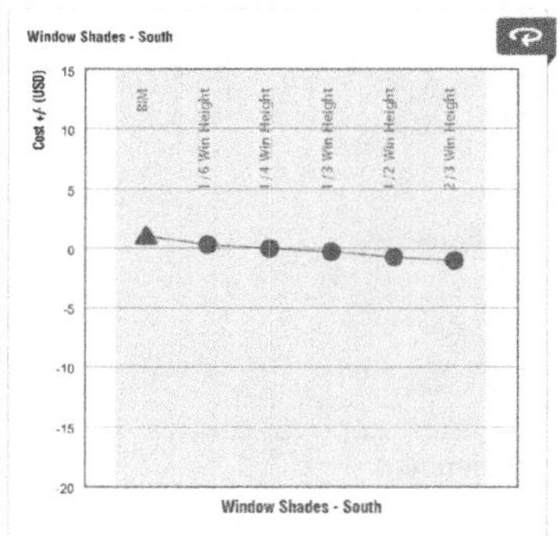
Operating schedule	Window shade—South

were higher at 33.06 AED/m2/year due to their less effective thermal regulation compared to courtyards.

Courtyards, by enhancing natural ventilation and air circulation, significantly reduced HVAC costs to approximately 14.69 AED/m2/year. Overall, courtyards provided a greater reduction in energy costs (24.24 AED/m2/year) compared to shading devices (6.61 AED/m2/year). While shading devices reduced lighting needs, courtyards had a more substantial impact on overall energy consumption, highlighting their effectiveness in passive design strategies for sustainable building practices and climate change mitigation.

By analyzing and comparing the two passive design solutions- shading devices and courtyards- and creating a stimulation that shows their effects on energy consumption in a villa located in Sharjah, United Arab Emirates, using Revit software, this research helps in the advancement of the subject of sustainable architecture as well as the use of passive design solutions.

Contributions:

1. The research provides a comprehensive analysis of the ways in which courtyards and shading devices impact energy use, including lighting efficiency, HVAC costs, and total energy consumption. Architects and planners can gain useful insights into the efficiency of these passive design techniques from this qualitative investigation.
2. With a focus on Sharjah, United Arab Emirates, the research discusses the unique opportunities and limitations of passive architecture in hot, arid climates. By placing the findings in a geographical context, the study can provide specific recommendations for sustainable building methods across the globe with similar climates.
3. The study offers architects and developers clear information on the best ways to design energy-efficient buildings by comparing and indicating that courtyards

Table 4 Comparative study between the three cases according to design parameters

Basic case (BC)	SBC	CBC
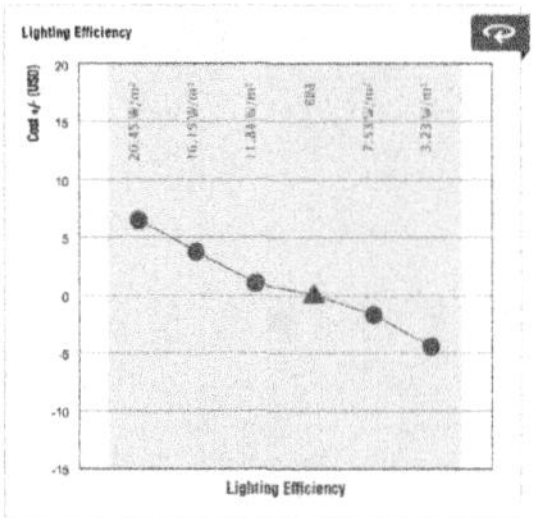	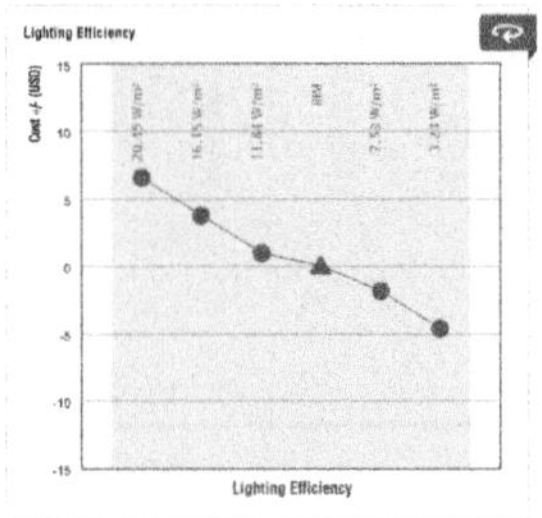	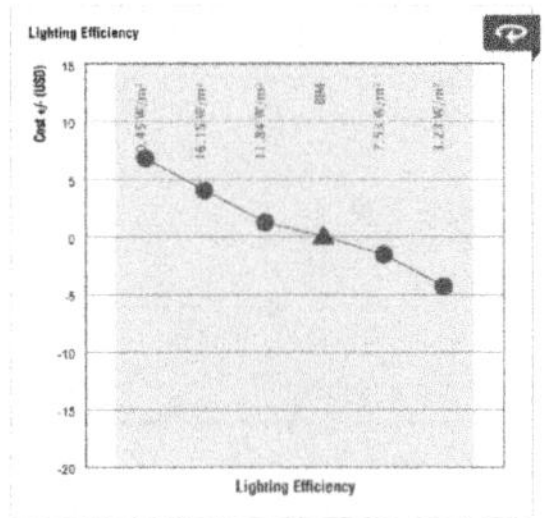
Lighting efficiency		
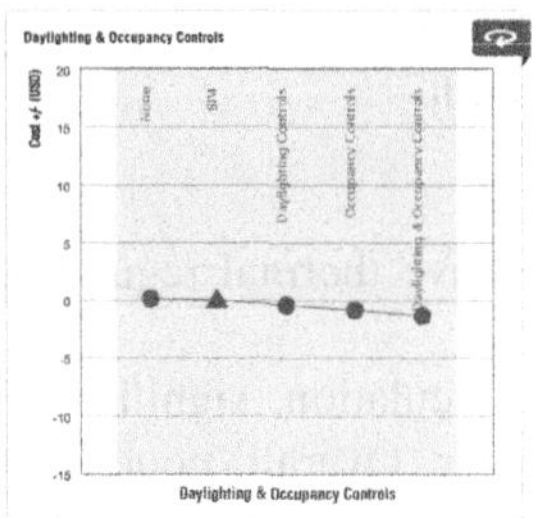	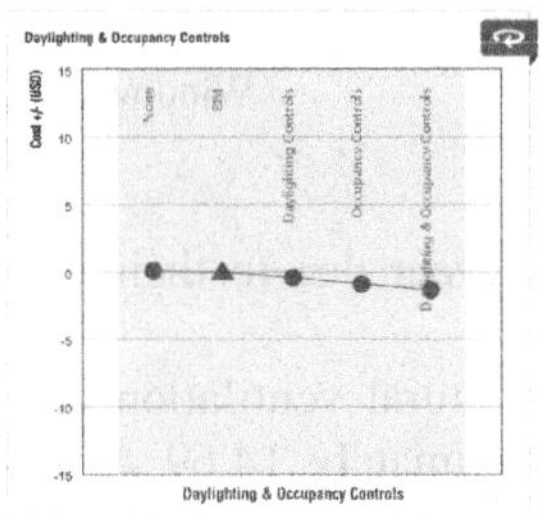	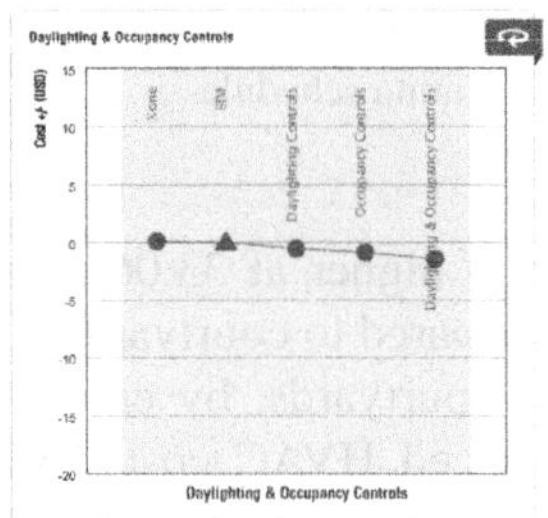
Daylighting and occupancy control		
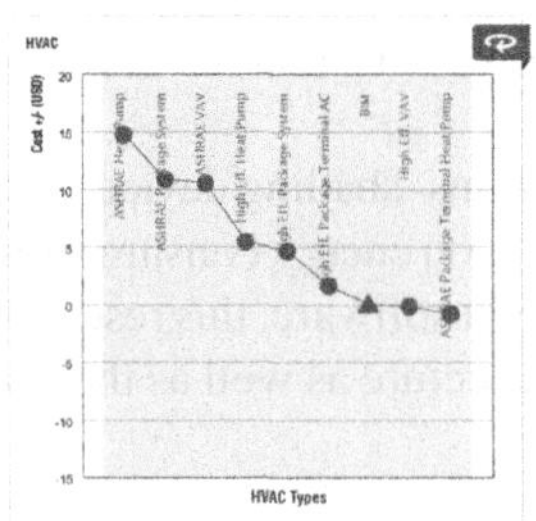	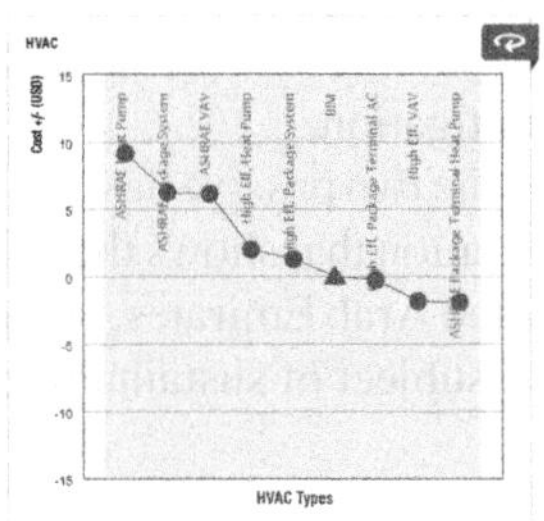	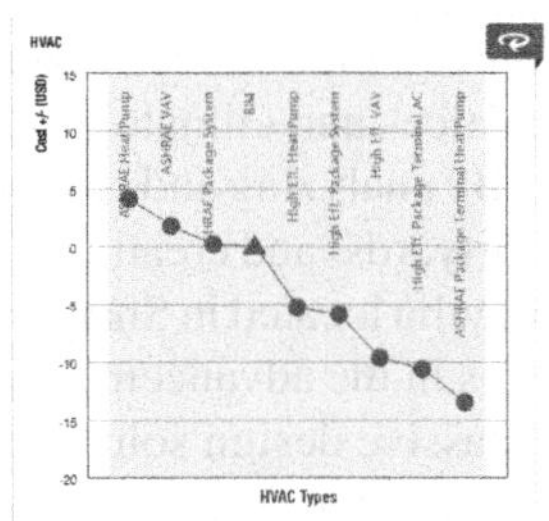
HVAC		
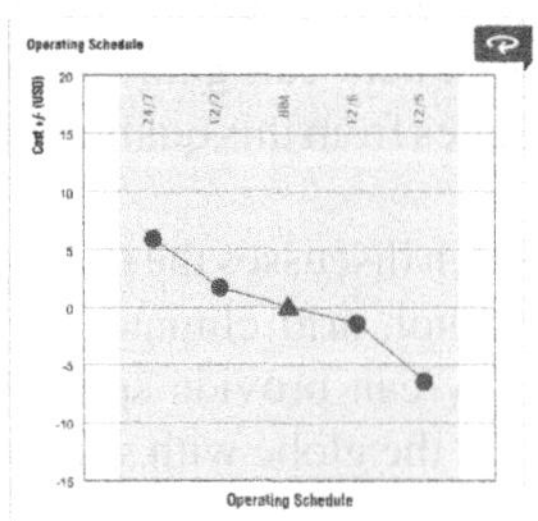	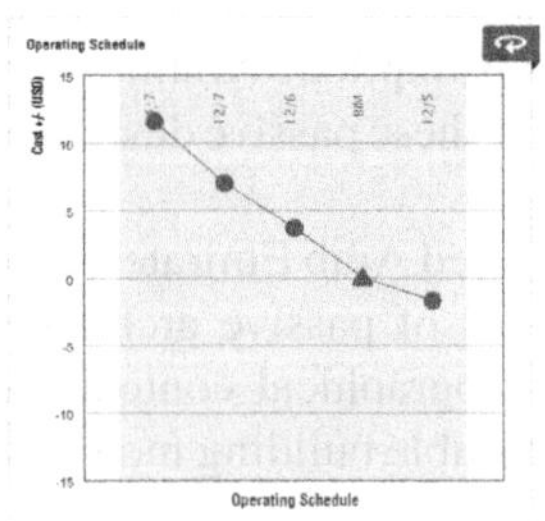	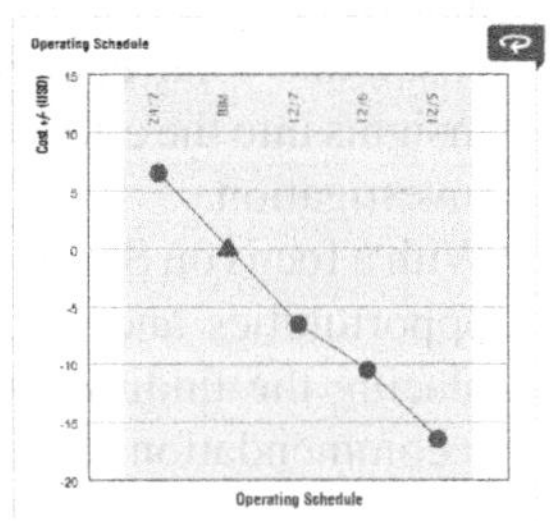
Operating schedule		

(continued)

Table 4 (continued)

Basic case (BC)	SBC	CBC
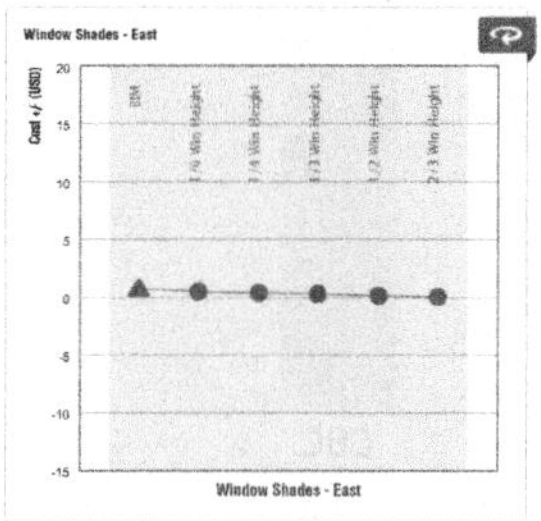	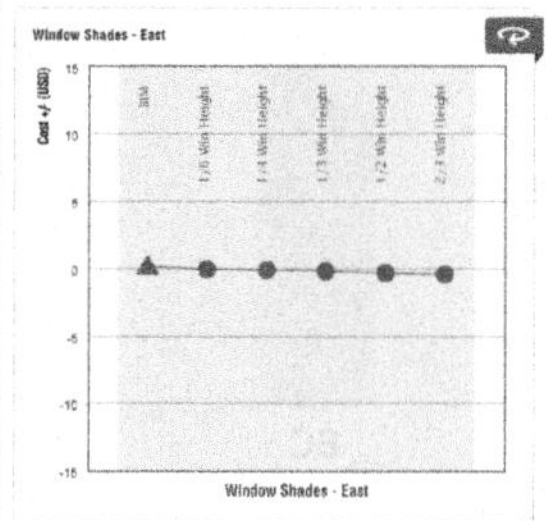	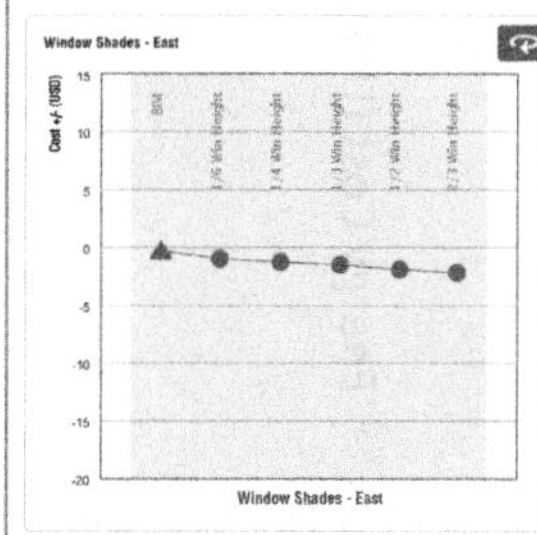
Window shades—East		
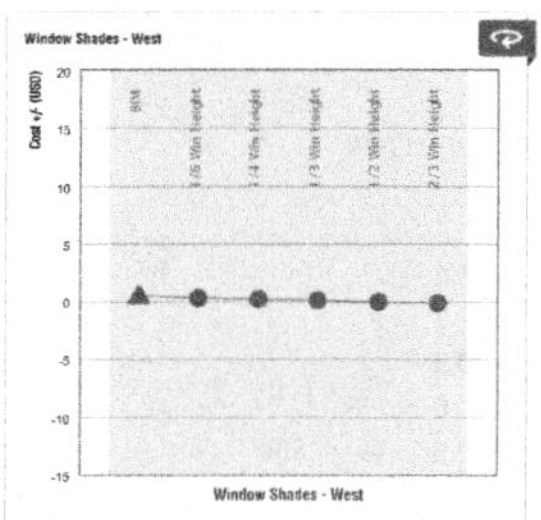	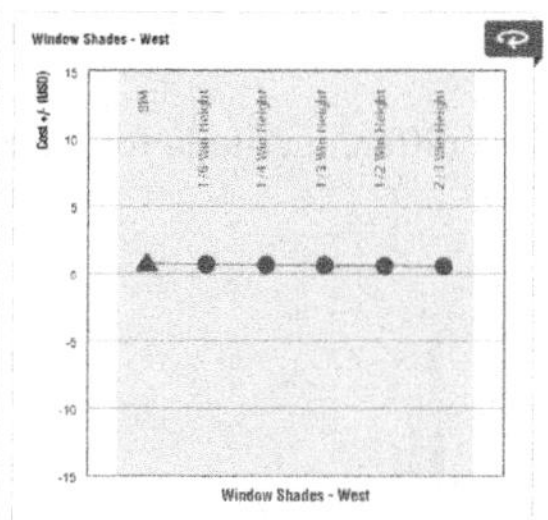	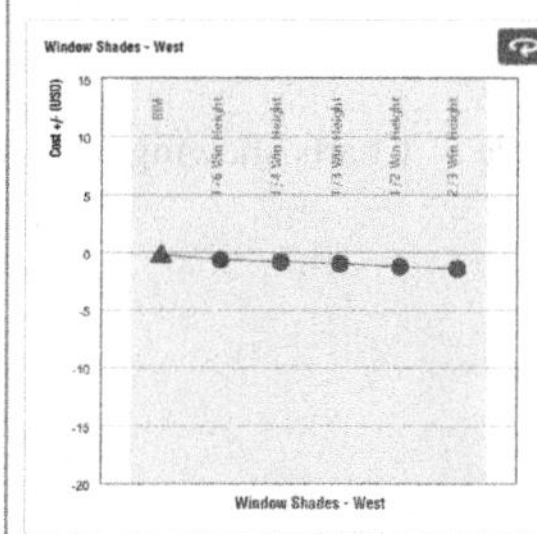
Window shade—West		
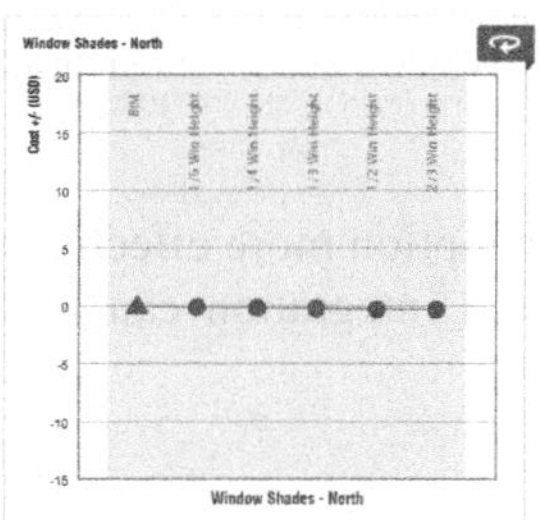	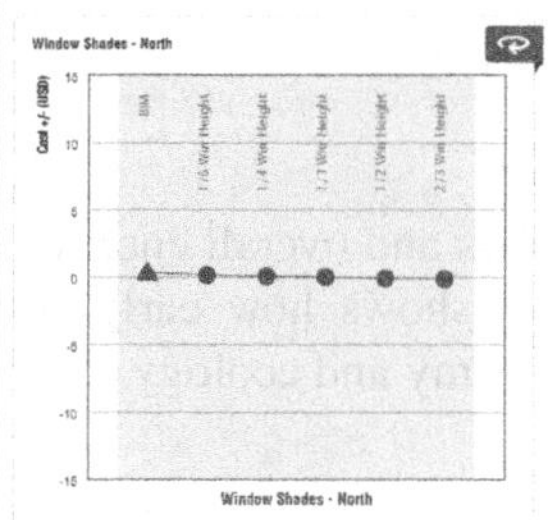	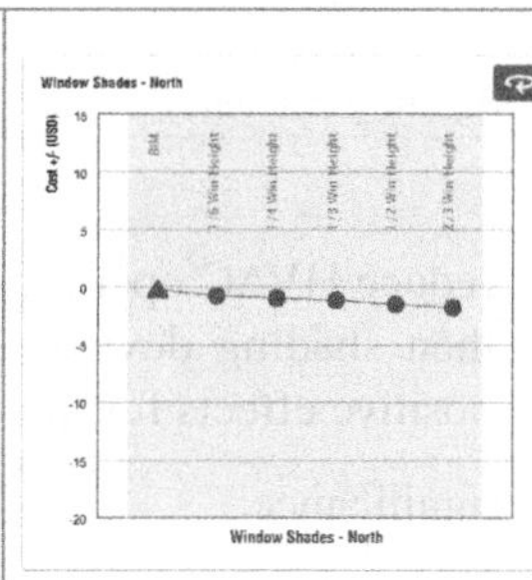
Window Shade—North		
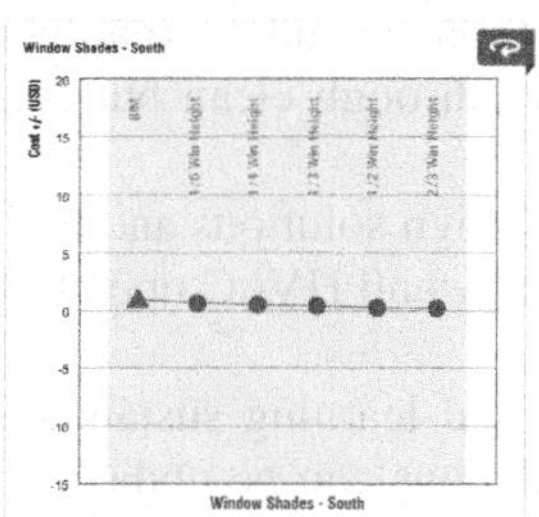	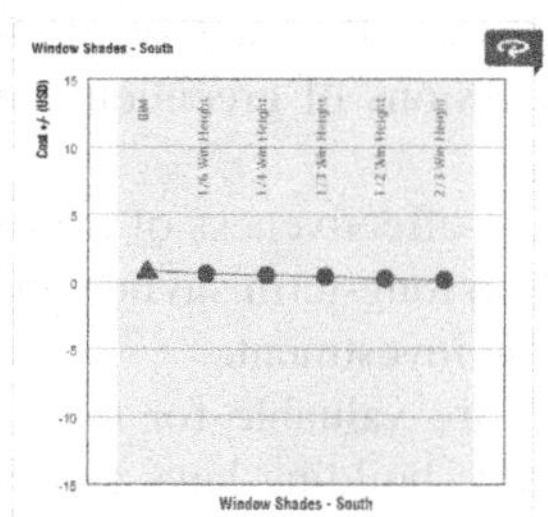	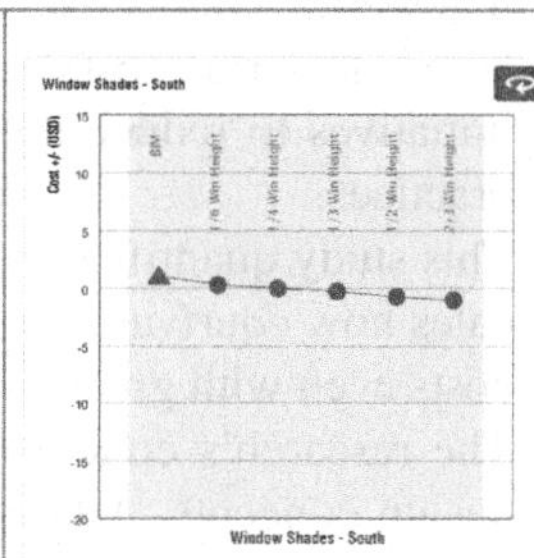
Window shade—South		

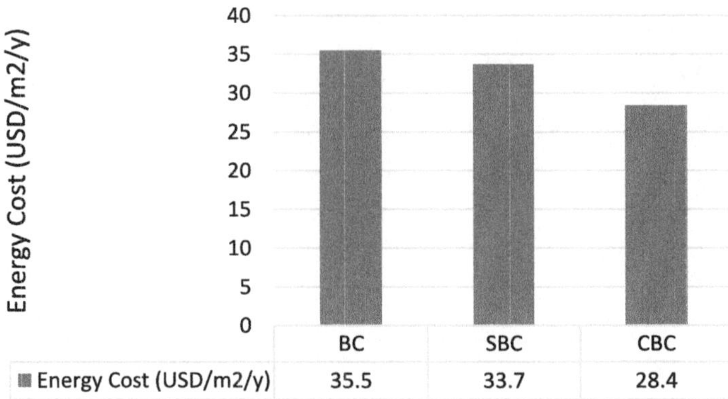

Fig. 5 Energy costs in different scenarios

Table 5 Charts showing the ranges of different criteria for the three scenarios

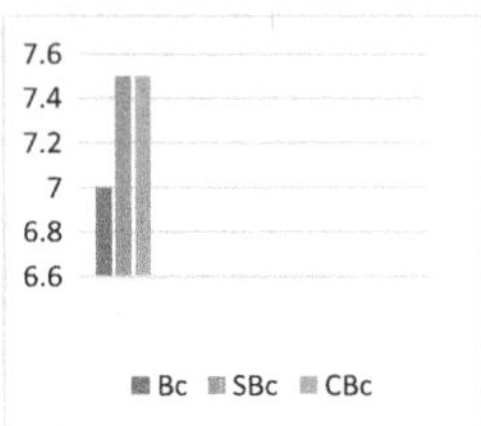	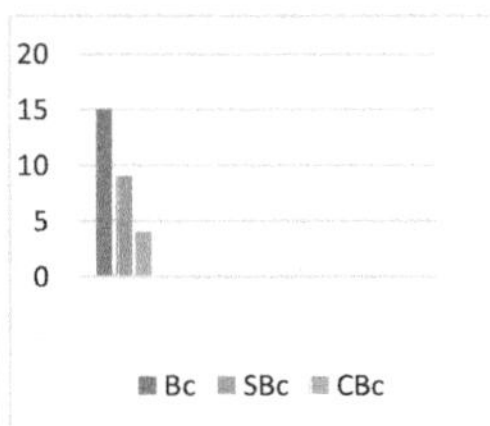	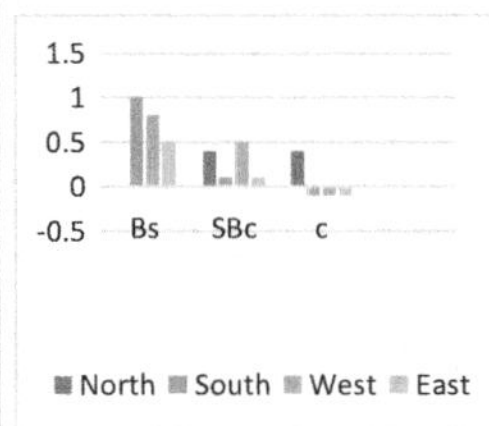
Daylighting efficiency	HVAC	Window shade—Four sides

reduce HVAC operating costs and overall energy consumption more effectively than shading devices. This shows how early courtyard integration can have positive effects for the economy and ecology.

Significance:

The results of this research highlight how passive design techniques are important in lowering energy use, especially with the help of courtyards. This does not only help in lowering costs but also supports sustainable development and international initiatives to reduce the emissions of greenhouse gases through clean building practices.
This study quantifies the cost-effectiveness of passive design solutions and indicates how courtyards result in long-term savings in energy and HVAC operation costs even with greater initial investment.
The research's conclusions are valuable for teaching and learning sustainable design concepts. Additionally, building laws and regulations can be updated to encourage the implementation of such design solutions on a larger scale for a more sustainable built environment.

References

Abbass JH, Taleb HM (2015) Applying passive design strategies: case study of an art gallery

Alothman, ARA and Mahmoud AH (2024) The effect of using horizontal shading elements on the energy efficiency of multistory residential buildings

Bektaş İ, Özköse A (2022) Investigation of housing projects for rural areas in terms of sustainability criteria with revit—the case of Kayseri. ICONARP Int J Archit Plan 10(2):551–574

Carvalho JP, Bragança L, Mateus R (2021) Sustainable building design: analysing the feasibility of BIM platforms to support practical building sustainability assessment. Comput Ind 127:103400

Drach JK-F (2014) Increasing ventilation by passive strategies: analysis of indoor air circulation changes through the utilization of microclimate elements

Friess KR (2017) A review of passive envelope measures for improved building energy efficiency in the UAE. Renew Sustain Energy Rev 72

Giovannini, Verso VRML, Karamata B, Andersen M (2015) Lighting and energy performance of an adaptive shading and daylighting system for arid climates

Ibrahim (2018) Eco-architecture VII: harmonization between architecture and nature

Konstantinou, Hoces AIP, Konstantinou T, Nata (2018) Environmental design principles for the building envelope and more: passive and active measures

Lee (2019) Heating, cooling, and lighting energy demand simulation analysis of kinetic shading devices with automatic dimming control for Asian countries

Mesloub, Ghosh A, Touahmia M, Albaqawy GA, Noaime E, Alsolami BM (2020) Performance analysis of photovoltaic integrated shading devices (PVSDs) and semi-transparent photovoltaic (STPV) devices retrofitted to a prototype office building in a hot desert climate

Mogili S, Avvari HV, Appecharla VK (2022) Energy consumption analysis of residential building using Autodesk Revit. In: IOP conference series: earth and environmental science, vol. 1086. IOP Publishing, p 012056

Salameh and Touqan B (2022) Traditional passive design solutions as a key factor for sustainable modern urban designs in the hot, arid climate of the United Arab Emirates

Taleb and AbuMoeilak L (2020) An assessment of different courtyard configurations in urban communities in the United Arab Emirates

Enhancing the Incident Response Capabilities of the UAE Critical Infrastructure Against Phishing Cyber Security Threat: An AI-Based Approach

Bushra Rwashdeh, Ahmed Awad, and Usman Butt

Abstract Phishing attacks pose a significant threat to critical infrastructure in the UAE. This study proposes an AI-driven approach to enhance incident response capabilities against such threats. The research adapts the Technology Threat Avoidance Theory (TTAT) to an organizational context and develops a machine learning model for phishing detection and response. This research contributes to both theory and practice by extending TTAT and creating an AI prototype tailored to the UAE's critical infrastructure needs. The study employs a design science approach, incorporating qualitative evaluations from UAE experts to refine the model. This research aims to strengthen organizational resilience and contribute to the UAE's national cybersecurity objectives.

Keywords Critical infrastructure · Incident response · AI · Phishing · Cybersecurity

1 Introduction

Cybersecurity is a paramount concern for protecting critical infrastructure in the UAE, which includes sectors like energy, transportation, telecommunications, finance, healthcare, and utilities from escalating cyber threats, particularly phishing attacks (Rajan, Ravikumar & Shaer 2017). Phishing attacks exploit human vulnerabilities to gain unauthorized access to sensitive systems supporting essential services, posing a serious threat to critical infrastructure organizations globally and in the UAE. Enhancing phishing incident response capabilities through an AI-driven approach is crucial.

This paper was prepared for RES606 Research Design and Planning, taught by Prof. Ahmed Awad and Prof. Usman Butt)

B. Rwashdeh (✉) · A. Awad · U. Butt
The British University in Dubai, Dubai, UAE
e-mail: 22000463@student.buid.ac.ae

K. Al Marri et al. (eds.), *BUiD Doctoral Research Conference 2024*, Lecture Notes in Civil Engineering 587, https://doi.org/10.1007/978-3-031-84371-6_16

Incident response is vital for mitigating cyber threats' impact, minimizing downtime, and ensuring business continuity. However, traditional manual incident response methods may be inadequate against constantly evolving phishing techniques. Leveraging AI and machine learning can automate detection, analysis, and response processes against phishing attacks (Verma & Ranga 2019). AI solutions can quickly identify patterns, and anomalies, and adapt to new phishing methods, minimizing potential impacts on critical infrastructure systems (Al-Mohannadi et al. 2016).

The UAE's National Critical Infrastructure Protection Program (NCIPP) aims to identify, prioritize, and protect critical infrastructure assets like energy, telecommunications, transportation, banking and finance, water supply, public health, emergency services, and government facilities from threats like cyberattacks (National Emergency Crisis and Disasters Management Authority [NCEMA] 2018).

2 Research Questions

Main Research Question:

How can the incident response capabilities be enhanced for detecting phishing attacks using AI?

Sub-Questions:

- What are the prevailing limitations in the current incident response mechanisms of critical infrastructure organizations in the UAE?
- How can advanced threat intelligence techniques be seamlessly integrated into incident response mechanisms through the application of cutting-edge AI?
- In what ways can the proposed incident response address the phishing attacks?
- How the proposed solution will inform the policymakers to improve security of Critical infrastructure against the phishing attacks?

3 Specific Aims

In response to the increasingly sophisticated cyber threats, critical infrastructure organizations in the United Arab Emirates (UAE) must enhance their defenses against malicious actors, especially in the context of phishing attacks. These deceptive attacks highlight the need for advanced incident response mechanisms tailored to the UAE's critical infrastructure challenges. To address this, a pioneering research initiative aims to leverage artificial intelligence (AI) for swift detection and effective mitigation of phishing attacks targeting critical infrastructure entities across the UAE (UAE Government 2023).

Aim: To develop an AI-driven solution that significantly enhances incident response capabilities, specifically targeting the detection and mitigation of phishing attacks within critical infrastructure organizations in the UAE.

Objectives:

The following objectives will help us to achieve the aim:

1. **Review the Current Scenario of Phishing Attacks on Critical Infrastructure:**
   - Conduct secondary research by analyzing existing data sources and literature related to phishing attacks on critical infrastructure organizations.
   - Identify key statistics, trends, techniques, and response mechanisms regarding phishing attacks.
   - Pinpoint limitations and gaps in existing incident response frameworks against phishing threats.
2. **Perform Thematic Analysis for Best Practices:**
   - Employ thematic analysis methodology to uncover recurring themes and patterns. Thematic analysis is a powerful method for identifying common themes and patterns (Castleberry et al. 2018).
   - Extract themes related to phishing threat avoidance, optimal response protocols, and recovery processes.
   - Outline organizational best practices and preparedness measures against phishing incidents.
   - Employ a qualitative methodology based on thematic analysis to validate the model's performance and effectiveness.
3. **Design and Develop AI Model for Enhanced Phishing Detection and Response:**
   - Analyze findings from secondary data analysis and thematic analysis.
   - Identify focus areas for enhancing existing machine learning models for phishing detection.
   - Incorporate findings into an improved ML model.
   - Integrate Technology Threat Avoidance (TTAT) strategies.
   - Develop an AI prototype focused on critical infrastructure phishing response.
4. **Validate and Refine AI Model:**
   - Test the prototype against simulated phishing attacks for evaluation.
   - Gather feedback data from implementation in real-world phishing incidents.
   - Continually refine the model to enhance the accuracy of phishing detection.

By carefully following these goals, this research aims to bring about a significant change in how we handle cybersecurity incidents. Specifically, we want to better protect critical infrastructure in the UAE by detecting and responding to phishing attacks more effectively.

4 Literature Review

4.1 Overview

Phishing attacks pose ongoing risks to critical infrastructure globally and in the UAE, resulting in unauthorized access, service disruptions, and financial harm. To combat these threats, enhancing incident response capabilities is crucial. The literature review emphasizes integrating AI-driven approaches to strengthen incident response within the UAE's critical infrastrucure.

The following table summarizes the key points from the literature review, covering various aspects related to cybersecurity phishing attacks on critical infrastructure, incident response models, the role of AI in incident response, challenges and limitations, and the research gaps.

Related Studies	Author	Key Findings
Introduction to Critical Infrastructure	(Orio et al. 2020)	- Critical infrastructure's vital role in societal functioning. - Necessity of incorporating security measures in development
	(Tiwari & Batra 2021)	- Complexity and interconnected nature of critical infrastructures - Vulnerabilities to various threats
Cybersecurity phishing attacks to Critical Infrastructure	(Knapp & Langill 2021)	- Sophisticated phishing techniques targeting SCADA and ICS systems
	(Arsene 2020)	Highlights spear-phishing campaigns targeting the petroleum and processing industry
	(Dragos 2017)	Analyzes the Crash Override malware used in the 2016 Ukrainian energy sector attack
	(Lee & Lim 2016)	Describes a phishing attack on Korea Hydro & Nuclear Power (KHNP) in 2014
	(Zetter 2016)	Describes the phishing-enabled attack that caused power outages in Ukraine in 2015
	(Lee et al. 2014)	Details the phishing attack on a German steel mill that caused physical damage

(continued)

(continued)

Related Studies	Author	Key Findings
	(MacKenzie 2012)	Discusses the SHAMOON malware attack on Saudi Aramco, initially thought to be phishing-enabled
Incident Response Models	(Chen et al. 2023)	- Frameworks for detecting, analysing, and responding to cyber-attacks. - Importance of systematic approaches
Current State of Incident Response in Critical Infrastructure	(Critical infrastructure protection & resilience - The EC proposes a … 2023)	Highlights disparities in incident response preparedness across different critical infrastructure sectors
	(Hartati & Muhammad 2023)	Lack of standardized incident response frameworks and processes hinders coordination and information sharing
	(Kleij et al. 2017)	Effective incident response is vital for protecting the economy and ensuring national resilience
AI in Incident Response for Critical Infrastructure	(Mohamed et al. 2023)	- AI-driven incident response capabilities. - Benefits of real-time threat detection and response
	(AlEroud & Alsmadi 2017)	- Challenges in adapting AI to CI environments. - Need for comprehensive evaluation
	(Larriva-Novo et al. 2020)	AI plays a crucial role in reinforcing protection against phishing cyber threats and ensuring infrastructure stability and security
	(Wu & Guo 2019)	LuNet, a hierarchical neural network combining CNN and RNN, achieves high detection rates for network intrusions with a low false positive rate by effectively extracting spatial and temporal features from network traffic data
Role of Artificial Intelligence in Cybersecurity phishing attacks	(Benavides-Astudillo et al. 2023)	Evaluated effectiveness of AI-based phishing detection methods

(continued)

(continued)

Related Studies	Author	Key Findings
	(Benavides-Astudillo et al. 2023)	Natural Language Processing techniques can analyze email content, URLs, and attachments to detect phishing patterns
	(Wang & Li 2021)	Balancing security and user privacy, as well as defending against adversarial attacks, are challenges in AI-based phishing detection
	(Wu et al. 2016)	Introduces innovative protective strategies against mobile phishing threats through a combination of behavioral analytics and automated learning systems for continuous protection
Critical Analysis and Synthesis: (strength and weakness)	(Mohamed et al. 2023)	AI algorithms can identify and prioritize potential threats, enabling well-informed risk mitigation and response strategies in the power sector
Challenges and Limitations	(Zeadally et al. 2020)	AI-based approaches rely on historical data and patterns, which may not capture evolving tactics of malicious actors
	(Demertzis et al. 2022)	Emphasizes the need for a balanced approach between AI capabilities and human oversight in cybersecurity
	(Aloul et al. 2020)	Highlights the lack of region-specific research and cybersecurity strategies for critical infrastructure in the Middle East
Research Gaps	(Mahmoud et al. 2015)	Identifies a research gap in exploring incident response capabilities within the Gulf region, despite the surge in cyber-attacks
	(Bethanyet al. 2024)	Highlights the need to study the integration of Large Language Models for large-scale phishing attacks targeting organizations

(continued)

(continued)

Related Studies	Author	Key Findings
	(Sarker et al. 2020)	- Neglect of AI-based defence against large-scale attacks. - AI limitations in adapting to evolving threats

4.2 Contribution and move forward

The integration of AI-driven phishing cybersecurity measures into critical infrastructure systems has the potential to significantly enhance incident response capabilities. This includes event prioritization, automated security analysis, and threat detection and prediction. The study contributes to the existing literature by surveying highly sophisticated attacks and emphasizing the prospects and challenges of using AI/ML in cybersecurity. It underscores the need for human oversight and ethical considerations. By addressing these limitations, the study aims to enhance the preparedness of critical infrastructure systems in the UAE. Stakeholders will benefit from automated phishing cybersecurity threat elements, enabling more effective decisions in incident response.

5 Research Methodology

The paper proposes an AI-based approach to enhance incident response capabilities against phishing cyber threats in the UAE, including testing and validating existing models and adapting them to the UAE's specific context. Through the utilization of AI, organizations can achieve real-time threat detection and automated response capabilities, promptly identifying and mitigating phishing cyber-attacks. The strengths and limitations of these AI techniques in the context of incident response for phishing cyber threats are also analyzed.

5.1 Theoretical Framework

The research aims to introduce a theoretical framework for enhancing phishing incident response in UAE's critical infrastructure. It will adapt the Technology Threat Avoidance Theory (TTAT) from individual to organizational context. The aim is to extend TTAT to address collective incident response challenges against phishing threats. This study represents the first attempt to apply TTAT at organizational level.

TTAT, validated for individual users, is being adapted to organizational settings in UAE's critical infrastructure. The research focuses on unique dynamics of collective phishing threat response in this context.

TTAT explains users' online security threat avoidance behaviours, emphasizing factors like self-efficacy and perceived costs of avoidance response. This framework underscores the importance of human oversight in ensuring the accuracy and reliability of AI-driven cybersecurity systems.

To empirically validate the organizational extension of TTAT, the research proposes:

1. Developing an Organizational TTAT (O-TTAT) model
2. Creating and validating a survey instrument for new constructs
3. Conducting quantitative validation using structural equation modeling

The theoretical contributions include:

1. Extending TTAT to the organizational level
2. Introducing organization-specific constructs
3. Developing a validated measurement tool for organizational threat avoidance behaviors

In exploring theoretical frameworks for this study, alternatives to TTAT such as Protection Motivation Theory (PMT) and the National Institute of Standards and Technology (NIST) Cybersecurity Framework were considered. PMT focuses on individual threat evaluation and coping mechanisms, lacking applicability to collective threat response at the organizational context.

Conversely, the NIST Framework provides industry standards for cybersecurity but overlooks understanding threat perceptions or improving incident readiness. TTAT, on the other hand, offers a relevant framework for modeling and enhancing phishing threat avoidance response in UAE critical infrastructure. By extending TTAT to an organizational context, this research can assess and improve readiness for phishing threats by considering shared threat perceptions, collective efficacy beliefs, and coordinated incident readiness. TTAT serves as an ideal foundation for modeling and influencing organizational behaviours and priorities based on threat avoidance attitudes and coping evaluations (Liang & Xue 2010).

Several studies have empirically supported the TTAT in various contexts (Wang et al. 2015) found that perceived threat of policy non-compliance significantly predicted higher compliance intentions among employees, confirming TTAT assumptions (Rocha Flores & Ekstedt 2016) used TTAT to assess how personal values influence online safety behaviours, demonstrating that value orientations shape threat avoidance responses. In the context of phishing threat avoidance, (Arachchilage et al. 2016) adapted TTAT to guide their study. They evaluated a game design framework through a mobile game prototype, aiming to enhance users' avoidance behaviours against phishing attacks. By incorporating TTAT, they provided a structured approach to understanding users' attitudes and behaviours related to phishing threat avoidance.

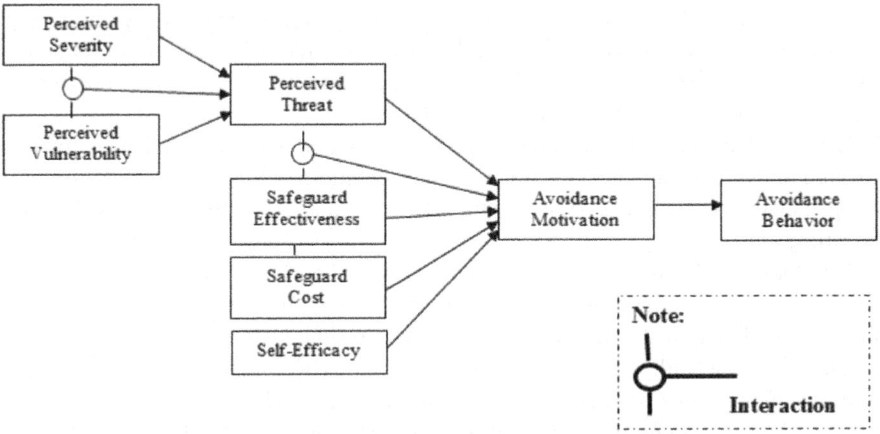

Fig. 1 Technology threat avoidance theory (TTAT), (Liang and Xue 2010)

Researchers have enhanced TTAT by adding rewards and costs to explore user motivations for security behaviors (Zahedi et al. 2020; Ifinedo 2021) further validated TTAT, highlighting perceived threat severity and susceptibility as key factors influencing employees' compliance with security policies (Fig. 1).

5.2 Research Approach and Justification

This study adopts a design science approach to develop an AI solution targeting the critical issue of insufficient phishing threat response capabilities within the UAE's critical infrastructure organizations. The design science methodology is particularly suitable for this research as it allows for the creation and evaluation of innovative artifacts to solve real-world problems (Hevner et al. 2004).

In the UAE, phishing poses a particularly high risk to the nation's critical information infrastructure, encompassing computer systems, networks, and assets supporting essential services such as electricity, water, transportation, financial services, and telecommunications (UAE Government 2023).

As illustrated in Fig. 2, the Approach Will Involve:

Collect data on existing phishing attacks targeting critical infrastructure from secondary sources like journal articles, research reports, and reputable platforms (e.g., PhishTank, CastleCops, Google Safe Browsing). Perform thematic analysis on the collected data to analyze secondary data to identify common trends, patterns, and techniques used in AI-based cyberattacks.

Create an ensemble AI model using findings from thematic analysis to enhance existing machine learning models for phishing threat detection and response. The model will combine Convolutional Neural Networks (CNNs), Recurrent Neural Networks (RNNs) with Long Short-Term Memory (LSTM), and Random Forest classifiers. Train the model on historical phishing attack data and synthetic data generated using Generative Adversarial Networks (GANs). Leverage the TTAT to align with established principles for mitigating IT risks. Map TTAT constructs to specific features and outputs of the AI model.

Utilize feedback data to enhance the AI model's accuracy and effectiveness through active learning and transfer learning techniques. Evaluate its performance, using metrics such as precision, recall, F1-score, and AUC, usability, and impact over time.

Conduct simulated phishing attacks in a controlled environment to validate the AI model's effectiveness. Iterate on the model design based on performance results and user feedback, following the design science research cycle.

The AI model, aligned with TTAT theory, aims to enhance phishing incident response capabilities for UAE's critical infrastructure. The integration of multiple machine learning techniques, diverse data sources, and a strong theoretical foundation provides a robust methodology for developing an effective AI-based solution to combat phishing threats (Fig. 2).

6 Conclusion

The research aims to enhance incident response to phishing threats in UAE's critical infrastructure using AI. This is vital for national cybersecurity and securing services like energy, water, and telecommunications. Phishing exploits human weaknesses, threatening sensitive systems. By using AI and cyber threat intelligence, the study seeks to improve response mechanisms and boost cyber resilience. Adopting the TTAT enhances organizational incident response. Through thematic analysis, AI model development, and validation, the research refines machine learning models for phishing detection and response, addressing current limitations and identifying best practices. It provides an AI-based prototype for improved phishing response and sets a standard for proactive defense against evolving threats.

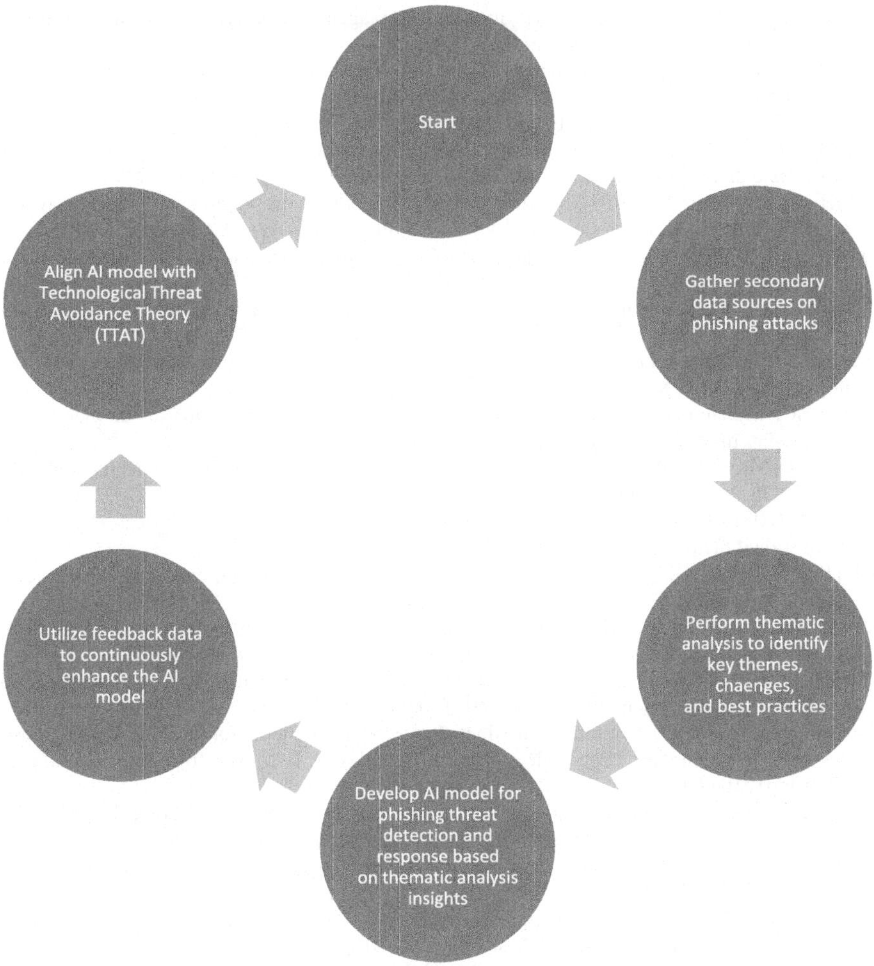

Fig. 2 Research methodology

References

AlEroud A, Alsmadi I (2017) Identifying cyber-attacks on software defined networks: an inference-based intrusion detection approach. J Netw Comput Appl 80:152–164. https://doi.org/10.1016/j.jnca.2016.12.024

Al-Mohannadi H, Mirza Q, Namanya A, Awan I, Cullen A, Disso JP (2016) Cyber-attack modeling analysis techniques: An overview. In: 2016 IEEE 4th international conference on future internet of things and cloud workshops (FiCloudW) (pp 69–76) IEEE

Aloul FA (2012) Phishing: a multifaceted challenge. Int J Comput Appl 52(5):1–4

Arachchilage NAG, Love S, Beznosov K (2016) Phishing threat avoidance behavior: an empirical investigation. Comput Hum Behav 60:185–197

Arsene (2020) Oil & gas spearphishing campaigns drop agent tesla spyware in advance of historic OPEC+ deal. Retrieved from https://bit.ly/2YyEcSa

Benavides-Astudillo E, Fuertes W, Sanchez-Gordon S, Nuñez-Agurto D, Rodríguez-Galán G (2023) A phishing-attack-detection model using natural language processing and deep learning. Appl Sci 13(9):5275. https://doi.org/10.3390/app13095275

Bethany M, et al (2024) Large language model lateral spear phishing: a comparative study in large-scale organizational settings. Available at: https://arxiv.org/abs/2401.09727

Castleberry A, Nolen A (2018) Thematic analysis of qualitative research data: is it as easy as it sounds? Curr Pharm Teach Learn 10(6):807–815. https://doi.org/10.1016/j.cptl.2018.03.019

Chen S et al (2023) Exploring the global geography of cybercrime and its driving forces. Available at: https://doi.org/10.1057/s41599-023-01560-x

Demertzis K, et al (2022) An ensemble transfer learning spiking immune system for adaptive smart grid protection. Retrieved from https://doi.org/10.20944/preprints202206.0076.v1

Dragos (2017) Crashoverride: analysis of the threat to electric grid operations. Retrieved from https://bit.ly/2KxWN8r

Hartati SC, Muhammad A (2023) Combating cybercrime and cyberterrorism in Indonesia. J HumItarian Innov 11(2). https://doi.org/10.18196/jhi.v11i2.15647

Hevner AR, March ST, Park J, Ram S (2004) Design science in information systems research. MIS Q 28(1):75–105. https://doi.org/10.2307/25148625

Ifinedo P (2021) Information systems security policy compliance: an empirical study of the motivational factors. Database Adv Inf Syst 52(1):70–104

Kleij Rvd, Kleinhuis G, Young AH (2017) Computer security incident response team effectiveness: a needs assessment. Front Psychol 8. https://doi.org/10.3389/fpsyg.2017.02179

Knapp ED, Langill JT (2021) Industrial network security: securing critical infrastructure networks for smart grid, SCADA, and other industrial control systems, 3rd edn. Syngress

Laplante P, Serebryakov S, Bennett JD (2020) Artificial intelligence and critical systems: from hype to reality. Retrieved from https://doi.org/10.1109/mc.2020.3006177

Larriva-Novo X, et al (2020) Efficient distributed preprocessing model for machine learning-based anomaly detection over large-scale cybersecurity datasets. Retrieved from https://doi.org/10.3390/app10103430

Lee RM, Assante MJ, Conway T (2014) German steel mill cyber attack. Ind Control Syst 30:62

Lee KB, Lim JI (2016) The reality and response of cyber threats to critical infrastructure: a case study of the cyber-terror attack on the Korea Hydro & Nuclear Power Co., Ltd. KSII Trans Internet Inf Syst 10(2):857–880

Liang H, Xue Y (2010) Understanding security behaviors in personal computer usage: a threat avoidance perspective. J Assoc Inf Syst 11(7):394–413

MacKenzie H (n.d.) Shamoon malware and SCADA security - what are the impacts? | Tofino industrial security solution. [Online]. Available at: https://www.tofinosecurity.com/blog/shamoonmalware-and-scada-security-%E2%80%93-what-are-impacts

Mahmoud R, Yousuf T, Aloul F, Zualkernan I (2015) Internet of things (IoT) security: Current status, challenges and prospective measures. In: 2015 10th international conference for internet technology and secured transactions (ICITST) (pp 336–341). London, UK. https://doi.org/10.1109/ICITST.2015.7412116

Mohamed N, Oubelaid A, Almazrouei KS (2023) Staying ahead of threats: a review of ai and cyber security in power generation and distribution. Int J Energy Environ Res. https://doi.org/10.37391/ijeer.110120

National Emergency Crisis and Disasters Management Authority (NCEMA) (2018) National critical infrastructure protection program (NCIPP). Retrieved from https://www.ncema.gov.ae/en/programs-andservices/national-critical-infrastructure-protection-program.aspx

Orio GD, Brito G, Maló P, Shrivastava A, Wirtz N, Monti A (2020) A cyber-physical approach to resilience and robustness by design. Int J Adv Comput Sci Appl 11(7). https://doi.org/10.14569/ijacsa.2020.0110710

Rajan AV, Ravikumar R, Shaer MA (2017) UAE cybercrime law and cybercrimes—An analysis. In: 2017 international conference on cyber security and protection of digital services (Cyber Security) (pp 1–6). London, UK. https://doi.org/10.1109/CyberSecPODS.2017.8074858

Rocha Flores W, Ekstedt M (2016) Shaping intention to resist social engineering through transformational leadership, information security culture and awareness. Comput Secur 59:26–44

Sarker H (2020) Cybersecurity data science: an overview from a machine learning perspective. Available at: https://doi.org/10.1186/s40537-020-00318-5

Tiwari A, Batra U (2021) Blockchain enabled reparations in smart buildings cyber physical system. Def Sci J 71(4):491–498. https://doi.org/10.14429/dsj.71.16454

UAE Government (2023) Critical information infrastructure protection (CIIP) Policy. Retrieved from: https://uaeun.org/press_release/uae-renews-commitment-to-strengthening-cyber-security/

Verma R, Ranga V (2019) Machine learning based approach for phishing detection. In: 2019 international conference on issues and challenges in intelligent computing techniques (ICICT) (pp 1–6). IEEE

Wang J, Gupta M, Rao HR (2015) Insider threats in a financial institution: Analysis of attack-proneness of information systems applications. MIS Q 39(1):91–112

Wu P, Guo H (2019) LuNet: a deep neural network for network intrusion detection. arXiv preprint arXiv:1909.10031. https://doi.org/10.48550/arXiv.1909.10031

Zahedi FM, Zahedi F, Zahedi F (2020) Security assurance behavior: modeling the simultaneous effects of rewards and punishments. Decis Support Syst 132:113227

Zeadally S et al (2020) Harnessing artificial intelligence capabilities to improve cybersecurity. Retrieved from. https://doi.org/10.1109/access.2020.2968045

Zetter K (2016) Everything we know about Ukraine's power plant hack. Retrieved from https://bit.ly/3eEnDJV

Examining Customer's Behavioral Intention Towards Chatbots Adoption—The United Arab Emirates Context

Heba Alkhoja

Abstract

Purpose The research aims to examine factors that could affect customer's behavioral intention towards the adoption of chatbots through a theoretical model based on Technology Readiness Index (TRI).

Design/methodology/approach PLS-SEM modeling technique using Smart-PLS is used to analyze online survey of 205 valid responses.

Findings obtained results indicated a significant impact of both innovativeness and trust on customer's hedonic motivation. Results also show that hedonic motivation has a direct significant impact on customer's behavior intention to adopt chatbots. Mediated role of hedonic motivation found to be significant where innovativeness and trust found to have significant effect on customer's behavior intention to use chatbots through hedonic motivation.

Research Implications This research helps marketers and business leaders in gaining more insights regarding customer's behavior intention toward the adoption of chatbots.

Research contribution The research plays important role in filling a current literature gap related to the adoption of chatbots, where the constructed theoretical model and the obtained results tested customer's behavior intention toward chatbots by considering the mediating effect of customer's hedonic motivation. Results contribute to the (TRI) by considering trust as a main independent factor and by testing hedonic motivation as a mediator. Moreover, originality of the research comes from its data context, where data is collected from customers in the UAE, and to the best of author's knowledge this is done for the first time in testing factors affecting chatbots from customer's perspective in the UAE.

Keywords Artificial intelligence · Chatbot · Behavioral intention

---

H. Alkhoja (✉)
College of Business Administration, Ajman University, Ajman, UAE
e-mail: 202111260@ajmanuni.ac.ae

K. Al Marri et al. (eds.), *BUiD Doctoral Research Conference 2024*, Lecture Notes in Civil Engineering 587, https://doi.org/10.1007/978-3-031-84371-6_17

1 Introduction

Artificial intelligence (AI) is the science of creating smart machines which can mimic human intelligence (Cheng et al. 2023; Yu et al. 2024). The adoption of AI is changing business and marketing industries (Huh et al. 2023; Mostafa and Kasamani 2022; Yu et al. 2024). One of the widely applied AI applications are chatbots (Blasco-Arcas et al. 2022; Hsu and Lin 2023). Chatbots are defined as computer programs which has the technical and technological ability to simulate conversations with human, via text or voice (Misischia et al. 2022). Previous scholars highlighted the importance of chatbots in business (Rese et al. 2020; Sheehan et al. 2020).

According to Bouhia et al. (2022), additional studies are still required to gain more understanding about customers' attitude and behaviours towards the adoption of chatbots. previous researchers stated that current literature needs more investigation about customer's interaction with chatbots and factors that affect their adoption (Misischia et al. 2022; Yu et al. 2024). Although the essential role of hedonic motivation has been highlighted as a main factor in investigating customers' acceptance and intention to use new technologies, however, current literature still need more investigation about its mediating effect (Parasuraman 2000; Vimalkumar et al. 2021). Furthermore, to the best of the author's knowledge this is the first research targeting data from users in the UAE as the geographical area to investigate user's behavioural intention to use chatbots. Considering the UAE as a source for data adds to the research value due to the multi-nationalities and the differentiations in customer's cultures living there which can give a clear general idea about customers behaviours.

Therefore, this research is going to fill a clear gap by investigating the following question: What factors can affect customer's behavioral intention when chatbots are adopted in marketing activities?

The research will test the effect of trust and innovativeness mediated by Hedonic motivation of consumers to adopt chatbots.

The constructed research model is built based on the Technology Readiness Index (TRI) where Innovativeness (INV) and Trust (TRU) are being considered as major factors that can affect customer's behaviour toward the adoption of chatbots and hedonic motivation (HMT) as a mediator. The main reason of choosing the TRI in formulating the theoretical model is the popularity of this index in business and marketing literature as an effective tool in measuring and understanding factors that affect customer's attitude and behaviour toward new technologies (Hasan et al. 2021; Parasuraman 2000).

2 Theoretical Background and Hypothesis Development

2.1 Behavioural Intention

Customer's behavioural intention refers to the chances of customer to adopt new technology (To and Trinh 2021). In business context understanding customer's behavioural intention towards the adoption of new technology plays a crucial role (Yu et al. 2024).

Due to the ability of chatbots in communicating effectively with human, it has been used for different customer's related services (Rese et al. 2020). Thus, Marketers and business decision makers need to obtain clear understanding of chatbots to improve their business (Yen and Chiang 2021).

2.2 Innovativeness and Hedonic Motivation

According to the TRI one of the important determinants that need to be considered when examining customer's willingness to adopt new technology is innovativeness (Parasuraman 2000). Innovativeness is defined as the degree to which users are ready to try innovative product or service (Lin and Hsieh 2012). Prior literature clearly indicated the positive impact of innovativeness on customer's hedonic motivation to use new technology (Liu et al. 2023). Thus, this research suggests the following relationship according to the research model which is shown in Fig. 1.

H1: Innovativeness has a significant positive effect on customer's hedonic motivation to adopt chatbot technology.

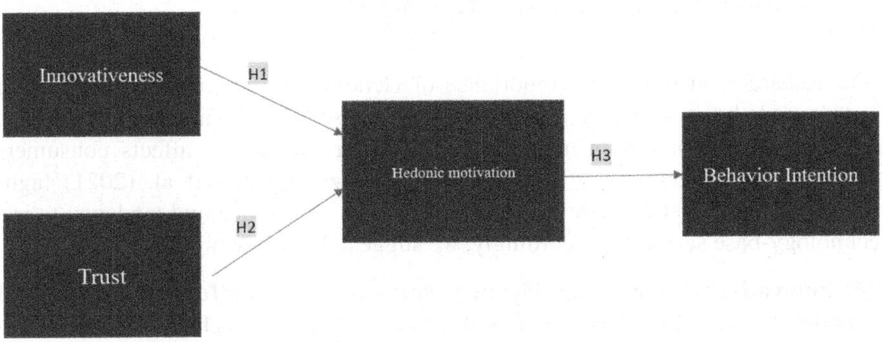

Fig. 1 Research theoretical model

2.3 *Trust and Hedonic Motivation*

Scholars stated that trust is one of the major factors that can directly affect customer's behaviour to adopt new technology (Hasan et al. 2021). Trust refers to customer's belief that a specific technology will operate and perform required tasks without causing them any harm (To and Trinh 2021).

Trust can affect customer's hedonic motivation to accept and use new technology (Komiak and Benbasat 2006; Yen and Chiang 2021). Accordingly, below hypothesis is proposed:

H2: Trust has a significant positive effect on customer's hedonic motivation to adopt chatbot technology.

2.4 *Hedonic Motivation and Behavioural Intention*

One of the major psychological factors that can affect customer's behavioural intention toward the adoption of new technology is the hedonic motivation (Vimalkumar et al. 2021). Referring to (TRI), hedonic motivation is essential when studying customer's' acceptance of new technology and it is defined as the extent of customer's satisfaction of a new technology (Parasuraman 2000). Thus, we suggest the following hypothesis:

H3: Hedonic motivation has significant positive direct effect on behavioural intention toward the adoption of chatbot.

2.5 *Hedonic Motivation as a Mediator*

Prior research illustrated the importance of Hedonic motivation in understanding customer's behaviours when a new technology is used (Cruz-Cárdenas et al. 2021). Hedonic motivation has been studied as a major factor that affects consumers tendency to use technology services, where Cruz-Cárdenas et al. (2021) highlighted its significant positive effect on consumers acceptance and tendency to use technology-base services. Accordingly, we suggest the following relations:

H4: Innovativeness has a significant positive significant effect on behavioural intention toward the adoption of chatbot technology through Hedonic Motivation.

Similarly, when consumers trust the adoption of new technology, and feel motivation and enjoyment towards it, their behaviour intention will be positively affected. Therefore, below relationship is being suggested:

H5: Trust has a significant positive significant effect on behavioural intention toward the adoption of chatbot technology through Hedonic Motivation.

3 Methodology

3.1 Data Collection Process

Quantitative research analysis was adopted in the study. Following the adopted data collection procedure based on available literature in marketing, survey method has been followed to collect relevant data (To and Trinh 2021; Yen and Chiang 2021). As a result, a questionnaire considering 16-scale items that directly measure the factors of the research has been developed. Each factor was measured using 4-scale items. The 16-scale items of the questionnaire were measured using 5-point Likert scale starting from (1 = strongly disagree) to (5 = strongly agree). Simple random sampling has been applied (James et al. 2019), where the questionnaire link has been distributed through social media platforms that includes WhatsApp and LinkedIn and Instagram. This method was chosen due to its ability in approaching variety random Internet users, A question (Are you living in the UAE?) was used to ensure all respondents are living in the UAE. This method is speed and economically suitable in collecting relative data (Sharma et al. 2021). A total of (205) valid responses were collected, and according to several research in marketing this sample size is considered as a sufficient and representative sample (Adam et al. 2021; Rhee and Choi 2020). Confidentiality has been obtained by providing a cover litter in the survey which clearly informed participants that their personal details will not be published, also participants were informed that collected data will be used for academic research purposes only.

Demographic data analysis results are indicated in Table 1. Majority of respondents were female with 62.4%. Most of the respondents falls in the age range of (31–45) years and majority of them holds a master's degree with 66.3%.

3.2 Data Analysis and Results

PLS-SEM using Smart-PLS was used to analyse collected data (Cheah et al. 2021; Hair et al. 2019b). Analysis included evaluating measurement and structural models (Hair et al. 2019a, b).

Table 1 Demographics results of respondents (n = 205)

Measure	Value	Frequency	Percent
Gender	Male	77	37.6
	Female	128	62.4
Age	Under 25 years	35	17.1
	30–25 years	32	15.6
	31–45 years	100	48.8
	Above 45 years	38	18.5
Education	Doctorate degree	10	4.90
	Master's degree	136	66.3
	Bachelor's degree	49	23.9
	High School	10	4.90

3.3 Evaluating Reflective Constructs

Measurement model helps in determining reliability and validity of the research data (Hair et al. 2019a). As a result, Table 2 indicates the values of outer loading (λ), composite reliability (CR), average variance (AVE) and Cronbach alpha:

According to Table 2, the obtained values of (CR) are far above the accepted level 0.7 which implies internal consistency of collected data (Hair et al. 2021). According

Table 2 Measurement model—construct reliability and validity

Construct	Item	λ	AVE	CR	Cronbach's α	rho_A
Behavioural intention	BI1	0.929	0.877	0.966	0.953	0.953
	BI2	0.948				
	BI3	0.954				
	BI4	0.914				
Hedonic motivation (HMT)	HMT1	0.903	0.86	0.961	0.946	0.948
	HMT2	0.927				
	HMT3	0.946				
	HMT4	0.932				
Innovativeness (INV)	INV1	0.892	0.805	0.943	0.919	0.920
	INV2	0.891				
	INV3	0.910				
	INV4	0.896				
Trust (TRU)	TRU1	0.907	0.824	0.949	0.929	0.930
	TRU2	0.924				
	TRU3	0.913				
	TRU4	0.886				

Table 3 Discriminant validity

Fornell-Larcker-discriminant validity				
	BI	HMT	INV	TRU
BI	0.840			
HMT	0.844	0.836		
INV	0.839	0.840	0.837	
TRU	0.838	0.847	0.848	0.848

to Hair et al. (2019a), values of Cronbach alpha need to be above 0.7, which is clearly satisfied in the obtained values of Table 2. The outer loading (λ) illustrated in the above table are all in the accepted range >0.5 (Hair et al. 2019a).

While Table 3 is illustrating the measures for discriminant validity adopting the Fornell-Larcker scale. All indicated measures are within the accepted range and all values are below the maximum accepted threshold level which is 0.85 (Henseler et al. 2015). The above results indicate that constructs are valid and reliable.

3.4 Evaluating Structural Model

Using the structural model (Hair et al. 2019a), hypothesized relations according to the constructed theoretical framework will be studied and indicated as shown in Table 4.

Table 4 clearly indicates that all hypotheses are accepted. Direct hypothesized relationships that include the first three hypothesis (H1–H3) are accepted. Results indicates that INV and TRU has positive direct effect on HMT, where INV affects HMT at ($\beta = 0.304$, $p < 0.01$) and TRU affects HMT at ($\beta = 0.61$, $p < 0.01$).

HMT also has a positive effect on BI at ($\beta = 0.884$, $p < 0.01$). Moreover, the mediating effect of HMT found to be significant at ($p < 0.01$), which means that both H4 and H5 are significant and accepted hypothesis. The effect of INV on BI through

Table 4 Hypothesis results using structural model

Parameter	β	(STDEV)	T-value	P value	Remarks
Direct effect					
H1: INV → HMT	0.304	0.085	3.589	0.000	Accepted
H2: TRU → HMT	0.61	0.081	7.557	0.000	Accepted
H3: HMT → BI	0.884	0.021	41.954	0.000	Accepted
Mediator effect					
H4: INV → HMT → BI	0.269	0.076	3.561	0.000	Accepted
H5: TRU → HMT → BI	0.539	0.073	7.393	0.000	Accepted

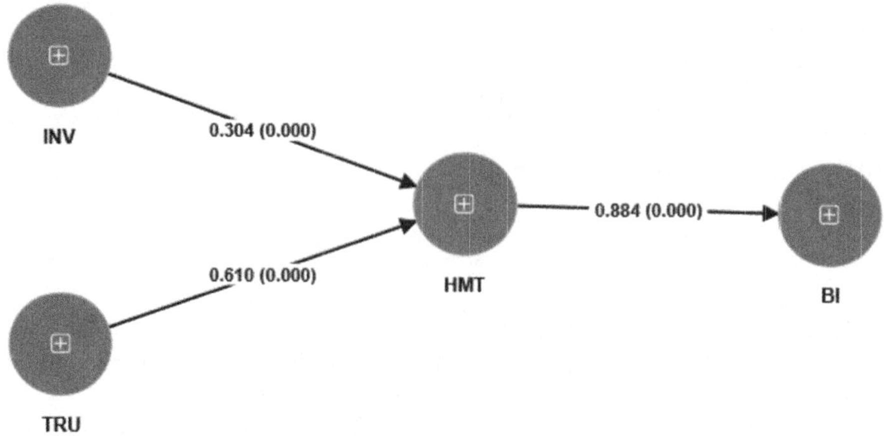

Fig. 2 Structural model of the research framework

HMT is significant at ($\beta = 0.269$, $p < 0.01$). TRU effect on BI through HMT has been found to be significant at ($\beta = 0.539$, $p < 0.01$) as shown in Fig. 2.

4 Discussion

The aim of the research is to examine factors that affect customer's behavioral intention when chatbot technology is used. Results clearly illustrated that innovativeness can affect customer's hedonic motivation significantly, this result is consistence with what was indicated by Cruz-Cárdenas et al. (2021). Results indicated the significant effect of trust on hedonic motivation which is consistence with previous studies (Mostafa and Kasamani 2022; To and Trinh 2021).

Results regarding the effect of trust on hedonic motivation is considered as one of the significant contributions of the research, as to the best of our knowledge, no other research tested this relation when chatbots are used. The direct impact of hedonic motivation on customer's behavioral intention found to be positive and significant. According to Cruz-Cárdenas et al. (2021), hedonic motivation affects technology-based consumers behavior positively. Our results are consistent with them but in the context of chatbot technology, which is another contribution of this research.

4.1 Theoretical and Practical Implications

Research theoretical implications are illustrated through the expansion of Technology Readiness Index by investigating the effect of trust and hedonic motivation which help in filling a gap in current literature. Such investigation plays an essential role in enriching current AI and chatbot literature.

Practical implications of this research are important for marketers and business decision-makers as it gives a clear idea of factors that affect customer's behavioral intention towards chatbots. Such results can help in developing business and marketing strategies (Balakrishnan et al. 2022; To and Trinh 2021). Business practitioners can use such studies in enhancing chatbots applicability in their services.

4.2 Limitation and Future Research

This research has several limitations, which can be considered as chances for future research. The study considered Trust and Innovativeness as main indicators to investigate customer's behavioral intention towards chatbots adoption, however, other factors related to customer's behavioral theories such as familiarity and attitude could be added to the model to gain more insights.

Second, although the sample size of (205) responses was enough for this research, however, increasing the sample size will help in generalizing results.

5 Conclusion

The research plays an important role in investigating major factors that affect customer's behavioral intention when chatbot are used, specifically in the UAE context. The main research question has been answered using PLS-SEM analysis technique.

The research's main objective was met successfully by obtaining these results which added valuable understanding and insights to current chatbot literature.

References

Adam M, Wessel M, Benlian A (2021) AI-based chatbots in customer service and their effects on user compliance. Electron Mark 31(2):427–445. https://doi.org/10.1007/s12525-020-00414-7

Balakrishnan J, Abed SS, Jones P (2022) The role of meta-UTAUT factors, perceived anthropomorphism, perceived intelligence, and social self-efficacy in chatbot-based services? Technol Forecast Soc Chang 180. https://doi.org/10.1016/j.techfore.2022.121692

Blasco-Arcas L, Lee HHM, Kastanakis MN, Alcañiz M, Reyes-Menendez A (2022) The role of consumer data in marketing: a research agenda. J Bus Res 146:436–452. https://doi.org/10.1016/j.jbusres.2022.03.054

Bouhia M, Rajaobelina L, PromTep S, Arcand M, Ricard L (2022) Drivers of privacy concerns when interacting with a chatbot in a customer service encounter. Int J Bank Market 40(6):1159–1181. https://doi.org/10.1108/IJBM-09-2021-0442

Cheah JH, Roldán JL, Ciavolino E, Ting H, Ramayah T (2021) Sampling weight adjustments in partial least squares structural equation modeling: guidelines and illustrations. Total Qual Manag Bus Excell 32(13–14):1594–1613. https://doi.org/10.1080/14783363.2020.1754125

Cheng CF, Huang CC, Lin MC, Chen TC (2023) Exploring effectiveness of relationship marketing on artificial intelligence adopting intention. SAGE Open 13(4). https://doi.org/10.1177/21582440231222760

Cruz-Cárdenas J, Guadalupe-Lanas J, Ramos-Galarza C, Palacio-Fierro A (2021) Drivers of technology readiness and motivations for consumption in explaining the tendency of consumers to use technology-based services. J Bus Res 122:217–225. https://doi.org/10.1016/j.jbusres.2020.08.054

Hair JF, Astrachan CB, Moisescu OI, Radomir L, Sarstedt M, Vaithilingam S, Ringle CM (2021) Executing and interpreting applications of PLS-SEM: updates for family business researchers. J Family Bus Strategy 12(3). https://doi.org/10.1016/j.jfbs.2020.100392

Hair JF, Risher JJ, Sarstedt M, Ringle CM (2019a) When to use and how to report the results of PLS-SEM. Eur Bus Rev 31(1):2–24. Emerald Group Publishing Ltd. https://doi.org/10.1108/EBR-11-2018-0203

Hair JF, Sarstedt M, Ringle CM (2019b) Rethinking some of the rethinking of partial least squares. Eur J Mark 53(4):566–584. https://doi.org/10.1108/EJM-10-2018-0665

Hasan R, Shams R, Rahman M (2021) Consumer trust and perceived risk for voice-controlled artificial intelligence: the case of Siri. J Bus Res 131:591–597. https://doi.org/10.1016/j.jbusres.2020.12.012

Henseler J, Ringle CM, Sarstedt M (2015) A new criterion for assessing discriminant validity in variance-based structural equation modeling. J Acad Mark Sci 43(1):115–135. https://doi.org/10.1007/s11747-014-0403-8

Hsu CL, Lin JCC (2023) Understanding the user satisfaction and loyalty of customer service chatbots. J Retail Consum Serv 71. https://doi.org/10.1016/j.jretconser.2022.103211

Huh J, Nelson MR, Russell CA (2023) ChatGPT, AI advertising, and advertising research and education. J Advert 52(4):477–482. https://doi.org/10.1080/00913367.2023.2227013

James TL, Wallace L, Deane JK (2019) Using organismic integration theory to explore the associations between users' exercise motivations and fitness technology feature set use appendix a developing the fitness technology feature sets. MIS Quart 43(1)

Komiak SYX, Benbasat I (2006) The effects of personalization and familiarity on trust and adoption of recommendation. MIS Quart 30(4)

Lin JSC, Hsieh PL (2012) Refinement of the technology readiness index scale: a replication and cross-validation in the self-service technology context. J Serv Manag 23(1):34–53. https://doi.org/10.1108/09564231211208961

Liu X, Zhang C, Wu J (2023) Explaining consumers' continuous purchase intention toward subscriber-based knowledge payment platforms: findings from PLS-SEM and fsQCA. Aslib J Inf Manag. https://doi.org/10.1108/AJIM-08-2022-0359

Misischia CV, Poecze F, Strauss C (2022) Chatbots in customer service: their relevance and impact on service quality. Procedia Comput Sci 201(C):421–428. https://doi.org/10.1016/j.procs.2022.03.055

Mostafa RB, Kasamani T (2022) Antecedents and consequences of chatbot initial trust. Eur J Mark 56(6):1748–1771. https://doi.org/10.1108/EJM-02-2020-0084

Parasuraman A (2000) Technology Readiness Index (TRI) a multiple-item scale to measure readiness to embrace new technologies. J Serv Res 2(4)

Rese A, Ganster L, Baier D (2020) Chatbots in retailers' customer communication: how to measure their acceptance? J Retail Consum Serv 56. https://doi.org/10.1016/j.jretconser.2020.102176

Rhee CE, Choi J (2020) Effects of personalization and social role in voice shopping: an experimental study on product recommendation by a conversational voice agent. Comput Hum Behav 109. https://doi.org/10.1016/j.chb.2020.106359

Sharma A, Dwivedi YK, Arya V, Siddiqui MQ (2021) Does SMS advertising still have relevance to increase consumer purchase intention? a hybrid PLS-SEM-neural network modelling approach. Comput Hum Behav 124. https://doi.org/10.1016/j.chb.2021.106919

Sheehan B, Jin HS, Gottlieb U (2020) Customer service chatbots: anthropomorphism and adoption. J Bus Res 115:14–24. https://doi.org/10.1016/j.jbusres.2020.04.030

To AT, Trinh THM (2021) Understanding behavioral intention to use mobile wallets in Vietnam: extending the tam model with trust and enjoyment. Cogent Bus Manage 8(1). https://doi.org/10.1080/23311975.2021.1891661

Vimalkumar M, Sharma SK, Singh JB, Dwivedi YK (2021) 'Okay google, what about my privacy?': user's privacy perceptions and acceptance of voice based digital assistants. Comput Hum Behav 120. https://doi.org/10.1016/j.chb.2021.106763

Yen C, Chiang MC (2021) Trust me, if you can: a study on the factors that influence consumers' purchase intention triggered by chatbots based on brain image evidence and self-reported assessments. Behav Inf Technol 40(11):1177–1194. https://doi.org/10.1080/0144929X.2020.1743362

Yu J, Dickinger A, So KKF, Egger R (2024) Artificial intelligence-generated virtual influencer: examining the effects of emotional display on user engagement. J Retail Consum Serv 76. https://doi.org/10.1016/j.jretconser.2023.103560

Exploring AI Conversational Chatbot UX Design: Insights from High School

Suha Khalil Assayed, Daniel Woods, Manar Alkahtib, and Khaled Shaalan

Abstract Advancements in state-of-the-art models and algorithms for chatbots have significantly driven the growth and evolution of human–computer interaction (HCI) in recent years. As a result, numerous authors are inspired to study the most effective interactive chatbot design that can maximize students' experiences. Despite the fact that high school is one of the most critical stage in a student's life, there is a lack of studies that focused on developing effective interactive high school advising chatbots. To address this current gap, this study aims to elicit the main effective features for high school advising chatbot. The authors in this study conducted a semi-structured qualitative interview with six high school students in UAE and the MAXQDA Analytics software—ver. 22.6.0 is implemented by processing the thematic analysis to study the findings. The results revealed that high school students recommended the same general interaction design characteristics that was derived from the previous systematic review with emphasizing more into accurate, reliable, and trustworthy short answers with having minimal conversational issues. In addition to that, emotions and empathic factors are less important for high school students.

Keywords Chatbot · Human-chatbot interaction · Education · HCI · High school · Qualitative study · User experience

1 Introduction

Human-computer interaction (HCI) is a multidisciplinary field that primarily combines psychology and behavioral science with computer and information systems (Chignell et al. 2023; MacKenzie 2024). It explores how people may efficiently communicate with computers to enhance user experiences and meet their needs. HCI

This paper was prepared for INF 611 & RES 600, taught by DR. Daniel Woods

S. K. Assayed (✉) · D. Woods · M. Alkahtib · K. Shaalan
The British University in Dubai, Dubai, UAE
e-mail: sassayed@gmail.com

K. Al Marri et al. (eds.), *BUiD Doctoral Research Conference 2024*, Lecture Notes in Civil Engineering 587, https://doi.org/10.1007/978-3-031-84371-6_18

plays an essential role in improving user experience by linking the humans' intentions with machine functions so as to make applications and devices more usable, affordable, helpful, efficient, and easy to learn (Assayed et al. 2023a; Shi et al. 2019; Van et al. 2020).

A chatbot can mimic human characteristics, particularly in responding to customer service inquiries in various sectors, such as education and healthcare. Based on their experiences and traits (Assayed et al. 2024a; Kim et al. 2023), this can influence users' intentions to adopt it. However, user perceptions of human-computer interaction (HCI) with chatbots vary. In the field of education, chatbots can be valuable pedagogical tools to support learning (Chen et al. 2023). They can answer students' inquiries instantly and effectively, while some are even designed with social characteristics to encourage and comfort students during their learning journey. Nevertheless, to maximize the benefits for students, researchers and developers need to study the effects of chatbot designs on student behavior.

Several studies have shown the importance of educational chatbots on students' experiences. For example, Kuhail et al. (2023) conducted a systematic review to study the educational chatbot from multiple dimensions, including the platform, interaction styles, and design approaches; according to the authors, chatbot interaction may be achieved by implementing characteristics like speech and visual fonts, which can help students perform better on educational assignments. Authors found that chatbots with personalities can significantly influence user engagement and satisfaction. This is achieved by tailoring the chatbot's personality to the specific audience. For instance, a university student would likely respond better to a more professional and informative chatbot persona, compared to a high school student who might benefit from a more interactive and engaging personality.

However, some researchers have studied the design features of a chatbot from different domains such as health sector (Kieffer et al. 2019), banking system (Gunathilaka 2022) and the educational sector (Burkhard et al. 2022; Assayed et al. 2023b), and it was found that most educational chatbots are focused on the teaching classes rather than advising students (Assayed et al. 2024b). It is widely acknowledged that high school is one of the most critical stages in a student's life; however, there are lack of research studies focusing on developing effective interactive high school advising chatbots to assist these students in helping them shape their futures and careers by selecting majors and universities which coincide with their passions and interests. To address this current gap, we aim to study the main effective features of the general chatbots design, which would be derived from the discussions of the systematic literature review that was conducted by Rapp et al. (2021). Afterward, we explore the ways to increase the students' experiences (UX) in using the advising chatbots by conducting a qualitative interview with six high school students. This aim to validate the derived general design factors for students in high schools. On this basis, we will explore the following research questions:

RQ1: Can the general chatbot interaction design characteristics identified by Rapp et al. (2021) be universally applied across all chatbot applications?

RQ2: What are the most effective characteristics of advising chatbot interaction design intended for high school students?

This paper is organized as follows: First, we introduced the readers to general background of chatbots themes. Next, the methodology is explained in Sect. 2, and the results and discussion are demonstrated in Sects. 3 and 4, respectively. Finally, the conclusions and future work are presented in Sect. 5.

1.1 Themes of HCI

Since this study aims to investigate how students define the effective design of advising chatbots in comparison to general factors that are derived from systematic reviews, we explore the most effective characteristics of chatbots which may affect users' experiences based on the main themes that have been derived by Rapp et al. (2021) as shown in Table 1.

2 Methodology

This study seeks to explore how students define the effective design of advising chatbots in comparison to general factors that have been derived from the systematic review as described in the previous section. Accordingly, a qualitative interview approach is conducted by interviewing six high school students from different schools where the results will be validated with general factors in order to propose the conceptual model for having an effective design for a student's-advising chatbot.

2.1 Data Collection

The data collected from six high school students, 50% of which are female. The sample includes students from different grades in high schools, where the majority of students come from grade 10. Table 2 shows the characteristics of the participants.

2.2 Sampling Technique

In this qualitative interview we used convenience sampling which pertains students who are currently in high schools, coming from the same geographic location, and who, ultimately, are more easily accessible for contact purposes.

1. Data collection instrument

Table 1 The chatbot interaction design characteristics based on themes derived by Rapp et al. (2021)

Themes of HCI from Rapp et al. (2021)	The impact on users' experiences according to Rapp et al. (2021)
Experiencing the chatbot	– Rapp et al. (2021) identified the satisfaction, engagement, and trust as the main keys that reference the user's experiences
	– Realistic expectation
	– When expectations are undefined, users seem to compare the chatbot with other technology services
	– Using emoji to engage users
	– Social desirable reaction
Conversational issues	– During the human-chatbot conversation, some misunderstandings can be produced due to either technical breakdown or other conversational problems
	– Poor language with spelling errors
	– Fail in understanding user's intention
Emotional experience and expression	– Since the chatbot is designed mainly to mimic human conversation, the emotions are fundamental for human-chatbot interaction
Humanness	– Humanness includes the essential characteristics that chatbots could deploy in order to appear as human-like agents
	– Giving the chatbot a name
	– Language style
	– Anthropomorphic cues
	– Social effects as human-like chatbots
	– Speech-based chatbots

Table 2 The description of the participants' characteristics

Student code	Grade	Gender
S1	10	F
S2	10	F
S3	10	F
S4	11	M
S5	12	M
S6	10	M

There are multiple methods of quantitative research who have been conducted by different authors. In this study, a one-to-one qualitative semi-structured interview is used as a master approach to collect data, which often includes open-ended questions which can have more flexibility for students to answer, with more variations to the questions.

2. Interview Process

The interview was conducted on May 2023 by using a virtual meeting via Zoom platform. Whereas, each participant was allowed to answer and discuss the questions within 10 min. Moreover, this study considered the procedures of meeting the ethical consideration by asking the students to sign a consent form to protect their rights, as well as, having the researchers remove any information that might reveal the identity of the students. Additionally, the records of students were stored in a safe place.

2.3 Data Analysis

In this study, authors applied the main phases of thematic analysis that are described by Braun and Clarke (2006). However, the MAXQDA Analytics software—ver. 22.6.0 is implemented to analyze the qualitative findings for processing the thematic analysis, and the below steps are phased accordingly:

1. Transcribing all interviews into transcripts of texts.
2. Identifying the common themes or ideas among all transcripts and coding them accordingly.
3. Mapping each section in the transcript with the matched common theme.

The phases of the thematic analysis in this study were conducted from the deductive perspectives, as we created pre-established codes that were developed before the beginning of data collection. These codes are derived from a previous systematic study about chatbot interaction design characteristics that was coordinated by Rapp et al. (2021).

3 Results

Since this study is mainly about students from high schools with ranging ages from 16 to 18, we needed to constitute the chatbot's design by using these particular students' perspectives. Therefore, all the semi-structured interviews with the (6) students are transcribed and coded precisely.

Interestingly, the word cloud chart in Fig. 1 visualizes the students' interview transcripts. In the word cloud, "chatbot" appears as the most prominent word, reflecting its importance in the student interviews. Other frequently mentioned terms include "advising," "students," "universities," and "humans".

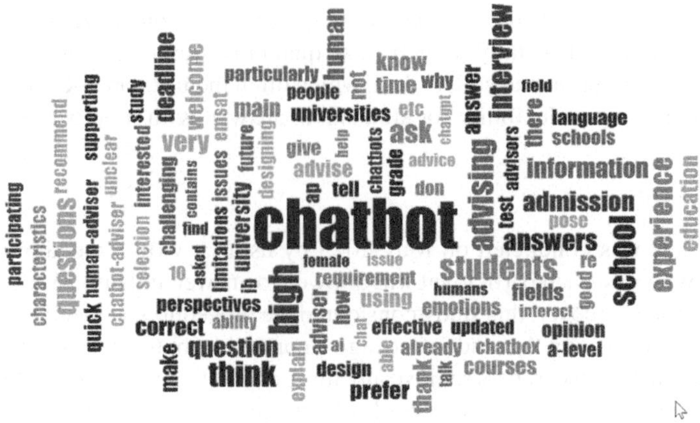

Fig. 1 The word cloud chart of students' interview transcripts

During the interview, students were asked to rank the most effective human-chatbot interaction factor that is considered the most important for them. Figure 2 shows the interview question; however, the results revealed that the conversational issue is the most important factor for them as illustrated in Fig. 3. This factor includes all issues that can cause negative impacts on understanding the chatbots, for instance, spelling and grammar errors, poor language and misunderstanding the users' requests.

In analyzing the data, we defined the common themes for all students' transcripts and coded them accordingly, as shown in Figure 4.

Conversational Issues

As presented in Fig. 3, most students ranked the conversational aspect as the most important factor among the chatbot-interaction design characteristics. This particular factor includes other sub-characteristic factors such as the ability of understand users' intention, poor languages and spelling errors.

9) Rank in order of the most effective human-chatbot interaction factors that you believe are the most important for you to use it? (1: Most important, 2: Important, 3: Less important)

- [ ] Fail in understanding user's intention, poor language with spelling errors
- [ ] Show emotions and empath in chatbot's conversations
- [ ] Including essential characteristics that chatbots could deploy it in order to appear as human-like agents (i.e. Giving the chatbot a name, Language style, Anthropomorphic cues, Speech-based chatbots)

Fig. 2 Question for ranking the most effective human-chatbot interaction factor

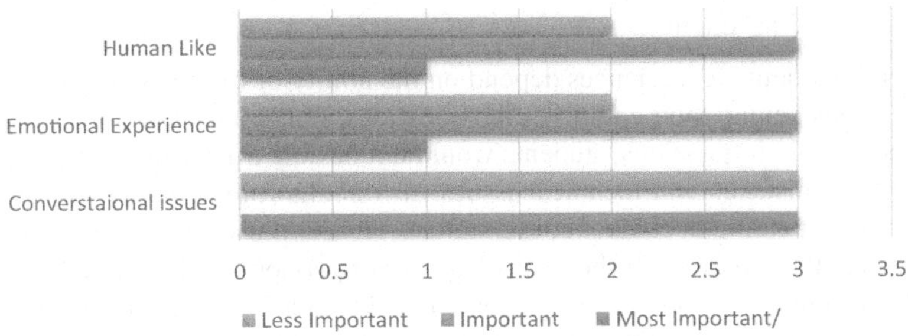

Fig. 3 The results from question of ranking the most important factor of chatbot interaction design

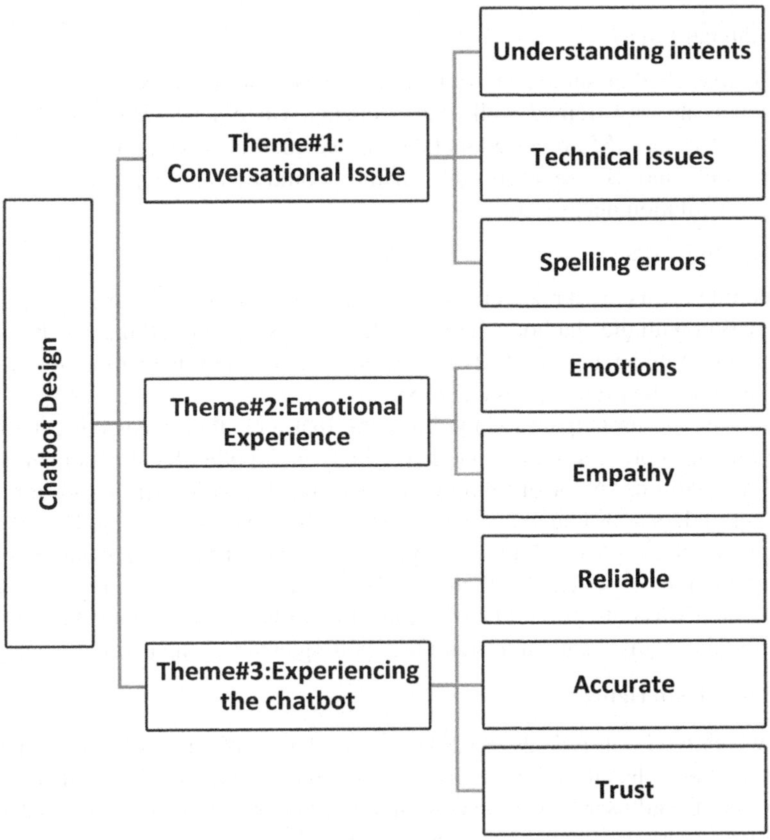

Fig. 4 The main themes and codes of chatbot design that derived from students' transcriptions

1. Understanding Intents.

In general, students' experiences depend on the ability of the chatbot to understand the intentions behind their enquiries and questions. For instance, if the chatbot fails to understand their questions, students would not be able trust using it. Nonetheless, the spelling mistakes and technical issues in ML algorithms can be the cause of the misunderstanding in chatbot conversations. Hence, if users realized that answers from the chatbot are not matched correctly in comparison to a human-adviser, they would certainly find it unreliable for use. For example, the participant (S5) has a concern in understanding the users' inputs as he said: "limitations to understanding the users' input and generating responses, as well as, the lack of ability to understand the input causes incorrect responses and irrelevant results"

2. Technical Issues

An effective chatbot should be trained and tested continuously. If the algorithms and models are not trained well, then students will not be able to understand a generative answer. (S5) reported that "a reliable chatbot should have more updated information", and (S3) said that "any chatbot would constantly have to be updated and its information needs to be up to date".

3. Spelling Errors

The spelling errors are a raised factor that can cause a misunderstanding during the conversation with the chatbot, where a participant (S3), reported that "a chatbot has a limitation in understanding texts with spelling errors and that would not motivate her to use it as she prefers to ask a human".

Similarly, the participant (S1) pointed out that a chatbot needs correct spelling and grammar in order to understand her, which on the other hand, a human-adviser would not require. "A lot of talking and chatting, I have to ask a question that a computer understands like including correct spelling and grammar, so I'd rather talk to a human-adviser "(S1). Interestingly, when we asked the participant (S1) if she had any previous experience with the chatbot, she answered "I actually did not even know what it was until you told me". Thus, this further confirms that she will prefer a human-adviser, as it can understand the spelling errors compared to the chatbot.

Emotional Experience

Only two high school students (S5, S3) expressed the preferences of having a human as adviser due to the lack of emotions and empathic in chatbots. (S5) said "I prefer to talk to a human adviser, as we can communicate better. He has the emotional aspect, so can help more, and know more about how to feel regarding college or a certain educational field, so I like to talk more with a human adviser "

Then again, (S3) said, "I would prefer to ask a human adviser as I can be able to interact with them better and ask them personalized questions".

Experience the Chatbot

Students' experiences depend on their knowledge, needs, and the level of their expectations in using the chatbot. The degree of users' experiences and expectations can affect their reliability, trusts and the accuracy in using the chatbots. The participants in this study have varieties in their experiences. For instance, (S1, S4) show more experiences than others in using the ChatGPT, (S1) reported that "I used a chatbot, like ChatGPT, when I stated studying AP computer science, and I use it in common apps, as well as, for applying to universities, and in writing essays. Similarly, (S4) explained his experience in using the chatbot "I use it for grammar detecting. There is one particular chatbot that I, and a majority of my colleagues have plenty of experience with, and it's ChatGPT. It works really well in terms of getting quick answers to simple questions. It also does put an effort into answering complicated questions involving educational fields".

1. Reliable and Accurate

The reliability of using the chatbot varied among students, for example, (S6) has a positive expectation in using the chatbot as he said, "It can be extremely beneficial to all students that are undergoing education, because it can easily supply enough details and statistics to allow students to choose which universities, colleges, or courses they want to experience". On the other hand, (S1) had recommended to ask a reliable experienced person for a short and understandable answer rather than ask a chatbot, she said "Personally I like talking to an experienced person that tells me the answer straight to the point, I don't like reading a lot but only if it is short and understandable".

Another participant has more expectations than only answering questions, (S2) explained her expectation with "Chatbots are used widely by students for help in school courses, but they can be advanced further by making them able to explain certain topics to students and answer any questions they may have". On the other hand, (S4) raised a concern of the reliability of the knowledge that is generated from the chatbot "Still, it often proves to be unreliable simply due to the limited knowledge it contains of a certain topic".

2. Trust

In order to improve the trust between students and chatbots, the interactive design should match with students' personalities. For instance, (S1, S4) show more their personalities during the interview, as they reported that they prefer not to talk more, so hence would opt for a quick and short answer for their questions, "I like talking to an experienced person that tells me the answer straight to the point" (S1). Conversely, (S4) has mentioned, "The chatbot would save the trouble of physically interacting with the human advisor, and would also save time in terms of that simply reading would be faster than talking." Moreover, (S3) feels that she will have more trust with chatbot in terms of privacy "I find it embarrassing to ask certain questions to human advisors".

Humanness

During the qualitative interview, students did not prefer to interact with chatbots that act similarly like human. For example, none of students showed interest to be answered by a specific human tone, or to deal with a chatbot that has a real human name. Some Students reported in general that they like to deal with humans. For instance, (S5) expressed the importance to interact with a human as they believe humans can understand them more compared to any chatbots, "I prefer to talk to a human adviser, we can communicate easier, he has emotional support and can help me more, (S5)".

4 Discussion

This study aims to understand high school students' requirements and expectations in designing an interactive advising chatbot, and accordingly the below two research questions are answered as follows:

RQ1: Can the general chatbot interaction design characteristics identified by Rapp et al. (2021) be universally applied across all chatbot applications?

Most of participants in this study focused on the same interaction design characteristics that are revealed from the study of Rapp et al. (2021), however, the level of importance varies based on students' personalities and experiences. Figure 5 demonstrates the frequencies of the main themes of the design characteristics, as it shows that the students' experiences (reliability, accuracy and trust) are the biggest concerns of students in interacting with the chatbots. Then, emotional and conversational factors come next. Interestingly, the humanness factors are not a priority for high school students. Indeed, students in this study expressed their concerns more in the quality and the accuracy of chatbot answers and conversations, rather than the special humanness interactive design. The reason behind these answers might be affected by students' situations. For instance, the interview conducted in this study was held during the final exams period. Besides that, students in this particular stage have higher stress levels regarding their future compared to other students; mainly in selecting the right majors and universities that fit with their interests, which in turn will explain their lack in concern for humanness, while emphasizing on the importance of accuracy of chatbot answers for facilitating their college education.

RQ2: What are the most effective and sustainable characteristics of advising chatbot interaction design intended for high school students?

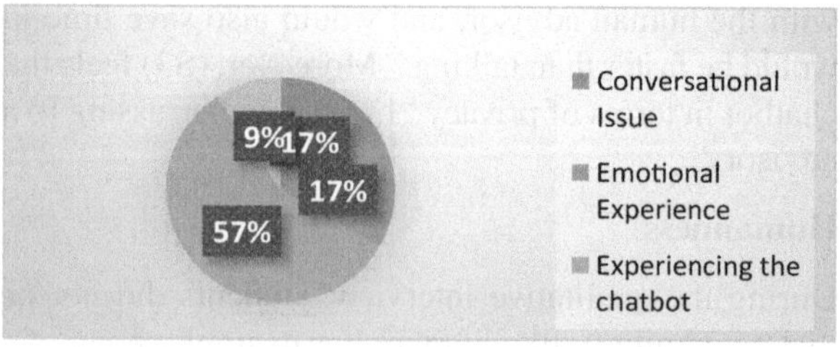

Fig. 5 The main themes frequencies of chatbot design that derived from students' transcriptions

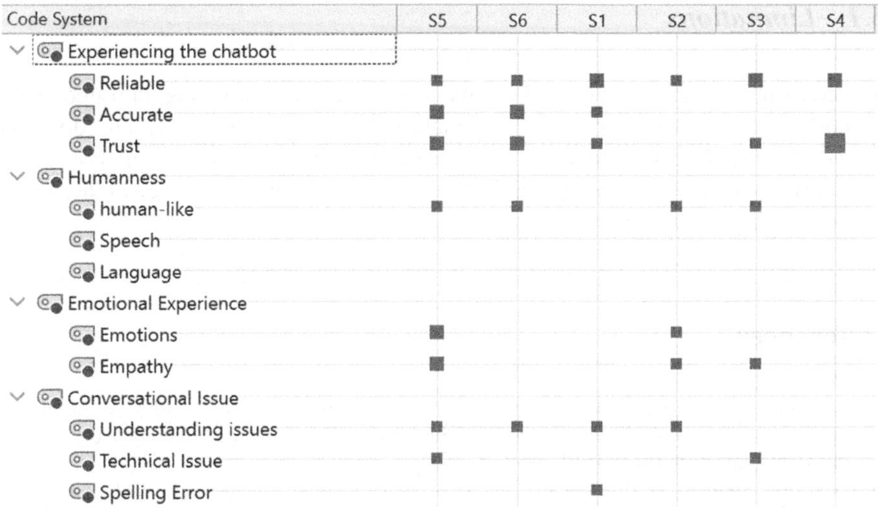

Fig. 6 The code matrix that visualize the most key ideas and themes that have the most impacts on students' expectations

According to this qualitative interview study, most students focused on the importance of chatbots to provide them with an understandable conversation, accurate, short answers, and reliable and trusted information as shown in Fig. 6. The code matrix visualized the key ideas and themes that have the most impact on students' expectations

5 Conclusion and Future Work

This study highlights the most important characteristics that high school students would recommend for having an interactive and effective design advising chatbot that can guide them towards their future by providing them with the deadline of university applications, admission tests, majors, etc. This study conducted semi-structured qualitative interview, revealing that students would recommend same general interactive design characteristics derived from the previous systematic review which emphasizes more on accuracy, yielding short answers, reliability and trust features with minimum conversational issues to avoid confusing them. On the other hand, emotions and empathic factors got fewer priorities for high school students. This result needs to be investigated more, since students are interviewed during their final exams, and students in this particular stage have higher stress levels compared to other students. However, in the future, quantitative studies with structural equation modeling (SEM) should be included to explore the impact of students' gender towards the empathic characteristics of human-chatbot interactions.

5.1 Limitations

The convenience sampling is used in this qualitative study, targeting a limited group of students from particular private international schools in the UAE, and does not include other public and local schools. Furthermore, the interviews were conducted during students' exams, during which students would likely experience higher levels of stress compared to other times.

References

Assayed SK, Alkhatib M, Shaalan K (2023a) Artificial intelligence based chatbot for promoting equality in high school advising. In: 2023 4th international conference on intelligent engineering and management (ICIEM). IEEE, pp 1–4

Assayed SK, Alkhatib M, Shaalan K (2023b) Advising chatbot for high school in smart cities. In: 2023 8th international conference on smart and sustainable technologies (SpliTech). IEEE, pp 1–6

Assayed SK, Alkhatib M, Shaalan K (2024a) Enhancing student services: machine learning chatbot intent recognition for high school inquiries. In: BUiD doctoral research conference 2023: multidisciplinary studies. Springer Nature Switzerland, Cham, pp 243–254

Assayed SK, Alkhatib M, Shaalan K (2024b) A systematic review of conversational AI chatbots in academic advising. In: BUiD doctoral research conference 2023. Springer, Cham, pp 346–359

Braun V, Clarke V (2006) Using thematic analysis in psychology. Qual Res Psychol 3:77–101

Burkhard M, Seufert S, Cetto M, Handschuh S (2022) Educational chatbots for collaborative learning: results of a design experiment in a middle school. International Association for Development of the Information Society

Chen Y, Jensen S, Albert LJ, Gupta S, Lee T (2023) Artificial intelligence (AI) student assistants in the classroom: designing chatbots to support student success. Inf Syst Front 25(1):161–182

Chignell M, Wang L, Zare A, Li J (2023) The evolution of HCI and human factors: integrating human and artificial intelligence. ACM Trans Comput Hum Interact 30(2):1–30

Gunathilaka C (2022) Exploring affordances of AI banking chatbots: towards understanding on how chatbots create value for customers

Kieffer S, Ko A, Ko G, Gruson D (2019) Six case-based recommendations for designing mobile health applications and chatbots. Clin Chim Acta 493:S29–S30

Kim J, Merrill K, Xu K, Collins C (2023) My health advisor is a robot: understanding intentions to adopt a robotic health advisor. Int J Hum Comput Interact 1–10

Kuhail MA, Alturki N, Alramlawi S, Alhejori K (2023) Interacting with educational chatbots: a systematic review. Educ Inf Technol 28(1):973–1018

MacKenzie IS (2024) Human-computer interaction: an empirical research perspective

Rapp A, Curti L, Boldi A (2021) The human side of human-chatbot interaction: a systematic literature review of ten years of research on text-based chatbots. Int J Hum Comput Stud 151:102630

Shi Q, Zhang Z, Chen T, Lee C (2019) Minimalist and multi-functional human machine interface (HMI) using a flexible wearable triboelectric patch. Nano Energy 62:355–366

Van NTT, Vrana V, Duy NT, Minh DXH, Dzung PT, Mondal SR, Das S (2020) The role of human–machine interactive devices for post-COVID-19 innovative sustainable tourism in Ho Chi Minh City, Vietnam. Sustainability 12(22):9523

Exploring the Effectiveness of Different Learning Methods: A Comparative Analysis

Mai Idrees Zaghab

Abstract This theoretical paper scrutinises the evolution of language teaching methods across the twentieth and twenty-first centuries, emphasising the continuous effort to enhance learning environments and outcomes. It investigates the Total-Physical Response (TPR), The Eclectic Approach, Task-Based Language Teaching (TBLT), and Suggestopedia. through a comparative analysis, the study aims to identify effective strategies for different learners, considering their styles and subject areas, each method is scrutinized based on its principles, techniques, strengths, and weaknesses, highlighting the need for adaptable methodologies that promote engagement and meaningful language acquisition. The findings and discussion offer insights into effective teaching methods, emphasizing personalized approaches to improve learning outcomes.

Keywords Total-Physical Response (TPR) · Suggestopedia · The eclectic approach · Task-Based Language Teaching (TBLT)

1 Introduction

Throughout the past century, there has been a concerted effort among educators to devise teaching strategies aimed at improving the learning process by fostering a positive, inclusive, and engaging environment. Although educators employed different techniques and approaches, all worked towards a mutual goal.

Teachers have their individual approaches to ensure that they have effectively transferred knowledge to all students, regardless of their academic levels, learning styles, or interests. It is the teacher's duty to utilise various methods and activities that will engage students and enable them to achieve the desired learning objectives

This paper was prepared for the Teacher and Professional Learning and Development Module, taught by Dr. Emad Abu-Ayyash.

M. I. Zaghab (✉)
Faculty of Education, The British University in Dubai, Dubai, UAE
e-mail: 21000915@student.buid.ac.ae

K. Al Marri et al. (eds.), *BUiD Doctoral Research Conference 2024*, Lecture Notes in Civil Engineering 587, https://doi.org/10.1007/978-3-031-84371-6_19

by the end of the lesson. Hence, educators need to recognise the diversity of their students and employ a range of instructional strategies to support their students' learning and development.

Although each method has shown its effectiveness during a particular time, it was eventually surpassed by another. Each method has its own strengths and weaknesses, which have been identified before and after its implementation. Based on these advantages and disadvantages, teachers were able to tailor various activities to suit the needs of their students and achieve optimal outcomes.

Educators' needs were the driving force behind the development of teaching methods. The grammar-translation method (GTM) and the Direct Method dominated the field for many years starting in the nineteenth century. However, in the 1950s and 1960s, psychologists and philosophers developed new methods and approaches to better serve learners. These included the Audio-Lingual Method (ALM), the Silent Way, and Total-Physical Response (TPR), which were designed to accommodate learners' different styles of learning.

During the 1970s and 1980s, the Eclectic Approach gained widespread popularity as it integrated the most beneficial features of all other methods into one comprehensive approach.

Towards the close of the twentieth century, educators embraced a new approach that placed greater emphasis on the students themselves. Task-based Language Teaching and Project-based Learning are two notable approaches in which students learn through inquiry, examination, and production. These methods have proven to be effective provided that there are sufficient resources available.

This paper seeks to investigate the effectiveness of different learning methods through a comparative analysis. The paper aims to identify which learning methods are more effective and efficient for different types of learners, including visual, auditory, and kinesthetic. It also seeks to identify which learning methods are most effective for different subject areas or levels of education, such as primary or secondary education. The paper will examine various learning methods and compare their effectiveness in terms of student learning outcomes, motivation, and engagement. It also aims to provide insights and recommendations to educators on the most effective methods for improving learning outcomes.

Every psychologist thinks his chosen approach or method is superior, but no one can claim that there is one best method. As a result, educators must adjust to their daily learning environment to ensure that their students are engaged and that their goal of imparting knowledge is successful. In this analytical paper, the researcher will set a comparison between four different language learning methods and attempt to identify one mixed method that can create an encouraging atmosphere for students to learn.

2 Literature Review

This section of the paper examines the conceptual framework, the theoretical framework, and the related studies.

2.1 *Conceptual and Theoretical Framework*

Four key concepts are to be investigated in this section: Total-Physical Response (TPR), Suggestopedia, the Eclectic Approach, and Task-Based Language Teaching (TBLT).

2.1.1 Total-Physical Response (TPR)

In his 1977 Learning Another Language Through Action, Asher identifies the Total Physical Response as a language teaching method that utilises learners' physical movements, senses, and emotions to facilitate language learning. This method is based on the idea that learners acquire a second language more effectively when associating it with physical movement.

TPR involves the teacher giving instructions in the targeted language and students responding with physical movements. The method is particularly useful for beginners or those who are learning a language for the first time, as it helps them learn through movements and repetition. For example, the teacher may give commands such as "walk to the door," or "open the window." The students are expected to follow these commands with the corresponding physical movement. As the lesson progresses, the teacher may give more complex instructions, such as "Pick up the book, walk to the door, and give it to your partner." This method provides students with an interactive way to learn basic vocabulary and sentence structures while promoting their listening and comprehension skills (Asher 2018).

One theory that underpins this method is:

1. Theory of Speech Act: Although Asher didn't relate his TPR to the Theory of Speech Act directly. Yet, he believed that language can be taught through the imperative structure if used comprehensively by a skilful educator to initiate requests and apologies. For example, in a language class, the teacher can initiate requests and apologies by saying: "George, Open the door," or "Beth, say sorry to your friend." This is comparable to what Austin (1975) and Searle (1969) highlighted in their theories, both believed that language is used to perform actions rather than convey information. By referring to the same example mentioned above, the teacher says: "George, open the door." This Locutionary act refers to the actual words used in the utterance. The illocutionary act refers to the speaker's intention and the effect of the utterance on the listener. This illocutionary act is

a request. As for the perlocutionary act, it depends on the listener's response, in this case, George opening the door.

While the Theory of Speech Act is concerned with the underlying intentions of the speaker, the TPR focuses on language acquisition through physical movement and gestures.

2.1.2 Suggestopedia

Suggestopedia is a language teaching method that was developed by the Bulgarian psychologist Lozanov in the 1970s. according to Lozanov (1978), Suggestopedia tackles the idea of relaxation and suggestion. he stated that learners' brains have vast capabilities for attaining knowledge if exposed to music and other relaxation techniques in their learning environment.

The theory that underpins the idea of suggestopedia is:

1. The Theory of Suggestion: Suggestopedia is created upon the idea that people can be influenced by the power of suggestion. It implies that learners are more likely to accept and act on suggestions if they are made in a positive and motivating environment. Lozanov, in his theory, referred to a state that is called concert pseudo-passiveness. According to Richards and Rodgers (2001), "This state is felt to be optimal for learning, in that anxieties and tension are relieved and power of concentration for new material is raised. Because the role of music is central in suggestopedic learning, it needs to be considered in somewhat more detail." (p. 321).

2.1.3 The Eclectic Approach

This approach "uses techniques and activities from a range of different language teaching approaches to suit the needs of individual students or groups of students" (Brown 2007, p. 41). It allows educators to adapt and combine any procedure or activity from any method depending on the students' needs. Griffiths (2012) stated that:

> Although the need to be aware of a variety of methods was acknowledged, several respondents also stressed the need to be able to choose methods appropriate to the needs of their students. Overall, the preference of the teachers in this study seemed to be for an eclectic approach to methodology, which leaves the teacher free to choose from a variety of methods in order to help their students achieve success in language learning. (p. 473).

The Eclectic Approach encourages teachers to select and adapt teaching methods and techniques from a range of different approaches based on the needs and goals of their learners. By drawing on a variety of theories and approaches, the eclectic approach aims to create a flexible and adaptable teaching approach that can be customized to meet the needs of different learners and situations.

Several theories underpin the aforementioned method, such as:

1. Behaviourism: Behaviourism emphasises the importance of positive and negative reinforcement in shaping behaviour. In language teaching, behaviourism is often associated with techniques such as drilling, pattern practice, and error correction, which aim to reinforce language patterns and structures through repetition and feedback. In the eclectic approach, behaviourist techniques are often used alongside other techniques from different language teaching approaches. For example, a teacher using the eclectic approach might combine behaviourist drilling exercises with cognitive techniques such as visualization, or constructivist techniques such as project-based learning (Richards and Rodgers 2014).
2. Constructivism: Both the eclectic method and constructivism share an understanding of the significance of individual differences and learning preferences. These approaches acknowledge the diversity of learners' experiences, knowledge, and skills, and consider them integral components in the teaching process. Additionally, both approaches recognize the active role of the learner in the learning process. They view learners as agents who actively construct their knowledge and understanding of the world. However, the eclectic method focuses on teaching language skills and structures, whereas constructivism emphasizes the learners' active participation in knowledge and understanding acquisition. Moreover, the eclectic method employs a teacher-centred approach in which the educator selects and modifies learning materials to meet the learners' needs. Constructivism takes a learner-centred approach, where the learners play a central role in the learning process, and the teacher facilitates the learning process by creating opportunities for exploration, discovery, and reflection (Richards and Rodgers 2001).

2.1.4 Task-Based Language Teaching (TBLT)

TBLT was developed as an alternative to the Direct Method, which focuses mainly on teaching vocabulary and grammar in isolation from real-world communication. TBLT is grounded in the belief that language is best learned through meaningful use of context. It involves providing students with practical tasks that require them to use the targeted language to complete the given tasks successfully. TBLT is commonly referred to as an approach, not a method, "it does not constitute one single methodology. It is a multi-faced approach, which can be used creatively with different syllabus types and for different purposes." (Leaver and Willis 2004, p. 3).

One main theory that has a strong connection to the TBLT is:

1. Constructivism: This theory suggests that learning occurs through the active construction of knowledge and learners need to be actively engaged in the learning process. In TBLT, learners are presented with real-world tasks that require them to use language to accomplish a goal, which encourages active learning. According to Richards and Rodgers (2014), TBLT and scaffolding share a common focus on promoting active language learning through learner-centred approaches. Both methods aim to help learners develop their language skills in a more genuine and authentic manner, rather than solely focusing on grammar or

vocabulary instruction. Additionally, both TBLT and scaffolding prioritize the importance of using language in context and its practical application in the real world.

2.2 Related Studies

2.2.1 Total-Physical Response (TPR)

According to Asher (1977), “the use of the body in association with language has been shown to be highly effective in learning.” (p. 5). TPR enables students to recall vocabulary and grammar structures more easily by associating them with physical actions and movements. Naiman et al. (1978) noted that “TPR is ideal for beginners or for those learning a new language in a classroom setting for the first time.” (p. 98). It is particularly effective for novices and young learners who may have difficulty understanding abstract language concepts. TPR is an engaging way to learn a language and can help to create a positive and interactive classroom environment. Larsen-Freeman and Anderson (2011) suggested that TPR “can be an enjoyable way to learn and foster a relaxed and interactive atmosphere in the classroom” (p. 333). However, some adversary studies claim that TPR may be less effective for more advanced learners who have already developed a solid understanding of the language. Richards and Rodgers (2001) commented that “the TPR approach is more appropriate for beginners or low-intermediate students than for advanced students.” (p. 98). Another view is that TPR may not be appropriate for all learners, particularly those who have physical limitations or who may not feel comfortable with physical movements; as Lightbown and Spada (2013) note, “TPR may not be appropriate for all learners, particularly those who are physically limited or who are uncomfortable with physical movements.” (p. 119). Some educators and learners may view the TPR method as childish or unprofessional, particularly in more formal or academic settings. Larsen-Freeman and Anderson (2011) stated that “some teachers and students may view TPR as childish or unprofessional, particularly in more formal or academic settings” (p. 333). Finally, TPR may not be suitable for all language learning contexts. Nunan (1991) argued that “TPR may not be suitable for all language learning contexts, particularly where a more communicative or task-based approach is preferred.” (p. 139).

2.2.2 Suggestopedia

Asher (1977) stated that suggestopedia can be effective in reducing anxiety as well as enhancing learning the language, and that allows them to learn more effectively. Lozanov (1978) argued that suggestopedia helps learners access their subconscious and learn more quickly and efficiently. He claimed that this is achieved through the use of music, which can help create a positive emotional state and a receptive mindset.

A study conducted by Hodges (1982) found that suggestopedia can be effective in improving language learning for college students. The study found that students taught using suggestopedia techniques scored significantly higher on a language proficiency test than those taught using traditional methods. However, critics of Suggestopedia argued that Suggestopedia is based on pseudoscientific principles and lacks empirical evidence to support its effectiveness. They argued that relaxation, music, and suggestion have a positive impact on learners but do not necessarily lead to improved learning outcomes (Swain 1981). Some researchers have criticized the suggestopedia method for its heavy reliance on the teacher and lack of learner autonomy. They argue that suggestopedia is overly teacher-centred and does not give learners enough control over their learning process (Littlewood 1984). Another study conducted by Cohen (1981) found that suggestopedia was no more effective in improving language acquisition than traditional teaching methods. The study found that while suggestopedia had a positive effect on learners' attitudes toward learning, it did not lead to significant improvements in language proficiency.

2.2.3 The Eclectic Approach

One of the strengths of the eclectic language teaching method is its flexibility. As Richards and Rodgers (2001) stated that the eclectic method allows language teachers to choose the most appropriate techniques and methods for their students and their learning goals. Moreover, The eclectic approach is personalized to suit the requirements of individual learners. McDonough and Shaw (1993) emphasize that teachers can tailor their teaching to cater to the unique needs of their students and utilize materials and activities that are pertinent and captivating to them. On the other hand, one criticism of the eclectic method is that it lacks coherence. As Nunan (1991) argued that "[t]he eclectic approach can be seen as a disjointed collection of techniques and methods without a coherent theoretical foundation, which can be confusing for both teachers and learners." (p. 42). Another criticism of the eclectic method is that it can lead to inconsistency in teaching. As Meddings and Thornbury (2009) noted that "[e]clectic language teachers may use different techniques and methods with different classes or even with different students, which can lead to inconsistency in teaching and may make it difficult to evaluate the effectiveness of the teaching." (p. 144).

2.2.4 Task-Based Language Teaching (TBLT)

TBLT encourages the use of language in real-life situations and tasks, which allows learners to develop their communicative competence by practicing using the language in meaningful contexts (Nunan 2004). It also provides learners with a sense of purpose and motivation because they work on tasks relevant to their needs and interests. TBLT encourages learners to take responsibility for their learning and to develop their own strategies for completing tasks (Ellis 2003). Yet, critics argue that TBLT may not provide learners with sufficient exposure to the target language because the focus is on completing tasks rather than language input (Long 2015). In addition, some learners may struggle with the task-based approach because they may lack the required language skills or may not be able to connect with the tasks that are presented (Swain 2005). Other critics argue that TBLT may not be practical in all contexts, particularly in large classes or where there is limited time available for language instruction (Willis and Willis 2007).

3 Findings and Discussion

Notably, the four teaching language approaches revolve around the same goal: to effectively help learners develop the new language rudiments in a motivating and engaging manner.

3.1 Total Physical Response Versus Suggestopedia

Both Total Physical Response and Suggestopedia have some similarities in terms of creating a positive learning environment and using creative activities to engage students, they also have some distinct differences in terms of their focus on physical movement versus creative activities, their approach to language acquisition, their target audience, and the settings in which they are commonly used.

Both methods highlight the importance of utilising physical movements and body language, music, and other creative activities to help students retain new vocabulary and grammar. For example, various classroom activities can assist students in acquiring new grammatical rules or vocabulary. (Simon says) is an effective activity that can help students learn new words while also encouraging physical movement. For instance, during a first-grade language lesson, the teacher can say, "Simon says: nod your head" or "Simon says: stamp your feet." This activity will introduce younger students to a new language; the different body parts, and the imperative structure. Another example is playing the popular song, "Head, Shoulders, Knees, and Toes," which enables students to learn new vocabulary related to body parts while also being engaged in a fun activity. Yet, such activities are often used with younger learners to boost their engagement but not necessarily attainment. Young learners can memorise the songs or sequences but can't use the language appropriately in a real-life context. Additionally, such activities are unsuitable for upper grades and big-group classes. Role-playing is considered an application of both methods, but still, can't be applied to every language lesson; mature students might consider it as childish or unprofessional as songs.

The effectiveness of a teaching method depends on various factors such as the learner's age, learning style, and proficiency level. Therefore, it is difficult to say which method is better. However, both TPR and Suggestopedia can be effective in certain contexts and for certain learners. It is important for teachers to choose a method that suits the needs of their learners and to adapt their teaching methods as necessary.

3.2 Eclectic Approach Versus Task-Based Language Teaching

Eclectic Language Teaching and Task-Based Language Teaching are two different approaches to teaching a second language. While the Eclectic Language Teaching method focuses on creating a personalized learning experience by combining different teaching techniques and approaches, the TBLT approach aims to develop students' communicative competence through task-based activities that simulate real-life situations.

Both methods prioritize the learners' needs and goals. The level of student-centeredness or teacher-centeredness of the Eclectic Method can vary depending on how it is implemented. It is flexible since it allows the teacher to adapt the lesson plan based on the students' interests. Still, the TBLT approach is more student-centered, with the teacher guiding students in completing a task in the targeted language. Both methods aim to make language learning relevant and meaningful to the learner. In the Eclectic method, activities can range from grammar drills, vocabulary exercises, and conversation practice, and can be adjusted to the needs of the students. For example, the teacher might use a grammar exercise such as fill-in-the-blank activities, sentence completion, or matching exercises to teach the past tense, followed by a role-play activity that encourages students to use the past tense in context. In contrast, TBLT activities revolve around completing a task verbally in the target language, such as planning a trip, ordering food at a restaurant, or writing a letter of complaint. However, the Eclectic method provides a flexible framework for incorporating a variety of teaching techniques and methods into the language classroom by utilizing a combination of activities that cater to different learning styles and preferences, students can have a more personalized and effective language learning experience. In contrast, the TBLT method encourages learners' autonomy and independence by developing students' communicative competence and providing opportunities for them to use the language in real-life situations.

In summary, the Eclectic Method and TBLT have similarities in their learner-centered approach and emphasis on communication, but differ in their level of structure and focus on different aspects of language learning.

4 Conclusion

In conclusion, it is important to design activities that are tailored specially to the student's needs, interests, and readiness. However, it is almost impossible to come up with one single activity that meets all students' learning styles. Therefore, educators are advised to develop a full-class plan that targets all learners and their preferences by incorporating different types of activities, providing opportunities for collaboration, considering the learning environment, using technology, and providing multiple modes of instruction.

References

Austin JL (1975) How to do things with words. Harvard University Press
Asher JJ (1977) Learning another language through actions: the complete teacher's guidebook. Sky Oaks Productions, Los Gatos, Ca
Asher JJ (2018) Total physical response: origins, developments, and implications. In: The Tesol encyclopedia of English language teaching. Wiley-Blackwell
Brown HD (2007) Principles of language learning and teaching, 5th edn. Pearson Education, White Plains, Ny
Cohen JM (1981) Suggestopedia: not a miracle cure. Lang Learn 31(1):129–140
Ellis R (2003) Task-based language learning and teaching. Oxford University Press, Oxford
Griffiths C (2012) Focus on the teacher. Elt J 66(4):468–476
Hodges CR (1982) Suggestopedia: a method for improving language proficiency in college students. Mod Lang J 66(2):133–139
Larsen-Freeman D, Anderson M (2011) Techniques and principles in language teaching, 3rd edn. Oxford University Press
Leaver M, Willis J (2004) Computer-assisted language learning. In Davies A, Elder C (eds) The handbook of applied linguistics. Blackwell Publishing
Lightbown PM, Spada N (2013) How languages are learned, 4th edn. Oxford University Press
Littlewood W (1984) Foreign and second language learning: language acquisition research and its implications for the classroom. Cambridge University Press
Long MH (2015) Second language acquisition and task-based language teaching. Wiley-Blackwell
Lozanov G (1978) Suggestology and outlines of suggestopedy. Gordon and Breach Science Publishers
Meddings L, Thornbury S (2009) Teaching unplugged: Dogme in English language teaching. Delta Publishing
McDonough J, Shaw C (1993) Materials and methods in ELT: a teacher's guide. Blackwell
Naiman N, Frohlich M, Stern HH, Todesco A (1978) The good language learner. Research in education series, no 7. Ontario Institute for Studies in Education
Nunan D (1991) Language teaching methodology: a textbook for teachers. Prentice Hall
Nunan D (2004) Task-based language teaching. Cambridge University Press
Richards JC, Rodgers TS (2001) Approaches and methods in language teaching (2nd edn.). Cambridge University Press
Richards JC, Rodgers TS (2014) Approaches and methods in language teaching. Cambridge University Press
Searle JR (1969) Speech acts: an essay in the philosophy of language, vol 626. Cambridge University Press
Swain M (1981) Suggestopedia: a critical analysis. Can Mod Lang Rev 37(1):72–83

Swain M (2005) The output hypothesis: theory and research. In: Hinkel E (edn) Handbook of research in second language teaching and learning. Lawrence Erlbaum Associates, pp 471–483
Willis J, Willis D (2007) Task-based language teaching: an overview. ELT J 61(1):4–15

Exploring the Efficacy of Adopting English as the Medium of Instruction in Moral Education: A Study of an American Curriculum School in Sharjah, UAE

Abdelhamid A. Khalil

Abstract Moral Education (ME) is a newly added subject to all schools' curricula in the UAE to promote tolerance and foster global principles, morals, and values shared by the overall community. Although schools are eligible to choose between English and Arabic as the language of instruction to teach ME, many schools which include mainly Arab students have opted for English as the medium of instruction (EMI) to teach ME. Therefore, the current study aims to investigate the policy of adopting EMI to teach ME at an American curriculum school in Sharjah, UAE. Moreover, eight English teachers and four members of the Senior Leadership Team were interviewed through three sets of semi-structured interviews. The study findings showed English language teachers' disagreement with the policy. Furthermore, the findings demonstrated that teachers' voices were fully disregarded while establishing the policy by the Senior Leadership Team.

Keywords Moral education · Medium of instruction · Curriculum · School policy

1 Introduction

Promoting the development of morally mature citizens has always been considered among the key responsibilities of schools. This is achieved through focusing on students' moral and character development (Althof and Berkowitz 2006). Morality is defined as the attitude towards life or the group of principles that if applied to certain contexts, will help to guide one's behaviours (Garrod and Bramble 1979). ME is the experience of assisting students to make better conclusions and make proper decisions that show an acceptable level of knowledge and understanding of the significance of moral ethics and values. Some researchers have identified positive results of integrating ME into the school curriculum (Berkowitz and Hoppe 2009);

A. A. Khalil (✉)
The British University in Dubai, Dubai, UAE
e-mail: abdelhamid.elmtiny@gmail.com

K. Al Marri et al. (eds.), *BUiD Doctoral Research Conference 2024*, Lecture Notes in Civil Engineering 587, https://doi.org/10.1007/978-3-031-84371-6_20

thus, it is significant to investigate the policy of teaching ME in English to Arab learners and state if it helps or hinders students understanding of ME.

1.1 The Research Problem

There are multiple challenges that the UAE society encounters. Among these issues is the growing concern regarding the arising immoral behaviours such as bullying, cheating, and violence within the school context, which make teaching moral skills and values indispensable (Bleazby 2020). In addition, a plethora of researchers have concluded that the impact of teaching ME at schools is reflected in reducing students' absenteeism rate, academic failure, and drop-out rates (Was et al. 2006).

Despite the benefits of implementing ME programs at schools, students do not always develop awareness of these initiatives. This could be the case with younger or primary school children who might be unaware of ME or who might regard it as inconspicuous (Revell 2002). Another critical issue associated with teaching ME in the UAE context is the choice of EMI by some schools whose student populations are mainly Arabs. Although many schools in the UAE have selected EMI to teach ME, there is a lack of literature regarding investigating the implementation of EMI policies to teach ME as a school subject, particularly in Sharjah. Besides, there are no studies investigating English teachers' and admins' perceptions of using EMI to teach ME and identify its impact on students' understanding of the ME subject.

1.2 Significance and Rationale

The rationale for conducting the present study originates from the controversial perceptions among educational leaders and teachers regarding choosing EMI to teach the ME curriculum despite the availability of the same curriculum using students' L1 (Arabic) (Tantawy 2020). In addition, the ME initiative of the UAE (2016) highlights the major role and responsibility of educators towards learners to develop them into open-minded, critical thinkers, and globally-valued citizens. As such, the current qualitative study is significant since it investigates the impact of employing EMI to teach ME to Arab learners.

1.3 Aims and Research Questions

To fulfil the study aims, the following research questions govern the scope of the study:

1. What are the perceptions of English language teachers of the policy of teaching ME in English?
2. What are the perceptions of the Senior Leadership Team regarding choosing English to teach ME at school?
3. What impact do English teachers perceive the adoption of EMI to have on students' understanding of ME?

2 Literature Review

This section elucidates the theoretical underpinning upon which the research is based and provides a review of previous studies related to the topic under investigation.

2.1 Theoretical Framework

Kohlberg's (1979) theory of Moral Cognitive Development (MCD) is the most effective in the realm of moral development psychology, and it is still widely applied and approved by educators and moral theorists all over the globe (Zhang and Zhao 2017). The reason behind the influential impact of Kohlberg's theory is that it is based on a combination of the beliefs of Dewey's Progressive Theory of Ethics and Piaget's Child Development Theory (Zhang and Zhao 2017). Another significance of the theory lies in the amount of challenge it poses against traditional ME practices and its urgency of reforming ME at schools (Schemrich 2003).

Stages of moral growth and development are regarded as the cornerstone of Kohlberg's theoretical system of moral development (Newton 1978). Furthermore, the key aspects of Kohlberg's theory are summarized as follows: first, moral and cognitive development are tightly connected, and the latter is the basic element leading to the establishment of moral development. Second, the level of moral development can only be parallel with the level of cognitive development and it can never exceed it by all means. Furthermore, the concept of morality in young and adult people is developed in some stages which can never be reversed.

2.2 Literature Review

2.2.1 The Role of School

Several studies have concluded that schools are responsible for students' moral development, yet an effective collaboration between parents and teachers will raise students' moral awareness and shape their moral behaviours (Bleazby 2020; Tantawy 2020). In the same aspect, Bleazby (2020) concludes that the increased amount of immoral or negative behaviours among students has made it vital to teach moral values through formal schooling. In a similar study conducted by Durmus (2019), the researcher states that becoming fully aware of moral values and advanced moral skills is best achieved through a curriculum of ME that is formally delivered by teachers in a school community.

2.2.2 Teachers' Voice of Educational Policies

A lot of education research assumes that there is a lack of teachers' indulgence in the decision-making process (Ingersoll 2007). Similarly, Hargreaves and Shirley (2011) concur that teachers' voice is always the final point of educational reform, and they are the last to speak or be involved in decision-making processes. They further view teachers as the objects of change rather than participants in making it happen. Ingersoll (2007) studies the reasons behind excluding teachers from policy-making processes and he concludes that policymakers regard teachers as the main cause of all issues that schools encounter nowadays. Other researchers argue against this issue by stating that teachers have to remain silent to maintain their power in their classrooms (McDonald 1986).

2.2.3 Challenges of Using EMI.

For the UAE to cope with the global market requirements and the competitive economic opportunities, the use of English has shifted from a second language (L2) to a medium of instruction in many schools. Several studies have investigated the impact of using EMI to teach academic subjects to Arab learners in general and UAE students in particular, and they have concluded that EMI is a great barrier to students' understanding and achievement in these subjects (Belhiah and Elhami 2015; Tantawy 2020; Troudi 2010). These findings are in harmony with those reached by Butzkamm (2003) who concludes that the power of using students' L1 to teach them content-related subjects is indispensable. By the same token, the use of students' L1 has both social and pedagogic functions that should be highly considered in a natural learning environment as confirmed by Tantawy (2020).

3 Methodology

3.1 Methodology and Approach

This study is based on the phenomenological qualitative research approach to investigate the perceptions of English teachers and SLT members regarding the EMI policy of teaching ME. Furthermore, the adopted research approach is the most feasible for the current study as it indicates the theory upon which knowledge and reality will be reached in light of answering the research questions (Troudi 2010).

3.2 Research Instruments and Data Collection Methods

Three sets of semi-structured interviews were used as data collection instruments to answer the three research questions of the present paper. The interviews were conducted face-to-face at school and each interview lasted between 15–20 min. Choosing the semi-structured interviews as data collection tools is effective in investigating teachers' perceptions of the EMI policy and deciding if their voices are heard while establishing the targeted policy. In addition, semi-structured interviews are crucial to collect data openly from the SLT members and reach deeper insights into the EMI policy establishment, implementation, and evaluation at school.

3.3 Participants and Sampling Procedures

Participants of the current study include eight English teachers who teach ME to middle and high school students and four SLT members who are responsible for establishing the EMI policy at school. English teachers are selected using convenience sampling due to the easy and convenient access to this specific population for data collection purposes. Additionally, purposive sampling is effectively employed with the SLT members as they possess knowledge and information about the EMI policy, that otherwise, cannot be obtained (Maxwell 2012).

3.4 Data Analysis

The qualitative data gathered from the semi-structured interviews were analysed using thematic analysis (TA) because it is a flexible approach that is easily adjusted based on the needs of individual studies so that it provides thorough, detailed, and rich data (Braun and Clarke 2012). Moreover, TA was employed to identify relevant themes that are useful to answer the three research questions governing the scope of the study. Therefore, using TA, data gathered from the semi-structured interviews were recorded, transcribed, and organized into relevant and meaningful themes for the purpose of the study.

4 Findings and Discussion

The findings of the study are reported according to the order of the research questions.

4.1 Teachers' Perceptions of the EMI Policy

While the majority of participants expressed their disagreement with teaching ME using the EMI policy, few participants emphasized their preference for using English to teach the ME curriculum. In this context, one participant mentions, "Teaching ME using a language of instruction rather than students' mother tongue (Arabic) is a challenging task since some moral concepts can hardly be explained using L2". However, another participant says, "Teaching ME through the world's most common language (English) would help students to acknowledge moral values not only within the UAE society but also on a worldwide scale".

Besides, participants identified several obstacles they face regarding the implementation of the EMI policy at school. Three participants argued that employing EMI to teach ME to Arab students is not very effective, particularly with those of limited academic performance since they are introduced to business, economic, and financial terminologies that are hard to explain in English. Also, the majority of participants concurred that the greatest obstacle to the EMI policy is the amount of preparation time it requires before each class.

Regarding teachers' voices and the role of the EMI policy at school, all participants expressed their frustration that their opinions were not considered nor discussed at the time of establishing the policy. Other participants, however, concluded that if they were to re-establish the policy, they would adopt students' L1 (Arabic) as the language of instruction to teach ME since it would serve the ME initiative's objectives.

4.2 SLT Perceptions of the EMI Policy

When they were asked about the causes behind opting for EMI rather than Arabic to teach the ME curriculum, some points emerged from interviewing the SLT. Two participants argued that being an American curriculum school, there are restrictions on the language of instruction from different bodies including Sharjah Educational Private Authority (SPEA) as well as international accreditation authorities. The other two participants concurred that the school is allowed to teach only Arabic and Islamic Studies using L1 (Arabic) and if they choose Arabic to teach ME as well, the school evaluation process will be affected.

Responding to a question about teachers' role in establishing the EMI policy at school, all participants reported that they were excluded from establishing the policy due to the mistrust of their intentions as well as their hidden agendas that they would work only to make their lives easy regardless of the school vision. Adding to this, two participants explained that teachers should not always take students' requirements for granted without considering the need to raise the standards of teaching as well as the necessity to challenge them and exceed their expectations. This explains why teachers' voices were disregarded as per the perceptions of the SLT.

Regarding a question about the evaluation of the EMI policy and the methods of dealing with new challenges, one participant reported that "The policy fits the context of an American curriculum school where English is the language of instruction starting from grade one except for Arabic and Islamic Studies. Also, there is a follow-up team of the admin staff that continually checks the implementation of the policy through student and teacher surveys as well as careful observation of classes".

4.3 Teachers' Perceptions of the Impact of the EMI Policy

Regarding teachers' perceptions of the impact of the EMI policy on students' understanding of ME, participants provided different reasons for and against the policy. For instance, three participants argued that spending the whole class time using EMI to teach ME results in partial achievement of the lesson objectives as well as limited understanding of the content of ME since some concepts are hard to explain in English. Two other participants maintained that integrating Arabic in teaching ME lessons proved to be very effective, that otherwise, students feel demotivated and disengaged as a result of their inability to interact with the content emotionally and behaviourally in English. Another participant confirmed the negative impact of the policy by reporting that EMI hinders students' cognitive and linguistic abilities as they have to decode the language before trying to understand concepts and topics taught theoretically.

4.4 Discussion

The findings of the current study revealed English teachers' disagreement with the EMI policy, which they are unwillingly forced to follow to the fullest. Furthermore, teachers are deprived of their rights to participate in the policy establishment even though they are responsible for implementing it. These findings are in harmony with those reached by Ingersoll (2007) in which teachers are viewed by stakeholders as objects rather than participants in policymaking. Acknowledging the reality that teachers' suggestions and views of the policy have never been considered adds to the doubts and uncertainty about the level of cooperation and professionalism that is found in the school environment to ensure the successful implementation of the EMI policy.

The results of the study also showed that the SLTs are the sole decision-makers at school, and they hold negative impressions against English teachers' intentions.

Furthermore, using words such as "hidden agendas" and "lack of trust" by the SLT members during the interview holds a negative connotation and it creates an uncomfortable educational environment where productivity and achievement are expected to be the outcome of such policy. These results are in line with what Simkins (2015) and Ellison et al. (2018) state about the sole and dominant practices of policymakers who consider themselves elite compared to the non-elite including teachers and students.

In terms of the impact of the EMI policy on students' comprehension of ME, the findings of the study demonstrated that teachers are worried that using English as the language of instruction instead of students' L1 (Arabic) leads to limited understanding and low achievement of ME concepts and values. In this context, most of the interviewed participants confirmed a low level of understanding of ME as a subject because of the language barrier. These findings are consistent with some other studies that concluded that using EMI forms a great barrier to students' understanding and achievement in academic subjects compared to delivering the same content in their mother tongue (Belhiah and Elhami 2015; Tantawy 2020).

5 Conclusions and Recommendations

5.1 Summary of the Paper and Key Findings

The purpose of the current study is to investigate the policy of teaching ME in English through the perceptions of English teachers and the SLT. The study also investigates the impact of the policy on students' comprehension of the ME curriculum. The study is significant as it provides deeper insights into the evaluation of the ME initiative of the UAE, especially when the curriculum is taught in English to Arab learners. The phenomenological qualitative research approach is chosen to serve the purpose of the study and get a deeper understanding of participants' perceptions of the EMI policy. Three sets of semi-structured interviews have been developed to collect data and answer the research questions.

The study findings revealed that English teachers hold negative perceptions of the EMI policy, and they are dissatisfied with it since their opinions have been disregarded and minimised to policy implementation only. Moreover, the findings showed that the SLT members are the only decision-makers at school, and they hold positive views regarding the EMI policy as compared to teachers' opinions. Furthermore, the findings demonstrated that students have very slow and limited performance due to the linguistic and cognitive challenges they face while studying ME using EMI.

5.2 Limitations

One limitation of the current study is the small number of participants involved, which makes it hard to generalize the findings even though many schools adopt the same EMI policy in the UAE. Another limitation is related to data triangulation which could have been improved by implementing another data collection tool to add to the validity and reliability of the study findings.

5.3 Recommendations

A key recommendation of the present study is to compare the results of students studying ME in English and of those studying the same curriculum in Arabic to identify which approach is more effective for teachers and students. Another recommendation is to modify the ME curriculum and include practical activities that would make the subject more meaningful and engaging to students.

5.4 Implications for Future Research

Based on the results of the present study, further research is required to explore students' perceptions of the EMI policy and its impact on their understanding of the ME curriculum. There is a vital necessity to use the quantitative approach to generate numeric data by including as many participants as possible for the sake of investigating the EMI policy in a wider context.

References

Althof W, Berkowitz MW (2006) Moral education and character education: their relationship and roles in citizenship education. J Moral Educ 35(4):495–518

Belhiah H, Elhami M (2015) English as a medium of instruction in the Gulf: when students and teachers speak. Lang Policy 14:3–23

Berkowitz MW, Hoppe MA (2009) Character education and gifted children. High Abil Stud 20(2):131–142

Bleazby J (2020) Fostering moral understanding, moral inquiry and moral habits through philosophy in schools: a Deweyian analysis of Australia's ethical understanding curriculum. J Curric Stud 52(1):84–100

Braun V, Clarke V (2012) Using thematic analysis in psychology. Qual Res Psychol 3(2):77–101

Butzkamm W (2003) We only learn language once. The role of the mother tongue in FL classrooms: death of a dogma. Lang Learn J 28(1):29–39

Durmus YT (2019) Early childhood education teachers' experiences on moral dilemmas and suggestions for morality and ethics in education course in faculty of educations. Int J Progress Educ 15(5):301–314

Ellison S, Anderson AB, Aronson B, Clausen C (2018) From objects to subjects: repositioning teachers as policy actors doing policy work. Teach Teach Educ 74:157–169
Garrod AC, Bramble GA (1979) Moral development and literature. Theory Pract 16(2):105–111
Hargreaves A, Shirley D (2011) The far side of educational reform. Can Teachers' Fed (NJ3)
Ingersoll RM (2007) Short on power, long on responsibility. Educ Leadersh 65(1):20–25
Maxwell JA (2012) Qualitative research design: an interactive approach. Sage publications
McDonald J (1986) Raising the teacher's voice and the ironic role of theory. Harv Educ Rev 56(4):355–379
Newton RR (1978) Kohlberg: implications for high school programs. Living Light 15(2):231–239
Revell L (2002) Children's responses to character education. Educ Stud 28(4):421–431
Schemrich CM (2003) Applying principles of Kohlberg's theory of moral development to classroom instruction, classroom discipline procedures, school-wide discipline procedures, district programs and community programs. Unpublished thesis. Graduate faculty of the University of Akron
Simkins T (2015) Leadership in education: 'what works' or 'what makes sense'? Educ Manage Adm Leadersh 33(1):9–26
Tantawy N (2020) Questioning the policy of teaching 'moral education through English as a foreign language in a private school in Dubai. Lang Teach Res Quart 16:71–88
Troudi S (2010) Paradigmatic nature and theoretical framework in educational research. TESOL Arabia Publications
Was CA, Woltz DJ, Drew C (2006) Evaluating character education programs and missing the target: a critique of existing research. Educ Res Rev 1(2):148–156
Zhang Q, Zhao H (2017) An analytical overview of Kohlberg's theory of moral development in college moral education in mainland China. Open J Soc Sci 5(8):151–160

Impact of Innovative Lighting Strategies on the Performance of Workplace Environments: Evaluation of Solar Tubes

Bana Eid and Wael Sheta

Abstract Most buildings in UAE rely on the daylighting as a sustainable source to provide good indoor lighting quality. This natural source will allow to save energy, improve the indoor ambiance, enhance the work productivity, provide good visual comfort as well as providing the physiological and psychological needs of humans. This paper covered the various techniques that are needed to promote daylighting in buildings, this includes literature review, present one international case study, and applying the learned concepts to the offices at The Abu Dhabi University campus. The different applied solutions of daylighting on the selected building were simulated using Dialux EVO software to study and analyse the improvements of lighting quality and the impact on energy consumption. The simulation phase has covered 4 strategies, strategy 1 is for solar tubes $\times 1$, size 10 in., strategy 2 is for solar tubes $\times 2$, size 24 in., strategy 3 is for solar tubes $\times 3$, size 24 in., and strategy 4 is for solar tubes $\times$ 4, size 24 in. The results showed that the offices in strategy 3 received 3 solar tubes with 24 in. each, and this have helped to enhance the amount of lux and achieve the requirements. The simulation results using solar tubes strategies have achieved the desired lux needed for indoor space with zero energy consumption as all artificial lights were switched off.

Keywords Daylighting · Light tubes · Solar tubes · Energy consumption · DIALux

This paper was prepared for the course lighting performance and strategies, taught by Dr. Wael Sheta.

B. Eid (✉)
Architecture and Sustainable Built Environment, The British University in Dubai, Dubai, UAE
e-mail: bana2610@hotmail.com; 20000005@student.buid.ac.ae

W. Sheta
Department of Engineering, The British University in Dubai, Dubai, UAE

K. Al Marri et al. (eds.), *BUiD Doctoral Research Conference 2024*, Lecture Notes in Civil Engineering 587, https://doi.org/10.1007/978-3-031-84371-6_21

1 Introduction

In office buildings and other types of buildings that generally operate during the daytime, a sustainable source of lighting is required to provide natural daylighting within the building, save energy and improve the overall ambiance. Previous studies have proven that daylight provides the best colour rendering, increases work productivity, good visual comfort, meets the physiological and psychological needs of humans, and saves energy. In addition, daylight reduces the heat wastage produced by artificial lighting as it causes less heat compared with artificial light for the same lighting amount if designed and integrated correctly. UAE, with its geographical location and climate, provides ample daylight. Although most high-rise buildings in UAE have entire glazed facades, the building's internal areas are still supplemented by artificial lighting due to improper design and integration of daylighting, resulting in high illuminance, non-uniform lighting distribution, and glare issues for internal spaces. Proper orientation of windows and skylights allows direct and diffused daylight to enter the building, producing the best combination of light entering the building while controlling glare. The selection and placement of windows and skylights should be determined by the amount of light needed and based on climate conditions (Majid et al. 2011). There is a need locally and globally to integrate daylighting as a more mainstream solution for satisfying customer lighting needs. This paper will study the various techniques used to promote daylighting by conducting a thorough literature review, reviewing one international case study, and applying the learned concepts to a locally case study building. The applied daylighting solution for the selected building will be simulated to study the improvement in lighting conditions and the impact on energy consumption.

2 Methodology

The research will be conducted in two stages, Stage 1 includes literature review to build current understating to daylighting principles and technologies used. Next, identify a daylighting technology to be studied under international and local context. Last, studying one international case study to make a case for local adoption. Stage 2 will cover the simulation of a local case study using the identified technology to observe performance improvement.

The identified daylight harvesting technology will be analysed using Dialux EVO to simulate the selected office building with the technology chosen for the study. Accordingly, the energy and lighting performance of the office building will be studied to draw comparisons between the conventional—artificially lit system and the proposed—daylighting system.

3 Literature Review

3.1 Daylighting General Concepts

The literature review was carried out to understand the basic concepts of daylights and various technologies and practices which support daylight harvesting. Daylighting in buildings refers to sunlight reflected into the building spaces through openings, influencing the quality and function of the Environment. As the building lighting loads reduce and operational efficiency increases, daylight harvesting encourages renewable energy sources and lowers energy consumption (Janda 2011). A good performance daylighting system initially requires a significant investment. However, if an integrated design approach is undertaken, the overall long-term savings make up for the initial costs spent on daylighting. In the same way, for example, when using Photo Voltaic solar cells resulting in savings in energy in a building in a certain payback period (Sharaf 2014). Various papers have explored the effects of daylighting on building performances. In one such report by Firas Sharaf (2014), he explores the impact of complimenting the existing artificial lighting setup by integrating daylighting. The author experimented with increasing the daylight factor with minimum change in the infrastructure of the present building by deploying better lighting control in areas with possible daylighting integration. Although simple solutions of integrating daylighting into the building are viable, they also pose challenges that make us explore indirect ways of daylight integration.

3.2 Introduction to Tubular Daylight Devices (Solar Tubes)

A solar tube or Sun tube is a tubular skylight that allows daylight received from the sun to be transported and delivered via a tube/pipe before being diffused into the area to be lighted. This indirect way of transmission of natural daylight prevents a considerable amount of thermal energy from being transferred during the transmission and distribution of sunlight, limiting the building's thermal heat gain. The design and operation principal are based on the basics of light propagation, using multiple reflections from highly reflective internal surfaces. These devices are also referred to as Tubular Daylight Devices. They are constructed of three components: a dome, a pipe, and a diffuser. The dome is generally a hemisphere made of clear plastic that captures sunlight and allows it into the pipe while keeping the rest of the environmental elements outside. Then there is the light transmission pipe, which is essentially a smooth cylinder with a highly reflective inside surface, to transmit the captured daylight into the building areas. Typically, the inner surface is either bare polished metal or a specially coated reflective sheet attached to the interior. The pipe travels through one or more transition zones, refracting light from the dome to the diffuser. Finally, the pipe ends in a diffuser which is a prismatic or frosted plastic sheet installed in the celling of the room. The diffuser distributes daylight evenly across the are being lit (Shuxiao et al. 2015) The aspect ratio, which is the ratio of the tube length to its diameter, is a crucial characteristic for the reflecting tube. Typically, the diffuser is linked to a light controller in order to adjust the intensity of the output light. The tubular daylight device can be used in conjunction with lighting controls that respond to the amount of sunlight available. In general, most energy saving building regulations advocate using intelligent daylight integrated lighting systems. (Balabel et al. 2021).

4 International Case Study

4.1 The Vine Offices

This case study is about an office building in the United States, shown in Fig. 1. For this project Four (4) SkyVault M74 with Amplifier units were bundled together with a bespoke diffuser to maximize daylight in this one-of-a-kind installation. As the LEDs had already been installed for both daylighting and night usage, the system was further modified with the addition of LEDs for lighting at night. The team intended to use daylight to infuse life into the hallways that are situated on the top floor of the building. A new problem presented itself because of the tangle of existing infrastructure, vents, and ductwork. The ultimate outcome was the installation of eight TDD 290DS devices. They selected the Spectra light Infinity tubing because of its reflectivity and high degree of proprietary use since they knew it would provide maximum light for lengthy tube runs with many 90° bends. With 290 Daylighting Systems installed, these internal corridors will have plenty of natural light. Commercial buildings almost always have runs of four to five feet in diameter. In this project, horizontal tube runs of 17 feet were required, and minimum light loss was attained. Instead of having to use more expensive temporary lighting during construction, TDD Systems installed daylighting in the interior areas, removing the need for it.

5 Analysis for Local Case Study

5.1 Site Selection

The selected building for simulation with the adoption of lighting tubes is the offices in Abu Dhabi University—Abu Dhabi Campus, UAE, as shown in Fig. 1. The University was opened in 2003 as the largest private University in the country and located on Al Ain highway road beside The British International School Abu Dhabi (Abu Dhabi University 2021).

The University consists of 2 floors for the classrooms and staff offices. In this paper, the focus will be on the staff offices on the 2nd floor. Figure 2 shows the

Fig. 1 Site photos of the vine offices. *Note* Solatube 2015

offices plan on the 2nd floor and shows the distribution of offices along with the circulation. Offices coloured with blue have access to natural daylight and outdoor view. The offices coloured with yellow are placed around the covered area with skylight and thus receive indirect daylight but with no exterior view. However, all the offices with white colours don't receive any natural daylight nor natural view to the outside. All corridors have artificial lights only with no access to any natural daylighting.

5.2 University Visit

A university visit was conducted, and the following pictures shows different spaces within the University. Reference to these pictures, all offices that have direct access to daylight have manual controlled curtains to control the amount of daylight entering the space and prevent glare issues. No shading elements were provided on the exterior façade of the University to control the building and adapt to the exterior Environment (Fig. 3).

Based on these pictures, fluorescent lighting is the primary type of light used in the University as artificial lighting. However, it has many disadvantages as it has sounds and buzz voice and flickering. Moreover, the artificial lights are controlled

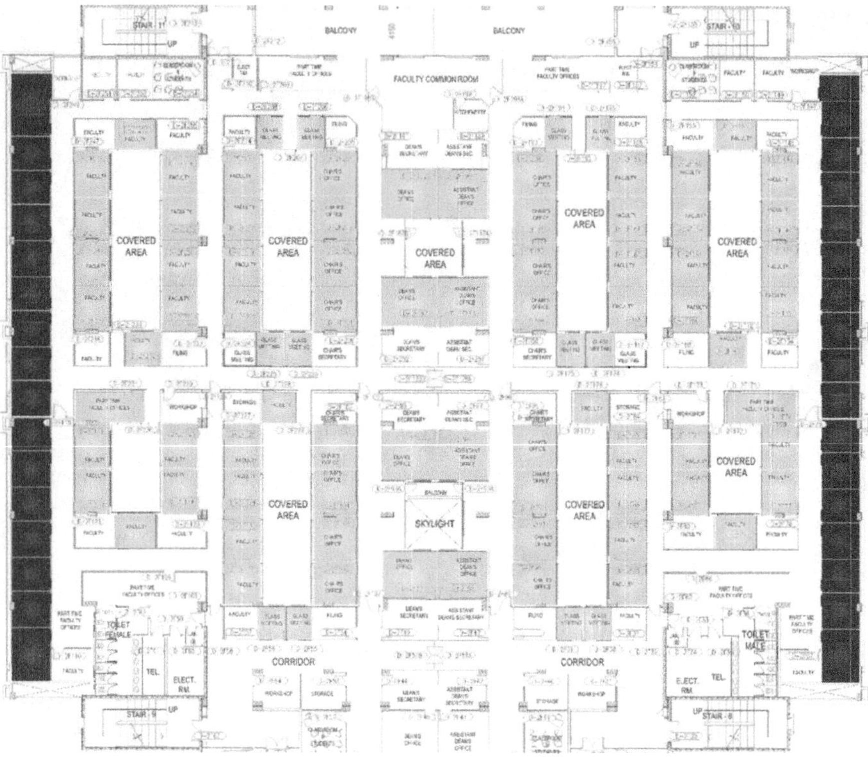

Fig. 2 Second-floor offices plan in Abu Dhabi University. *Note* The Author 2021

manually by the switches in each room as there are no occupancy sensors to control any lights.

6 Model Set-Up on DIALux

Part of second floor plan was modelled on DIALux, as shown in Fig. 4. In the construction section, the following sections were set and determined at the first stage to apply the simulations on the model: the site location of building, storey and building construction, apertures, façade elements, determine the spaces, room elements, roof, ceilings, furniture and objects, and materials. Next stage was setting the lights, in this section the luminaires were selected and distributed as per the actual case at site. For the base case scenario, 2 simulations were applied, the first is daylight simulation and the second is daylight with artificial light. Next, simulations were applied on 4 different strategies using the solar tubes (10 and 24 in.) to achieve the best results in Lux amount and energy efficiency, as the following:

1. Multi offices room with no windows, thus no natural daylighting.

2. Office room located around the covered area with skylight allowing for indirect daylight.

3. & 4. Office rooms with large windows allowing for daylighting and controlled by curtains.

5. Office room with no windows, thus no natural daylighting.

6. Student meeting room with no windows, thus no daylighting.

7. Corridor with no windows, thus no daylighting.

Fig. 3 Site visit pictures for different spaces at the University. 1. Multi offices room with no windows, thus no natural daylighting. 2. Office room located around the covered area with skylight allowing for indirect daylight. 3, 4. Office rooms with large windows allowing for daylighting and controlled by curtains. 5. Office room with no windows, thus no natural daylighting. 6. Student meeting room with no windows, thus no daylighting. 7. Corridor with no windows, thus no daylighting. *Note* The Author 2021

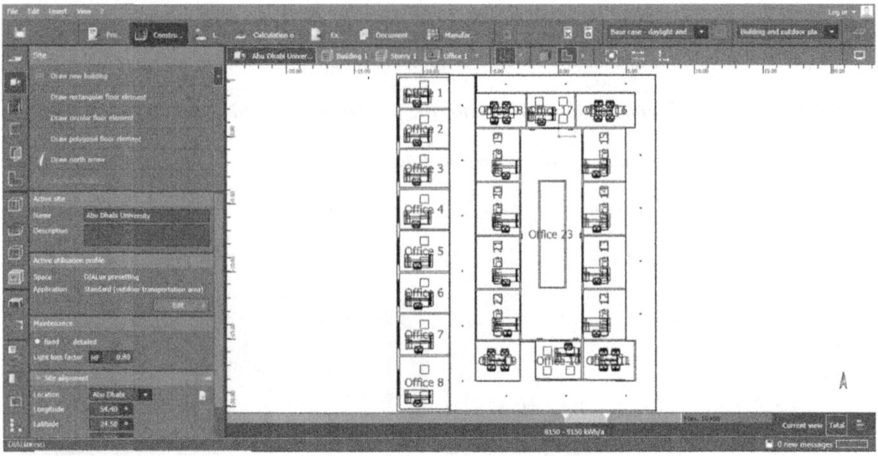

Fig. 4 Modelling the base case scenario on DIALux. *Note* The Author 2021

1. Solar tubes × 1 (10")
2. Solar tubes × 2 (24")
3. Solar tubes × 3(24")
4. Solar tubes × 4 (24")

7 Simulation and Analysis

7.1 Base Case Scenario: Daylight

In the base case scenario, all the windows and the type and quantity of all artificial lights were set in the model. Next, the simulation was applied while all the artificial lights switched off. The results show that all offices have direct access to natural daylight while there are 4 offices do not have any windows, and thus do not have access to daylight. After that, the simulation was done again with switching all artificial lights on, as shown in Figs. 5 and 6. To move forward with the simulation and concentrate the lighting strategy analysis, 2 offices with no windows were selected to apply the four simulations on. The offices are office 9 and 11. The base case scenario for office 9 and 11 with artificial lights show that the available lights did not meet the minimum requirements of 500 lx for work plane in each office as shown in Fig. 7. However, both offices don't have access to daylight. In the following stages and strategies, the simulation was done by adding solar tubes to enhance the amount of lux of each office and reduce the energy consumption.

The catalogues in DIALux did not have the option of installing solar tubes to the model, for that reason the skylights were used as an indication of solar tubes in the

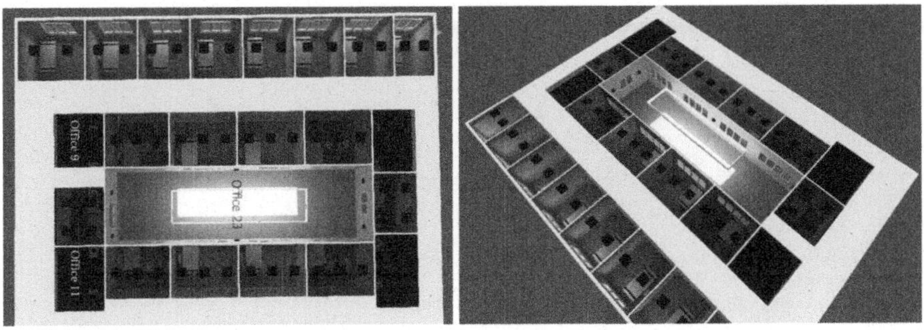

Fig. 5 Base case scenario with daylighting only, date: June 15th, 2021, at 12:00 pm. *Note* The Author 2021

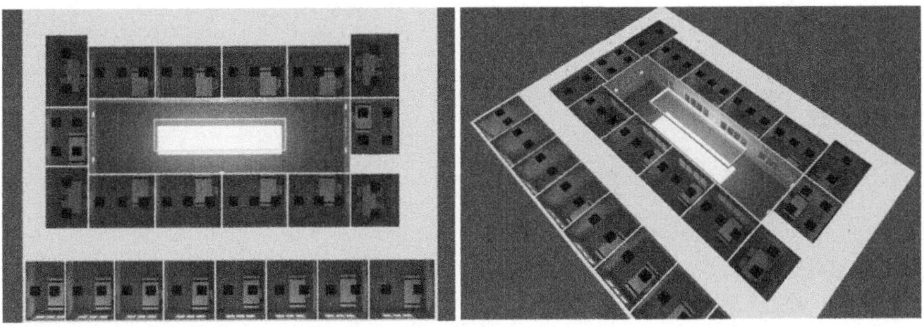

Fig. 6 Base case scenario with daylighting and artificial light, date: June 15th, 2021, at 12:00 pm. *Note* The Author 2021

project. Four different strategies were applied, and the simulations were done for each strategy on 3 times per day (9:00 am, 12:00 pm, and 3:00 pm), and in two days of the year (Jan 15th, and June 15th), as per LEED requirements.

7.2 Strategy 1: Solar Tubes × 1 (10")

In this strategy 1 solar tube with size 10 in. was added to each of the two offices in order to monitor the amount of natural light the solar tube can add to the space. The results show that providing 1 solar tube of 10 in. and switch off all artificial lights in the office is not enough to achieve the minimum requirements of 500 lx in the offices. Figure 8 shows the results of analysis for both offices in Jan at 12:00 pm, office 9 achieved 23 lx while office 11 achieved 19.9 lx which both are very low.

Building 1 · Storey 1 · Office 9

Summary

Results

	Symbol	Calculated	Target	Check	Index
Workplane	$\bar{E}_{perpendicular}$	447 lx	≥ 500 lx	×	S2
	g_1	0.47	-	-	S2
Visual task areas	$\bar{E}_{Task\ Area}$	589 lx	≥ 500 lx	✓	S6
	$g_{1\ Task\ Area}$	0.77	≥ 0.60	✓	S6
	$\bar{E}_{Surrounding\ area}$	386 lx	≥ 300 lx	✓	S7
	$g_{1\ Surrounding\ area}$	0.68	≥ 0.40	✓	S7
	$\bar{E}_{Background\ area}$	-	≥ 100 lx	×	S5
	$g_{1\ Background\ area}$	-	≥ 0.10	-	S5
Consumption values	Consumption	240 kWh/a	max. 350 kWh/a	✓	
Lighting power density	Room	13.36 W/m²	-	-	
		2.99 W/m²/100 lx	-	-	

Building 1 · Storey 1 · Office 11

Summary

Results

	Symbol	Calculated	Target	Check	Index
Workplane	$\bar{E}_{perpendicular}$	396 lx	≥ 500 lx	×	S4
	g_1	0.46	-	-	S4
Visual task areas	$\bar{E}_{Task\ Area}$	515 lx	≥ 500 lx	✓	S9
	$g_{1\ Task\ Area}$	0.76	≥ 0.60	✓	S9
	$\bar{E}_{Surrounding\ area}$	356 lx	≥ 300 lx	✓	S10
	$g_{1\ Surrounding\ area}$	0.71	≥ 0.40	✓	S10
	$\bar{E}_{Background\ area}$	-	≥ 100 lx	×	S8
	$g_{1\ Background\ area}$	-	≥ 0.10	-	S8
Consumption values	Consumption	240 kWh/a	max. 400 kWh/a	✓	
Lighting power density	Room	11.25 W/m²	-	-	
		2.84 W/m²/100 lx	-	-	

Fig. 7 Simulation result for base case artificial lights in offices 9 and 11. *Note* The Author 2021

Work planes

Properties	Ē (Target)	E_{min}	E_{max}	g_1	g_2	Index
Workplane (Office 9) Perpendicular illuminance (adaptive) Height: 0.800 m, Wall zone: 0.000 m	23.0 lx (≥ 500 lx) ×	4.46 lx	97.5 lx	0.19	0.046	S2
Workplane (Office 11) Perpendicular illuminance (adaptive) Height: 0.800 m, Wall zone: 0.000 m	19.9 lx (≥ 500 lx) ×	3.22 lx	97.2 lx	0.16	0.033	S4

Fig. 8 Simulation result for 1 solar tube of 10 in. for offices 9 and 11. *Note* The Author 2021

Work planes

Properties	Ē (Target)	E_{min}	E_{max}	g_1	g_2	Index
Workplane (Office 9) Perpendicular illuminance (adaptive) Height: 0.800 m, Wall zone: 0.000 m	468 lx (≥ 500 lx) ×	169 lx	828 lx	0.36	0.20	S2
Workplane (Office 11) Perpendicular illuminance (adaptive) Height: 0.800 m, Wall zone: 0.000 m	467 lx (≥ 500 lx) ×	178 lx	827 lx	0.38	0.22	S4

Fig. 9 Simulation result for 2 solar tubes of 24 in. for offices 9 and 11. *Note* The Author 2021

7.3 Strategy 2: Solar Tubes × 2 (24")

Reference to strategy 1, the size of solar tube was small based on the amount of light provided by it. Therefore, the size of solar tubes in strategy have increased to reach 24 in. The two offices were designed to receive 2 solar tubes with 24 in. size to enhance the amount of lux provided in the space. Figure 9 shows the simulation result in Jan at 12:00 pm where office 9 have achieved 468 lx, and office 11 achieved 467 lx, which is higher than the amount of lux in strategy 1 but still did not meet the requirements.

7.4 Strategy 3: Solar Tubes × 3(24")

For strategy 3, the size of solar tubes remains the same, but we added 1 more solar tube to reach 3 solar tubes in each office. Figure 10 shows the results of simulation done in Jan at 12:00 pm, office 9 achieved 654 lx, and office 11 achieved 642 lx. This strategy has achieved the amount of lux needed in the office space with zero energy consumption as no artificial lighting were used.

Work planes

Properties	Ē (Target)	E_{min}	E_{max}	g_1	g_2	Index
Workplane (Office 9) Perpendicular illuminance (adaptive) Height: 0.800 m, Wall zone: 0.000 m	654 lx (≥ 500 lx) ✓	282 lx	1104 lx	0.43	0.26	S2
Workplane (Office 11) Perpendicular illuminance (adaptive) Height: 0.800 m, Wall zone: 0.000 m	642 lx (≥ 500 lx) ✓	316 lx	987 lx	0.49	0.32	S4

Fig. 10 Simulation result for 3 solar tubes of 24 in. for offices 9 and 11. *Note* The Author 2021

Work planes

Properties	Ē (Target)	E_{min}	E_{max}	g_1	g_2	Index
Workplane (Office 9) Perpendicular illuminance (adaptive) Height: 0.800 m, Wall zone: 0.000 m	866 lx (≥ 500 lx) ✓	414 lx	1182 lx	0.48	0.35	S2
Workplane (Office 11) Perpendicular illuminance (adaptive) Height: 0.800 m, Wall zone: 0.000 m	867 lx (≥ 500 lx) ✓	402 lx	1201 lx	0.46	0.33	S4

Fig. 11 Simulation result for 4 solar tubes of 24 in. for offices 9 and 11. *Note* The Author 2021

7.5 Strategy 4: Solar Tubes × 4 (24")

Although strategy 3 has achieved the target of lux amount and zero energy consumption, this strategy was applied to measure the difference in lux between the two strategies. Four solar tubes with 24 in. size were added to each office to run the analysis. Figure 11 shows the results of simulation where office 9 received 866 lx and office 11 received 867 lx, which both are high and above the requirements with zero energy consumption.

8 Results and Findings

As the chosen offices were used during the day only, the focus of this analysis was to get benefit of natural daylight while minimizing the energy consumption. Strategy 1 showed that the 10 in. size of solar tube is very small to achieve the targeted lux in the spaces. However, in the following strategies the size of solar tubes was 24 in. which allowed for more natural daylight to enter the offices. Table 1 summarize the outcomes of the simulations that were done for 4 strategies, 3 times per day, and in 2 days of the year. Strategy 3 had the best simulation results among the other strategies. The offices in strategy 3 received 3 solar tubes with 24 in. each, and this have helped to enhance the amount of lux and achieve the requirements.

9 Conclusion

To conclude, Solar tubes as a lighting strategy is very important to consider while designing a project as they provide a clean natural daylight to the indoor spaces. Installing solar tubes to the buildings allow for daylight to access the building while minimizing the amount of thermal heat gain entering the space. In this paper, the focus was on applying solar tubes in office using DIALux software and switch off all artificial light to reduce the amount of energy consumption. However, the simulation results using solar tubes strategies have achieved the desired lux needed for indoor space with zero energy consumption as all artificial lights were switched off. The DIALux software was limited in terms of selecting type of glass used in skylight to control the light and heat transfer through it, but overall, it did the job and showed us the improvements in the lux when changing the solar tubes layout and sizes. The main disadvantages of solar tubes are that the received natural light is not as bright as the light coming from skylight. Moreover, the solar tubes can't be open for ventilation, not as normal windows that can be open and ventilate the space. In addition, the more storeys in the building the less effective natural light from solar tubes provided. This is because the increased depth of solar tubes will reduce the amount of light transferred through the pipe. All these points should be considered while designing any project with solar tube lighting strategies.

Table 1 Summary of the simulations done for the 4 strategies of solar tubes

		Office—9				Office—11			
		15th Jan		15th Jun		15th Jan		15th Jun	
		Task area illumination	Electrical power	Task area illumination	Electrical power	Task area illumination	Electrical power	Task area illumination	Electrical power
Base case	Artificial lighting (CFL)	447.00	60 W	–	–	396.00	60 W	–	–
Config 1	Solar Tubes × 1 (10″)								
	9:00 AM	23.00	0.00	23.00	0.00	19.90	0.00	19.90	0.00
	12:00 PM	23.00	0.00	23.00	0.00	19.90	0.00	19.90	0.00
	3:00 PM	16.30	0.00	25.50	0.00	44.00	0.00	19.30	0.00
Config 2	Solar Tubes × 2 (24″)								
	9:00 AM	468.00	0.00	468.00	0.00	467.00	0.00	467.00	0.00
	12:00 PM	468.00	0.00	468.00	0.00	467.00	0.00	467.00	0.00
	3:00 PM	468.00	0.00	468.00	0.00	467.00	0.00	467.00	0.00
Config 3	Solar Tubes × 3 (24″)								
	9:00 AM	654.00	0.00	654.00	0.00	642.00	0.00	642.00	0.00
	12:00 PM	654.00	0.00	654.00	0.00	642.00	0.00	642.00	0.00
	3:00 PM	654.00	0.00	654.00	0.00	642.00	0.00	642.00	0.00
Config 4	Solar Tubes × 4 (24″)								
	9:00 AM	866.00	0.00	866.00	0.00	867.00	0.00	867.00	0.00
	12:00 PM	866.00	0.00	866.00	0.00	867.00	0.00	867.00	0.00
	3:00 PM	866.00	0.00	866.00	0.00	867.00	0.00	867.00	0.00

Note The Author 2021

References

Balabel A, Alwetaishi M, Abdelhafiz A, Issa U, Sharaky IA, Shamseldin AK, Al Surf M, Al-Harthi M (2021) Potential of Solatube technology as passive daylight systems for sustainable buildings in Saudi Arabia. Alex Eng J. https://www.sciencedirect.com/science/article/pii/S1110016821003628

Janda KB (2011) Buildings don't use energy: people do. Archit Sci Rev 54(1):15–22

Majid S, Shemirani M, Gholam H, Memarian NS, Hashem Nejad H, Vaziri V (2011) Investigating the behaviors of the elementary school students in reference to factors associated with daylight. Asian Soc Sci 7(3)

Mohapatra BN, Kumar MR, Mandal SK (2020) Analysis of light tubes in interior daylighting system for building. Indonesian J Electr Eng Comput Sci 17(2):710

Sharaf F (2014) Daylighting: An alternative approach to lighting buildings sustainable energy in buildings view project. J Am Sci 10(4)

Shuxiao W, Jianping Z, Lixiong W (2015) Research on energy saving analysis of tubular daylight devices. Energy Procedia 78:1781–1786. https://reader.elsevier.com/reader/sd/pii/S1876610215020378?token=9E4FC088BAA80A26A61244B2F8116857BDF774CEBBEB02AC3C9AD312FD4B7DA59F12ED77029D0E309D89C65E098EB6BE&originRegion=eu-west-1&originCreation=20210702084205

Toledo GE, Pelegrini AV, Heemann A (2016) Design parameters for solar light pipes in the Brazilian context. Int J Sustain Eng 9(4):251–258

Yun GY, Hwang T, Kim JT (2010) Performance prediction by modelling of a light-pipe system used under the climate conditions of Korea. Indoor Built Environ 19(1):137–144

Zhang X, Muneer T, Kubie J (2002) A design guide for performance assessment of solar light-pipes. Light Res Technol 34(2):149–168

References

[illegible]

Investigating Indoor Air Quality in UAE Restaurants: A Case Study

Priji Biju, Bashaer Saumer, and Wael Sheta

Abstract Particulate matter, oxides of carbon and volatile organic compounds released during cooking can have a harmful impact on the health of employees and clienteles of restaurants. Hence it is crucial to monitor and maintain a healthy indoor air quality (IAQ) of restaurants and this study intends to assess the parameters of indoor air quality of a busy restaurant in Dubai with the intention to offer recommendations to enhance the indoor environment of the restaurant. Following a preliminary assessment and quantitative walk-through inspection of the kitchen and dining area, a survey was administered to the restaurant's staff and patrons in order to determine quality of interior environment. The responses received for the employees' survey emphasised the areas of concern as well as the variables that needed to be assessed and monitored in order maintain the indoor air quality at desired levels. Using the compatible instruments, parameters such as $PM_{2.5}$, PM_{10}, Temperature, Humidity, Interior Lighting, CO_2, CO, and Total VOCs were measured to evaluate the indoor air quality. While analysing the results it was observed that cooking activities in the kitchen are producing particulate matter to an undesired level and augmentation of ventilation is required to maintain a healthy working atmosphere for the employees. Customer response was satisfactory on the quality of indoor environment of the restaurant as per Likert 5-point analysis.

Keywords Indoor air quality · Particulate matter · VOC · Ventilation

This paper was prepared for Advanced Indoor Air Quality and Climate module, taught by Dr. Wael Sheta.

P. Biju (✉) · B. Saumer · W. Sheta
Faculty of Engineering and IT, The British University, Dubai, UAE
e-mail: 23001165@student.buid.ac.ae

K. Al Marri et al. (eds.), *BUiD Doctoral Research Conference 2024*, Lecture Notes in Civil Engineering 587, https://doi.org/10.1007/978-3-031-84371-6_22

1 Introduction

Climate and ecosystems are closely related to air quality, which has drawn more attention from scientific, political, and environmental institutions. Poor air quality, both indoors and outdoors, can cause a host of detrimental health problems, such as headaches, biliousness, skin irritation, kidney failure, and cancer. As humans spend 90% of their time indoors, built environment's occupants and guests should have access to a clean, healthy interior space (Cincinelli & Martellini 2017). Adverse effects on health due to interior air pollutants are higher than that of exterior air pollutants. Indoor air quality eco system consists of occupants and their activities, natural and mechanical ventilation systems, building envelopes and its environmental settings and time of exposure (Burroughs & Hansen 2020). According to U.S. Army research, the nation's air quality related health issues are estimated to cost 15 billion dollars per year in direct medical expenses and around 150 million missed workdays.

2 Factors Affecting Indoor Air Quality

Tighter and insulated interior spaces with central ventilation systems are being built as a result of design philosophy of minimal energy consumption. Such environments will significantly affect the indoor air quality due to any ventilation system malfunctions, occupant activities and built environment of space. Pollutants that are dispersed in the outdoor air also have an impact on the quality of the indoor environment (Marć et al. 2018). According to the US Environmental Protection Agency, sources (internal and external), HVAC design, pollutant pathways, and occupants are the major aspects that contribute to the development of IAQ issues. Malfunctioning equipment, building components or furnishings, housekeeping and maintenance activities, occupant activities, contaminated outdoor air, soil gases like radon, fuel leakages, pesticides and moisture are some of the sources. The HVAC system is the primary conduit or pathway for air flow in buildings and distribution of contaminants is influenced by the interaction of the air with parts of the building, furnishings and people inside. Owing to individual differences in sensitivity, one person may become adversely affected by an IAQ issue and different people may react differently to a single indoor air pollutant or issue (Epa & Environments Division 1991).

Climate change can also have an impact in the indoor environment due to greater heat transfer and more polluted air with augmented concentration of ozone, fine particles, nitrogen oxides and volatile organic compounds from outdoors. Increased indoor air temperature can aggravate the pollutant emission from building, chemical reactions and partitioning between gases and fine particles. Growing demand of heating and cooling due to climate change is demanding change in insulation materials, heat protection and reducing the use of open windows and fans. The extreme events happening in the nature like floods, hurricanes and wildfires are causing outdoor as well as indoor pollutants such as molds, fungi, carbon monoxide and

Table 1 Indoor air contaminants, sources and related health impacts

Contaminants	Source	Health impacts
Endotoxins	Contaminated humidifiers, stored food wastes, lower ventilation rates	Asthma, reduced lung function
Wetness and mold	Water system leak, food waste	Asthma, upper respiratory illness
CO_2, Nitrous Oxide particulates, various hydrocarbons	Combustion of solid fuels	Pulmonary cancer, pulmonary diseases
Carbon monoxide	Outdoor air, bio fuels, cigar smoke	Headache, nausea, fatigue
Nitrogen Dioxide (NO_2)	Combustion of fossil fuels	Respiratory illness
Pesticides	Contaminated soil, storage	Respiratory illness
Formaldehydes (-HCHO)	Internal finishes, cigarette smoke	eye, nose, throat irritation, asthma, bronchitis, and possible carcinogen
VOCs	Cigar smoke, drycleaned materials, room deodorizers, paints, carpets	Asthma, bronchial hyper-reactivity
Ultra fine particles	Frying and combustion activities	Heart and lung disease

Source Mannan & Al-Ghamdi 2021

suspended particles (Mansouri et al. 2022). The classification of various pollutants in the interior environment and their health effects are briefed in Table 1.

3 Prevailing Standards

When at least 20% of the occupants of a building exhibit symptoms of illness for a duration of two weeks or more, it is defined as "sick building syndrome" (Bas n.d. 2004). Various organizations, including WHO, US EPA, and ASHRAE have established guidelines and threshold values for various IAQ parameters, acknowledging IAQ as a multidisciplinary phenomenon (Mannan & Al-Ghamdi 2021). Table 2 provides some of the standards acknowledged by different organisations and those required to meet Dubai green building regulations.

Table 2 Prevailing standards for indoor air quality

Parameter	WHO	US EPA	ASHRAE	Dubai GBRS
Carbon dioxide	10 mg/m3	800 ppm	1000 ppm	800 ppm
Carbon monoxide	8.6 ppm	9 ppm	6 ppm	9 ppm
Formaldehyde	0.2 mg/m3	920 µgms/m3		< 0.08 ppm
Total VOC	200µgms/m3	< 200µgms/m3	100µgms/m3	< 300 µgms/m3
PM_{10}	150 µgms/m3	150 µgms/m3	150 µgms/m3	150 µgms/m3
$PM_{2.5}$	25 µgms/m3	35 µgms/m3	35 µgms/m3	
Relative humidity	30% - 60%	30% - 50%	30% - 60%	30% - 60%

Source Green Building Regulations & Specifications, n.d.; Mannan & Al-Ghamdi 2021;Shirin et al. 2013

4 Indoor Air Quality Maintenance Strategies

Ensuring and enhancing Indoor Air Quality necessitates a holistic approach that combines technology with various essential strategies. Even though there isn't a single, agreed-upon definition of what constitutes "good" indoor air quality (IAQ), ASHRAE and USEPA recommends strategies like managing contaminants at source, dilution by improving ventilation with clean air and use of air purifiers (AHRAE, USEPA et al. 2009).

5 Air Quality of Restaurants and Cafes

Cooking effluents are a major source of environmental pollution whether it is interior or exterior space. Over 20% of $PM_{2.5}$ in the atmosphere has been attributed to cooking oil fumes (COFs). Researches have indicated that COF exposure is linked to respiratory conditions like rhinitis, asthma, emphysema, aberrant lung function, and an increased risk of lung cancer death in non-smoking employees of hotels and restaurants. Burning solid biomass fuels releases a number of pollutants that are harmful to human health, including carbon monoxide (CO), particulate matter (PM), some hydrocarbons and volatile organic compounds (Sharma & Jain 2019). As professionals from the International Agency for Research on Cancer categorized compounds emitted during food preparation as a probable human carcinogen group 2A, designing, monitoring and managing of quality restaurant and kitchen environment is imperative. Kitchen ventilation need to be given more attention as it is the key to solving air quality problems in restaurant environments (Marć et al. 2018). Malfunctioning of the refrigeration system can increase moisture levels and enhance the growth of mold and mild dew and cleaning or disinfecting the dining areas with chemicals having high VOC content release pollutants. Hence, the IAQ hotspots in a restaurant are kitchen, refrigeration unit, the HVAC system and the dining area.

6 International Case Studies—Air Quality Management in Restaurant Environments

Enhancing restaurant air quality can have a positive economic impact by boosting customer confidence. Restaurant owners should prioritize making sure that patrons feel completely safe and at ease within their establishment's four walls by keeping emissions to a minimum. And in the area of restaurant air quality monitoring and improvement, this has opened the door for additional studies and research.

A study at a university cafeteria of Izmir Institute of Technology, Turkey on the impact of margarine-based cooking in the IAQ revealed that VOCs and aldehydes increased by 1.50 folds and remained for one and half hours after cooking and $PM_{2.5}$ increased by threefold and PM_{10} increased by 6.6-fold when margarine was used for cooking (Sofuoglu et al. 2015). Investigation of the cause of many non-smoking women working in Chinese barbeque restaurants ailing from lung cancer pointed out to the high concentrations of carbon monoxide from burning of charcoal in table-top barbeque stations (Zhang et al. 2017). Researchers at University of Florida examined the IAQ of open kitchen restaurants and observed that mean values of invisible pollutants like $PM_{2.5}$ and PM_{10} are 113.1 μg/m3 and 548.1 μg/m3 respectively during their study (Chang et al. 2021). 17 restaurants in Penang Island of Malaysia were studied by the researchers at Universiti Technologi, MARA to evaluate the impact of different cooking methods on IAQ and was observed that frying increases PM concentrations and boiling increase temperature, relative humidity, CO and CO2 levels (Nazli et al. 2023).

Above case studies displayed the impact of cooking effluents in the air quality parameters of restaurant environments and the importance of monitoring of particulate matter levels and other IAQ parameters to ensure healthy conditions within.

7 Case Study in Dubai, UAE

A modern Asian fine dining restaurant, located within a 5-star hotel in Downtown, Dubai. The restaurant having a seating capacity of more than 200 diners and is having an open kitchen as well as a closed one to prepare the pan -Asian menu (Figs. 1 and 2).

8 Methodology

A preliminary investigation and quantitative walk-through inspection were done at the restaurant, with permission from the management. Observations were made on the general layout, finishes, floor and walls for any dampness or molds, natural ventilation systems, HVAC layout, open kitchen activities, equipment and exhaust

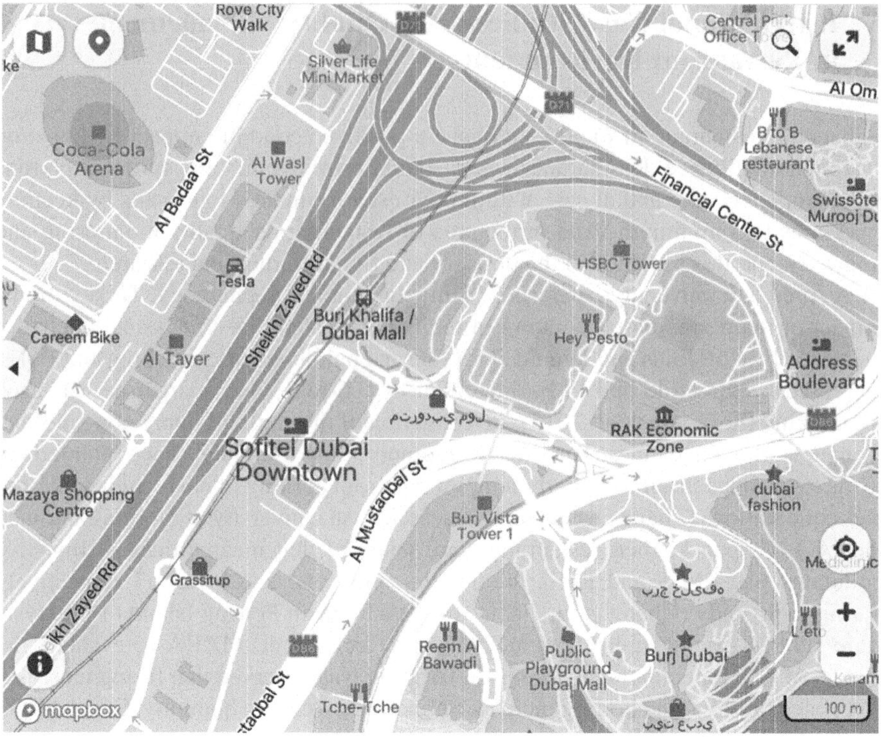

Fig. 1 Location and photograph of the restaurant under study. *Source* (Mapcarta 2024)

system of kitchen. This was followed by a diagnostic survey among the staff to understand whether long exposure in the indoor environment of the restaurant was causing any discomforts to them.

The results of the survey illustrated as shown in Fig. 4 indicated that primary concerns are insufficient light, health issues like headache, eye and nose irritation, nausea and fatigue. The survey results indicated the parameters to be examined and temperature, relative humidity, CO, CO_2, $PM_{2.5}$, PM_{10} and TVOC and aldehydes were measured using the instruments CO_2 meter CO240, FP-30 HCHO detector, HD450 light meter, Thermo-Anemometer 45,158 and VT 9IN 1 air quality detector.

9 Results and Discussion

The sampling locations chose for the survey is marked in Fig. 3 as X1 to X6 and observations were taken during off-peak hours (1 pm to 3 pm) on a working day and peak hours (8 pm to 10 pm) on a weekend. The results are summarized as shown below in Tables 3 and 4.

Fig. 2 Main dining area of the restaurant. *Source* (Authors 2024)

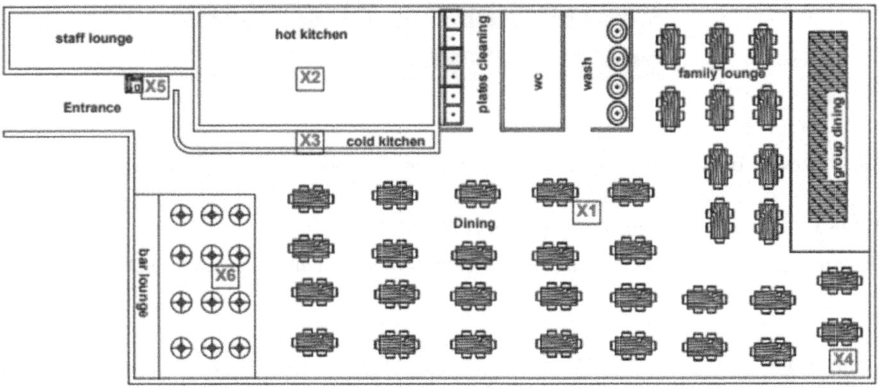

Fig. 3 Layout of the restaurant with sampling locations marked (X1 -X6 are sampling locations). *Source* (Authors 2024)

Air Quality Index (AQI) is calculated as per EPA standards, using Air Quality Index equation:

$$\text{Air Quality Index (AQI)},\ \text{Ip} = [\text{IHi}-\text{ILo}/\text{BPHi}-\text{BPLo}](\text{Cp}-\text{BPLo}) + \text{Ilo} \quad (1)$$

where:

Ip = AQI of pollutant p,

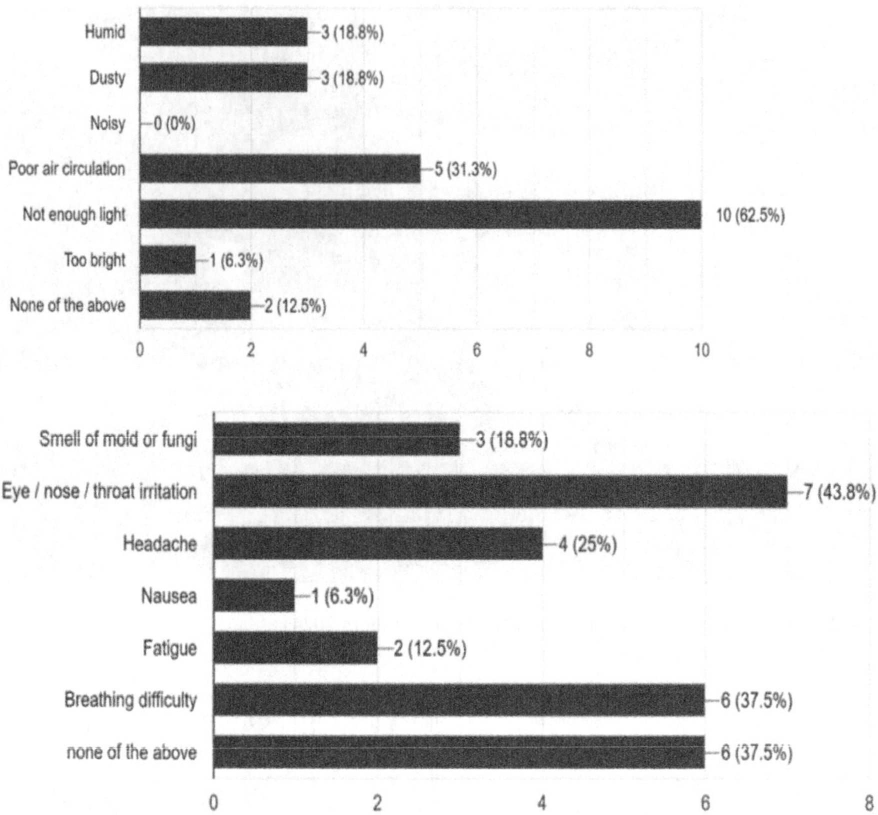

Fig. 4 Results of employee response survey. *Source* (Authors 2024)

Table 3 Observations during off peak hours (T1)

Parameters	Observed values at different locations in the restaurant during T1							Dubai GBC threshold
	Unit	X1	X2	X3	X4	X5	X6	
Temperature	C	24	26	24	24	24	24	≤ 25
Relative Humidity	%	60	65	60	60	60	60	30–60
Light	lux	92	450	165	260	163	260	≥ 220 -Rest ≥ 550-Kitchen
Formaldehyde	mg/m3	0.012	0.01	0.012	0.006	0.004	0.008	≤ 0.08
$PM_{2.5}$	µg/m3	30	47	28	29	22	22	≤ 35
PM_{10}	µg/m3	47	1 74	43	43	34	32	≤ 150
CO_2	ppm	428	408	407	407	418	409	≤ 800
CO	ppm	1	1	1	1	2	2	≤ 9
TVOC	mg/m3	0.022	0.016	0.022	0.012	0.012	0.012	≤ 0.3

Source (Authors 2024)

Table 4 Observations during peak hours (T2)

Parameters	Observed values at different locations in the restaurant during T2							Outer air
	Unit	X1	X2	X3	X4	X5	X6	
Temperature	C	24	26	24	24	24	24	25
Relative Humidity	%	50	65	50	50	50	50	48
Light	lux	92	450	158	128	160	159	310
Formaldehyde	mg/m3	0.009	0.01	0.009	0.009	0.007	0.008	0.005
$PM_{2.5}$	µg/m3	26	123	35.5	22	40	23	24
PM_{10}	µg/m3	36	187	47	38	65	49	35
CO_2	ppm	451	553	464	523	406	438	413
CO	ppm	2	2	2	2	2	2	3
TVOC	mg/m3	0.015	0.016	0.017	0.015	0.014	0.024	0.012

Source (Authors 2024)

Cp = observed concentration of pollutant p,
BPHi = concentration breakpoint i.e. greater than or equal to Cp,
BPLo = concentration breakpoint i.e. less than or equal to Cp,
IHi = AQI value corresponding to BPHi,
ILo = AQI value corresponding to BPLo

In this case study observed values of $PM_{2.5}$, PM_{10} and CO were considered for the AQI calculation (See Table 6). The calculated values of AQI using Eq. (1) and resulting colour categorization of sampling locations are given in Table 6. From the calculated values it can be seen that the restaurant is having moderate indoor air quality, cold kitchen and reception which is located close to cold kitchen are in the orange category and hot kitchen is in the unhealthy category as per indoor air measurements observed during this study. The main contributing parameter for this is the presence of higher concentrations of $PM_{2.5}$.

Results were then plotted to represent the levels of indoor air quality parameters against the threshold values. Figures 5, 6 and 7 show the trends of temperature, humidity, $PM_{2.5}$, PM_{10} Interior lighting, CO_2, CO, formaldehyde and TVOC trends in the restaurant.

From Figs. 5, 6 and 7, it can be observed that temperature and humidity are at near threshold limits, interior lighting is below minimum required and values of $PM_{2.5}$ and PM_{10} crossing the threshold values at hot kitchen, cold kitchen and reception areas. It can also be observed that the concentrations of particulate matter are increasing by more than 150% in the hot kitchen due to peak time activities and there is an increase in the concentration of $PM_{2.5}$ in the cold kitchen area and reception area during peak hour. Even though the concentration of other parameters also varies, values are within the threshold limits.

Along with the measurement of IAQ parameters, a customer survey was also conducted and responses were analyzed. Indoor Air Quality questionnaire was

Table 5 Break points for AQI

These breakpoints							.. is equal this AQI	.. and this category
O3(ppm) 8-h	O3(ppm) 1- hour	$PM_{2.5}$ (μgm/m3) 24-h	PM_{10}(μgm/m3) 24-h	CO(ppm) 8-h	SO_2 (ppb) 1-h	NO_2 (ppb) 1-h	AQI	
0.000–0.054		0.0–12.0	0–54	0.0–4.4	0–35	0–53	0–50	Good
0.055–0.070		12.1–35.4	55–154	4.5–9.4	36–75	54–100	51–100	Moderate
0.071–0.085	0.125–0.164	35.5–55.4	155–254	9.5–12.4	76–185	101–360	100–150	Unhealthy for sensitive people
0.086–0.105	0.165–0.204	55.5–150.4	255–354	12.5–15.4	186–304	361–649	151–200	Unhealthy
0.016–0.200	0.205–0.404	150.5–250.4	355–424	15.5–30.4	305–604	650–1249	201–300	Very unhealthy
	0.405–0.504	250.5–350.4	425–504	30.5–40.4	605–804	1250–1649	301–400	Hazardous
	0.505–0.604	350.5–500.4	505–604	40.5–50.4	805–1004	1650–2049	401–500	Hazardous

Source Aqi-Technical-Assistance-Document-Sept2018, n.d

Table 6 AQI values for the sampling locations inside the restaurant

Sampling station	Sampling location	AQI calculated		AQI category	
		T1	T2	T1	T2
X1	Restaurant area	89	80	Moderate	Moderate
X2	Hot kitchen	129	186	Unhealthy/ sensitive groups	Unhealthy
X3	Cold kitchen	84	101	Moderate	Unhealthy/ sensitive groups
X4	Under AC duct	87	72	Moderate	Moderate
X5	At reception desk	72	112	Moderate	Unhealthy/ sensitive groups
X6	At bar lounge	72	74	Moderate	Moderate

Note (Authors 2024)

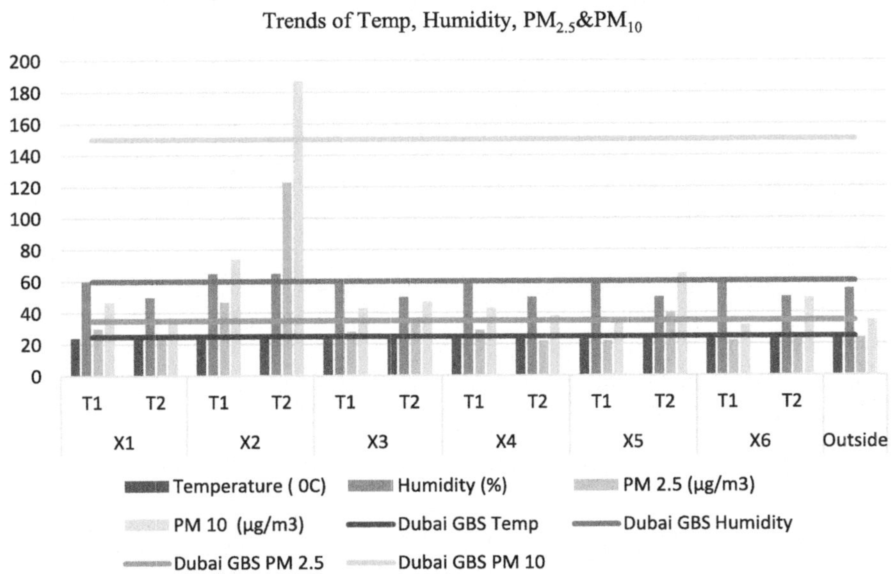

Fig. 5 Trends of indoor air quality parameters inside the restaurant. *Source* (Authors 2024)

prepared in a Likert scale and QR code for the same was distributed among the customers for feedback on the indoor environment of the restaurant. A 5-point Likert scale analysis was done to construe customer responses and it was observed that the customers were satisfactory about the indoor environment of the restaurant. The standard deviation varies from 0.77 to 1.52. This implies that the variation in customer opinion is moderate.

Even though employees have expressed some discomfort during the working hours, customer responses were mostly in the "agree" or "strongly agree" category

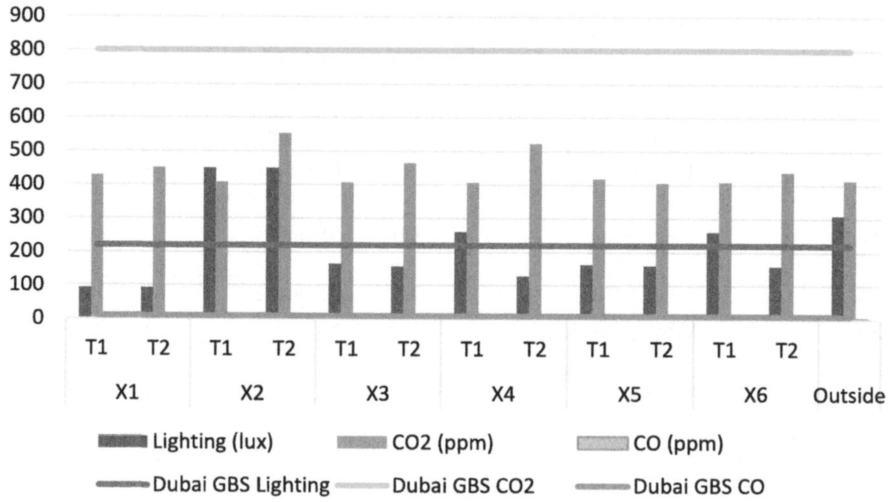

Fig. 6 Trends of interior lighting, CO_2 & CO inside the restaurant. *Source* (Authors 2024)

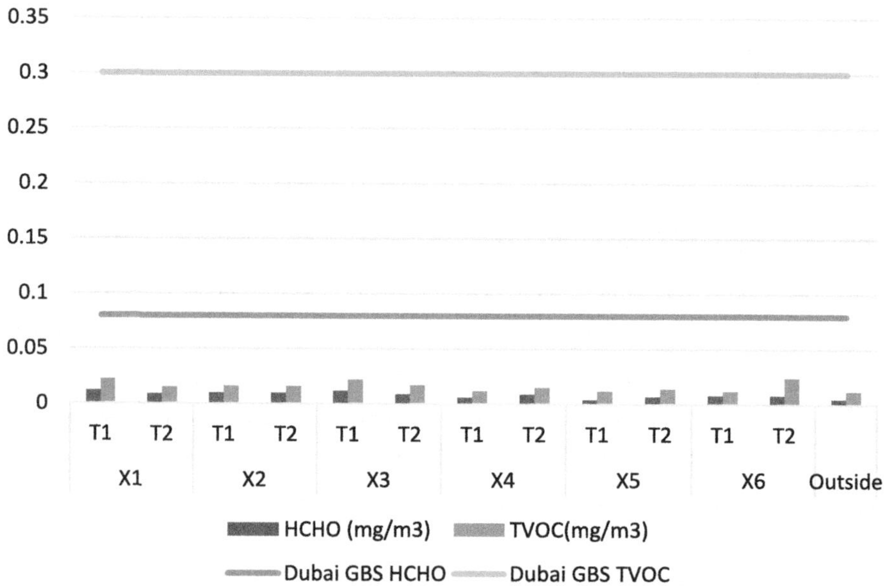

Fig. 7 Trends of HCHO &TVOC inside the restaurant. *Source* (Authors 2024)

regarding the comfort in the restaurant and around 90% of the customers has opined that they are satisfied about the indoor air quality of the restaurant. Shorter exposure of the customers in the restaurant environment compared to the employees might be the probable reason for the same. Observations have shown that the restaurant environment is in the moderate category and the area near cold kitchen is unhealthy for sensitive occupants. Particulate matter which is an invisible pollutant being the reason for these categorizations of indoor environment in this case study, mitigatory measures needs to be taken for the same. Insufficient ventilation can cause PM emissions from cooking to accumulate in the kitchen and spread into the dining room. High-capacity ventilation systems are therefore essential in commercial kitchens for the safety of kitchen staff as well as diner patrons. As temperature, humidity, $PM_{2.5}$ and PM_{10} were found to be more than desirable limits, dilution strategy with increased ventilation is required in the kitchen to maintain the IAQ parameters within desirable range.

10 Conclusion

The restaurant environment which was providing intangible dining experience for the customers with clean interiors and no visible discomforts is found to be having particulate matter concentrations above threshold limits. This has resulted in higher values of Air Quality Index and the restaurant area is mostly in the moderate air quality category. The open kitchen area and the locations close to it are found to be unhealthy for sensitive occupants due to peak-hour activities. Hence increased ventilation in the hot kitchen, air-purifiers in the restaurant area, increased interior lighting for a better dining experience for the customers and well-being of employees and indoor plants in the restaurant area are recommended as a mitigatory measure. Monitoring of effectiveness of the ventilation system and concentration of indoor air quality parameters on a regular basis is also recommended.

References

AHRAE, USEPA, The American Institute of Architects, US Green Building Council (2009) Indoor air quality guide best practices for design, construction, and commissioning

Alameri, A. (n.d.) DHCR HSE food safety policy & procedure and guideline department: HSE DHCR HSE food safety procedure

Aqi-Technical-Assistance-Document-Sept2018, n.d

Bas E (n.d.) Indoor air quality guide for facility managers

Burroughs HE, Hansen SJ (2020) Managing indoor air quality, fifth edition. Taylor & Francis

Chang H (Sean), Capuozzo B, Okumus B, Cho M (2021) Why cleaning the invisible in restaurants is important during COVID-19: a case study of indoor air quality of an open-kitchen restaurant. Int J Hosp Manag 94. https://doi.org/10.1016/j.ijhm.2020.102854

Cincinelli A, Martellini T (2017) Indoor air quality and health. Int J Environ Res Public Health 14(11). MDPI. https://doi.org/10.3390/ijerph14111286

Epa U, Environments Division I (1991). Indoor Air Facts No. 4 Sick Building Syndrome.

Green Building Regulations & Specifications. (n.d.)

https://www.who.int/news-room/fact-sheets/detail/household-air-pollution-and-health. (n.d.)

Indoor air quality guide best practices for design, construction, and commissioning. (n.d.). www.ashrae.org/publicationupdates

Mannan M, Al-Ghamdi SG (2021) Indoor air quality in buildings: A comprehensive review on the factors influencing air pollution in residential and commercial structure. Int J Environ Res Public Health 18(6):1–24. MDPI AG. https://doi.org/10.3390/ijerph18063276

Mansouri A, Wei W, Alessandrini JM, Mandin C, Blondeau P (2022) Impact of climate change on indoor air quality: a review. Int J Environ Res Public Health 19(23). MDPI. https://doi.org/10.3390/ijerph192315616

Mapcarta (2024) https://mapcarta.com/

Marć M, Śmiełowska M, Namieśnik J, Zabiegała B (2018) Indoor air quality of everyday use spaces dedicated to specific purposes—a review. In Environ Sci Pollut Res 25(3):2065–2082. Springer Verlag. https://doi.org/10.1007/s11356-017-0839-8

Nazli SN, Ul-Saufie AZ, Abd Razak A, Elbayoumi M (2023) Assessment of indoor air quality in small and medium food industries and effects towards perceived IAQ symptoms. Sustainability (Switzerland) 15(5). https://doi.org/10.3390/su15054065

Sharma D, Jain S (2019) Impact of intervention of biomass cookstove technologies and kitchen characteristics on indoor air quality and human exposure in rural settings of India. Environ Int 123:240–255. https://doi.org/10.1016/j.envint.2018.11.059

Shirin E, Bindu H, Reddy MV (2013) Indoor air quality in commercial kitchens. Int J Sci Res 5:2319–7064. https://doi.org/10.4103/0019

Sofuoglu SC, Toprak M, Inal F, Cimrin AH (2015) Indoor air quality in a restaurant kitchen using margarine for deep-frying. Environ Sci Pollut Res 22(20):15703–15711. https://doi.org/10.1007/s11356-015-4762-6

Zhang Z, Zhao Y, Zhou M, Tao P, Li R (2017) Measurement of indoor air quality in chinese charcoal barbecue restaurants. Procedia Eng 205:887–894. https://doi.org/10.1016/j.proeng.2017.10.088

Zhao Y, Liu L, Tao P, Zhang B, Huan C, Zhang X, Wang M (2019) Review of effluents and health effects of cooking and the performance of kitchen ventilation. Aerosol Air Qual Res 19(8):1937–1959). AAGR Aerosol and Air Quality Research. https://doi.org/10.4209/aaqr.2019.04.0198

Investigating the Causes of Food Waste in the Hospitality Sector in Dubai-UAE. Toward Implementing Food Waste Prevention Strategies

Reham Zeyad Sadeh and Bassam Abu Hijleh

Abstract One of the major environmental problems worldwide is food waste and loss. In Dubai, an Emirate in the United Arab Emirates, this problem is most significant in the hospitality sector including restaurants, hotels and food services, at the consumption level. The efforts to create campaigns and initiatives are notable, however, the theoretical part of finding the actual causes of this problem is still unobtrusive. This research focuses to fill this gap and find the main causes of food waste (FW) especially in the hospitality sector of Dubai and suggests prevention strategies for food services to follow. It used a mixed-mode methodology by spreading two surveys for food services guests and employees, then interviewing expert-positioned people in the emirate to verify the data. It found that customers' main causes of FW were taking big portions and lack of awareness. The causes of FW from food services were mainly their over-production and wrong forecasting for numbers. Interviewees suggested strategies to prevent FW. Thus, the research combined all results to come up with a FW reduction fundamental plan for food services to follow.

Keywords Food waste in Dubai · Food waste prevention · Causes of food waste

Food trends, food exhibitions, new technologies…etc. can affect the amount of FW negatively or positively depending on our use. To benefit from it and lower FW, people must know what is causing this waste and prevent it accordingly. United Arab Emirates (UAE) is one of the countries with a high number of food festivals, and food services and is considered a tourist hub, making it very prone to FW; which is food thrown after consumption, usually in homes and hospitality sectors mainly in developed countries; to show a sophisticated economic position, (Filimonau & De Coteau 2019).

This paper was prepared for the author's dissertation in the M.Sc. Sustainable Design of the Built Environment, supervised by Prof. Bassam Abu Hijleh.

R. Z. Sadeh (✉) · B. A. Hijleh
Faculty of Engineering and IT, The British University in Dubai, Dubai, United Arab Emirates
e-mail: 20001523@student.buid.ac.ae

K. Al Marri et al. (eds.), *BUiD Doctoral Research Conference 2024*, Lecture Notes in Civil Engineering 587, https://doi.org/10.1007/978-3-031-84371-6_23

The UAE is following a set of national and international goals to fight FW such as "zero hunger" by 2051 and halving FW per capita by 2030. This research focuses on theoretically finding the causes of FW in the food services of Dubai, one of UAE's seven emirates, and suggesting a fundamental plan for hospitality sectors to follow to prevent larger amounts of FW as the theoretical part is lacking.

1 Research Motivation (Significance of the Study)

The UAE FW problem is affecting all three sustainability pillars: environmental, economic, and social. Water In UAE is lost by around 790 m3/capita/year, the highest in the Near East and North Africa (NENA) countries, (Berjan et al. 2018), much of it is on farms. Monetary losses of FW cost around 282 million dirhams (about 77 million US dollars) in Dubai alone yearly, and around 6 billion dirhams yearly in the UAE (about 1.63 billion US dollars), (Sewatkar & Al Khayyat 2022).

2 Theoretical Underpinning (Literature Review)

Despite the high per capita number for FW in UAE, 224 kg per year (Kohli 2021), and the high percentage of food loss and waste in the Arabic region, around 44%. The literature available in UAE about FW is lacking and the available papers are old. The paper by Pirani & Arafat (2016) concluded some causes for FW in the emirate of Abu Dhabi such as forecasting, the food varieties and the serving style and times of serving.

On the other hand, UAE has many initiatives and companies that are concerned about FW in hospitality or other sectors. The main ones are the UAE Food Bank, the Ministry of Climate Change and Environment (MOCCAE) and the Emirates Foundation which created the "Ne'maa initiative" in 2022 or "Blessing Pledge" that aims to encourage reducing FW by all the UAE communities, (Sewatkar & Al Khayyat 2022), and Dubai Municipality which is responsible for food safety programs. Besides, the (UAE Food Waste Pledge) for big hotels to lead by example such as Armani Hotel Dubai, (winnow n.d.).

3 Materials and Methodology

3.1 Surveys

This research used a mixed-mode methodology to collect the data. It used surveys to target two groups of people in the hospitality industry, the guests, and the employees. This method is easy to construct, distribute and analyze. Followed by interviewing the responsible people to verify the results. The two surveys were divided into four groups of questions: demographic, awareness, causes of FW and prevention strategies questions. The guests' survey was completed by 361 people and the employees' survey by 91 employees. The study concluded special variables that may affect FW in the hospitality sector were asked about. Such as the season, weekdays, level of service, type of service…etc.

3.2 Interviews

The interviews' main aim was to verify the data collected from the surveyed participants. Thus, experts were interviewed, working in the governmental, private (business), and nonprofit or social sectors of Dubai, Table 1.

Table 1 The personal information of the interviewed people

	Date of interview	Name of interviewee	Position of interviewee	Name of institute
1	30th Sept 2022	Mr. D	Procurement manager of Food &Beverage	Major hotels' company in Dubai and the Middle East
2	7th Oct 2022	Ms. Tatiana Abella	Founder and managing director of Goumbook	Goumbook
3	10th Oct 2022	Dr. Sayed Essam Al-Hashmi	Head of the Food Trade Control Section at the Food Safety Department	Dubai Municipality (DM)
4	14th Oct 2022	Mr. Kym Barter	Vice president of F&B at Atlantis	Atlantis Resorts
5	19th Oct 2022	Ms. Amal Al-Ahmadi	Head of R&D in the ministry, Food security portfolio, and member of the general for the Emirates Food Security Council	Ministry of Climate Change and Environment
6	03rd Nov 2022	Ms. Maitha AlKaabi	Food safety and hygiene manager	Dubai World Trade Center (DWTC)
7	03rd Nov 2022	Ms. Habiba AlMara'shi	Co-founder and Chairperson of EEG	Emirates Environmental Group (EEG)

The interviews were semi-structured with open-ended questions ranging between nine to thirteen depending on their role.

4 Results and Discussion

In this chapter, the guests' survey (N = 361) and the employees' survey (N = 91) data were transferred to Microsoft Excel and SPSS to be analyzed and represented as tables.

4.1 Hospitality Guests' Survey

4.1.1 Demographics

Most of the hospitality sector guests surveyed were aged between 20 years (yrs.) to 60 years. The salaries ranged from 8–33% for each category. The contribution of the males was double the females, and the expats were the most participating group. The most visited type was (a la carte) restaurants, 68.7%, with mid-level as the top-visited level.

4.2 Hospitality Employees' Survey

4.2.1 Demographics

The majority (72.5%) of this survey's participants were aged from 20 to 39yrs, with similar experience levels. The sample mainly consisted of employees working in "A la carte" services (50.5%), between high and mid-level restaurants (45.1% and 44%), respectively.

4.3 Data Analysis and Discussion

4.3.1 FW Awareness Comparison

The current level of FW awareness in participants was investigated. Three groups of questions were asked. They resulted in high awareness in the first and third groups, and low awareness in the second. However, according to Jones (2017), saving money affects people's consciousness more than the environmental impact. The overall employees' awareness has higher levels than guests, Fig. 1.

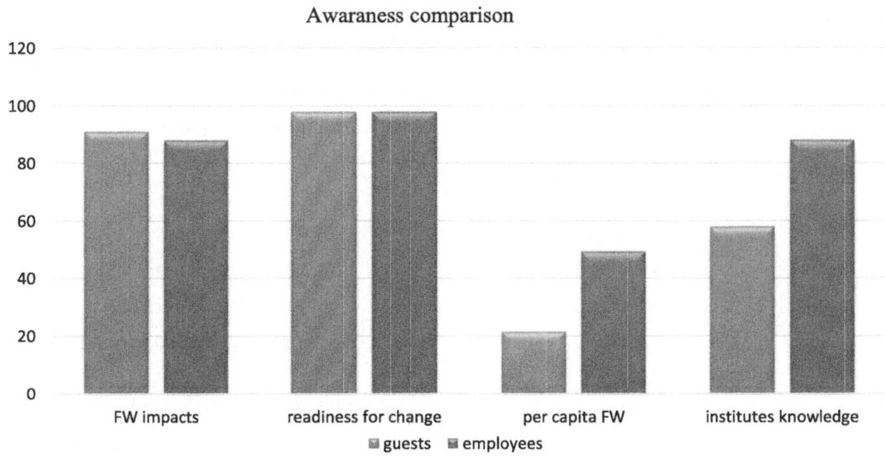

Fig. 1 Awareness percentages comparison between guests and employees

4.3.2 Causes of Food Waste Comparison

Causes of FW can be categorised into causes from customers' behaviours, food services, and contributing factors. The FW factors studied in this paper were the months, days of the week, gender, and type of food service.

Winter months in Dubai, "December, January, and February", were agreed on as the most months contributing to FW. While an increase of FW reached around 60% during the Holy month of Ramadan that year, (dcce 2019). While weekends were agreed on as a significant second contributor to FW.

The Buffet type was chosen as the most contributing type to FW, caused by people's lack of awareness, taking more quantities than needed, destroying the dishes…etc. besides its policies.

Table 2 highlighted colours, such as yellow, showing the causes of FW from food services from the points of view of guests and employees at similar agreement levels and green and blue colours for similar causes but at different agreement levels.

The guests and the employees agreed that many different causes of FW from customers have similar levels of importance, Table 3. Generosity in over-ordering was confirmed by Baig et al. (2019) in KSA to be most contributing to FW, above 70% in households.

The top agreed-on cause of FW for both studies by Pirani & Arafat (2016) and this study was the faults in forecasting and planning, Table 2. They disagreed that the food varieties have a significant effect on it. This disagreement might be attributed to the different perceptions of the participants and the time of publishing the paper.

Table 2 Ranking the causes of FW from food services from the point of view of guests and employees

Causes/Services	Guests	Employees
1	Poor planning or forecasting	Poor planning or forecasting
2	Food presentation and preparations	Over-production
3	Storage and expiration dates	Customer plate waste
4	Over-production	Depends on mealtime
5	Menu's many choices	Storage and expiration dates
6	Unqualified cooking skills/staff errors	Cooking skills/staff errors
7	–	Food presentation and preparations in a special
8	–	Low meal prices & many choices

Table 3 Ranking of the causes of FW from customers by the opinions of guests and employees

Causes/Customers	Guest	Employees
1	Taking big portions	Offering big portions/big plate size
2	Generosity and wealth in over-ordering	More varieties of food and menu choices—I order more
3	Lack of awareness	Lack of awareness
4	I do not like the taste/unexpected quality	Generosity and wealth in over-ordering
5	Tourist behaviors—try as many available options as possible	Tourist behaviors—try most options as possible
6	More varieties of food and menu choices—I order more	They do not like the taste
7	Un-wanted food sides like decoration items of vegetables and sauces/food type	Low prices of food
8	Not declare about allergies/religious restrictions	Not declare about allergies/religious restrictions
9	Low prices of food	–
10	I do not see it as a big problem	–

4.3.3 FW Prevention Strategies Comparison

To prevent kitchen waste; chefs must be aware of the differences between "Best before" and "used by". Mabaso & Hewson (2018) insisted on the importance of using a FIFO (first-in-first-out) system in storing food items, which is used in 95.6% of the food services surveyed in this paper. Moreover, over 93% of the sample confirmed having periodical monitoring and maintenance for the storage systems.

In comparison, both groups agreed on three out of five strategies but not in the same ranking, Table 4.

Table 4 The ranking of the efficient FW prevention strategies from the guests' and employees' perception

Strategy	Guests	Employees
1	Encouraging doggy bags/takeaway bags	Workshops for staff to improve skills
2	Better storage and handling	Better storage for food items
3	Encouraging messages/menu design	Study customer behaviors/portion sizes
4	Study customer behaviors/portion sizes	Exact forecasting and planning
5	Better forecasting and planning	Technology, ex. Smart scales like Winnow
6	Fruit to root/use all parts of food	Incentives for staff and guests to reduce food waste

Most of the employees' sample refused to let the customers pay extra charges for wasting, at 74.7%. However, weighted buffet types are used as a solution for buffet restaurants FW, (Eckert Matzembacher et al. 2020) as people are more cautious about wasting when paying money.

5 Interviews Data Collection and Analysis

The data collected previously needed to be verified by professionals, (Table 1), with different roles in Dubai's society to confirm the main results and come up with a plan for suitable prevention strategies to be implemented.

All interviewees agreed that the FW problem is prominent in Dubai, especially in the hospitality sector relating it to Dubai as being a tourist hub confirmed in Sect. 1 and needs to be measured and managed.

They agreed that the top leading cause of FW from customers was the lack of awareness shown in their greed to take bigger portions. Followed by the cultural habits of being generous. Moreover, Dr Essam added hygiene problems in some labour areas. While the hotel services' managers added cancelling orders as a main cause.

They agreed on overproduction and generosity as the top leading causes of FW from food services. Followed by wrong planning for guests as there are no strict reservation policies. Four interviewees agreed that the awareness of chefs was a significant problem.

All interviewees disagreed on eliminating buffet-type restaurants which cause high amounts of FW while applying reduction strategies is enough. This type gives a level of luxury that guests are waiting for. Besides, having buffets especially at breakfast time and during Ramadan Month is more convenient.

Dubai Municipality started the law to fine food services on waste per ton in 2022. Mr. Kym confirmed being affected by this law as it pushed his resort to use new technologies in their restaurants, such as Winnow Solutions, where they noticed a

47% reduction in FW in four months. This is studied by Filimonau et al. (2021) and called the 'pay-as-you-throw' method. However, Mr. Essam suggested applying the law on small services as well, as their number in the Emirate is very high.

The interviewees agreed that the awareness in the hospitality sector is developing as many of them have their sustainable practices mainly coming from their company's origins. Besides, the aim enhances their financial status. However, MOCCAE, Goumbook and EEG do awareness campaigns and workshops among individuals and companies. Mr. Kym and Ms. Maitha confirmed the importance of changing people's eating habits to a healthy lifestyle to lower FW as portions are calculated and as Coudard et al. (2021) explained, may save up to 28% of vegetables in the hospitality sector.

Governmental institutes confirmed using their platforms for the same purpose such as the "Ne'ma initiative". All interviewees confirmed that a higher level of awareness is found in the youth and that the work on their awareness is intense; in school campaigns, workshops and curricula as they are the future leaders. Moreover, a paper in the UAE done on university students by Yagoub et al. (2022) confirmed that 57% of their respondents agreed that young people waste more food, showing the importance of focusing on their awareness.

All the interviewees confirmed the importance of technology use to reduce FW in food services, such as the Winnow smart scale system, proven to lower FW in major hotels in Dubai. It measures the quantity of wasted food and categorises it by type using a camera to avoid cooking it in the future. Followed by reducing portion/plate sizes in the services and promoting individual sizes in catering services. This strategy was proven to reduce about 20% of FW, (Al-Obadi et al. 2022).

Most of the interviewees agreed on the strategy of sourcing locally to extend the shelf life of food items as it adds to the wasted days in importing, transporting, and getting certifications. Ms. Amal and Dr. Essam confirmed working on reducing strict safety policies on imported foods within safe limits. Ms Habiba added that in their 4th-panel discussion of 2022, they aimed primarily to address the effective use of packaging to eliminate FW and support a more circular economy.

The three food-related services managers encouraged using menu engineering. This is done using controlled portions and favoured choices of people and upcycling items like sauces to be used for different meal options on the menu. The latter was confirmed by Okumus (2019) and Martin-Rios et al. (2018).

Three interviewees agreed that adding encouraging messages on menus, screens, and between buffet items is an efficient prevention strategy that is being used by Goumbook in Dubai. In addition, cooking more live sections as in Hilton Hotels, not introducing cut fruits and using a governmental ranking (scoring) system help in FW prevention.

All interviewees confirmed the importance of the involvement of the government. Their efforts include raising awareness, implementing policies, and suggesting prevention strategies. Half of the interviewees confirmed that laws and legislation are fundamental. In contrast, others expressed that it must start from inside the person with the governmental legislation as support. While forcing a plan for all food services to follow to reduce their FW was agreed on by all to be a bit challenging.

6 Conclusion

6.1 Research Proposal

In alignment with the world trend of "Target, Measure, and Act" this basic plan is suggested to be used by food services with tailoring to what suits their service, Fig. 2.

7 Conclusion

The FW impacts affecting the environmental, social, and economic pillars are causing damage worldwide. It is in urgent needs to implement FW prevention and recovery strategies. UAE is one of the highly affected countries by this problem, as it has one of the highest per capita numbers of FW per year and ranks second in wasting fresh water globally from FW. Efforts are being made to control this problem but theoretical studies on root causes are needed.

This research investigated the causes of FW in the hospitality sector in the Emirate of Dubai, aiming to suggest prevention strategies to reduce it, as source reduction is the top problem solution. It used a mixed-mode methodology for data collection. It used surveys for the guests attending the food services and the service employees. The results showed that the main causes of FW from customers were taking big portions and lack of awareness. While causes of FW from food services were mainly their over-production and wrong planning and forecasting for numbers. While winter season, weekend days, and buffet type were the most FW-increasing factors.

Interviewing seven of the highly positioned experts in Dubai was an important step to verify the results of the survey and suggest a prevention plan. Interviewees agreed on many FW prevention strategies such as using smart scale technology, reducing portion/plate sizes…etc. The study used these strategies along with the surveys' verified ones to come up with a FW reduction base plan for food services to follow. It briefly includes quantifying the FW, offering workshops and spreading encouraging messages, this was tested by Pinto et al. (2018) to reduce about 15% in the consumption index. In alignment with department communications. Using menu engineering was verified by interviewees as important. It included strategies like using vegetarian options which may reduce up to 28% of vegetables in the hospitality sector, (Coudard et al. 2021), and reducing plate size results in a 20% reduction of FW, (Al-Obadi et al. 2022). 14% of food is lost between harvesting to retail and sourcing locally will eliminate this problem (Globalagriculture 2019). Finally, offering incentives for services to reduce FW such as financial discounts and higher governmental scoring. Chalak et al. (2018) found in their study on 33 countries that using a verified legislative framework, awareness campaigns and financial incentives will considerably lower FW by about 17.6%, 21.3% and 14.3%, respectively. Generally, it was studied by Champions 12.3 that hotels are estimated to save $7 for every $1 paid for reducing FW investment (Lim 2019).

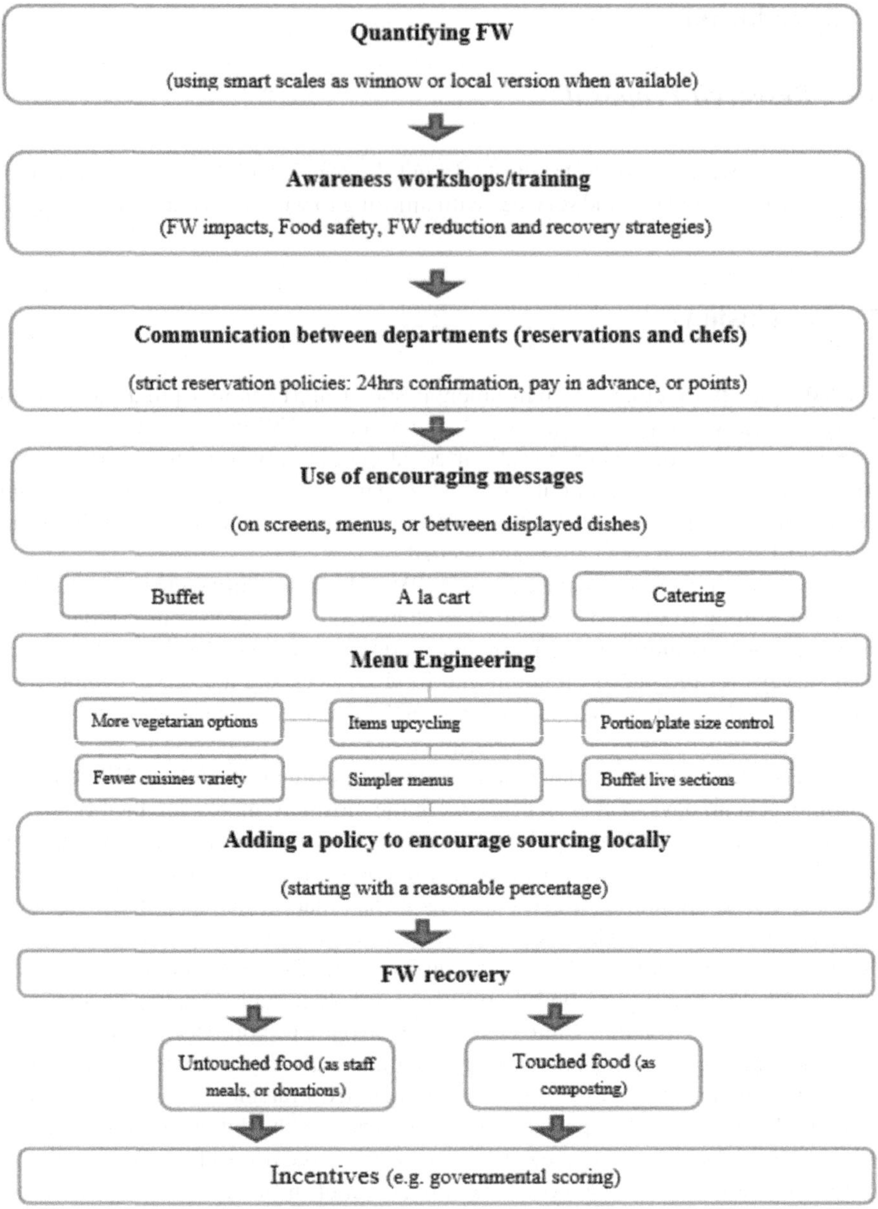

Fig. 2 Food services fundamental plan for FW prevention

Recommendations focus mostly on raising awareness by offering public and specific workshops for hospitality sectors, using a strict reservation system and emphasizing mentality change as not to treat the "guest as a king", but rather convince them to be "environmentally responsible", as also mentioned in Mabaso & Hewson (2018). Starting Shopping Festivals and Ramadan Holy month with awareness campaigns through social media, videos on screens, flyers…etc. Besides, extensive efforts to introduce FW education in universities and at all school grade levels.

Recommendations for governmental organizations to lower food safety's strict policies and increase shelf-life expiry dates in safe ranges. Besides, suggesting a governmental mobile/tablet application for food services to self-report their waste and send daily data while offering them incentives to suggest real solutions. Moreover, incentivizing small food services and branched ones to donate their extra food such as offering transportation with nominal fees, outdoor fridges…etc.

References

Al-Obadi M, Ayad H, Pokharel S, Ayari MA (2022) Perspectives on food waste management: Prevention and social innovations. Sustain Prod Consumption Elsevier 31:190–208

Baig MB, Al-Zahrani KH, Schneider F, Straquadine GS, Mourad M (2019) Food waste posing a serious threat to sustainability in the Kingdom of Saudi Arabia—a systematic review. Saudi J Biol Sciences Elsevier 26(7):1743–1752

Berjan S, Capone R, Debs P, Bilali HEl (2018) Food Losses and Waste: A Global Overview with a Focus on Near East and North Africa Region. International Journal of Agricultural Management and Development [online]. [Accessed 22 June 2021]. Available at: www.ijamad.iaurasht.ac.ir

Chalak A, Abou-Daher C, Abiad MG (2018) Generation of food waste in the hospitality and food retail and wholesale sectors: lessons from developed economies. Food Security. Springer Netherlands 10(5):1279–1290

Coudard A, Corbin E, de Koning J, Tukker A, Mogollón JM (2021) Global water and energy losses from consumer avoidable food waste. J Clean Prod. Elsevier Ltd 326

Eckert Matzembacher D, Brancoli P, Moltene Maia L, Eriksson M (2020) Consumer's food waste in different restaurants configuration: a comparison between different levels of incentive and interaction. Waste Manag. Elsevier Ltd 114:263–273

Filimonau V, De Coteau DA (2019) Food waste management in hospitality operations: a critical review. Tour Manag. Elsevier Ltd 71:234–245

Filimonau V, Nghiem VN, Wang L en (2021) Food waste management in ethnic food restaurants. Int J Hosp Manag. Pergamon 92:102731

Globalagriculture (2019) FAO: 14% of the world's food is lost between harvest and retail [online]. [Accessed 11 November 2022]. Available at: https://www.globalagriculture.org/whats-new/news/en/33821.html

Kohli SA (2021) Ramadan 2021: UAE aims to cut food waste by half by 2030—News. Khaleej Times [online]. [Accessed 9 May 2022]. Available at: https://www.khaleejtimes.com/ramadan/ramadan-2021-uae-aims-to-cut-food-waste-by-half-by-2030

Lim A (2019) 5 ways to reduce commercial buffet food waste. winnow solutions [online]. [Accessed 11 November 2022]. Available at: https://blog.winnowsolutions.com/reduce-buffet-food-waste

Mabaso CH, Hewson D (2018) (PDF) Employees' perceptions of food waste management in hotels [online]. [Accessed 14 September 2022]. Available at: https://www.researchgate.net/publication/336916093_Employees%27_perceptions_of_food_waste_management_in_hotels

Martin-Rios C, Demen-Meier C, Gössling S, Cornuz C (2018) Food waste management innovations in the foodservice industry. Waste Manag. Elsevier Ltd 79:196–206

Okumus B (2019) How do hotels manage food waste? evidence from hotels in Orlando, Florida. J Hosp Mark Manag. Routledge

Our food is damaging the environment | dcce. (n.d.) [online]. [Accessed 21 June 2021]. Available at: https://dcce.ae/press_releases/our-food-is-damaging-the-environment/

Pinto RS, Pinto RM dos S, Melo FFS, Campos SS, Cordovil CM dos S (2018) A simple awareness campaign to promote food waste reduction in a University canteen. Waste Manag. Pergamon 76:28–38

Pirani SI, Arafat HA (2016) Reduction of food waste generation in the hospitality industry. J Clean Prod. Elsevier Ltd 132:129–145

Sewatkar S, Al Khayyat H (2022) Launching a "Blessing Pledge" to spur stopping food waste—SK TODAY'S NEWS June 9, 2022. SK Todays news [online]. [Accessed 14 August 2022]. Available at: https://sktodaysnews.com/09/06/2022/world/launching-a-blessing-pledge-to-spur-stopping-food-waste/

winnow (n.d.) How hotels in the middle east are leading the fight against food waste three hotels in UAE show you how to drive revenue and reduce costs

Yagoub MM, Al Hosani N, Alsumaiti T, Kortbi O, Alshehhi AA, Aldhanhani SR, Albedwawi SA (2022) University students' perceptions of food waste in the UAE

Leveraging Network Traffic Byte-Streams for Machine Learning Based Early Botnet Attack Detection

Rajesh Thomas, Suleiman Yerima, and Khaled Shaalan

Abstract Botnet attacks can overwhelm networks and severely affect the availability of services. Anomaly based detection techniques using machine learning are effective against zero-day attacks. However, they require complex data preprocessing and feature extraction which can affect the early detection of botnet attacks. In this paper we propose a novel approach, for early detection of botnet attacks using machine learning models that learn from byte representation of raw network traffic flows. The study departs from the traditional approach of network-based intrusion detection which relies on flow statistics and other hand-crafted features. We discuss our framework which includes light weight network traffic pre-processing, transformation, and model training. We used the CTU-13 dataset to evaluate the proposed byte-based botnet detection system. Our results show that byte-based representation can provide an effective and ultra lightweight means of developing network intrusion detection systems that can match the performance of traditional approaches, while also enabling early detection of botnet attacks. In our experiments we achieved accuracy of 99.9% consistently across different byte stream sizes for the Decision Tree and Logistic Regression classifiers.

Keywords Botnet detection · Machine learning · Early detection · Network intrusion detection · Byte-stream representation

R. Thomas (✉) · S. Yerima · K. Shaalan
Faculty of Engineering and IT, The British University, Dubai, UAE
e-mail: rajeshthomas@gmail.com

S. Yerima
e-mail: suleiman.yerima@buid.ac.ae

K. Shaalan
e-mail: khaled.shaalan@buid.ac.aem

K. Al Marri et al. (eds.), *BUiD Doctoral Research Conference 2024*, Lecture Notes in Civil Engineering 587, https://doi.org/10.1007/978-3-031-84371-6_24

1 Introduction

Threat actors have been increasingly using botnets to conduct attacks against critical network infrastructure (Live Botnet Threats Worldwide—Spamhaus Technology, n.d.).

A Botnet is a network of compromised computing devices known as bots that are controlled by the hacker or botmaster. With the proliferation of IoT devices, the attack surface area has widened to include vulnerable IoT devices in addition to traditional computing devices. Once a device has been compromised it becomes part of the botnet and communicates with a Command and Control (CnC) using various communication protocols such as IRC, http and P2P among others. Botnets are used to execute different types of attacks such as Distributed Denial of Service (DDoS), ransomware, data exfiltration, phishing campaigns, etc. Therefore, detecting botnet attacks is of great importance.

1.1 Motivation and Aim

Traditional signature-based detection methods are not effective against the advanced and evolving botnet threat. Machine Learning based detection models are better at detecting these advanced threats including zero-day attacks. However, there are considerable overheads associated with the data pre-processing and feature extraction which may impact early detection of botnet attacks. Additionally, domain knowledge expertise is required to extract or design meaningful features. Early detection techniques with less overhead can be efficient and serve as a first point of detection in a layered security approach.

1.2 Objectives of the Study

In this research, we investigate the effectiveness of detecting botnet attacks by leveraging information that machine learning models can learn from raw bytes extracted from network flows. We are motivated by the potential to enable earlier detection of botnet attacks if the models can utilize bytes from packet flow, compared to existing approaches which are largely based on packet and flow-based statistics. Therefore, the main objective of this study is to develop and evaluate botnet detection approaches based on learning from byte representation of network traffic flows with machine learning.

1.3 Contribution

The notable contributions of our research are the following.

1. An extensible and customizable framework for representing network traffic flows as raw bytes and related formats.
2. This study aims to answer the question: ‘how many bytes do we need to extract from each flow of packets to enable the machine learning models to effectively distinguish botnet attack flows from normal flows?’. Therefore, we present the analysis of the performance of different total bytes extracted from the packet flow in relation to their corresponding botnet attack detection rates.
3. We present the investigation of a new approach to detect botnet attacks from network flows using raw bytes representation, unlike existing research which is based mainly on the use of statistical packet and/or flow-based features such as packet rates, packet inter arrival times, number of bytes per flow/packet, etc.

2 Related Work

Most of the works related to botnet detection are based on hand-crafted statistical features extracted from packets or network flows. Compared to these established traditional techniques, packet byte stream analysis and other byte representations for early attack detection is an emerging field. Here we discuss similar studies that make the use of packet bytes. Fox (2023) used byte and image bitmap formats extracted from traffic flows for detecting anomalous network flows through classical (Random Forrest and kNN) and deep learning techniques (CNN and MLP) across multiple datasets. Different datasets were used for training and testing.

The study in this paper differs from Fox (2023) in several ways. Unlike their approach, where the emphasis is on the first 2 packets of only TCP flows found within IPv4 packets, our method considers all flows that fall within the defined maximum bytes, irrespective of the number of packets. Our method is also independent of the transport layer protocol and considers both IPv4 and IPv6 packets. Furthermore, since our method directly extracts the bytes, it could work well in cases of low-rate botnet attacks.

Taheri et al. (2018) proposed a deep learning IoT botnet detection by using traffic data to image conversion on the CTU-13 dataset. Although we used the same dataset in our research, our methodology is completely different.

Wang et al. (2017) proposed a malware classification model using representation learning. Their approach used raw traffic-based images representing different combinations of the network layers. Hu et al. (2023) proposed a network classification system for encrypted traffic using long short-term memory (LSTM) and convolutional neural network (CNN) using multiple datasets. They considered the first 784 bytes to create a 28×28 pixel image for traffic classification rather than botnet detection. Agrafiotis et al. (2022) used CNN in their research along with their IDS-Vision Toolkit. Their framework transforms the raw network traffic from the CIC IDS-2017 dataset into grayscale images. Unlike our proposed approach, the existing studies have significant overheads that could prevent them from being effectively used for early botnet attack detection in practice.

3 Proposed Approach

Existing approaches based on network flow statistics such as average packet size, flow duration, number of IP hops, interpacket arrival times, etc. require considerable pre-processing and feature reduction techniques. Such techniques may not be ideally suitable for early attack detection. Our proposed approach to a lightweight framework for botnet detection as shown in Fig. 1, has minimal processing overhead and is independent of feature selection. We discuss this approach in the following sub sections.

3.1 Botnet Detection Framework

The botnet and normal traffic captures were converted into bi-directional flow using the SplitCap utility and our custom python program. This process extracts the bidirectional flow information from the full packet capture (FPC) file. Our methodology is different from statistical flow processing techniques which convert a FPC into binary format representation.

The parser submodule extracts the information from the traffic flow files. This is a hex byte representation of the packets belonging to a particular flow. It captures the details of the protocols from Layer2 (L2) to the application layer along with the payload. Our technique for byte extraction is different from the statistical flow generation methods that usually capture the summary of the flow.

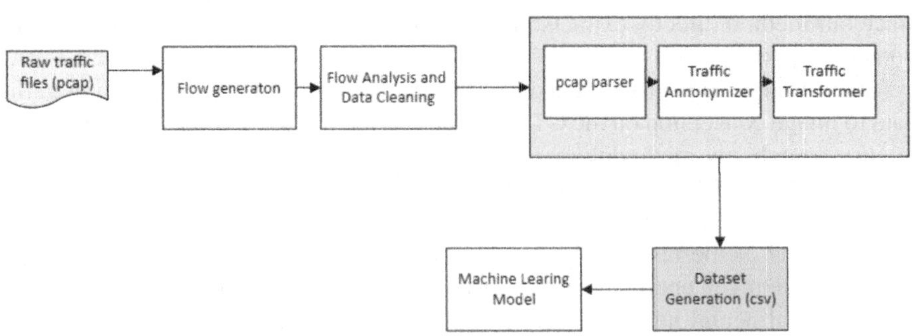

Fig. 1 Bot detection framework

The anonymizer module will mask the L2 mac address and the L3 IPv4 or IPv6 addresses if the flow contains L3 information. Anonymizing prevents the bias from associating the botnet characteristics to an IP address. Signature based detection techniques use IP addresses to detect malware with limited success. Moreover, this is not effective against IP spoofing attacks.

The transformer submodule transforms the hex byte stream into a format suitable to be represented in our dataset. Each pair of hex characters in the stream is considered as a byte and transformed into an integer range of 0–255. If a flow does not contain the maximum bytes to be extracted, it is padded with 0 values to obtain uniform records.

3.2 Byte Extraction and Validation

A sample of the botnet network traffic used in our experiments is show in Fig. 2. The highlight is over an example packet number 4 showing layer details along with the packet payload. Each layer from L2 to L5, considering the 5-layer TCP/IP stack, has its own meta data (headers). All this information along with the payload is a representation of the network traffic that forms the basis of our framework for botnet detection.

Figure 3 highlights the details of the example packet. The information for each layer can be seen in the hex byte stream and ASCII format on the right-hand side pane in Wireshark.

We used the dpkt python library (dpkt PyPI, n.d.) for reading the packets from the pcap file. The dpkt module is efficient for low level packet reading and is also known to be efficient especially for large flow files. Our framework can optionally

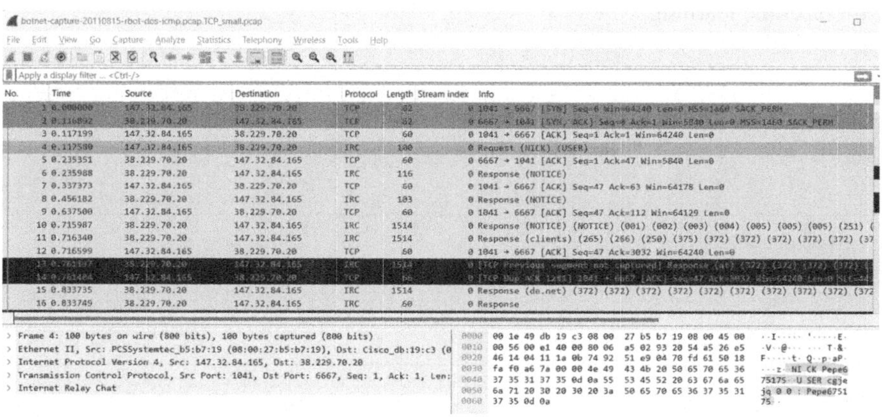

Fig. 2 Botnet packet in Wireshark

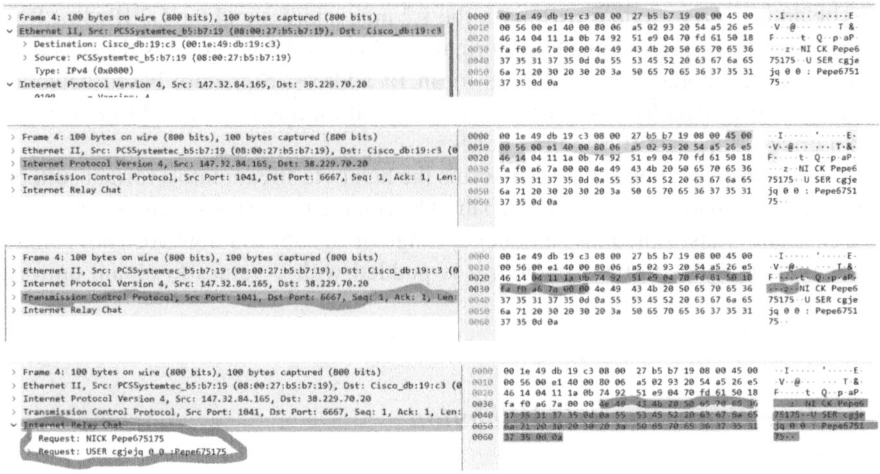

Fig. 3 Details of Ethernet, network, transport and application layers for a sample botnet packet

use the scapy module and is also integrated with tshark utility which can be used for future analysis and validation.

Figure 4 part [a] shows the sample data read from the botnet network traffic and part [b] shows the corresponding anonymized hex byte stream representation.

We validated our network traffic bytes extraction algorithm by comparing it with the hex dump obtained in Wireshark. Figure 5 shows the exact match of the hex byte stream extracted through our algorithm and that of the hex dump in Wireshark. This verifies that all the information from the TCP/IP layers are captured in the byte stream format by our framework. We then convert each pair of hex characters or nibbles into a decimal representation in the range 0–255 that is suitable to be used with machine learning.

```
Ethernet(dst=b'\x00\x1eI\xdb\x19\xc3', src=b"\x08\x00'\xb5\xb7\x19",
data=IP(len=86, id=225, df=1, ttl=128, p=6, sum=42242, src=b'\x93 T\xa5',
dst=b'&\xe5F\x14', opts=b'', data=TCP(sport=1041, dport=6667, seq=1955746281,
ack=74513761, flags=24, win=64240, sum=42618, opts=b'', data=b'NICK Pepe675175\r\
nUSER cgjejq 0 0 :Pepe675175\r\n')))
```

[a] botnet traffic

```
b'00000000000000000000000008004500005600e140008006a502000000000000000004111a0b749
251e90470fd615018faf0a67a00004e49434b20506570653637353137350d0a555345522063676a65
6a7120302030203a50657065363735313735 0d0a'
```

[b] hex byte stream

Fig. 4 Comparison of botnet traffic read and corresponding byte stream

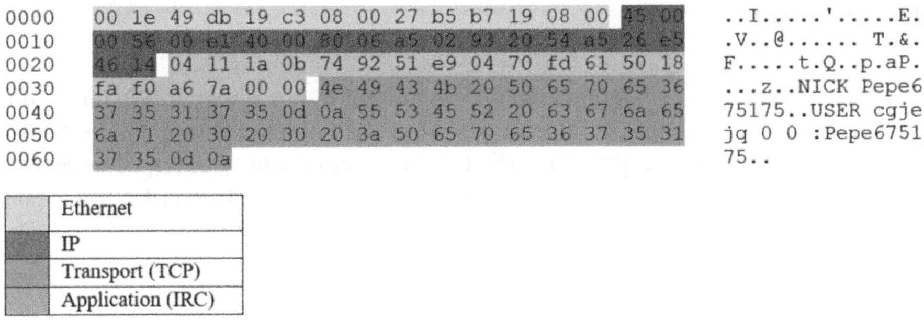

Fig. 5 Byte extraction validation with Wireshark

3.3 Byte Sizes Choices

We experimented with 3 different choices of the total bytes extracted from the network traffic. We started with the first 64 bytes and gradually increased this to 256 bytes and finally to 784 bytes. The choice of 784 bytes was influenced by similar studies such as (Hu et al. 2023; Wang et al. 2017; Agrafiotis et al. 2022).

In our system, the byte extraction starts with the first packet of the network traffic flow and continues till the maximum defined bytes are extracted. Depending on the transport layer protocol, botnet and attack types, these packets could convey significant and interesting information about the characteristics of the network traffic.

Our choice of considering the extracted data in terms of bytes rather than number of packets is reasoned by the outcome of obtaining uniform records with less padding for analysis. The dataset used in our experiments contains a mix of small and large flows. Although the total number of bytes extracted does represent a certain number of packets in the network flow, it would be difficult to ascertain this due to inconsistency in the number of packets in the network flow. Moreover, our rationale is that in case of encrypted traffic or large flows, considering all packets will lead to inefficient and inconsistent large number of features across the extracted flows.

4 Experiments

In this section we describe the details of our experiments for different total bytes extracted from the traffic flow. We conducted multiple experiments to generate the different byte features csv files by varying the maximum bytes parameter. Our chosen values are 64, 256 and 784 bytes.

4.1 Dataset

The CTU botnet dataset (García et al. 2014) is used in our research. We chose this dataset as it contains real traffic captures from the botnet malware family mixed with normal and background traffic. The CTU dataset matches our requirements for a realistic botnet dataset containing raw traffic captures of different bot types and protocols.

4.2 Training and Testing Approach

We extracted the bytes from the generated flow and saved the data in csv format. Each row in our dataset represents a network traffic flow. This is the byte representation of the packets converted into numerical format suitable for ML analysis.

Our dataset contains 115,666 instances. Of these 71,066 belong to bot and 44,600 are normal traffic instances. This proportion is similar to the related research utilizing this dataset.

We used the k-fold cross validation (k = 10) and ran different classifiers. The k-fold cross validation is known to be effective against overfitting (Wong & Yeh 2020).

5 Results

The results for 64 bytes, 256 bytes and 784 bytes are summarized in Tables 1, 3 and 5. Similarly the training and testing times are summarized in Tables 2, 4 and 6 respectively. All times are in seconds.

The confusion matrix for some of the classifiers is shown in Fig. 6.

6 Findings and Discussion of Results

We start our discussion with the number of bytes that were used for botnet detection and their effectiveness for each classifier. We achieved very encouraging results by varying the total number of bytes extracted from the network flow. For the MNB classifier, there is an increase in the accuracy moving from 64 to 256 bytes and then a decrease in the accuracy to 85.1% for 784 bytes implying that the 256 bytes case may be optimal for this classifier. However, further experiments may need to be conducted to confirm the exact value or range of the optimal bytes. A decrease in the accuracy is noticed for the GNB classifier as the bytes extracted are increased. GNB may not be suitable for large flow extractions. The DT, KNN and LT classifiers

Table 1 Classification metrics for maximum traffic size of 64 bytes

Classifier	Precision normal	Precision bot	Recall normal	Recall bot	Accuracy	F1
Multinomial naive Bayes (MNB)	0.916	0.875	0.783	0.955	0.889	0.887
Decision tree (DT)	0.999	0.999	0.999	0.999	0.999	0.999
K-nearest neighbor (KNN)	0.999	0.999	0.999	0.999	0.999	0.999
Gausian naive Bayes (GNB)	0.93	0.999	0.999	0.953	0.97	0.97
Logistic regression (LR)	0.999	0.999	0.999	0.999	0.999	0.999

Table 2 Training and testing times for 64 bytes

Classifier	Training Time	Testing Time
Multinomial naive Bayes (MNB)	0.112	0.003
Decision tree (DT)	0.762	0.003
K-nearest neighbor (KNN)	0.01	33.513
Gausian naive Bayes (GNB)	0.192	0.027
Logistic regression (LR)	6.286	0.004

Table 3 Classification metrics for maximum traffic size of 256 bytes

Classifier	Precision normal	Precision bot	Recall normal	Recall bot	Accuracy	F1
Multinomial naive Bayes (MNB)	0.874	0.999	0.999	0.91	0.944	0.945
Decision tree (DT)	0.999	1	0.999	0.999	0.999	0.999
K-nearest neighbor (KNN)	0.999	0.998	0.998	0.999	0.999	0.999
Gausian naive Bayes (GNB)	0.916	1	1	0.943	0.965	0.965
Logistic regression (LR)	0.999	0.999	0.999	0.999	0.999	0.999

Table 4 Training and testing times for 256 bytes

Classifier	Training Time	Testing Time
Multinomial naive Bayes (MNB)	0.547	0.01
Decision tree (DT)	2.373	0.006
K-Nearest neighbor (KNN)	0.013	32.228
Gausian naive Bayes (GNB)	0.718	0.103
Logistic regression (LR)	2.77	0.007

Table 5 Classification metrics for maximum traffic size of 784 bytes

Classifier	Precision normal	Precision bot	Recall normal	Recall bot	Accuracy	F1
Multinomial naive Bayes (MNB)	0.731	0.9765	0.97	0.776	0.851	0.853
Decision tree (DT)	0.999	1	0.999	0.999	0.999	0.999
K-nearest neighbor (KNN)	0.997	0.983	0.973	0.998	0.988	0.988
Gausian naive Bayes (GNB)	0.776	0.999	0.999	0.819	0.889	0.891
Logistic regression (LR)	0.999	0.999	0.99	0.999	0.999	0.999

Table 6 Training and testing times for 784 bytes

Classifier	Training Time	Testing Time
Multinomial naive Bayes (MNB)	2.540	0.025
Decision tree (DT)	6.962	0.021
K-nearest neighbor (KNN)	0.049	96.179
Gausian naive Bayes (GNB)	56.685	1.03
Logistic regression (LR)	0.805	0.113

achieve consistent accuracies across all the 3 cases. The results of our experiment show that the highest accuracy of 99.9% is achieved through the Decision Tree and Logistic Regression classifiers. Using a smaller number of bytes as features achieves good results. However, the case for large number of bytes opens opportunities for further research involving deep learning. As the maximum bytes that are extracted from the traffic increases there is higher computational overhead due to the large number of features. This can be verified from the 784 bytes features testing and training times which are lower than that of the 256 and 64-bytes versions.

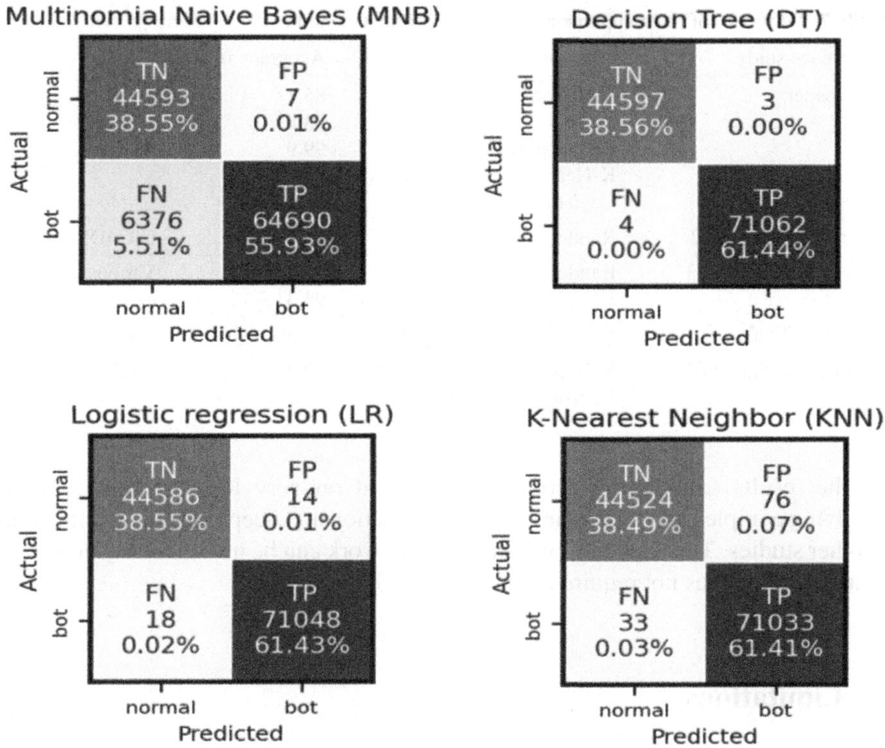

Fig. 6 Confusion matrix for classifiers using 256 bytes features

The results obtained in our study are better compared to those of traditional machine learning models reported in other works. For example, (Agrafiotis et al. 2022) achieved 98.5% accuracy using Random Forrest based on grayscale images. (Fox & Boppana 2023) achieved balanced accuracy of 99.9% using Random Forrest based on the first 2 packets of the flow on the IDS17 dataset. Their kNN model achieved 94.3% balanced accuracy and F1 score of 93.4% as compared to ours which has an accuracy and F1 scores of 99.9%. (Taheri et al. 2018) achieved 78.56% accuracy for the model based on Logistic regression. They employed the same dataset as our study although our methodology has significantly low overhead and in our case the LR model achieved significantly higher accuracy of 99.9% for all byte cases. (Moorthy & Nathiya 2022) achieved accuracies of 82.38 and 92.21 using KNN and Decision Tree based on 70:30 testing and training using the CTU dataset. which is significantly lower than our results. Table 7 provides a comparison of our results with the related works.

All the afore mentioned works have substantial overhead of pre-processing and feature extraction except in the case of (Fox & Boppana 2023) where the overhead is significantly less although it is based on the number of packets as compared to our technique of a defined maximum bytes.

Table 7 Summary of the comparison of results with related studies

Reference study	Classifier	Accuracy in %	Dataset
This paper	Multinomial Naive Bayes (MNB)	85.1	CTU
	Decision Tree (DT)	98.8	
	K-Nearest Neighbor (KNN)	99.9	
	Logistic regression (LR)	99.9	
Agrafiotis et al. 2022	Random Forrest	98.5	CIC IDS-2017
Fox & Boppana 2023	Random Forrest	99.9	Various
	K-Nearest Neighbor (KNN)	94.3	
Taheri et al. 2018	Logistic regression (LR)	78.56	CTU
Moorthy & Nathiya 2022	K-Nearest Neighbor (KNN)	82.38	CTU
	Decision Tree (DT)	92.21	

The results from our experiments are based on very low overhead without involving complex pre-processing, feature extraction and deep learning as compared to other studies. The results show that our framework can be used for early detection of attacks and does not require domain knowledge expertise.

7 Limitations

The experiments were conducted on the CTU botnet dataset. Our results are for binary classification and the performance of our framework for multi class classification needs further work.

8 Conclusion and Future Work

Our experiments show promising results and consistent performance across different byte sizes of the network traffic flow. The network byte stream can be extracted and represented directly as features for enabling a lightweight detection system for botnet attacks.

For future work we plan to conduct further analysis on network traffic based on more granular total byte sizes using other datasets to study their influence on the detection rates. We also plan to extend our framework for image representation of the network traffic from the traffic bytes. Finally, we plan to further investigate byte representations for efficient feature extraction and Deep Learning.

References

Agrafiotis G, Makri E, Flionis I, Lalas A, Votis K, Tzovaras D (2022) Image-based neural network models for malware traffic classification using PCAP to picture conversion. ACM Int Conf Proceeding Ser. https://doi.org/10.1145/3538969.3544473

dpkt PyPI. (n.d.). Retrieved April 10, 2024, from https://pypi.org/project/dpkt/

Fox G, Boppana RV (2023). On early detection of anomalous network flows. https://doi.org/10.1109/ACCESS.2017.DOI

García S, Grill M, Stiborek J, Zunino A (2014) An empirical comparison of botnet detection methods. Comput & Secur 45:100–123. https://doi.org/10.1016/j.cose.2014.05.011

Hu F, Zhang S, Lin X, Wu L, Liao N, Song Y (2023) Network traffic classification model based on attention mechanism and spatiotemporal features. Eurasip J Inf Secur 2023(1). https://doi.org/10.1186/s13635-023-00141-4

Live botnet threats worldwide—Spamhaus Technology (n.d.). Retrieved March 12, 2024, from https://www.spamhaus.com/threat-map/

Moorthy RSS, Nathiya N (2022) Botnet detection using artificial intelligence. Procedia Comput Sci 218:1405–1413. https://doi.org/10.1016/j.procs.2023.01.119

Taheri S, Salem M, Yuan JS (2018) Leveraging image representation of network traffic data and transfer learning in botnet detection. Big Data Cogn Comput 2(4):1–16. https://doi.org/10.3390/bdcc2040037

Wang W, Zhu M, Zeng X, Ye X, Sheng Y (2017) Malware traffic classification using convolutional neural network for representation learning. Int Conf Inf Netw 712–717. https://doi.org/10.1109/ICOIN.2017.7899588

Wong TT, Yeh PY (2020) Reliable accuracy estimates from k-fold cross validation. IEEE Trans Knowl Data Eng 32(8):1586–1594. https://doi.org/10.1109/TKDE.2019.2912815

References

[illegible]

Memory Enhancing Strategies to Support Learning and Cognitive Functioning in Students with Neurodevelopmental Disorders

Iman S. Sabih and Ahmed Kuyini

Abstract The article aimed to assess and combine methods that improve memory in students with neurodevelopmental disorders to support academic outcomes. The focus is on identifying practical ways to enhance memory, learning, and cognitive development that can be integrated into educational practices. The literature review approach was designed to analyze empirical studies and theoretical analyses to related memory strategies within cognitive psychology and neuroscience. Employing a systematic review format, several articles were collated and analyzed, the results show that personalized memory strategies, along with regular and structured practice, significantly improve learning and retention in students with neurodevelopmental challenges. Techniques like visualization, association, and spaced repetition are especially helpful in enhancing both short-term and long-term memory. These results have important implications for educators and practitioners, indicating a need for adaptable educational frameworks that integrate memory enhancement techniques. This has the potential to elevate academic performance and cognitive development in students with neurodevelopmental disorders. The significance of this study lies in its ability to impact educational methods and policy development by highlighting evidence-based, non-pharmacological interventions.

Keywords Memory enhancement · Neurodevelopmental disorders · Educational strategies · Mnemonic devices · Cognitive functioning

---

I. S. Sabih (✉) · A. Kuyini
Department of Psychology, Faculty of Education, The British University, Dubai, UAE
e-mail: 23002533@student.buid.ac.ae

A. Kuyini
e-mail: ahmed.abubakar@buid.ac.ae

K. Al Marri et al. (eds.), *BUiD Doctoral Research Conference 2024*, Lecture Notes in Civil Engineering 587, https://doi.org/10.1007/978-3-031-84371-6_25

1 Introduction

1.1 Background to the Study

Neurodevelopmental disorders present significant obstacles within educational settings, profoundly impacting cognitive functions, academic attainment, and daily living skills. Difficulties with memory processing and retention are particularly prominent across a range of these disorders, further complicating the learning process. For instance, individuals with Autism Spectrum Disorder often experience challenges with episodic and prospective memory, hindering their ability to generalize learned skills (Bowler et al. 2014). Similarly, dyslexia, frequently rooted in phonological processing issues (Snowling 2000), can negatively impact both working and long-term memory, affecting reading and writing abilities (Swanson & Jerman 2006). Intellectual Disability presents notable limitations in intellectual functioning and adaptive behavior, with memory, attention, and executive function deficits posing significant barriers to learning (Boyle et al. 2010). While not classified as a neurodevelopmental disorder, slow learners also exhibit below-average cognitive abilities, encountering difficulties with memory retention and information processing that impede their academic progress (Reis & McCoach 2000). Addressing these unique memory and learning challenges is crucial for fostering the inclusion and academic success of all students.

1.2 Statement of Problem

Students with neurodevelopmental disorders, including Autism Spectrum Disorder, Dyslexia, and Intellectual Disability, often experience significant memory deficits that hinder their learning and academic achievement (Mastropieri & Scruggs 1998). While mnemonic and retrieval-based strategies, such as the method of loci and visual imagery, hold promise for improving memory function (Bakken 2017; Mastropieri & Scruggs 1998; Bellezza 1981), research on their specific application and effectiveness for these student populations remains limited.

1.3 Purpose

This research aims to synthesize existing literature and explore the effectiveness of various memory enhancement strategies for students with ASD, Dyslexia, and ID. The study will focus on identifying evidence-based practices and understanding their applicability to the unique cognitive profiles of each disorder.

1.4 Research Questions

1. What memory enhancement strategies, specifically mnemonic and retrieval-based techniques, have been studied in relation to students with ASD, Dyslexia, and ID?
2. What is the relative effectiveness of these memory enhancement strategies in promoting learning and cognitive functioning for students with ASD, Dyslexia, and ID?
3. How can these strategies be effectively implemented in educational settings to support students with ASD, Dyslexia, and ID?

1.5 Rationale for the Study

This research is crucial for addressing the gap in current literature regarding effective memory interventions for students with neurodevelopmental disorders. By synthesizing existing knowledge and identifying best practices, this study will provide valuable insights for educators, practitioners, and policymakers to develop targeted interventions that cater to the unique needs of these students, ultimately promoting their academic success and overall well-being.

2 Methodology

In this literature review, focused on the past decade's publications (2014–2024), we targeted research addressing memory enhancement in students with autism, dyslexia, intellectual disability, and slow learners, emphasizing studies that explored interventions affecting memory, cognition, and academic performance. Comprehensive searches across platforms like PubMed, PsycINFO, ResearchGate, and Google Scholar utilized specific keywords to ensure thorough coverage. Specific keywords such as "memory enhancement strategies," "mnemonic training," "retrieval practice," "memory interventions", "memory performance," "cognitive function," "academic achievement," "learning outcomes", "neurodevelopmental disorders," "autism," "dyslexia," "intellectual disability," and "slow learners" are used in the search strategy. Screening processes, from eliminating duplicates to detailed assessments of titles, abstracts, and full texts, ensured the relevance of studies included. In this phase, further pertinent studies are identified by manually reviewing bibliographies and tracking citations. Studies were considered eligible for inclusion based on the following criteria:

- **Publication Date**: Published between January 2014 and December 2024.
- **Population**: Focused on students (ages 5–18) diagnosed with autism, dyslexia, intellectual disability, or categorized as slow learners.

- **Intervention**: Investigated the effects of specific memory enhancement strategies or interventions.
- **Outcomes**: Included measures of memory performance, cognitive function, and/or academic outcomes.

On the other hand, studies were excluded based on the following criteria: Primarily focused on pharmacological interventions., lacked a specific memory enhancement component, investigated populations with co-occurring conditions that could significantly confound the results, were reviews, dissertations, or conference abstracts without sufficient methodological detail.

Data extraction and analysis highlighted key attributes (e.g., authors, publication year, study design), participant details (e.g., sample size, demographics), intervention details (e.g., memory enhancement approach, duration, intensity), and outcome measures (e.g., memory performance, academic attainment) leading to the synthesis of themes and insights pertinent to memory strategies in these student groups. Extracted data were synthesized using a narrative synthesis approach. We identified key themes related to the effectiveness of different memory enhancement strategies, factors influencing intervention outcomes, and gaps in the existing literature. These findings are then presented coherently, allowing for a comprehensive discussion of the evidence on memory enhancement strategies for students with neurodevelopmental disorders.

3 Findings

Cognitive psychology has established that memory enhancement strategies can significantly aid in the encoding, storage, and retrieval of memories in individuals with and without neurodevelopmental disorders. Popularized methods like mnemonic devices utilize semantic encoding and elaborative rehearsal, using acronyms, imagery, and locations, as in the method of loci, to facilitate recollections via associative connections. These techniques help to acquire vocabulary and make the retrieval process easier, faster, and more self-directed (Farrokh et al. 2021). Memory is integral to the mastery of vocabulary and grammar, consisting of short-term and long-term segments. Educational goals focus on shifting knowledge from short-term to long-term storage, ensuring its long-lasting preservation. (Farrokh et al. 2021). Mnemonic devices involve verbal or visual cues that link new information with existing knowledge, enhancing its storage and recall. These strategies are particularly beneficial for students with lower proficiency levels who typically engage in tasks requiring memorization. (Farrokh et al. 2021).

Researchers such as Raugh and Atkinson, and Pressley et al. found that mnemonic devices, like the keyword method, are effective in the process of learning second language vocabularies, in both short-term and long-term retention contexts (Farrokh et al. 2021). Visual mnemonics, particularly, offer stable memory storage and direct retrieval paths proving beneficial for vocabulary development (Farrokh et al. 2021).

While verbal strategies may be comparatively more effective in some cases, overall evidence suggests that mnemonic techniques are valuable for enhancing memory in language learners across different age groups (Farrokh et al. 2021; Dunlosky et al. 2013). Loci is a memory technique that has been proven to enhance memory consolidation, facilitating the transfer of information from short-term to long-term memory, Moreover, it is associated with enhanced connectivity between the neocortex and the hippocampus, suggesting improved consolidation of memory (Drigas et al. 2022).

In cognitive psychology, spaced repetition and elaborative encoding are regarded as essential for enhancing memory by effectively utilizing the spacing effect and connecting new information with prior knowledge, respectively. These methods, alongside retrieval practice and metacognitive strategies like self-monitoring, play a significant role in improving memory and learning outcomes for students with neurodevelopmental disorders. Such research-supported strategies, mentioned by Drigas et al. 2022 tap into the brain's abilities for pattern recognition and information consolidation.

The relationship between learning and memory is closely tied to how we concentrate and organize information. Logan's "An Instance Theory of Attention and Memory" (2002) explains that learning and memory are intricately connected to attention and the organization of information, with attention facilitating memory retention by focusing on pertinent cues. This understanding can be applied through practical strategies to support students with learning challenges. Special education research underscores the significance of memory enhancement techniques customized to address the distinctive learning requirements of individuals with conditions such as autism, dyslexia, intellectual disabilities, or those classified as slow learners.

Structured Learning Logs, as promoted by Banikowski & Mehring, promote memory consolidation through increased emphasis on repetition and retrieval. This aligns with Logan's framework, which emphasizes the importance of organization and attention in memory formation. Additionally, incorporating mental imagery, a key element of the dual-coding technique, can significantly benefit learners, particularly those with cognitive delays, by providing multiple retrieval paths for information (Banikowski & Mehring 2017; Drigas et al. 2022).

Activating Prior Knowledge: To support students with learning disabilities, it is essential to connect new information to pre-existing knowledge, facilitating memory development as indicated by Graham & Bellert (2004). Breaking learning materials into smaller units, a concept related to Logan's principles on the importance of categorization and attention i optimizing short-term memory capacity and facilitating retrieval (Logan 2002). Systematic repetition, as suggested in the phrase "use it or lose it," supports memory maintenance by regularly activating relevant brain area (Drigas et al. 2022).

Hermans et al. note that Obsessive–Compulsive Disorder is characterized by a significant lack of cognitive confidence and mistrust in memory as a result of repetitive checking behaviors (Hermans et al. 2007). This is particularly notable in individuals with learning disabilities, as they may also encounter metacognitive difficulties, wherein they struggle to rely on their cognitive processes, such as memory. Hence, the adoption of learning strategies that promote metacognitive trust and reduce over-reliance on repetition could be more advantageous in enhancing cognitive abilities.

3.1 Some Ways to Improve Learning and Memory Are

Personalized Learning Materials: Tailoring materials to students' personal experiences can increase engagement and improve memory (Hermans et al. 2007; Banikowski & Mehring 2017).

Limited and Structured Checking: Implementing limited and structured checking mechanisms can avoid the decline of memory confidence that excessive checking can cause. This adapts Hermans et al.'s findings to the educational environment. Banikowski and Mehring (2017) discuss techniques that enhance memory, such as effective response mechanisms. Although not directly connected to "limited and structured checking," some principles of structuring learning tasks could potentially be adapted to develop effective checking mechanisms.

Metacognitive Training: Tailoring materials to students' personal experiences can increase engagement and improve memory (Hermans et al. 2007). Techniques for activating learners, such as reciprocal teaching/learning, align with the idea of metacognitive training by allowing learners to monitor and regulate their cognitive processes (Banikowski & Mehring 2017).

Supportive Feedback Mechanisms: Gradual and supportive feedback mechanisms that confirm correct responses can build cognitive confidence. This technique can utilize insights on OCD and memory distrust. Immediate feedback on learning, as mentioned by Banikowski and Mehring (2017), is one way to support memory confidence, which could be adapted from insights on OCD and memory.

Graham and Bellert highlight the need to assess, stimulate, and enhance pre-existing knowledge in students with learning disabilities, including those with dyslexia. This basic knowledge helps to connect new information to what the student already knows, leading to better understanding and retention of the material.

Burack et al. (2016) suggest a more dynamic interaction between the individual, their brain development, and the environment for students with autism. They suggest that emphasis on the fluidity of attention is critical, and that attentional biases are shaped by personal experience and rewarding outcomes. As a result, it is vital to adopt approaches that consider the individual's history, motivation, preferences, and cultural background.

Banikowski and Mehring recommend several methods for enhancing memory and learning. These include organizing and rehearsing the material, using graphic organizers, mnemonic devices, spaced practice, and imagery to improve comprehension and retention. These techniques are especially useful for students with intellectual disabilities or slower learners since they provide visual and organizational aids that simplify complex information.

The literature emphasizes that for students with developmental and learning challenges, including background knowledge, personalized learning approaches, and memory enhancement strategies can significantly improve cognitive functioning and learning outcomes (Sachdeva et al. 2015). These methods emphasize the significance of tailored educational strategies that address the unique needs and capabilities of every student, moving beyond a standardized, universal approach.

Husain and Mehta (2011) reported on pharmacological treatments. While the finding does not specifically address non-pharmacological strategies or learning support for students with neurodevelopmental disorders, it provides insights into drugs used for cognitive enhancement as part of a comprehensive approach. Husain and Mehta's findings, discuss the use of drugs targeting the noradrenergic and dopaminergic systems, such as methylphenidate and atomoxetine, which are commonly prescribed for ADHD. They also discuss the targeting of neuromodulatory systems such as the cholinergic, dopaminergic, noradrenergic, and serotonergic pathways by various cognitive enhancers. They acknowledge that the impact of these drugs on cognitive enhancement tends to be modest and varies significantly between individuals. Therefore, personalized medicine and careful monitoring of treatment outcomes are recommended.

Dresler et al. (2013) discuss Non-pharmacological cognitive enhancement processes to improve memory function which are considered safer options. These approaches include mnemonic training, proper diet, physical activity, sleep, meditation, brain stimulation, and computer training. Although there are no direct comparative efficacy studies between pharmaceutical and non-pharmacological cognitive enhancers, non-pharmacological methods provide a feasible, less risky alternative to pharmaceuticals for memory enhancement. Nevertheless, additional evidence is required to support recommendations for their use.

4 Discussion

In this literature review, the significance of strategies for enhancing memory in cognitive psychology is highlighted. These strategies can aid in the cognitive functioning and learning of students with neurodevelopmental disorders. They provide various methods and techniques aimed at improving the processes of memory encoding, storage, and retrieval, which are often impaired in individuals with slow learners, ID, autism, and dyslexia.

Research has shown that the use of mnemonic techniques, including the method of loci and visual imagery, can improve the encoding and retrieval of memory in individuals with ID (Boyle et al. 2010; Higbee 2001) and slow learners (Reis & McCoach 2000). Mnemonic devices can be helpful in memorization, but they may not always lead to long-term retention. Summarization and imagery use for text learning have been shown to help some students on some criterion tasks, yet the conditions under which these techniques produce benefits are limited (Dunlosky et al. 2013).

To improve memory, retrieval practice is a powerful technique that should be incorporated into a comprehensive learning strategy. Repeated retrieval itself has been shown to be a powerful mnemonic enhancer, producing large gains in long-term retention compared to repeated studying" (Dresler et al. 2013; Roediger & Karpicke 2006). Retrieval practice involving self-testing or retrieval exercises are known to improve memory performance of individuals with dyslexia (Kang et al. 2007) and autism (Bowler et al. 2014) by strengthening memory traces through repeated retrieval attempts.

Karpicke and Bauernschmidt's research suggests that the regularity with which retrieval practice is conducted plays a crucial role in determining long-term memory retention. This points towards the potential benefits of implementing spaced repetition interventions for students. By incorporating increasingly longer intervals between recall attempts, strategies could be developed to improve the consolidation of memory. Although no particular schedule for the relative spacing of retrieval practice was found to be superior over another, the clear implication is that repetition over time plays a vital role in enhancing learning.

Cognitive enhancement methods that don't rely on medication, like memory training, regular physical activity, balanced diet, adequate rest, and mindfulness meditation, can strengthen memory functions and contribute to general health and wellness. However, there is a need for systematic validation of their effectiveness (Dresler et al. 2013). The use of mnemonic strategies can be effective in enhancing memory, especially for easily learnable material, but their efficacy may decrease with complexity. Mnemonic strategies have been shown to be effective for retaining easy-to-learn material in students with neurodevelopmental disorders (Dresler et al. 2013).

Brain stimulation methods such as transcranial magnetic stimulation and transcranial direct current stimulation (tDCS) hold promise for enhancing cognition and boosting performance in memory-related tasks among healthy individuals. Non-pharmacological methods for improving cognition offer alternative therapies to drug treatments. There are various approaches and uses for these methods. One type of training focuses on teaching techniques for encoding and retrieving information from working memory, which has been shown to benefit populations ranging from children with Down syndrome to older adults. "The training paradigms usually teach effective approaches to encode, register, and retrieve from WM. Such methods of 'strategy training' have been successfully used in children with Down syndrome to supplement specific WM deficits" (Sachdeva et al. 2015). This suggests that memory retention may be improved, and cognitive decline may be slowed. Strategy training was reported to slow the decline of, and perhaps improve, WM in older adult populations (Sachdeva et al. 2015).

Cognitive remediation therapy has been demonstrated to be effective across various cognitive domains in dementias and neuropsychiatric conditions. "Cognitive Remediation therapy aims to improve cognitive processes in dementias and neuropsychiatric disorders. CRT expects that changes will maintain over a follow up period and will translate into real world benefit" (Sachdeva et al. 2015). In addition, cognition-enhancing effects in healthy persons and improvements in tasks related to memory have been observed with the use of brain stimulation methods like transcranial magnetic stimulation and transcranial direct current stimulation (tDCS). However, the exact mechanisms and long-term effectiveness of these techniques require further investigation. Developed originally for therapeutic use in psychiatry and neurology, some brain stimulation methods have shown to enhance cognitive functions in healthy individuals (Sachdeva et al. 2015).

The nuanced application of non-pharmacological cognitive enhancement approaches means that individual cognitive capacities and contextual relevance must be considered when treating memory impairments. Non-pharmacological cognitive enhancement strategies offer promising ways to improve memory for various populations. However, ongoing research is necessary to fully understand their impact and optimize their application.

Elaborative encoding techniques, such as generating personal associations and examples, promote deeper processing of information (Bower et al. 1979). This process aids in merging new data with pre-existing knowledge and experiences, bolstering understanding and the retention of memories. Individuals with ASD may experience difficulties with memory processing, particularly affecting their ability to retain social information such as faces and social scenes. However, despite these challenges, they can remember intricate details for longer periods, which suggests that their memory profile is unconventional. This disparity in memory capabilities may be related to the effectiveness of memory encoding strategies. The findings of a study by Williams et al. (2005) align with emerging theories that suggest that autism's underlying neurobiological implications involve intricacies in the localized circuitry of the neocortex. To address these challenges, personalized memory enhancement techniques may be necessary to cater to the complex information processing requirements of individuals with ASD. Adcock and Cuvo have emphasized that instructional modifications utilizing enhanced stimulus control procedures can significantly enhance the learning and memory performance of children with ASD, providing practical applications for educators to scaffold learning experiences that are tailored to the unique cognitive mechanisms characteristic of ASD. Therefore, integrating structured learning cues and individualized strategies could potentially address the deficits in processing complex social information and enhance memory retention for this population. Although strategies for enhancing memory present hopeful options for aiding students with neurodevelopmental disorders, accounting for individual variances is essential. Tailoring interventions to address the distinct requirements of every student is critical. Moreover, implementing these strategies in educational settings may require collaboration among families, educators, and special education professionals to ensure consistency and effectiveness across different contexts.

Additionally, Bower et al. (1975) found that using both verbal and visual information improves memory recall. This aligns with multimedia learning and dual coding theories, which also support the use of multiple representations for better comprehension. Combining information in multiple layers enhances conceptualization and recall, which can be useful for educators and learners.

Future studies ought to concentrate on evaluating the enduring impact of interventions aimed at memory enhancement on educational attainment, practical results, and life quality in those with neurodevelopmental disorders. Longitudinal studies are necessary to evaluate the sustained impact of these interventions and identify factors that contribute to their success. Finally, research should explore the feasibility and scalability of implementing memory enhancement strategies in inclusive classrooms and special education settings, considering resource constraints and practical considerations.

5 Conclusion

Cognitive psychology research provides examples of effective techniques to enhance memory which can be useful for students with neurodevelopmental disorders to improve their cognitive abilities and learning. Educators and practitioners can use evidence-based strategies such as elaborative encoding, mnemonic devices, and retrieval practice to address specific memory deficits in slow learners, individuals with autism, dyslexia, and ID. However, there is a need for more research to confirm the effectiveness of these strategies in real-life educational settings and to identify the best approaches for providing personalized support. Improving educational practices and outcomes for students with neurodevelopmental disorders can promote their academic success and well-being, which can be achieved by addressing these research gaps.

References

Bakken JP (2017) Mnemonic strategies: helping students with intellectual and developmental disabilities remember important information 2(3). https://doi.org/10.19080/gjidd.2017.02.555587

Banikowski AK, Mehring TA (2017) Strategies to enhance memory based on brain research. Univ Kans 32(2)

Bellezza FS (1981) Mnemonic devices: classification, characteristics, and criteria. SAGE Publ 51(2):247–275. https://doi.org/10.3102/00346543051002247

Bower GH, Karlin MB, Dueck A (1979) Comprehension and memory for pictures. Mem Cognit 7(1):2–14

Bowler DM, Gaigg SB, Gardiner JM (2014) Effects of related and unrelated context on recall and recognition by adults with high-functioning autism spectrum disorder. Neuropsychologia 63:107–113

Boyle CA, Boulet S, Schieve LA, Cohen RA, Blumberg SJ, Yeargin-Allsopp M, Kogan MD (2010) Trends in the prevalence of developmental disabilities in US children, 1997–2008. Pediatrics 127(6):1034–1042

Burack JA, Russo N, Kovshoff H, Fernandes TP, Ringo J, Landry O, Iarocci G (2016) How I attend—Not how well do I attend: rethinking developmental frameworks of attention and cognition in autism spectrum disorder and typical development. J Cogn Dev 17(4):553–567

Dresler M, Sandberg A, Ohla K, Bublitz C, Trenado C et al (2013) Non-pharmacological cognitive enhancement. Elsevier BV 64:529–543

Drigas A, Mitsea E, Skianis C (2022) Metamemory: metacognitive strategies for improved memory operations and the role of VR and mobiles. Multidiscip Digit Publ Inst 12(11):450–450

Dunlosky J, Rawson KA, Marsh EJ, Nathan MJ, Willingham DT (2013) Improving students' learning with effective learning techniques. Psychol Sci Public Interes 14(1):4–58

Graham L, Bellert A (2004) Difficulties in reading comprehension for students with learning disabilities. In: Wong B (ed) Learning about Learning Disabilities. Academic Press, pp 251–279

Farrokh P, Vaezi H, Ghadimi H (2021) Visual mnemonic technique: an effective learning strategy. GiST Educ Learn Res J 23:7–32. https://doi.org/10.26817/16925777.1042

Hermans D, Engelen U, Grouwels L, Joos E, Lemmens J, Pieters G (2007) Cognitive confidence in obsessive-compulsive disorder: distrusting perception, attention and memory. Behav Res Ther 46(1):98–113. https://doi.org/10.1016/j.brat.2007.11.001

Higbee KL (2001) Your memory: how it works and how to improve it, 2nd edn. Pearson Education, Upper Saddle River, NJ

Husain M, Mehta MA (2011) Cognitive enhancement by drugs in health and disease. Trends Cogn Sci 15(1):28–36. https://doi.org/10.1016/j.tics.2010.11.002

Kang SHK, McDermott KB, Roediger HL III (2007) Test format and corrective feedback modify the effect of testing on long-term retention. Eur J Cogn Psychol 19(4–5):528–558

Logan GD (2002) An instance theory of attention and memory. Am Psychol Assoc 109(2):376–400

Mastropieri MA, Scruggs TE (1998) Enhancing school success with mnemonic strategies. SAGE Publ 33(4):201–208. https://doi.org/10.1177/105345129803300402

Reis SM, McCoach DB (2000) The underachievement of gifted students: what do we know and where do we go? Gifted Child Quarterly 44(3):152–170

Roediger HL, Karpicke JD (2006) Test-enhanced learning: taking memory tests improves long-term retention. Psychol Sci 17(3):249–255

Sachdeva A, Kumar K, Anand KS (2015) Non-pharmacological cognitive enhancers—current perspectives. JCDR Res Publ PrivE LteD

Snowling MJ (2000) Dyslexia, 2nd edn. Blackwell Publishing, Oxford, UK

Swanson HL, Jerman O (2006) Math disabilities: a selective meta-analysis of the literature. Rev Educ Res 76(2):249–274

Williams DL, Goldstein G, Minshew NJ (2008) Impaired memory for faces and social scenes in autism: clinical implications of memory dysfunction. Arch Clin Neuropsychol 23(2008):699–704

Regulatory and Governmental Drivers of Eco-Innovation

Ali Al Hammadi and Edward Ochieng

Abstract This paper Aimed at exploring the role of innovation in CSR through environmental regulation and government support. The paper used a six stages systematic literature review on CSR and Eco-innovation to examine the relationship between various factors effecting the innovation within CSR. The findings showcase a close relation between innovation and CSR, manifesting in various fields including Eco-innovation. It was found that Environmental regulation has a significant effect on Eco-innovation products and processes. The literature also revealed that government support positively impacts Eco-innovation. The paper concluded that corporate social responsibility (CSR), implemented through innovative projects, which is mainly driven by environmental regulation coupled with government support, has a major impact on the success and sustainability of an organization. Finally, the paper presented three practical implications.

Keywords CSR · Innovation · Eco-innovation · Corporate social responsibility

1 Introduction

Eco-innovation is a key component of sustainable development and corporate social responsibility (CSR), and it's driven by government and regulatory influences. The interplay between CSR and innovation has been highlighted in previous studies as being crucial for addressing both social and environmental issues (Wu and et al. 2023). Eco-innovation is a novel approach to addressing environmental concerns and promoting sustainable practices. In this context corporations have a multifaceted social responsibility, which includes shareholders, consumers and employees.

This paper was prepared for Evolutionary themes module, taught by Prof. Edward Ochieng.

A. Al Hammadi (✉)
Faculty of Business and Law, The British University in Dubai, Dubai, UAE
e-mail: A23sky@gmail.com

E. Ochieng
Department of Project Management, Dubai, UAE

K. Al Marri et al. (eds.), *BUiD Doctoral Research Conference 2024*, Lecture Notes in Civil Engineering 587, https://doi.org/10.1007/978-3-031-84371-6_26

While the role of government policies and regulatory frameworks in fostering Eco innovation remains underexplored. This paper is an attempt to fill the research gap, by looking at how government and regulatory drivers affect eco-innovation. It will also contribute to the larger discourse on corporate responsibility and sustainable development.

2 Research Question

This paper aims to examine the role that governmental regulations, policies and incentives plays in Eco innovation within organizations through. To that end, the research objectives are: To Evaluate the Impact of regulations and government support on Eco Innovation. The paper also asks the following **Research Questions**:

RQ1 How does government support influence Eco Innovation

RQ2 How might environmental regulation influence Innovation?

3 Literature Review

3.1 Innovation

Innovation is the ability of a human being to transform an idea into a reality and push the boundaries of what's possible (Romano 1990). It is the creation of new ideas, their translation into new products or processes, as well as the dissemination of these products and processes to potential users (Dziallas & Blind 2019).

Widely in Economics and policy discussions to cover a wide range of activities, it describes the process of putting new ideas into practice. It is a practical way to achieve the new goals that an organization wants (Stoneman 2022). Simply put, "innovation" is a new approach to something. It can refer to incremental or radical changes in products, processes, organizations, or thinking (Wu et al. 2023).

3.2 Corporate Social Responsibility

Corporate social responsibility (CSR) is a difficult concept to grasp. one of its first definitions state that corporate social responsibility is an obligation of businessmen in taking actions, making decisions, and framing policy in a way that achieves objectives and gives due consideration to values of society (Carroll & Shabana 2010).

It can be defined as the voluntary commitment by companies to act ethically, to contribute to economic growth, and to improve the quality-of-life for their employees, the local community, as well as society in general (Velte 2022).

Corporate social responsibility is also a strategy for businesses in developing economies to partner with governments and provide the essential infrastructure amenities needed for industrialization (Yoopetch et al. 2023).

3.3 *Eco-Innovation*

Combining CSR with innovation improves financial performance and is the best strategy for businesses looking to improve their financial performance as well as being socially responsible (Costa & Fonseca 2022).

CSR can also be improved through innovation by implementing research and development on environmental protection, employee training, and systematic stakeholder survey (Ubrežiová et al. 2020) furthermore, CSR-driven innovations can be used to redefine business models, meet compliance requirements, and create win–win scenarios for all stakeholders (Journal 2018).

CSR-driven innovations are defined as a process of innovation that is aimed at designing a profitable service or product in a user-friendly and innovative way, which benefits the environment and the society (Journal 2018). One of the main innovations within CSR is Eco-Innovation.

The field of eco-innovation may be young, but it is a growing concern among policy makers, academics, and practitioners. Eco innovation is defined as the reduction in negative environmental impacts, while providing new products and process as an advantage to both the customer and business (Hojnik & Ruzzier 2016).

It is a CSR-driven field that looks at the exploitation of the opportunities for innovative products to tackle critical environmental issues (Ubrežiová et al. 2020). This paper defines Eco-innovation as the introduction of any new product, process, organizational change, or marketing solution that reduces the consumption of natural resources, including materials, energy, and water (Hojnik & Ruzzier 2016).

The growing demand of environmental protection, both globally and locally, has led to an increase in environmentally friendly products, services, and techniques (Arranz et al. 2020).

Eco-innovation helps create a sustainable environment by developing green technologies and improving organizational performance.

The European Commission's Green Deal includes initiatives like the Circular Economy Action Plan and Biodiversity Strategies for 2030. These are all linked to digital technology and aim to drive a circular economy (Doni & Fiameni 2024).

Analysis of patent data reveals that environmental technologies and technologies for climate change mitigation have made significant advances, especially in energy production, transmission, and distribution. These technologies are in line with international environmental policies (Hlioui & Yousfi 2022).

3.4 Theoretical Underpinning

Research in this field is mostly based on resource-based theory and institutional theory where competitive advantage and regulation are the dominant factors.

Several studies have explored the relationship between environmental regulations and eco-innovation. The common drivers of product Eco innovation, process innovation, organizational Eco innovation, environmental R&D, and cost-savings seem to be regulations, market-pull factors, EMS, and cost-savings, as well as being positively related to company size (Hojnik & Ruzzier 2016).

Environmental concerns are either driven by external pressures, such as increased governmental regulation or stakeholders, or by the recognition of how it can improve performance and competitiveness through cost reduction (Díaz-García et al. 2015). Furthermore, Literature has shown that eco-innovations and environmental innovations differ from each other in terms of externalities, drivers for their introduction, and the like. They are primarily triggered by regulation.

As for the factors driving Eco innovation, Competitive pull and regulation are the main factors recognized by the literature (Arranz et al. 2020). Many countries have implemented regulations to protect ecosystems. These regulations are imposed on individuals, businesses, and organizations. These environmental regulations have forced companies to find innovative solutions to preserve the environment and stay complaint (Kesidou & Wu 2020).

This can be observed for example, in the rising appeal of electric cars. Which can be considered a product on Eco innovation which is stemmed from regulations on reducing carbon emissions (Sierzchula et al. 2012).

Furthermore, (Kesidou & Wu 2020) found that Firms located in provinces that have stricter pollution standards produced more green patents. Additionally, Liao and Tsai (2019), has emphasized that environmental policies obligate businesses to implement eco innovations, and investors prefer industries with low pollution. This aligns with the Porter Hypothesis that Environmental regulation can reduce pollution through the promotion of innovative technologies (van Leeuwen & Mohnen 2017).

Environmental regulations can be a key factor in generating public concern, shifting consumption habits towards eco-friendly options, and ultimately driving the demand for eco innovations by businesses (Han & Chen 2021). As evidenced by numerous studies (Fernández et al. 2021), competition is a major factor driving companies to adopt eco-innovation. The regulations imposed by the government on businesses to protect the environment and the growing demand for eco-friendly goods have prompted the development of innovative technologies to improve environmental performance (Han & Chen 2021).

The pressure of competitors who are implementing environmental strategies within the same industry also plays a crucial role in motivating businesses to pursue Eco innovation (Geng et al. 2021).

The government can also play an important role in eco-innovation (Han & Chen 2021) note that the government plays a key role in facilitating eco-innovation.

The government can help this process by improving the knowledge and technological capacity of its human resources. Governments have adopted various policies and programs to encourage eco-innovation. Ecolab net, an international partnership network aimed at encouraging eco-innovative behaviors, is one example.

Indonesia's government recognizes that eco-innovation is a necessary foundation for sustainable industrial growth. In Central and Eastern European nations, government tax policies and non-tax policy indirectly affect circular economy (CE), through eco-innovation (Arthur et al. 2023). In China, the relevant government departments provide cash assistance, tax incentives, environmental protection investments, and other forms of support to businesses. This promotes innovation in green technology research and developments.

The Chinese government also implemented green finance policies based on place, including pilot zones for green financing reforms and innovations. This is to encourage green innovation among manufacturing enterprises these policies and initiatives show the government's commitment to promoting eco-innovation.

These strategies include efficient resource management, the reduction of waste, and the integration into corporate culture and decisions processes of sustainable principles (Campbell 2018).

4 Methodology and Approach

The researcher used systematic literature review to understand the current research state on the CSR and innovation. A literature review is an objective-driven process that identifies, evaluates, and interprets existing literature using a reproducible and explicit design (Kraus et al. 2020). This allows for an inductive research style and the development significant models/theories from the literature data (Wolfswinkel et al. 2013)

This review was conducted on six stages, stage 1 is to identify a benchmark for the review (Wolfswinkel et al. 2013) this means that the articles reviewed would be in English, focuses on innovation and CSR and are peer reviewed.

Second stage involves conducting the search, which was done through google scholar and BUID library. Key words included CSR and innovation, emerging technologies in CSR etc.

Third stage is the refinement of the search results where all articles that were deemed no relevant were removed. Stage four is where certain dominant themes are identified and lead to the proposal of the conceptual framework.

The next stage involves presenting and synthesising the literature, followed by discussing the findings, implications, and future research direction if any.

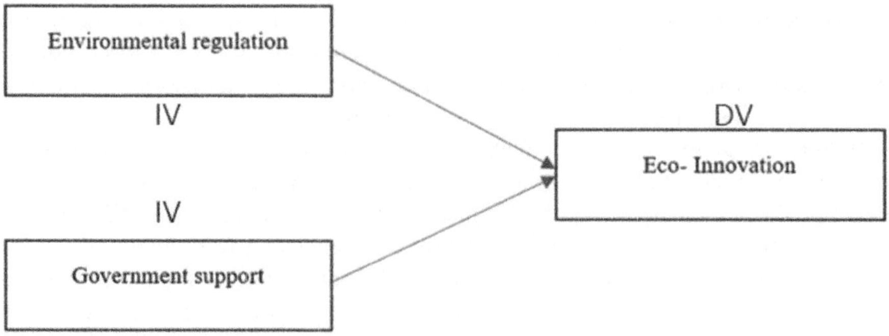

Fig. 1 Conceptual framework

5 Conceptual Framework

The literature indicates that regulation plays a role in encouraging Eco innovation (Geng et al. 2021; Fernández et al. 2021) where Environmental regulations that are stricter can encourage a competitive business environment with an emphasis on sustainability.

The relationship between eco-innovation and environmental regulation is complex and indicates that there's an optimal level of regulatory that promotes eco innovation (Costantini 2022).

Additionally, Government support can have an impact on eco-innovation through promoting enterprise growth and improving their ability to innovate (Arthur et al. 2023) Government-to-firm subsides, reduced firm tax burdens and increased provisions for environmental infrastructure can enhance the effectiveness of environmental regulations in promoting Eco innovation (Wang et al. 2023).

Tax and non-tax policies of the government can indirectly influence circular economies (CE) through eco-innovation by firms (Zhang 2023). These Tax incentives from the government encourage construction companies and suppliers to collaborate on green building projects (Zeng et al. 2022).

Therefore, the following can be conceptualized (Fig. 1).

6 Discussion

As defined by Carroll & Shabana (2010), Velte (2022) among others, corporate social responsibility reflects an extension of the business objectives, beyond profit maximization, to include ethical conduct and contribution to economic development, as well as enhancement of quality-of-life for a variety of stakeholders. This expansion in corporate goals reflects a shift of business to a holistic approach where economic, environmental, and social considerations are integrated.

The integration of CSR in companies' core strategy and value propositions reflects a wider trend where businesses are embracing their role of agents of positive transformation, with innovation as the means and end of that transformation.

Furthermore, Eco- innovation that is driven by regulation is now increasingly viewed as a powerful motive for organizations. This relationship highlights the potential of CSR-driven strategy to promote both technological and processes innovations, thus contributing to a firm's competitive pull.

Eco-innovation is a complex interaction between policy, innovation, and industry. The role of regulatory structures, market dynamics, and competitive pressures are all important factors. This dynamic highlight the need for a multi-stakeholder strategy to foster eco-innovation.

7 Conclusion and Recommendations

This paper examined existing research to get a better understanding of the role that regulation and governmental support plays in Eco-Innovation.

Businesses are under pressure to help the environment and society. Corporate Social Responsibility, or CSR, is a concept business cannot ignore. Innovating, or finding new and better ways to do things, is key for companies who want to be socially responsible and sustainable.

When businesses combine CSR with innovation, they are more likely to be successful, grow faster and have a positive effect on society and environment (Díaz-García et al. 2015). Eco-Innovation is about companies doing their part to help the environment and society and looking beyond the bottom line to consider how business impacts employees, communities, and the environment.

Following the Review on innovation, Corporate Social Responsibility, and eco-innovations, several recommendations can be made to businesses that want to incorporate these principles into their sustainable success strategies.

Eco-innovation should be a core element in the corporate strategy. It is not enough to allocate resources, but also to integrate these values in the culture, goals, and decision-making process of an organization.

Innovation can be used to promote corporate responsibility in many ways, such as developing eco-friendly products and using technology to improve social outcomes. It can not only help the environment and society, but also create new markets and opportunities.

Eco-Innovation has evolved, and now is about integrating ethical and environmental concerns into business operations. Businesses can solve social and environmental problems in new ways by finding innovative solutions to old issues. Using new technology can, for example, help businesses reduce waste, lower emission, and improve community well-being.

The government is also acting, establishing rules to encourage companies to be more responsible. These regulations can encourage companies to innovate, and to find new ways to meet the standards without sacrificing profits. Laws requiring lower emission levels, for example, could lead to cleaner manufacturing technologies.

This study relied on existing literature from the sources mentioned in the methodology chapter and did not use primary date, which may not capture all recent developments in Eco-Innovation. In addition, the author No access to proprietary or unpublished data. In order to fill the gaps addressed in this paper, more research needs to be conducted in countries which have strict environmental regulations and support environmental programs.

References

Arranz N, Arroyabe M, Li J, Fernandez de Arroyabe JC (2020) Innovation as a driver of eco-innovation in the firm: an approach from the dynamic capabilities theory. Bus Strat Environ 29(3). https://doi.org/10.1002/bse.2448

Arthur EE, Darfo-Oduro RKA, Gyamfi S, Sein YY, Stejskal J, Prokop V (2023) Tax and non-tax policies towards the finance of sustainable economy: the mediating role of eco-innovation. In: Economics and finance readings: selected papers from asia-pacific conference on economics & finance. 2022. https://doi.org/10.1007/978-981-99-1979-6_11

Campbell JL (2018) Reflections on the 2017 decade award: corporate social responsibility and the financial crisis. Acad Manag Rev 43(4). https://doi.org/10.5465/amr.2018.0057

Carroll AB, Shabana KM (2010) The business case for corporate social responsibility: a review of concepts, research and practice. Int J Manag Rev 12(1). https://doi.org/10.1111/j.1468-2370.2009.00275.x

Costa J, Fonseca JP (2022) The impact of corporate social responsibility and innovative strategies on financial performance. Risks 10(5). https://doi.org/10.3390/risks10050103

Costantini V (2022) Environmental regulation as a stimulus for eco-innovation. Elgar Encycl Econ Knowl Innov. https://doi.org/10.4337/9781839106996.00021

Díaz-García C, González-Moreno Á, Sáez-Martínez FJ (2015) Eco-innovation: insights from a literature review. Innov: Manag, Policy Pract 17(1). https://doi.org/10.1080/14479338.2015.1011060

Doni F, Fiameni M (2024) Can innovation affect the relationship between environmental, social, and governance issues and financial performance? empirical evidence from the STOXX200 index. Bus Strat Environ 33(2). https://doi.org/10.1002/bse.3500

Dziallas M, Blind K (2019) Innovation indicators throughout the innovation process: an extensive literature analysis. Technovation 80–81. https://doi.org/10.1016/j.technovation.2018.05.005

Fernández S, Torrecillas C, Labra RE (2021) Drivers of eco-innovation in developing countries: the case of Chilean firms. Technol Forecast Soc Chang 170. https://doi.org/10.1016/j.techfore.2021.120902

Geng D, Lai K, Hung, Zhu Q (2021) Eco-innovation and its role for performance improvement among Chinese small and medium-sized manufacturing enterprises. Int J Prod Econ 231. https://doi.org/10.1016/j.ijpe.2020.107869

Han MS, Chen W (2021) Determinants of eco-innovation adoption of small and medium enterprises: an empirical analysis in Myanmar. Technol Forecast Soc Chang 173. https://doi.org/10.1016/j.techfore.2021.121146

Hlioui Z, Yousfi O (2022) CSR and innovation: two sides of the same coin. Corp Soc Responsib. https://doi.org/10.5772/intechopen.94344

Hojnik J, Ruzzier M (2016) What drives eco-innovation? a review of an emerging literature. Environ Innov Soc Trans 19. https://doi.org/10.1016/j.eist.2015.09.006

Journal I (2018) Corporate social responsibility (CSR) driven innovation & opportunities for sustainable international business. Int J Bus Soc Res 8(3)
Kesidou E, Wu L (2020) Stringency of environmental regulation and eco-innovation: evidence from the eleventh five-year plan and green patents. Econ Lett 190. https://doi.org/10.1016/j.econlet.2020.109090
Kraus S, Breier M, Dasí-Rodríguez S (2020) The art of crafting a systematic literature review in entrepreneurship research. Int Entrep Manag J 16(3). https://doi.org/10.1007/s11365-020-00635-4
Liao YC, Tsai KH (2019) Innovation intensity, creativity enhancement, and eco-innovation strategy: the roles of customer demand and environmental regulation. Bus Strategy Environ 28(2):316–326
Romano CA (1990) Innovation. The promise and the perils for nursing and information technology. Comput Nurs 8(3)
Sierzchula W, Bakker S, Maat K, Van Wee B (2012) Technological diversity of emerging eco-innovations: a case study of the automobile industry. J Clean Prod 37. https://doi.org/10.1016/j.jclepro.2012.07.011
Stoneman P (2022) Defining innovation. Elgar Encycl Econ Knowl Innov. https://doi.org/10.4337/9781839106996.00014
Ubrežiová I, Sokil O, Janošková M (2020) Corporate social responsibility as a part of the strategies for business sustainability based on the innovative activity development. Strat Bus Sustain Collab Econ. https://doi.org/10.4018/978-1-7998-4543-0.ch009
van Leeuwen G, Mohnen P (2017) Revisiting the porter hypothesis: an empirical analysis of green innovation for the Netherlands. Econ Innov New Technol 26(1–2). https://doi.org/10.1080/10438599.2016.1202521
Velte P (2022) Meta-analyses on corporate social responsibility (CSR): a literature review. Manag Rev Q 72(3). https://doi.org/10.1007/s11301-021-00211-2
Wang J, Hu S, Zhang Z (2023) Does environmental regulation promote eco-innovation performance of manufacturing firms?—empirical evidence from China. Energ 16(6). https://doi.org/10.3390/en16062899
Wolfswinkel JF, Furtmueller E, Wilderom CPM (2013) Using grounded theory as a method for rigorously reviewing literature. Eur J Inf Syst 22(1). https://doi.org/10.1057/ejis.2011.51
Wu R, Zhang J, Yu Y, Jasimuddin SM, Zhang JZ (2023) The impact of value cocreation on CSR innovation and economic performance. Sustain (Switzerland) 15(5). https://doi.org/10.3390/su15054008
Yoopetch C, Nimsai S, Kongarchapatara B (2023) Bibliometric analysis of corporate social responsibility in tourism. Sustain (Switzerland) 15(1). https://doi.org/10.3390/su15010668
Zeng X, Li S, Yin S, Xing Z (2022) How does the government promote the collaborative innovation of green building projects? an evolutionary game perspective. Buildings 12(8). https://doi.org/10.3390/buildings12081179
Zhang X (2023) How does government support promote the relationship between ESG performance and innovation? J Innov Dev 3(2). https://doi.org/10.54097/jid.v3i2.9391

Risk Identification of Electric Vehicle Infrastructure Projects: A Systematic Review

Yifei Sheng, Kamal Qazi, and Amir Rahbarimanesh

Abstract Introduction: Carbon emissions and energy poverty have prompted the British government to invest in the electric vehicle industry. Despite financial support, nearly half of the charging infrastructure projects in the transportation sector experience cost overruns. Methods: With the aim of identifying risk factors within electric vehicle infrastructure projects, a review on the risk identification of electric vehicle infrastructure projects is conducted using preferred reporting items for systematic reviews and meta-analysis (PRISMA) approach and classification analysis of literature. This resulted in the evaluation of 104 critical articles collected from two online databases (Scopus and Web of Science). Findings: This research identified 24 risk factors during electric vehicle infrastructure construction process and classified them based on PESTLE model. The research findings indicate that the public–private partnership models have evolved from initial employment systems to collaborative approaches. Inflation, which can directly result in soaring prices, will trigger a series of problems including worker strikes, public opposition, and unsatisfactory services. Conclusion: These risk factors do not operate in isolation; rather, they dynamically influence each other. In the future, interview and survey will be conducted to identify the strategies which minimize the risk in electric vehicle infrastructure projects.

Keywords Risk management · Electric vehicle infrastructure · Systematic review

1 Project Context

Excessive extraction of fossil fuels leads to energy poverty (Butler 2022), climate change (McCarthy et al. 2012; Howarth et al. 2018) and air pollution (Hawkes 2015). According to National Statistics, 13.4% of households (3.26 million) in the UK are

This paper was prepared for 8th BUiD Doctoral Research Conference 2024, taught by University of Manchester.

Y. Sheng (✉) · K. Qazi · A. Rahbarimanesh
Faculty of Science and Engineering, The University of Manchester, Manchester, UK
e-mail: yifei.sheng@postgrad.manchester.ac.uk

K. Al Marri et al. (eds.), *BUiD Doctoral Research Conference 2024*, Lecture Notes in Civil Engineering 587, https://doi.org/10.1007/978-3-031-84371-6_27

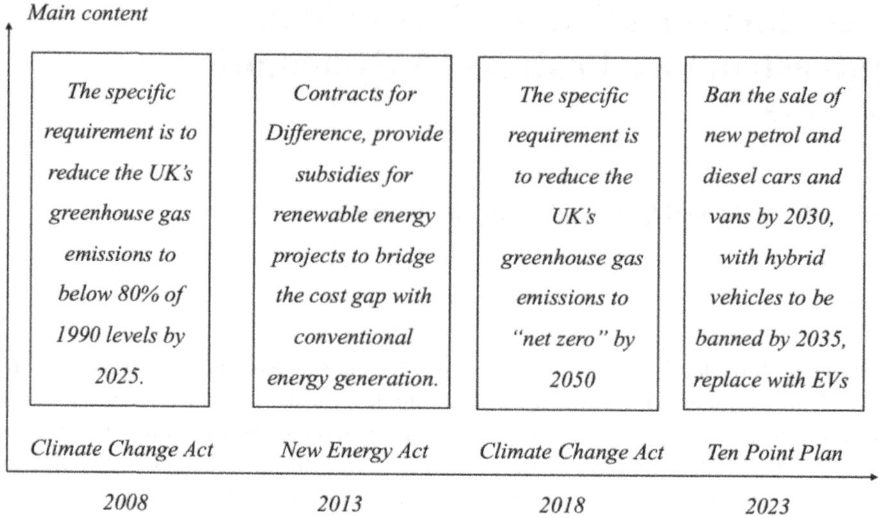

Fig. 1 The historical evolution of electric vehicle policies in UK. *Note* Author's own work 2023

suffering from fuel poverty. To meet this challenge, the UK government raised the target to "net zero by 2050" in 2019 based on *Paris Agreement* and *United Nations Framework Convention on Climate Change* (Fig. 1).

As indicated by the United Nations Environment Programme (2022), 14% of global carbon emissions stem from the transportation industry. Cheung (2022) concluded that electric vehicles (EVs) can contribute to decarbonization up to a certain level. The number of EVs on UK roads is projected to reach 5 million by 2029, 10 million by 2034, and 15 million by 2040. However, the current supply state of charging infrastructure is far from adequate. More than half of UK residents have not purchased EVs due to the unavailable charging stations (Emodi et al. 2023). Jones et al. (2023) also raised concerns among respondents regarding availability of facility and technical support. In addition, there exists an uneven geographical distribution of charging points as indicated in Fig. 2, which is a problem of insufficient supply and demand.

The number of bankruptcies among manufacturing enterprises remained stable and even experienced a slight decline since 2012 as illustrated in Fig. 3. However, this trend began to reverse following the impact of Covid-19 on the real economy. Moreover, owing to the increased inflation and increased interest in the UK, an increase in liquidations happened.

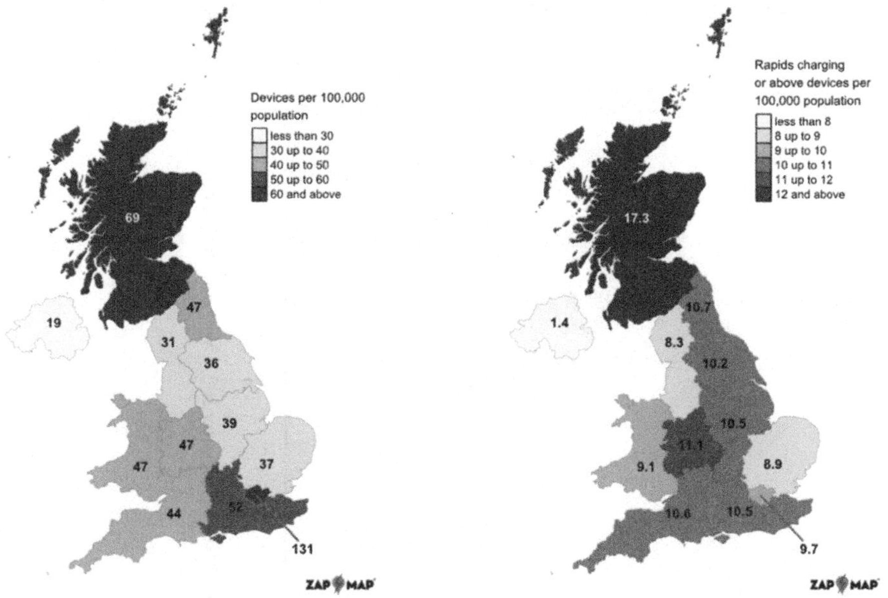

Fig. 2 Charging device and rapid charging distribution in the various cities of UK. *Note* Department for transport−GOV.UK (2023)

2 Supply Chain of Electric Vehicle Infrastructure in the United Kingdom

The British government has set clear deadlines for electric vehicle penetration by 2030 and for carbon emission targets by 2050. According to the Iron Triangle of project management, when time and quality are constraints, progress can only be ensured through the budget adjustments. Mousavi et al. (2023) argued that deviations from project progress could directly result in increased costs. If the actual expenses are underestimated, the project may encounter a funding shortfall (Siemiatycki 2018). Flyvbjerg & Bester (2021) concluded that more than half of these public engineering projects experience cost overruns.

The complex supply chain structure of electric vehicle infrastructure will bring more risks (Lefeng et al. 2020). Lithium-ion battery suppliers face dual threats from upstream and downstream (Shi et al. 2023), including the instability of chemical raw materials and the volatility of the electric vehicle market. Jadidi et al. (2020) believed that this uncertainty will directly affect the cash flow of the project operation. If a project that cost over budget is not refinanced in time, dividends can be cut, or even cash flow can be disrupted (Taha et al. 2022), especially for the megaproject (Flyvbjerg 2017) (Fig. 4).

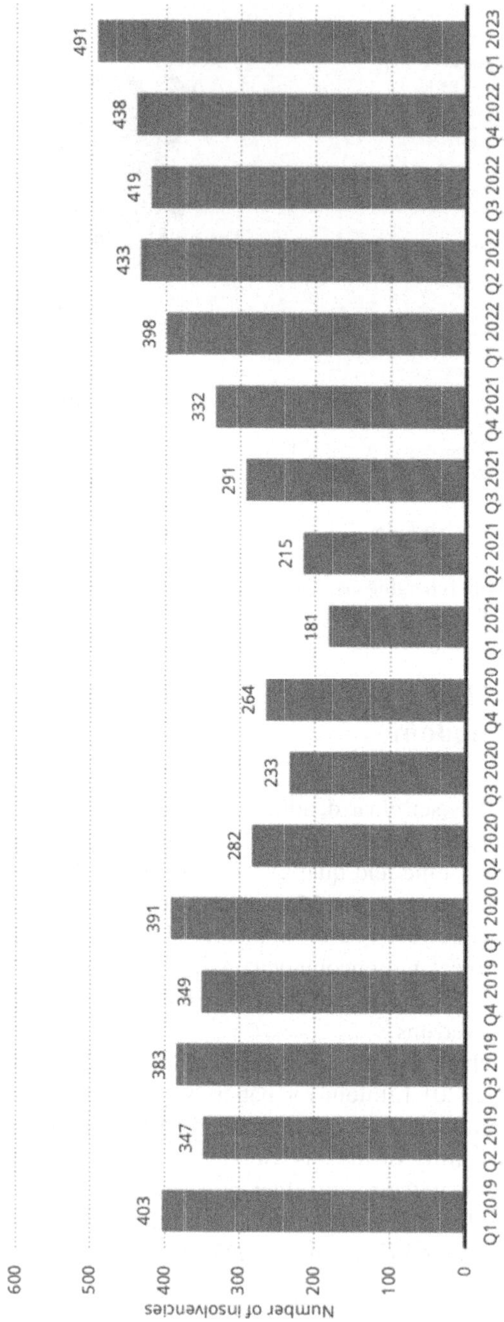

Fig. 3 Number of manufacturing company insolvencies in England and Wales from 1st quarter 2019 to 1st quarter 2023. *Note* The Insolvency Service-GOV.UK. (2023)

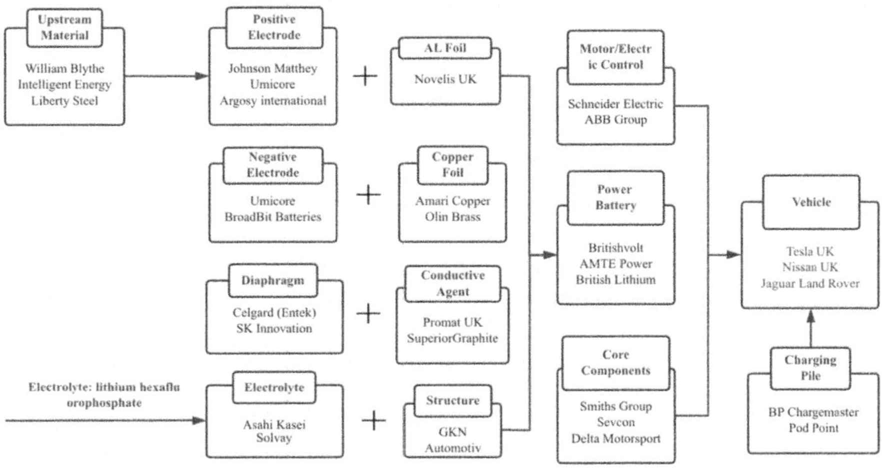

Fig. 4 Electric vehicle supply chain and related enterprises in UK. *Note* Author's own work

3 Systematic Literature Review

The aim of a systematic review is to identify all empirical evidence that fits the prespecified inclusion criteria to answer a particular research question or hypothesis. By using explicit and systematic methods when reviewing articles and all available evidence, bias can be minimized, thus providing reliable findings (Snyder 2019).

3.1 Search Keywords

The literature search was conducted using two online databases: Scopus and Web of Science, owing to their technical prowess, diversity, and size, especially in industrial sustainability, advanced technology, and manufacturing sector. The searching keywords are as follows:

(TITLE-ABS-KEY ("electric vehicle charg*") AND TITLE-ABS-KEY ("risk identification" or "risk assessment" or "risk management")) AND PUBYEAR > 2013 AND PUBYEAR < 2025 AND (LIMIT-TO (DOCTYPE, "ar")) AND (LIMIT-TO (PUBSTAGE, "final")) AND (LIMIT-TO (SRCTYPE, "j")) AND (LIMIT-TO (LANGUAGE, "English")).

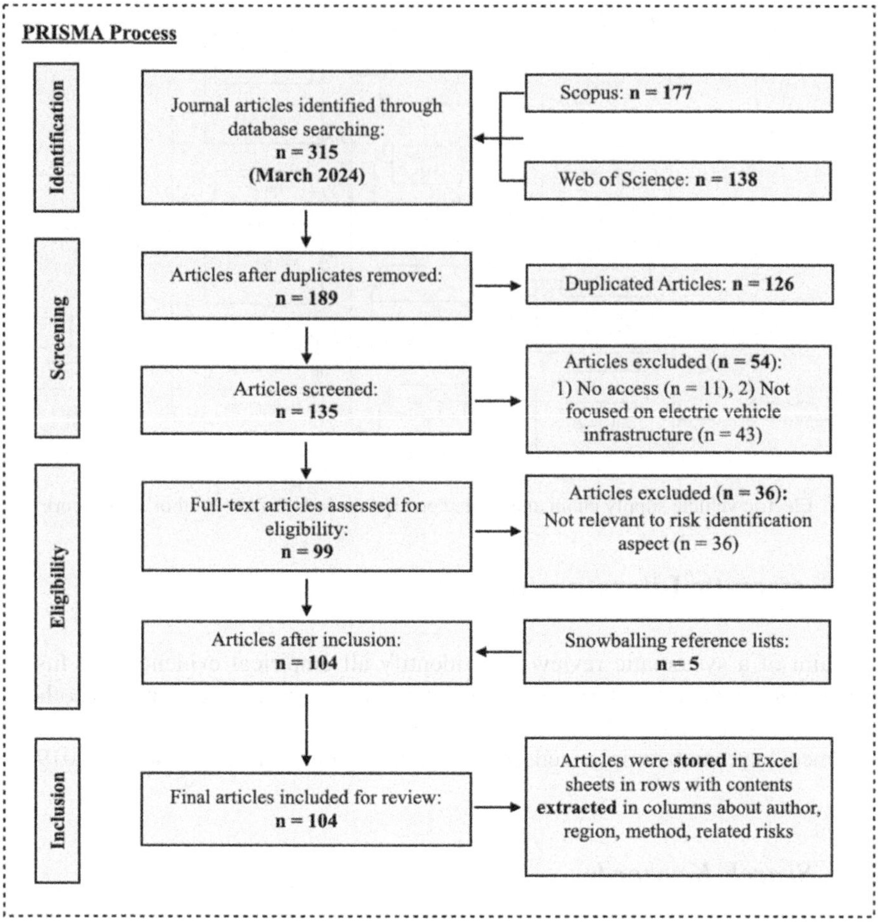

Fig. 5 PRISAM process of the systematic review. *Note* Author's own work

3.2 Inclusion and Exclusion Criteria

Scopus and web of science have different representations of various filter indicators (Fig. 5 and Table 1).

3.3 Visualization of Co-Occurrence

The next step is to bring the cross combination of keywords into different retrieval software, such as "Scopus" and "Web of Science". After screening process, 104

Table 1 Screening criteria for searching literature

	Scopus	Web of science
Year	2014 to 2024	2014 to 2024
Document type	Article	Article
Source type	Journal	Web of science index: science citation index expanded (SCI-EXPANDED) or social sciences citation index (SSCI) or emerging sources citation index (ESCI)
Publication stage	Final	Published
Language	English	English

peer-reviewed articles were selected through PRISMA. The filtered documents will be imported into VOSviewer (Fig. 6).

To conclude, the demanding topics are mainly focus on "battery management systems", "energy management", "power transmission", "energy storage", and "safety and health problem". Most of risk stems from the technology's uncertainty. Whether a technology that has not been tested by the market have investment value is a question worth considering.

The research selected 9 documents with high influence and citations from 315 documents for detailed analysis (Tables 2 and 3).

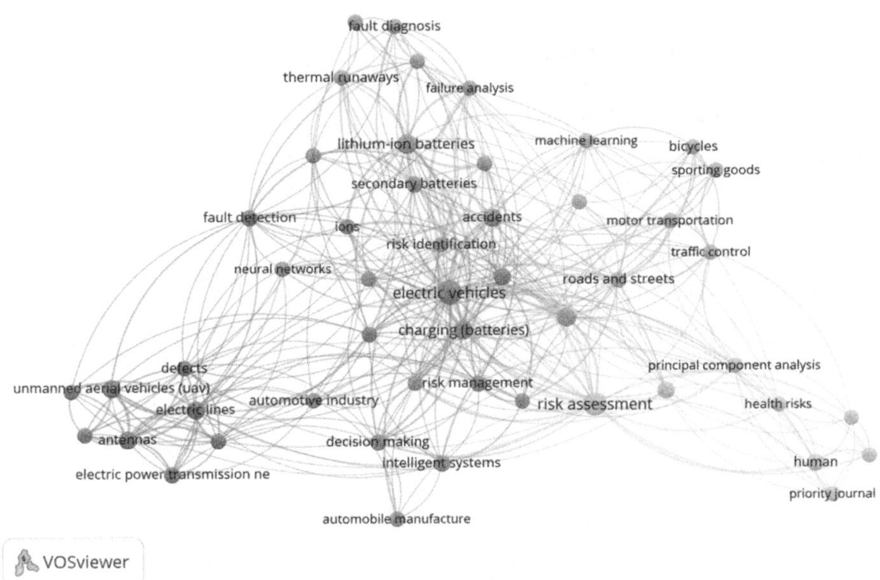

Fig. 6 Network visualization of electric vehicle charging projects and risk identification

Table 2 Risk identification for electric vehicle projects

Type	Risk factors	References									Total
		1	2	3	4	5	6	7	8	9	
Political	Government revokes project permit			*		*			*		3
	Graft and corruption	*		*				*			3
	Changes in tax policy			*	*	*			*	*	5
	Lack of government subsidies	*				*	*		*		4
	Expropriated land resources	*	*	*		*		*			5
Economic	Raw material price fluctuations				*	*	*			*	4
	Construction or operation cost overrun			*	*	*		*	*	*	6
	Inflation and interest rate volatility	*	*	*	*	*	*				6
	Unattractive investment projects	*	*	*	*	*		*		*	7
Social	Unsatisfactory charging service	*	*				*	*	*	*	6
	Labor strike or unavailable	*	*		*	*			*		5
	Public opposition			*	*	*					3
	Product obsolescence and unsalable	*	*				*		*		4
Technical	High R&D cost			*	*	*		*	*	*	6
	Lack of expertise				*			*		*	3
	Substandard construction quality	*	*					*	*	*	5
	Grid overload operation						*		*	*	3
Environment	Heavy metals pollution	*		*			*	*	*	*	6
	Noise pollution	*	*			*	*	*		*	6
	Emissions problem	*	*	*		*	*	*			6
Legal	Break environmental laws		*		*		*	*	*		5
	Break land planning act	*		*	*		*			*	5
	Consumption and data security issues		*		*	*		*	*	*	6
	Contract conflict		*		*	*			*		4

Table 3 Main reference list of the identified risks

No	Authors	Time
1	Cebrian & Kagan	2010
2	Zhao et al.	2013
3	Chou & Pramudawardhani	2015
4	Bel et al.	2017
5	Liu & Wei	2018
6	Yan et al.	2020
7	Featherman et al.	2021
8	Hajeforosh et al.	2022
9	Mousavi et al.	2023

4 Findings and Discussion

The 24 risk factors can be classified into the following six main categories. The two largest proportion on risk factors are economic and technological (22%), followed by political impact (19%), and the other three parts (legal, social, and environment) each account for about 12%.

4.1 Political Risks

These issues need to be considered during the collaboration between the public and private sectors. Due to the transformation of the public–private partnership model, the project's dependence on the government has weakened (Shu et al. 2023) (Fig. 7).

4.2 Economic Risks

The public project estimates involved in the cost–benefit analysis process are often inaccurate and biased (Flyvbjerg & Bester 2021), which may cause insufficient cash flow. Flyvbjerg & Bester (2021) conducted a comparative analysis between budget and actual cost with a sample of 2062 government projects. The research found that half of the projects were cost overrun, and the largest proportion was road facilities, such as electric vehicle infrastructure (Table 4).

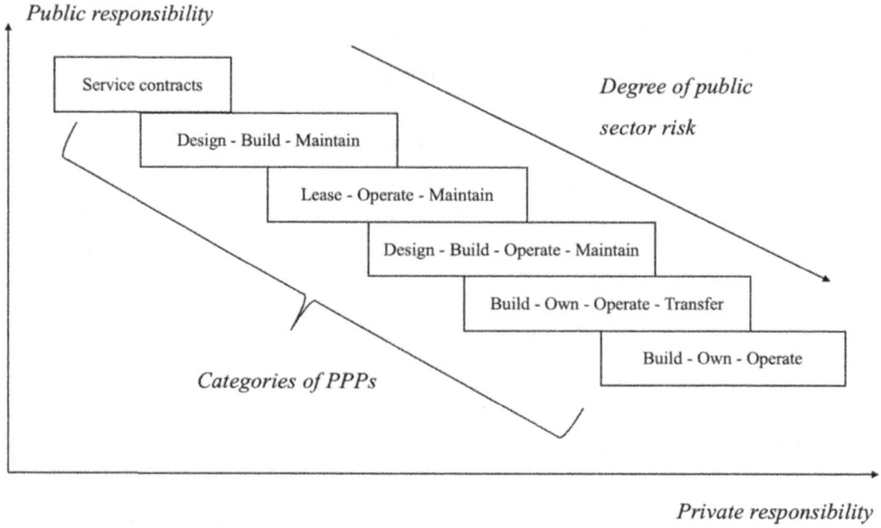

Fig. 7 Changes of the public–private partnership model in the UK

Table 4 Cost and benefit estimates for different types of public investment projects

Investment type	Cost overrun (A/E)			Benefit overrun (A/E)			p**
	N	Average	p*	N	Average	p*	
Dams	243	1.96	< 0.0001	84	0.89	< 0.0001	< 0.0001
BRT	6	1.41	0.031	4	0.42	0.12	0.007
Rail	264	1.4	< 0.0001	74	0.66	< 0.0001	< 0.0001
Tunnels	48	1.36	< 0.0001	23	0.81	0.03	0.015
Power plants	100	1.36	0.0076	23	0.94	0.11	0.0003
Buildings	24	1.36	0.00087	20	0.99	0.77	0.01
Bridges	49	1.32	0.00012	26	0.96	0.099	< 0.0001
Roads	869	1.24	< 0.0001	532	0.96	< 0.0001	< 0.0001
Total	1603	1.39/1.43a	< 0.0001	786	0.94/0.83a	< 0.0001	< 0.0001

Note Flyvbjerg & Bester (2021)

4.3 Social Risks

As delineated in the survey report from the Department for Transport (2022), in the UK, only one in three residents are satisfied with charging infrastructure services (Fig. 8).

In addition to service quality concerns, labour strikes severely impact the construction of charging piles. Public opposition is a social activity like labour strikes. Labour

Fig. 8 Service satisfaction of UK residents with public electric vehicle charging infrastructure. *Note* Department for transport (2022)

strikes are primarily initiated by workers, whereas public opposition involves residents or customers. Both are social movements caused by residents' dissatisfaction. The consequences are public resentment and failure to adapt a positive role.

4.4 Technological Risks

The main technologies of electric vehicle charger include smart grid (Medved et al. 2023), renewable energy storage (Tomar et al. 2023), and battery management system (Lipu et al. 2022). Electric vehicle charging piles are integrated into the power grid, resulting in grid overloading or instability. This uncertainty could lead to service interruptions, expensive repair cost, and contract breach. Medved et al. (2023) intelligently regulates grid power distribution to prevent the overload by detecting the charging current in smart parking and photovoltaic power distribution.

Battery technology is the critical factor for the electric vehicle development. Positive and negative electrode materials are added and combined with different chemical elements, such as copper, lead, mercury, nickel, and lithium. The price of lithium hydroxide has increased to seven times its previous level (Schmidt & Staffell 2023). The increase in raw material cost has led to one time rise in the ternary batteries (NMC) prices, and one and half times increase in the lithium iron phosphate battery packs (LFP) (Fig. 9).

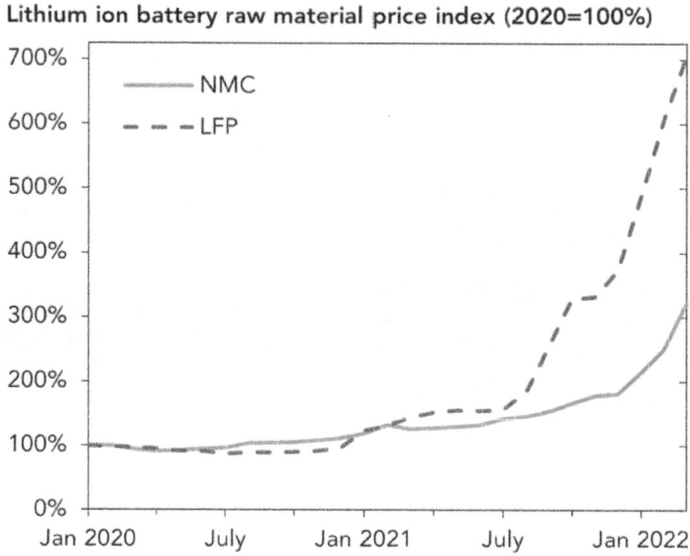

Fig. 9 Price fluctuation chart of lithium raw materials (Schmidt & Staffell 2023). *Note* Schmidt & Staffell (2023)

In conclusion, the estimation of electric vehicle energy battery prices frequently lack accuracy owing to the complex chemistries and designs. Technological breakthroughs in lithium-oxygen batteries, solid batteries, and silicon anode materials could disrupt the market competition. Advancements in battery technology may diminish the compatibility of existing charging stations, or even require reconstruction.

4.5 Environmental Risks

Environmental pollution can bring severe ecosystem consequences, possibly leading to the forced suspension of operations, public backlash, community health damage, and legal repercussions, especially in urban areas (Surya et al. 2020). Emission, material, and noise are three main pollution concerns in the construction or operation. The production and installation of these charging stations necessitate various metal and plastics materials. Their production process will involve carbon emissions and environmental pollution. In 2010, EU had not established comprehensive regulations or standards on the environmental impact of charging piles. As a result, several equipment may not have been subject to strict environmental safety assessments. The production phases of batteries, electronic control systems, cables and interfaces could all bring metal pollution to the environment. Moreover, the construction team maybe generate noise that reaches disturbing levels, which could potentially

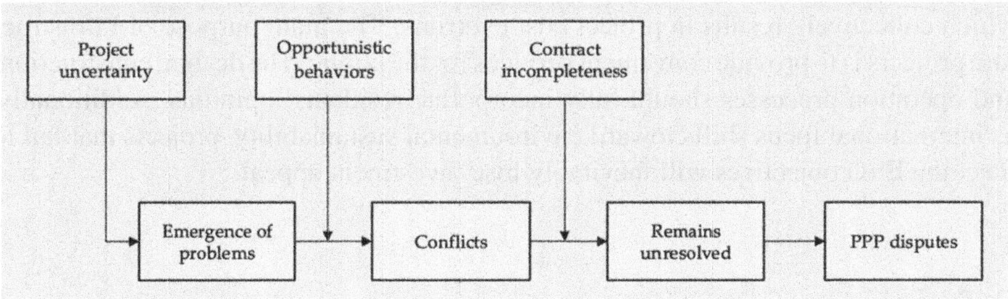

Fig. 10 The mechanism of conflict in PPP projects (Zheng et al. 2022)

result in residents' complaints. In conclusion, these issues underscore the importance of responsible management and adherence to quality and environmental standards. Studies on corporate environmental responsibility in heavy pollution industries reveal a strategic correlation between environmental stewardship and financial performance (Yang et al. 2020). In addition, Gubeladze & Pavliashvili (2021) calls for sustainable ecosystem management and optimal resource use.

4.6 Legal Risks

Conflicts, litigation, and contractual disputes can arise among multiple stakeholders in the PPP. Zheng et al. (2022) concluded that the probability of conflicts occurrence in public–private partnership project is as high as 94% through Fault Tree Analysis. Lawsuits arising from conflicts can result in property damage, or even bankruptcy. Such conflicts may not only arise during the construction phase, but also persist through the operation and service phases (Fig. 10).

The structure of this contract model is design–build–finance–Operate–Maintenance (DBFOM). Contractors with a package contract can achieve full lifecycle management, minimizing the time and cost overruns risk.

5 Conclusion & Recommendations

The study identified 24 risk factors in electric vehicle infrastructure projects through systematic literature review. These risk factors were divided into six key aspects, including policy, economy, society, technology, environment, and law. The degree of public sector intervention has gradually decreased. Moreover, the UK's high standing in global government transparency indices suggests that the risk caused by governmental corruption is minimal. In addition, financial or economic issues, such as inflation, cost fluctuations, and interest spread, can form a closed-loop structure, which collectively results in project cost overruns. The main purpose of infrastructure projects is to provide convenient services for the public. The design, construction, and operation processes should fully incorporate residents' opinions. Additionally, as international focus shifts toward environmental sustainability, projects that fail to meet the ESG objectives will inevitably lose investment appeal.

References

Butler C (2022) Energy poverty, practice, and policy. Palgrave Macmillan, Cham, Switzerland. https://directory.doabooks.org/handle/20.500.12854/91293

Cheung WM (2022) A scenario-based approach to predict energy demand and carbon emission of electric vehicles on the electric grid. Environ Sci Pollut Res Int 29(51):77300–77310. https://doi.org/10.1007/s11356-022-21214-w

Department for transport (2022) Electric vehicle charging research-survey with electric vehicle drivers. Department for transport, London. https://assets.publishing.service.gov.uk/media/628f5603d3bf7f037097bd73/dft-ev-driver-survey-summary-report.pdf

Department for transport-GOV.UK (2023) Electric vehicle charging device statistics: January 2023. Department for transport-GOV.UK. https://www.gov.uk/government/statistics/electric-vehicle-charging-device-statistics-january-2023/electric-vehicle-charging-device-statistics-january-2023#about-this-release

Emodi NV, Akuru UB, Dioha MO, Adoba P, Kuhudzai RJ, Bamisile O (2023) The role of internet of things on electric vehicle charging infrastructure and consumer experience. Energies (Basel) 16(10):4248. https://doi.org/10.3390/en16104248

Flyvbjerg B, Bester DW (2021) The cost-benefit fallacy: why cost-benefit analysis is broken and how to fix it. J Benefit-Cost Anal 12(3):395–419. https://doi.org/10.1017/bca.2021.9

Flyvbjerg B (2017) The oxford handbook of megaproject management. Oxford University Press, Oxford, UK. https://doi-org.manchester.idm.oclc.org/10.1093/oxfordhb/9780198732242.001.0001

Gubeladze I, Pavliashvili N (2021) Agroecological sustainability, green economy, and business model of financial integration. World Sci 4(4):1–8. https://doi.org/10.1007/s10644-022-09408-1

Hawkes N (2015) Air pollution in UK: the public health problem that won't go away. BMJ 350:h2757. https://doi.org/10.1136/bmj.h2757

Howarth C, Morse-Jones S, Brooks K, Kythreotis AP (2018) Co-producing UK climate change adaptation policy: an analysis of the 2012 and 2017 UK climate change risk assessments. Environ Sci Policy 89:412–420. https://doi.org/10.1016/j.envsci.2018.09.010

Jadidi H, Firouzi A, Rastegar MA, Zandi M (2020) Bayesian updating of solar resource data for risk mitigation in project finance. Sol Energy 207:1390–1403. https://doi.org/10.1016/j.solener.2020.07.096

Jones B, Nguyen-Tien V, Elliott RJR (2023) The electric vehicle revolution: critical material supply chains, trade, and development. World Econ 46(1):2–26. https://doi.org/10.1111/twec.13345

Lefeng S, Shengnan L, Chunxiu L, Yue Z, Cipcigan L, Acker TL (2020) A framework for electric vehicle power supply chain development. Util Policy 64:101042. https://doi.org/10.1016/j.jup.2020.101042

Lipu MSH, Mamun AA, Ansari S, Miah MS, Hasan K, Meraj ST, Abdolrasol MGM, Rahman T, Maruf MH, Sarker MR, Aljanad A, Tan NML (2022) Battery management, key technologies, methods, issues, and future trends of electric vehicles: a pathway toward achieving sustainable development goals. Batteries (Basel) 8(9):119. https://doi.org/10.3390/batteries8090119

McCarthy MP, Harpham C, Goodess CM, Jones PD (2012) Simulating climate change in UK cities using a regional climate model, HadRM3. Int J Climatol 32(12):1875–1888. https://doi.org/10.1002/joc.2402

Medved D, Bena L, Oliinyk M, Dzmura J, Mazur D, Martinko D (2023) Assessing the effects of smart parking infrastructure on the electrical power system. Energies 16(14):5343. https://doi.org/10.3390/en16145343

Mousavi SA, Hafezalkotob A, Ghezavati V, Abdi F, Mobarra R (2023) Sustainable construction project of electric vehicle charging stations: a risk-based hybrid decision-making approach. J Clean Prod 402:136565. https://doi.org/10.1016/j.jclepro.2023.136565

Schmidt O, Staffell I (2023) Monetizing energy storage-a toolkit to assess future cost and value. Oxford University Press, Oxford. https://doi.org/10.1093/oso/9780192888174.001.0001

Shi Y, Feng Y, Zhang Q, Shuai J, Niu J (2023) Does China's new energy vehicles supply chain stock market have risk spillovers? evidence from raw material price effect on lithium batteries. Energy 262:125420. https://doi.org/10.1016/j.energy.2022.125420

Shu X, Smyth S, Haslam J (2023) Understanding absences and ambiguities of post-decision project evaluation in the UK's PPPs: drawing from the sociology of ignorance. Account, Audit, & Account J 37(1):363–392. https://doi.org/10.1108/AAAJ-03-2020-4451

Siemiatycki M (2018) The making and impacts of a classic text in megaproject management: the case of cost overrun research. Int J Project Manage 36(2):362–371. https://doi.org/10.1016/j.ijproman.2016.07.003

Snyder H (2019) Literature review as a research methodology: an overview and guidelines. J Bus Res 104:333–339. https://doi.org/10.1016/j.jbusres.2019.07.039

Surya B, Sutanta H, Rudiarto I (2020) Environmental pollution control and sustainability management of slum settlements in Makassar city, South Sulawesi, Indonesia. Land 9(9):279–279. https://doi.org/10.3390/land9090279

Taha G, Sherif A, Badawy M, El Kafhali S, El Kafhali S (2022) Overall cost overrun estimate in residential projects: a hybrid dynamics approach. Appl Comput Intell Soft Comput 2022:1–17. https://doi.org/10.1155/2022/2285971

The Insolvency Service-GOV.UK (2023) Commentary-monthly insolvency statistics May 2023. The Insolvency Service. https://www.gov.uk/government/statistics/monthly-insolvency-statistics-may-2023/commentary-monthly-insolvency-statistics-may-2023

Tomar A, Viral R, Asija D, Rao UM, Sarwar A (2023) Smart grids for renewable energy systems, electric vehicles, and energy storage systems. CRC Press, Boca Raton. https://doi-org.manchester.idm.oclc.org/10.1201/9781003311195

United Nations Environment Programme (2022) Emissions Gap Report 2022: the closing window-climate crisis calls for rapid transformation of societies. United Nation. https://wedocs.unep.org/20.500.11822/40874

Yang X, Zhang Y, Wang Y (2020) Internal control quality, enterprise environmental protection investment and finance performance: an empirical study of China's a-share heavy pollution industry. Int J Environ Res Public Health 17(17):6082. https://doi.org/10.3390/ijerph17176082

Zheng X, Liu Y, Jiang J, Su N (2022) Quantitative assessment of the likelihood of disputes in PPP projects using fault tree analysis. Buildings (Basel) 12(3):384–384. https://doi.org/10.3390/buildings12030384

Stakeholder Management in Construction Projects: A Systematic Literature Review

Lantian Liu, Amir Rahbarimanesh, Mohamed Abadi, and Kun Wang

Abstract Stakeholder management (SM) is critical in project management. However, SM was less effective in the construction industry. Given the critical role and complexity of Stakeholder Management in Construction Projects (SMCP), this research has used a systematic literature review approach to select 56 articles for bibliometric and content analysis. Findings of this research provide a clear understanding of the state-of-the-art study on SMCP. These include communication as one of the major topics, the trend of this area is to integrate sustainability and BIM in SMCP. In addition, this study identified critical challenges in SMCP: (1) Dynamic management of stakeholders. (2) Increasing interest in the construction industry to promote digital transformation in SMCP. (3) Integrate sustainability into SM. Although the limited search scope reflects the overall development of SMCP research trends, it may overlook some research not included in them. Overall, this comprehensive review of SMCP provides valuable insights for researchers, giving them a detailed view of previous studies and it recommends future research agenda.

Keywords Stakeholder management · Construction projects · Systematic literature review

1 Introduction

"Stakeholders" was first defined as "those groups who can affect or are affected by the achievement of an organisation's purpose" (Freeman 1984). Stakeholder Management (SM) is an iterative process that identifies individuals/groups that could impact or be impacted by the project, analyses their needs and impact, and then develops proper strategies for engaging project stakeholders effectively to support project decisions and implementation (PMI 2017). SM is a significant concept in the context of projects (Aaltonen et al. 2008). In the past few years, the interest in SM has increased.

L. Liu (✉) · A. Rahbarimanesh · M. Abadi · K. Wang
Department of Engineering Management, School of Engineering, The University of Manchester, Manchester, UK
e-mail: lantian.liu@manchester.ac.uk

K. Al Marri et al. (eds.), *BUiD Doctoral Research Conference 2024*, Lecture Notes in Civil Engineering 587, https://doi.org/10.1007/978-3-031-84371-6_28

Littau et al. (2010) conducted a comprehensive review of stakeholder theory within leading project management journals and concluded that stakeholder plays a critical role in successful project management, with the construction industry being one of the main industries in this context. Many studies discussed improving stakeholder identification and analysis, such as the proposed saliency model (Olander & Landin 2008; Mok et al. 2017). Several studies explored Critical Success Factors (CSFs) that influence SM (Yang & Shen 2015; Mashali et al. 2023). However, though project managers tried various methods to improve SM, there were still some failed projects (Bulğan & Taş 2023). Reasons include having multiple stakeholders, which means more conflict rather than consensus (Loosemore 2006). The diversity of tasks and various parties involved in construction projects also undergo many changes and uncertainty (Kumar et al. 2024; Yang & Shen 2015). Previously published studies are limited to investigating one stage of SM process or focusing on finding CSFs. Up to now, little attention has been given to the overview of Stakeholder Management in Construction Projects (SMCP) research.

This research aims to fill above gap and provide a comprehensive overview of SMCP. Therefore, this study formulates three research questions:

RQ1: What are the major themes discussed within SMCP?
RQ2: What is the research trend of SMCP in the last 5 years?
RQ3: What are the critical challenges of SMCP?

By addressing the three research questions, this research demonstrates the state-of-the-art study on SMCP and provides potential opportunities to encourage further research in this area.

2 Methodology

2.1 Systematic Literature Review (SLR)

This research used an SLR approach, which differs from conventional reviews. SLR is a scientific, transparent, and reproducible literature review method, and it has been described as gold standard among reviews (Snyder 2019). It helps find the most significant gaps contributing to theory development (Tranfield et al. 2003). PRISMA is a preferred reporting item for systematic reviews and meta-analysis, which shows a transparent and clear literature selection process (Moher et al. 2010). Figure 1 shows the PRISMA process.

2.1.1 Stage 1: Identification

Identifying relevant literature by searching the keywords ("stakeholder management" OR "management of stakeholder") AND ("construction project" OR "infrastructure

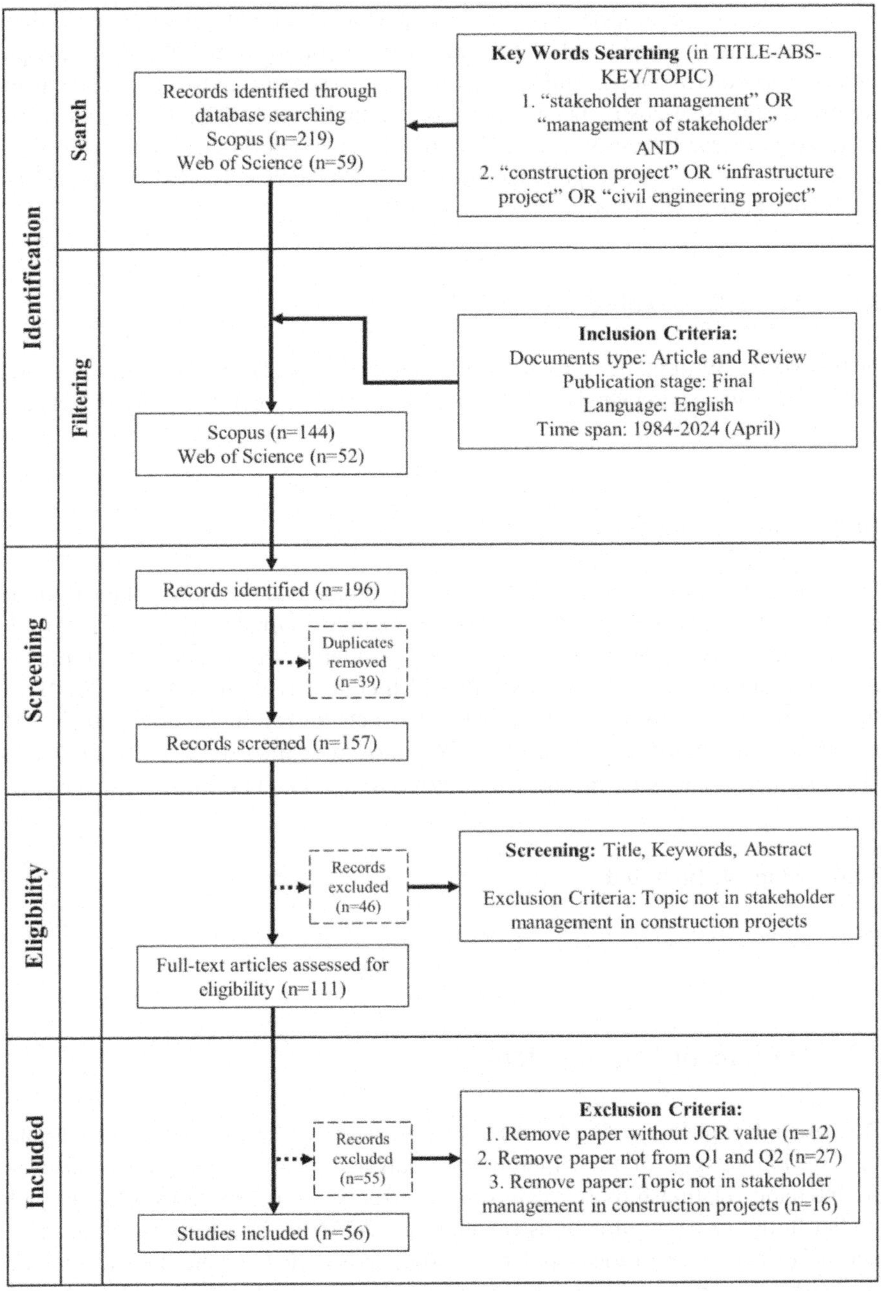

Fig. 1 PRISMA process for this systematic review

project" OR "civil engineering project" on databases "Web of Science Core Collection (WoS)" and "Scopus". The peer-reviewed literature in the WoS and Scopus databases cover research achievements in many research fields and represent the largest databases of abstracts and citations (Alaloul et al. 2022). The initial search retrieved 59 research from WoS and 219 from Scopus. Filtering was done by limiting the search to peer-reviewed published articles and reviews in English between 1984 and 2024 (April), which led to 196 articles left.

2.1.2 Stage 2: Screening

At this stage, 39 duplicate articles were removed first. Then, the title, keywords, and abstract were screened to remove unrelated papers, which led to 111 articles remaining.

2.1.3 Stage 3: Eligibility

The Journal Citation Report (JCR) provides journals that only meet their rigorous selection criteria, which locate the most influential journals (Clarivate 2023), 12 papers without JCR values were removed. Considering that papers from reputable and cited journals increase the credibility of literature (Knight & Nurse 2020), the articles are selected from highly academic journals ranked in the first quartile (Q1) and second quartile (Q2). 27 papers were removed as they were not from Q1/Q2. The final step is checking full texts to remove 16 unrelated papers.

2.1.4 Stage 4: Included

There are 56 articles finally selected for review.

2.2 Analysis of Selected Articles

This research comprised of bibliometric analysis and content analysis. Co-occurrence analysis is one type of the bibliometric analysis, usually been used to discover the thematic relationships by analysing the frequency of the keywords and help predict future research directions (Donthu et al. 2021). Next section demonstrates a co-occurrence map of the popular keywords in SMCP. Content analysis includes grouping the themes of selected articles and finding the trends and challenges in SMCP research.

3 Results

3.1 Findings of Co-Occurrence Analysis

The network map of keyword co-occurrence analysis is shown below in Fig. 2. The size of nodes reflects the frequency of their occurrence, larger nodes indicate that the keyword appears more frequently (Yu et al. 2020). Also, the Table 1 shows the most frequently keywords present the popular topics, objects, and methodology in SMCP research.

Figure 3 shows the overlay visualisation of the keywords co-occurrence map of the selected 56 papers. From the distribution of popular keywords along the temporal perspective to understand the trend of SMCP research. The colours range from dark blue (older publications) to yellow (latest publications) of the overlay map demonstrating that SMCP research shifted their attention to 'sustainability' and 'building information modelling' in recent years, showing researchers started to combine their thinking of SM with sustainability and digitalisation.

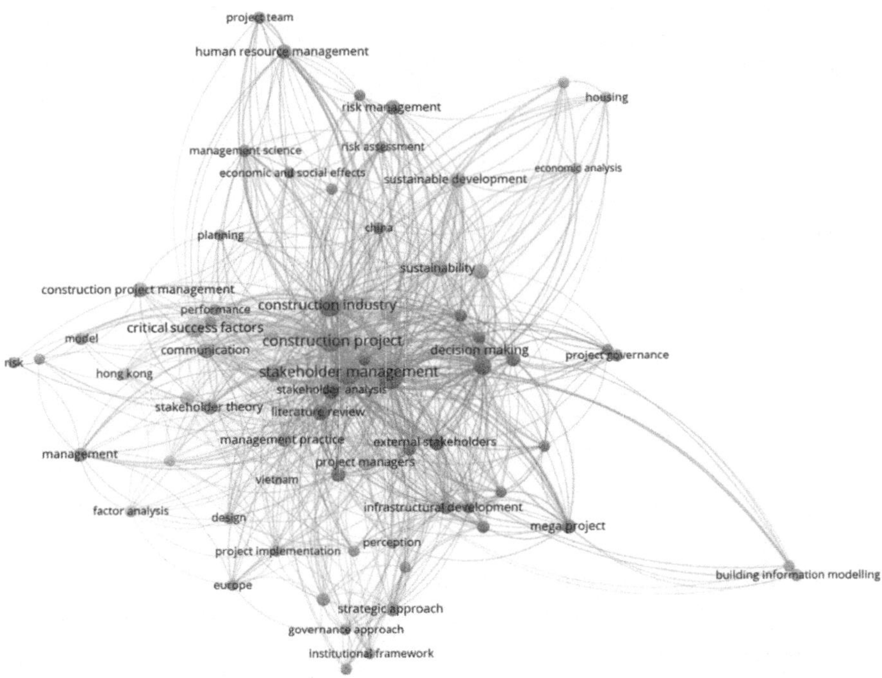

Fig. 2 Network map of keyword co-occurrence via VOSviewer

Table 1 Popular keywords in SMCP research

Popular topics	Popular research objects	Popular research methodology
Stakeholder analysis	Construction project	Literature review
Stakeholder engagement	Infrastructure project	Systematic literature review
Communication	Mega project	Surveys
Critical success factors		Case studies
Building information modelling (BIM)		
Sustainability		

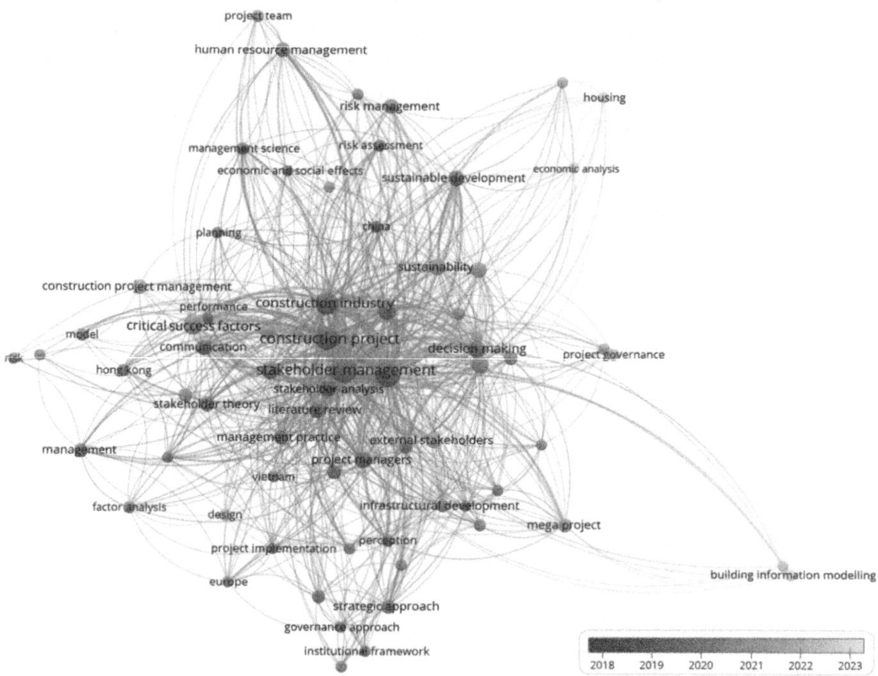

Fig. 3 Overlay visualization of the keywords co-occurrence via VOSviewer

3.2 *Findings of Content Analysis*

3.2.1 Thematic Analysis

Table 2 outlines the distribution of publications via research topics and periods. The first four themes align with SM process as defined in the PMBOK guide (PMI 2017). ‘Entire Process’ refers to the comprehensive SM process emphasised in the article. ‘Others’ indicates studies without considering SM process.

Table 2 The distribution of publications across identified research themes

Research theme	Period year				
	2005–2008	*2009–2012*	*2013–2016*	*2017–2020*	*2021–2023*
Identify and analyse stakeholders	2	3	1	7	4
Plan strategies for stakeholder engagement	0	0	0	0	0
Manage stakeholder engagement	1	0	1	3	2
Monitor stakeholder engagement	1	0	0	1	1
Entire process	3	2	5	4	7
Others	0	0	2	4	2

Identify and Analyse Stakeholders

Identifying and analysing stakeholders is a popular topic in SMCP research. Stakeholder analysis is used to determine how external stakeholders' needs and concerns can be addressed and to evaluate the potential implications of failing to meet these needs (Olander & Landin 2008). How to efficiently identify and analyse stakeholders is the first research direction that many scholars consider, and a variety of models or tools have been developed, such as saliency framework (Aaltonen et al. 2008), stakeholder network (Mok et al. 2017). Previous studies focus on the identification and analysis of individual attributes of stakeholders, without considering the interrelationships between stakeholders (Mok et al. 2017). Focus on selecting the best stakeholders does not guarantee project success (Dadpour et al. 2022). Yang et al. (2011) stated that no method is perfect for identifying and prioritising stakeholders, it is best to integrate multiple approaches in need.

Plan Strategies for Stakeholder Engagement

No study refers to this theme because the plan stage depends more on the practical scenario.

Manage Stakeholder Engagement

Communication has become a focus among scholars. El-Diraby and Wang (2005) proposed that one-way communication should be transformed into two-way communication; that is, the person in charge of the project cannot only convey information to the public but should constantly listen to the opinions of the public for better feedback. Yang et al. (2010) had the same conclusion from their case study the project was delayed 8 years due to one-way communication. During the management phase of communication, stakeholders' interpretations are complex and unpredictable (Collinge & Harty 2014). A transparent and inclusive SM program can help build better communication (Di Maddaloni & Derakhshan 2019).

Monitor Stakeholder Engagement

The importance of monitoring stakeholders has been realised. Chinyio and Akintoye (2008) highlighted the salience of stakeholders must be continuously monitored, while methods of engaging with them are required to be varied based on their dynamism. Yang et al. (2018) also indicated that project teams have recognized the importance of continuity in managing stakeholders in megaprojects and recommend promoting ongoing stakeholder engagement.

Entire SM Process

As for the studies focusing on entire SM process. Mok et al. (2015) pointed out that existing methods are mainly aimed at the project's early stages and lack a comprehensive management model covering the entire project life cycle. SM frameworks based on different theories have been proposed (Gaur & Tawalare 2022; Yang & Shen 2015) to provide systematic guidance for the entire SM process. Ngampravatdee et al. (2023) explored the communication issues throughout the project SM process and believed effective communication builds a bridge between stakeholders.

Significantly, 8 articles primarily investigated the CSFs in SMCP. Those papers are grouped into the 'entire SM process' because CSFs listed are involved in different stages. For example, Yang et al. (2010) listed key factors such as proper identification of stakeholders, understanding stakeholders' needs and constraints, and effective communication with stakeholders.

Others

Papers in the 'other' group focus on other aspects. Zheng et al. (2016) provided a literature review on Social Network Analysis (SNA) to offering insights. Xia et al. (2018) integrated SM with risk management, advocating for a combined method to consider risks and stakeholders concurrently. Xue et al. (2022) explored collaborative organizational citizenship behaviour in construction projects, suggesting the selection of partners based on mutual trust and prior collaboration experience.

Future Areas and Recommendations

It can be suggested that future areas focus more on the "plan" and "monitor" stages as they got little attention. This study recommends that researchers continue to integrate different topics (e.g., risk management) within SM so that they have a broader perspective to obtain ideas for enhancing SM.

3.2.2 Realise the Critical Challenges in SMCP

Dynamic Management of Stakeholder

Collinge (2020) stated that projects are distinguished by stakeholders' dynamics and the evolving nature of relationships. Stakeholder influence strategies are dynamic throughout the project lifecycle as stakeholders' roles change over time, and different actions are adopted regarding the changing project environment (Mok et al. 2015; Yang & Shen 2015). The stakeholder "saliency model" is the earliest proposed SM model (Mitchell et al. 1997), but cannot reflect the dynamically changing attitudes of stakeholders (Di Maddaloni & Davis 2017). Regular network analysis of stakeholder issues and relationships throughout the project lifecycle will help address dynamic changes in the project environment (Mok et al. 2017). Yang et al. (2018) noted that stakeholder engagement practices have developed to the point that the dynamics of stakeholder demand can be monitored by project teams. Although many authors mentioned the dynamic nature in SM, no article has focused on discussing dynamics among the 56 articles reviewed. The possible reason is the significant time and practical experience required to study dynamics. And the stakeholders are diverse and complex, it is difficult to achieve a stable monitoring process.

Increasing Interest in the Construction Industry to Promote the Digital Transformation in SMCP

Prebanić and Vukomanović (2021) conducted a SLR of digital transformation in SM and emphasised that although there are many technologies and tools for digital management of stakeholder (e.g., Darzin, BIM), there are few examples of successful implementation. The construction industry is slow when it comes to implementation of new digital ways of working, as evidenced by the fact that construction is recorded as one of the least digitised industries (Prebanić & Vukomanović 2021). Although the use of digital technology for SM is gradually becoming an emerging trend, there is still a lack of motivation to achieving digitalisation in SMCP.

Integrate Sustainability into SM

Scholars believe that appropriate SM contributes to the sustainable development of projects (Di Maddaloni & Derakhshan 2019). Safapour et al. (2019) discussed the relationship between stakeholders and sustainable development in projects under the perspective of COVID-19 pandemic and highlighted that efficient SM can help achieve sustainable development goals. Khalilzadeh et al. (2023) pointed out, "there is currently a huge gap in the construction industry in providing sustainable stakeholder engagement strategies and maximising stakeholder engagement levels".

4 Conclusion

This study systematically analysed SMCP research and identified major themes, including identify and analyse stakeholders, stakeholder engagement, communication and CSFs. It also identified that the recent literature is interested in exploring sustainability and BIM in SMCP. The critical challenges in SMCP are: (1) Dynamic management of stakeholder. (2) Increasing interest in the construction industry to promote the digital transformation in SMCP. (3) Integrate sustainability into SM. These findings demonstrate the state-of-the-art study on SMCP and provide clear directions for the scholars who focus on SMCP field.

5 Limitations and Future Work

This research is limited to using Scopus and WoS only, thus restricting the coverage of the reviewed literature, although these two databases represent the most extensive databases (Alaloul et al. 2022). Moreover, this research focused only on construction projects. The findings are limited to a specific industry, which may not work for general SM. Future work needs to expand the scope of literature review and explore the identified gaps to enhance SM in construction projects and even for other sectors.

References

Aaltonen K, Jaakko K, Tuomas O (2008) Stakeholder salience in global projects. Int J Project Manage 26(5):509–516

Alaloul WS, Alzubi KM, Malkawi AB, Al Salaheen M, Musarat MA (2022) Productivity monitoring in building construction projects: a systematic review. Eng Constr Archit Manag 29(7):2760–2785

Bulğan J, Taş EF (2023) Future opportunities for stakeholder management in construction projects

Chinyio EA, Akintoye A (2008) Practical approaches for engaging stakeholders: findings from the UK. Constr Manag Econ 26(6):591–599

Clarivate (2023) 2022 journal imapct factor. J Cit Rep. Web of Science
Collinge W (2020) Stakeholder engagement in construction: exploring corporate social responsibility, ethical behaviors, and practices. J Constr Eng Manag 146(3):04020003
Collinge WH, Harty CF (2014) Stakeholder interpretations of design: semiotic insights into the briefing process. Constr Manag Econ 32(7–8):760–772
Dadpour M, Shakeri E, Nazari A, Moghadas Nejad F (2022) Optimizing the portfolio of road project stakeholders with cost and risk approach. Int J Constr Manag 1–12
Di Maddaloni F, Davis K (2017) The influence of local community stakeholders in megaprojects: rethinking their inclusiveness to improve project performance. Int J Project Manage 35(8):1537–1556
Donthu N, Kumar S, Mukherjee D, Pandey N, Lim WM (2021) How to conduct a bibliometric analysis: an overview and guidelines. J Bus Res 133:285–296
El-Diraby T, Wang B (2005) E-society portal: integrating urban highway construction projects into the knowledge city. J Constr Eng Manag 131(11):1196–1211
Freeman RE (1984) Strategic management: a stakeholder approach. Cambridge university press
Gaur S, Tawalare A (2022) Investigating the role of BIM in stakeholder management: evidence from a metro-rail project. J Manag Eng 38(1):05021013
Khalilzadeh M, Kebriyaii O, Rezaei R (2023) Identification and selection of stakeholder engagement strategies: case study of an Iranian oil and gas construction project. Int J Constr Manag 23(3):484–494
Knight R, Nurse JR (2020) A framework for effective corporate communication after cyber security incidents. Comput Secur 99:102036
Kumar V, Singh R, Pandey A (2024) Multiple stakeholders' critical success factors scale for success on large construction projects. Asian J Civ Eng 25(2):1691–1705
Littau P, Jujagiri NJ, Adlbrecht G (2010) 25 years of stakeholder theory in project management literature (1984–2009). Proj Manag J 41(4):17–29
Loosemore M (2006) Managing project risks. In: The management of complex projects: a relationship approach. pp 187–204
Di Maddaloni F, Derakhshan R (2019) A leap from negative to positive bond. A step towards project sustainability. Adm Sci 9(2):41
Mashali A, Elbeltagi E, Motawa I, Elshikh M (2023) Stakeholder management challenges in mega construction projects: critical success factors. J Eng, des Technol 21(2):358–375
Mitchell RK, Agle BR, Wood DJ (1997) Toward a theory of stakeholder identification and salience: defining the principle of who and what really counts. Acad Manag Rev 22(4):853–886
Moher D, Liberati A, Tetzlaff J, Altman DG, Group P (2010) Preferred reporting items for systematic reviews and meta-analyses: the PRISMA statement. Int J Surg 8(5):336−341
Mok KY, Shen GQ, Yang J (2015) Stakeholder management studies in mega construction projects: a review and future directions. Int J Project Manage 33(2):446–457
Mok KY, Shen GQ, Yang RJ (2017) A network theory-based analysis of stakeholder issues and their interrelationships in large construction projects: a case study. Int J Constr Manag 17(3):210–227
Ngampravatdee C, Gharehbaghi K, Hosseinian-Far A, Tee KF, McManus K (2023) Strategic initiatives for large transport infrastructure planning: reinforcing sustainability in urban transportation through better stakeholder engagement. Sustain 15(18):13912
Olander S, Landin A (2008) A comparative study of factors affecting the external stakeholder management process. Constr Manag Econ 26(6):553–561
PMI (2017) A guide to the project management body of knowledge (PMBOK Guide), 6th edn. Project Management Institute, Inc.
Prebanić KR, Vukomanović M (2021) Realizing the need for digital transformation of stakeholder management: a systematic review in the construction industry. Sustain 13(22):12690
Safapour E, Kermanshachi S, Kamalirad S, Tran D (2019) Identifying effective project-based communication indicators within primary and secondary stakeholders in construction projects. J Leg Aff Disput Resolut Eng Constr 11(4):04519028

Snyder H (2019) Literature review as a research methodology: an overview and guidelines. J Bus Res 104:333–339

Tranfield D, Denyer D, Smart P (2003) Towards a methodology for developing evidence-informed management knowledge by means of systematic review. Br J Manag 14(3):207–222

Xia N, Zou PX, Griffin MA, Wang X, Zhong R (2018) Towards integrating construction risk management and stakeholder management: a systematic literature review and future research agendas. Int J Project Manage 36(5):701–715

Xue H, Sun T, Zheng J (2022) Organizational elements and collaborative organizational citizenship behavior in off-site construction projects: configurational approach. J Constr Eng Manag 148(10):04022108

Yang RJ, Shen GQ (2015) Framework for stakeholder management in construction projects. J Manag Eng 31(4):04014064

Yang J, Shen GQ, Drew DS, Ho M (2010) Critical success factors for stakeholder management: construction practitioners' perspectives. J Constr Eng Manag 136(7):778–786

Yang J, Shen GQ, Bourne L, Ho CMF, Xue X (2011) A typology of operational approaches for stakeholder analysis and engagement. Constr Manag Econ 29(2):145–162

Yang RJ, Jayasuriya S, Gunarathna C, Arashpour M, Xue X, Zhang G (2018) The evolution of stakeholder management practices in Australian mega construction projects. Eng Constr Archit Manag 25(6):690–706

Yu Y, Li Y, Zhang Z, Gu Z, Zhong H, Zha Q, Yang L, Zhu C, Chen E (2020) A bibliometric analysis using VOSviewer of publications on COVID-19. Ann Transl Med 8(13)

Zheng X, Le Y, Chan AP, Hu Y, Li Y (2016) Review of the application of social network analysis (SNA) in construction project management research. Int J Project Manage 34(7):1214–1225

The Effectiveness of the Universities Entrepreneurship Ecosystem in Shaping the Students' Entrepreneurial Career Choices: Evidence from UAE Context

Rania Gamal El Din Sheir and Husam-Aldin N. Al-Malkawi

Abstract This study investigates the factors that influence the establishment and functioning of the university entrepreneurship ecosystem (UEE) and its impact on the student entrepreneurial career choice. This research utilises the case-study methodology to conduct an evidence-based analysis of the current theoretical framework. It aims to validate the role of its elements through empirical experience, which is gathered from semi-structured interviews with 7 respondents from three private universities in Dubai. The results shows that the business incubators programs have a greater favourable influence on student entrepreneurship compared to official courses and various tiers of networks can facilitate the flow and interchange of resources. In addition, the findings reveal that culture has a crucial role in shaping and fostering individual entrepreneurial behaviour and their clustering in the UEE. Furthermore, the UEE's creation and progress are significantly influenced by strong entrepreneurial vision and enduring commitments, particularly in terms of leadership.

Keywords Entrepreneurial programs · Career choices · Theory of planned behaviour · University entrepreneurship ecosystem culture · Industry and government collaboration

1 Introduction

Since the early 21st century, entrepreneurship has been pivotal in addressing global issues and driving economic growth. Research has increasingly focused on the entrepreneurial environment (EE), which boosts local economic and non-economic performance through start-ups (Art 2018). However, there is a lack of empirical studies on factors influencing career decisions, with most focusing on demographic and economic trends (Douglas and Shepherd 2002; Feldman and Bolino 2000).

This paper was prepared for PhD dissertation, taught by Prof. Husam-Aldin Al-Malkawi.

R. G. E. D. Sheir (✉) · H.-A. N. Al-Malkawi
Faculty of Business and Law, The British University in Dubai, Dubai, UAE
e-mail: 20183139@student.buid.ac.ae

K. Al Marri et al. (eds.), *BUiD Doctoral Research Conference 2024*, Lecture Notes in Civil Engineering 587, https://doi.org/10.1007/978-3-031-84371-6_29

Universities play a crucial role in these contexts by identifying and capitalizing on opportunities (Audretsch 2014), but they are often bureaucratic and inflexible. Kirby et al. (2011) suggest that universities need to adopt an entrepreneurial mindset and innovate in connecting education and research. There has been an increase in initiatives to provide entrepreneurial support to faculty, staff, and students.

In the UAE, entrepreneurship is promoted to increase youth self-employment, with many institutions requiring courses on entrepreneurship and innovation. This study aims to assess the impact of entrepreneurship education on student goals and the university entrepreneurship ecosystem (UEE) in the UAE. It will provide insights for curriculum developers, institutions, and policymakers. The study focuses on understanding entrepreneurial practices in universities and evaluating the impact of these practices on students' entrepreneurial inclinations and career decisions using the theory of planned behaviour. The outcomes are expected to help educational institutions improve entrepreneurship programs, particularly in the UAE.

2 Theoretical Background

The Theory of Planned Behaviour (TPB) proposed by stands out among all other theories in academic literature. The TPB proposes that it can be used to interpret entrepreneurial intentions and reactions toward starting business ventures. Since its conception 20 years ago, TPB has been essential in understanding human behaviour. It has been used to tackle a variety of behaviours with success. There have been few attempts to employ TPB in complicated behaviours like managerial decision-making. This is especially true when an institute internationalizes. Additionally, the TPB exhibits a strong entrepreneurial background and has been widely acknowledged by various research scholars. Furthermore, the theory is compatible with academic research focusing on the impact of entrepreneurship education on individual intentions. The TPB offers divergent aspects of entrepreneurship irrespective of its nature. For example, Iglesias-Sanchez et al. (2015) used Ajzen's influential model (1991) with the TPB theory to study students' intentions at Malaga University.

3 Methodology

The intricate and interrelated nature of entrepreneurship research necessitates the use of appropriate research methodologies (Crowther et al. 2018). According to, researchers should possess an epistemological framework. The qualitative technique promotes the investigation of a phenomenon by comprehending the intricacies of its processes

This paper employs an exploratory case study approach for three private universities across Dubai to explore the impact of entrepreneurial programs on the entrepreneurship career choices of students in the UAE context (Takahashi and

Araujo 2020). These private universities were selected because they exhibit good reputations in terms of successful entrepreneurship models, have business incubators, and strong entrepreneurship and innovation initiatives.

In this regard, three different interview questionnaires were developed for students, incubation managers, and higher management to assess the effectiveness of entrepreneurial programs for the career choices of students. For this purpose, the researcher interviewed seven participants from the universities of the UAE.

4 Data Analysis

The collected data from the participants was analysed through the thematic analysis technique, which has been extensively used in the literature. This is a qualitative research approach proposed by Braun and Clarke (2012) that encounters 6-step procedures and requires familiarisation with the data, coding procedures, making of themes, review of themes, finalising themes, and reporting of results. The thematic analysis identified the main elements of the university entrepreneurship ecosystem and key themes that emerged from the interviews, indicating the impact of entrepreneurship programs, networks, incubators, government, and industrial support. Following the steps, the study developed 11 themes that are presented in the following section.

5 Findings

The researcher identifies the fact that every respondent exhibits unique insights, and thus every participant was considered a source to generate themes. Therefore, this study considered the session with every respondent and generated different themes following a thematic analysis approach.

Table 1 demonstrates the findings of the discussion with participants who were associated with higher management and faculty members in two different universities. This session helped to end with the theme named "University Entrepreneurial Culture".

Table 2 expresses the finding of the discussion with participants from students and faculty members who declare that the universities play a crucial role in fostering entrepreneurship.

Table 3 provides a concise summary of the participants' motives for starting a business and their reasons for choosing self-employment. The participants describe multiple causes or circumstances that motivated their decision to pursue entrepreneurship.

Table 4 Summarizes the participants responses toward the challenges hinder the students to choose the entrepreneurial career choice and start their businesses

Table 1 University entrepreneurial culture

Theme	Summary	Responses
Strategic emphasis on entrepreneurship	Promotes entrepreneurship as a viable career choice	**Respondent 1**: It's encouraged opening the students come up with different ideas, innovative ideas from even the faculty members because you know, I even sometimes I hate to call entrepreneur as a career. When we are talking about entrepreneurs, real career and entrepreneurial mindset is very important from university perspective to have it as a good component in their culture."
Nurturing of entrepreneurship culture	Creates an environment where students feel comfortable taking risks and starting their own businesses	**Respondent 2**: "We have a dedicated incubation centre. We have rolled out many businesses, including the current transportation on demand by RTA. The software was written by our student. Another notable project is Thalia, which means 'bag' in Hindi. A student has made Snickers out of used poly bags, selling them through resellers and requesting returns for recycling. This demonstrates our strong entrepreneurial philosophy and focus on our incubation centre."

Table 5 provides a summary of the participants' comments, which state that entrepreneurship is fostered by government support and university-industry partnership.

6 Discussion

The present research focused on examining entrepreneurial education at the university level, considering the perceptions of students, incubator managers, and higher management. The study assessed the inclination of respondents toward entrepreneurship in the foreseeable future. The analysis conducted in this study focuses on the examination of several aspects related to entrepreneurial education and the university entrepreneurial ecosystem (UEE), including the understanding of necessary procedures and the practicalities associated with being a stakeholder at the university level (Gurrero et al. 2020).

This study is based on the premise that leadership, entrepreneurial culture, networks, and extracurricular activities all play a role in building an entrepreneurial ecosystem at the university. Our primary source for making this assumption is the current body of research in the UEE. According to Art (2018), the network is crucial in constructing the UBEE. Put another way, the university acts as a link between the campus and the surrounding community. Furthermore, incubators play a crucial role

Table 2 Role of universities in fostering entrepreneurship

Theme	Summary	Responses
Entrepreneurship program	Integrates entrepreneurship into all programs, from year one to graduation	**Respondent 1** "We said, let's not teach a course, let's make it a journey. For each program, we identified courses called innovation and entrepreneurship nodes. From day one, we teach students basic innovation concepts like idea prototyping. Later, we teach them entrepreneurship skills, guiding them to bring an idea, problem, or challenge from industry and work on it in their project." **Respondent 4** I used to have a passion for this field. The entrepreneurship program at my university helped me think beyond engineering and technical aspects to the business side. As engineers, we usually focus on technical details. The university gave me an entrepreneurial mindset to convert a lab product into a market-ready product."
Role of incubators in supporting entrepreneurship	Can provide mentorship, consultation, IP registration, network access to finance and investors	**Respondent 5** "Incubators are essential for entrepreneurs and start-ups, providing services like mentorship, consultation, IP registration support, and access to finance and investor networks." **Respondent 4** "It's good to have this incubator. Each student is passionate about something, but some push themselves to follow this passion, while others are lazy. Incubators help by putting you on the right track or inspiring you to create a business out of something you love."

in fostering entrepreneurship by supporting students to develop their startups through mentorship, technical support, funding support, and IP registration, in addition to their essential role in shaping the entrepreneurial mindset.

This research focused on examining entrepreneurial education at the university level, considering the perceptions of students, incubator managers, and higher management. The study assessed the inclination of respondents toward entrepreneurship in the foreseeable future. The analysis conducted in this study focuses on the examination of several aspects related to entrepreneurial education and the university entrepreneurial ecosystem (UEE), including the understanding of necessary procedures and the practicalities associated with being a stakeholder at the university level (Gurrero et al. 2020).

The practical implications of this research, which examines the impact of entrepreneurship education on entrepreneurial intention and entrepreneurial career

Table 3 Motivation to entrepreneurial career choice

Theme	Summary	Responses
Family background & successful entrepreneurs	Fosters an environment where there kids are encouraged to take risks and start their own businesses	**Respondent 6 (Student who started his own enterprise)** "Actually, my parents are both entrepreneurs, and I'm really inspired by my mother. The way she handles and manages the business is very creative. She's the biggest motivator in my life, so I take inspiration from her." **In contrast Respondent 7** "Another student who started his own business claims, 'To be honest, my family sometimes encourages me, sometimes not, because the journey is long, and they don't see instant results. They wonder if I'm doing the right thing or benefiting. On the other hand, I got inspired by a close friend who is a successful entrepreneur. He gave me the mindset, passion, and smart work ethic. It's important to have at least one person believing in and pushing you, which is enough to keep you going.'"
Entrepreneurship Rewards	Advantages of entrepreneurship and flexibility in working in something you are passionate about	**Respondent 7 (students who started his own business)** "So as I said, being an entrepreneur is doing what you love the most, which is the fun part of a business. You're not going to do something you hate or aren't good at because you chose this journey. You're choosing something you belong to, are good at, and love to do." **Respondent 5** "The first interesting thing in entrepreneurship is you get to be your own boss. Second, you get to make an impact faster." **Respondent 6** "Yeah, I believe it's more of an advantage because more doors open up and you're more open to opportunities. For me personally, I really love and enjoy what I'm doing. The fact that it's helping society gives back to me. So, I feel it's more of an advantage."

choice, will provide insights into the need for entrepreneurship programs, their organization, and instructional approaches. The findings also indicated that the assistance provided to entrepreneurs by the surrounding environment (such as mentors, government, and financial institutions) might have an impact on their decision to pursue an entrepreneurial career. This study also confirmed the influence of an individual's entrepreneurial family background on their choice of an entrepreneurial career.

Table 4 Challenges that hinder the entrepreneurial career choice

Theme	Summary	Responses
Access to finance	Entrepreneurial projects need some fund for developing the Idea, prototypes, licensing and IP registration	**Respondent 6** "I'm not facing any problems personally. I was prepared for pitching by university mentors and got to meet investors. Fundraising isn't a problem, but the skills and capabilities needed to raise funds are crucial. Many entities offer accessible funding, and you can easily access institutions or angel investor networks to secure funds for your project." **Respondent 7** "The fund was a problem at the beginning, but I succeeded in raising it through a government-managed crowdfunding platform. I learned about this platform through my university, which partnered with this government entity."

(continued)

Table 4 (continued)

Theme	Summary	Responses
Access to technical support	Developing specific projects need technical support and expertise	**Respondent 5** "I was developing a website for my project and needed technical and IT support to create the platform. However, the university's incubator only provided ideas, not the necessary technical support. My challenge was the lack of knowledge to create or program the website. This technical support was crucial at this stage." **Respondent 7** "As for my personal experience, my main challenge is the technical support, as my product is purely technical—a battery connected to circuit boards. In our region, we lack a strong technical foundation in areas like electric or civil engineering. The university tried to support with lab equipment and assistance, but it wasn't enough for the product itself. Developing the product mainly involved having a place, space, and equipment, but there was no expertise."
Entrepreneurial mindset	Students are most inclined towards getting job in public & private sector	**Respondent 6** "As far as I've seen, the new generation is more inclined towards entrepreneurship, experimenting with starting small businesses. If they fail, they often choose jobs for a more relaxed life, preferring to learn and work for MNCs or institutions. Universities traditionally channel students towards job-seeking, internships, and comfort zones." **Respondent 1** "They see entrepreneurship as a tough journey and a unique way of thinking that not everyone has. Many students prefer permanent jobs in the private or public sector over entrepreneurial careers, possibly because they find entrepreneurship hard."

Table 5 The university-government-industry collaboration

Theme	Advantage	Quotation
Government support	Helps to create a more favourable environment for start-ups to thrive	**Respondent 1 (Faculty Higher Management)** "On the long term, more start-ups and local ideas will boost the country's economy. The government plays a crucial role, benefiting the community, local people, and founders. A proper structure supported by big entities through the incubator role will ensure success."
Industry partnerships	Provides students with access to real-world experience and mentorship from entrepreneurs	**Respondent 2 (Faculty Member)** "We have nine successful entrepreneurs from industry who teach students how to build their business. At the end of the training, these entrepreneurs examine the students' work. Successful ideas are reviewed by the university's spin-outs and start-ups committee, which then supports these ideas." **Respondent 5 (student)** "Yes, I believe it's essential because it helps young entrepreneurs absorb more and collaborate with the government and other entities, inspiring and motivating them to join the entrepreneurship journey. My university has an incubator that links us to specific incubators interested in our industry to push us forward."

7 Conclusion

The research was carried out at three private universities in Dubai. The research findings in this study, like other qualitative studies, have limited generalizability. This study utilized interactive interviews to identify the primary components that influence the establishment and functioning of the UEE based on empirical evidence. The findings of this study serve as a research basis for future quantitative research. To broaden its scope, it is necessary to conduct a comparative analysis of the UEE at other higher education institutions across the UAE. The findings of this study reveal the role of universities' entrepreneurial ecosystems in fostering entrepreneurship and how the entrepreneurship programs, culture, incubators, and extracurriculars are shaping students' entrepreneurial career choices. These factors may vary if conducted at a different university, but this does not diminish the value and quality of the research. It is necessary to verify additional factors and mechanisms that influence the formation and operation of the UEE in order to provide more empirical support for future theoretical development.

References

Ajzen I (1991) The theory of planned behavior, organizational behavior and human decision processes. Cited Hansen 50:93–114

Art S (2018) Universities and the entrepreneurship ecosystem. Research Gate. https://www.researchgate.net/publication/325310578_Universities_and_the_Entrepreneurship_Ecosystem
Audretsch DB (2014) From the entrepreneurial university to the university for the entrepreneurial society. J Technol Transf 39:313–321
Braun V, Clarke V (2012) Thematic analysis. American Psychological Association
Crowther P, Orefice C, Beard C (2018) At work and play: business events as entrepreneurial spaces. Int J Entrep Innov 19(2):90–99
Douglas EJ, Shepherd DA (2002) Self-employment as a career choice: attitudes, entrepreneurial intentions, and utility maximization. Entrep Theory Pract 26(3):81–90
Feldman DC, Bolino MC (2000) Career patterns of the setf-etnployed: career motivations and career outcomes. J Small Bus Manage 38(3):53–68
Guerrero M, Urbano D, Gajón E (2020) Entrepreneurial university ecosystems and graduates' career patterns: do entrepreneurship education programmes and university business incubators matter? J Manag Dev 39(5):753–775
Kirby DA, Guerrero M, Urbano D (2011) The theoretical and empirical side of entrepreneurial universities: an institutional approach. Can J Adm Sci 28:302–316
Takahashi ARW, Araujo L (2020) Case study research: opening up research opportunities. RAUSP Manag J 55:100–111

The Impact of Distance Learning on Identifying Primary Students with Specific Learning Disorder (SLD) During COVID-19 in Two Schools in Ajman City in the UAE

Mona Abuhattab and Eman Gaad

Abstract The purpose of this study is to highlight the impact of distance learning on identifying primary students with specific learning disorders (SLD) during the COVID-19 pandemic in two private schools in Ajman city in the UAE. The present paper is a small-scale study that employed the qualitative design method to gather responses to the research question, depending on semi-structured interviews. A thematic analysis was used to analyze data collected from 101 teachers, parents, special education teachers, and coordinators. In light of the results, the study revealed that schools faced obstacles, which hindered the effective identification during distance learning. Findings referred these barriers to lack of knowledge and awareness about students with SLD, lack of instructions, resources, and data, as well as lack of assessment reliability and validity and poor student evaluation. In addition, there was lack of communication and coordination among teachers, special education teachers, coordinators, and parents. Lastly, the absence of a team of experts and trained teachers in the field played an important role. The study wrapped up with a list of recommendations for effective identification in an online environment such as training teachers on inclusive education, establishing a well-articulated and efficient policy for identification methods in the schools, and raising awareness within the school community. Stakeholders and policymakers are urged to examine the problem from different perspectives to overcome all the barriers hindering effective identification. Finally, various factors posed limitations to the study, including restricted access to schools, restrictive rules of social interaction during the pandemic, and having to rely solely on semi-structured interviews. The significance of this study arises from the sacristy of research, at least locally, and the increasing numbers of cases of learning difficulties, as well as the lack of awareness among all stakeholders on dealing with such cases.

Keywords Distance learning · Identification · COVID-19 · Students with a specific learning disorder

M. Abuhattab (✉) · E. Gaad
The British University in Dubai, Ajman, UAE
e-mail: monaabuhattab@gmail.com

K. Al Marri et al. (eds.), *BUiD Doctoral Research Conference 2024*, Lecture Notes in Civil Engineering 587, https://doi.org/10.1007/978-3-031-84371-6_30

1 Introduction

Since the 1990s, and due to the influence of technology, the education system, globally, has witnessed significant changes (Sivin-Kachala and Bialo 1994). Worldwide, many educational institutions have utilized multiple models of online education such as hybrid, blended, and distance learning. (Palvia et al. 2018). More recently, because of unprecedented health crises, schools over the globe migrated to fully online learning (Tavares et al. 2021).

The story originated in December 2019 when schools throughout the world shut down due to the outbreak of the contagious COVID-19 pandemic—a health crisis that began in Wuhan Province, People's Republic of China, and has grown into a pandemic that resulted in lockdowns in most parts of the world. The impact of the pandemic has extended to all fields of life, affecting not only human health, but also education and the economy (Tirivayi et al. 2020).

The outbreak of the COVID-19 pandemic has negatively disrupted the education systems in the most extensive way in human history. Nearly 1.6 billion learners in more than 200 countries were affected, making up 94% of the student population of the world (López-Vargas et al. 2021).

Consequently, the COVID-19 pandemic has forced schools to turn to distance learning and deliver classes online, on an untested and unprecedented scale (Tartavulea et al. 2020). This caused obstacles for students, teachers, and parents who were not qualified initially to move online. However, it was the most appropriate solution to carry on education during the lockdown (Pokhrel et al. 2021).

During the COVID-19 outbreak, the UAE was one of the first countries to apply distance learning. It successfully shifted thousands of students online almost overnight (The United Arab Emirates government portal 2020). The UAE government adopted several education policies to effectively meet the challenges that may face schools during distance education (The United Arab Emirates government portal 2020). Nevertheless, schools had limited time to plan and less clarity about what may happen next, leading to a flow of learning full of complexities and limitations (Gül Özüdoğru 2021).

For example, special education field was hugely impacted throughout this transformation. The schools experienced several obstacles related to the consistency of providing services for the students who experience special educational needs and disabilities (SEND) (Ayda et al. 2020), particularly, students with undiagnosed learning disabilities who may be missing out on services and supports during distance learning. Also, there was an unprecedented challenge on how to continue or begin a special education identification assessment during schools closure (Tremmel P, Myers R).

2 Literature Review

2.1 *Identification of Specific Learning Disorder (SLD) at School Level*

The concept of identification, rooted in Freud's ideas, refers to associating oneself with others' characteristics. In education, it involves identifying whether a student has a disability and planning appropriate services. This systematic process includes both formal and informal assessments to determine the type of learning disability (Ahmad and Hanifah 2015). According to Gresham (2002), identification entails collecting, reviewing, and using information to make educational decisions, especially regarding students' eligibility for special education. SLD identification is crucial, as it helps in addressing students' needs and developing tailored educational programs.

Pierangelo and Giuliani (2002) outlined three common referral methods for SLD identification: teacher concerns about performance or behavior, injury or treatment, and parental requests for evaluation. Lerner (2000) added that the process involves several stages: pre-referral, formal referral, assessment, eligibility determination, case discussions, and developing an Individualized Education Plan (IEP).

In the UAE, the Ministry of Education outlines a six-step process: pre-referral, referral, evaluation, developing and implementing IEPs, monitoring progress, and issuing certificates of completion. Schools may use assessments and universal screening to identify students at risk for SLD early on. Before formal referral, teachers typically initiate pre-referral interventions to provide support and minimize the need for special education services. If these interventions are unsuccessful, formal evaluation begins, often involving intelligence tests, academic history, classroom observations, and more.

Parental involvement is essential throughout the process. However, the lack of uniform criteria for distinguishing SLD from other learning challenges makes the process complex and varied across schools worldwide, contributing to confusion among educators and families.

3 Methodology

This study was carried out in Ajman city in the UAE; two primary schools located in the same region were selected in this study. A total of 58 teachers, 7 SEN teachers, 2 SENCOs, and 34 parents from the selected schools were interviewed. Primary students were chosen because SLD are more obvious in this stage, and the signs are more common than others at different grades.

The main purpose of the paper was to examine the impact of distance learning on identifying primary students with SLD during COVID-19 in two primary schools in Ajman city, UAE. To establish this, a qualitative research design was used to answer the questions of what, when, where, who and how that are of relevance to

Table 1 Population per school

Participates	School 1	School 2
Teachers	30	28
SEN teachers	3	4
SENCOs	1	1
Parents	16	18
Total	50	51

the research problem. Comprehensively, the qualitative research approach enables examining peoples' experiences in detail by using a set of research methods such as in-depth interviews and content analysis (Moser and Korstjens 2018). Within this approach, information is obtained from participants who are directly involved in the issue. Considerations of factors that shape a certain population group are also kept in check, including their values, behaviors, attitudes, and shared social context.

The targeted groups in this study are primary school teachers, parents, special education teachers, and SENCOs from two schools in Ajman city. To gather wealth of information in relation to the research questions, 58 teachers, 7 SEN teachers, 2 SENCO, and 34 parents were selected (Table 1).

This study explored the impact of distance learning on identifying students with specific learning disabilities (SLD) in two American curriculum schools in Ajman, serving over 1,000 students with special educational needs units. Informants, including teachers, SEN teachers, SENCOs, and parents, were selected based on their experience with SLD students and interviewed individually using a semi-structured format to gather in-depth insights. Interviews, lasting about 30 min, were conducted via telephone and Zoom due to COVID-19, with platforms tested for accessibility. Consent forms were obtained via email, and interviews were recorded with participant approval. While reading body language was challenging, the method enabled effective data collection.

3.1 Data Presentation

In the interview guide, 16 questions were for teachers, 16 questions were for SEN teachers and SENCOs, and another 16 questions were for parents. The questions focused particularly on attitudes, knowledge, and perspectives concerning students with SLD and the barriers to their identification. From the outset, it has been important to reconsider the ongoing controversy about how identification should proceed and the ambiguity of definitions. Therefore, terms such as learning difficulties, learning disorders, and learning disabilities were used if participants were unfamiliar with the term SLD.

Data from multiple sources were subjected to content analysis and coded into themes. Four major themes were identified after content analysis: (1) Awareness of the concept of SLD among teachers and parents, (2) Identification procedures, (3) barriers to effective identification, and (4) collaboration among teachers, parents, and special education specialists during distance learning.

3.1.1 Awareness of the Concept of SLD

Awareness in School 1

Teachers' Responses

The data collected from School 1 revealed that many teachers understood what the term SLD referred to. However, they lacked information on the causes of SLD. They all cited different causes, with some attributing them to genetic conditions or poor nutrition.

They were also opposed to the inclusion of students with SLD in the regular classroom, arguing that they lacked the necessary knowledge and training to identify and support students with SLD in online settings. Teachers stated that they were too busy due to distance learning and did not have time to support students with SLD. As a result, it is quite possible that these students were treated the same as others because teachers lacked sufficient knowledge on this field.

Awareness in School 2

Teachers' Responses

Responses suggested that there is a lack of awareness of SLD among teachers. There is evidence that some of them do not view SLD as disabilities; rather, they see them as problems caused by negligence or ineffective pedagogical practices. However, there were attempts to raise teachers' awareness through workshops to develop positive attitudes toward students with SLD. Teachers mentioned that the school's SENCO organized two workshops on SEND students during distance learning.

Parents' Responses from Schools 1 and 2

Data collected from schools 1 and 2 showed that parents lacked both the conceptual knowledge on the symptoms of SLD and the appropriate guidelines for dealing with their child's problem. Parents also presented negative attitudes and reactions to their child's diagnosis of SLD.

Apparently, parents from both schools do not associate their children's school problems with learning disorders. In order to avoid this stigma, such families often deny their children have SLD by associating it with being naughty or spoiled. In addition, parents often lack sufficient knowledge and information on SLD, resulting in a lack of or delay in identifying the need for action. As a result, parents refuse to acknowledge that their children have learning problems when teachers or others point them out. Also, when a child is diagnosed with a disability, parents generally exhibit negative attitudes toward the problem; these could range from denying the problem, rejecting the diagnosis, or blaming themselves for the situation.

3.1.2 Identification Procedures in School 1

Teachers' Responses

When School 1 participants were asked about their approach to identifying students with SLD during distance learning, they indicated that they use direct observation and checklists. In addition, they said we bring to the attention of school specialists any student who shows poor performance in academic subjects as well as in behavior.

Teachers reported that they usually try to help these students before referring them to specialists. They use interventions and several strategies within the classroom, such as direct and individual support. Teachers said that some interventions may work better than others, and they agreed that a single intervention or approach is not sufficient to address the complexity of learning disorders. However, some teachers added that it is not their job to identify these students because they are not able to address their needs and are not equipped to help them. Therefore, they had difficulty identifying these signs before and during distance learning.

3.1.3 Identification Procedures in School 2

Teachers' Responses

Just as in School 1, teachers in School 2 relied on direct observation to identify students with SLD. However, in collaboration with SEN teachers, they developed pretests to assist them in identifying students suspected of having SLD during distance learning. The teacher used a simple administered test to determine the student's level of performance. If the results indicate that a student may have SLD, further assessment will be conducted by the school's SENCOs or SEN teachers. However, teachers stated that the distance learning assessments are not valid to evaluate the level of the students. The data collected from teacher show that it is difficult to ensure certain quality criteria in distance learning since academic integrity must be ensured.

SEN Teachers' and SENCOs' Responses in School 1 and School 2

SENCOs at both schools reported following a systematic process for identifying students with SLD, both before and during distance learning. This includes informal referral, intervention, formal referral, multidisciplinary assessment, eligibility decisions, and IEP development. SENTs agreed, emphasizing the importance of gathering data from multiple sources, including classroom performance, to make informed decisions. At School 1, teachers are trained to report concerns to special education staff, who then provide targeted interventions before formal assessment. At School 2, SENTs introduced pre-administered tests to help identify SLD students during distance learning, noting that school closures negatively impact student achievement, especially for SEND students. When intervention is insufficient, parents are informed and consent to assessments by a team of professionals. Both schools found it challenging to implement these steps during distance learning.

3.1.4 Barriers to Effective Identification During Distance Learning in School 1 and School 2

Teachers' Responses

Teachers in the selected schools agreed that distance learning makes it difficult to identify students with SLD for several reasons:

(a) Lack of interest and engagement: Teachers reported that students did not take distance learning seriously, seeing home as a place to rest and play. This made it difficult for teachers to assess struggling students and maintain engagement, especially without parent support.
(b) Lack of effective instruction: Teachers found it challenging to deliver high-quality, differentiated instruction in virtual environments, impacting their ability to identify students needing special education.
(c) Lack of data: Teachers struggled to gather and interpret data during virtual teaching, complicating assessments.
(d) Validity and reliability issues: Assessments lacked validity, reliability, clarity, and were vulnerable to technical issues, making it difficult to accurately evaluate students.
(e) Lack of parent collaboration: Teachers blamed parents for not creating a supportive learning environment or helping students, with some even refusing to acknowledge their child's learning difficulties.
(f) Lack of resources and training: Teachers felt unqualified to teach students with SEND, citing inadequate training, outdated curricula, and a lack of strategies to support these students.
(g) Lack of expertise and collaboration: Teachers emphasized the need for more guidance from specialists and ongoing training to effectively support SLD students, noting a shortage of qualified special education teachers.

SEN Teachers' and SENCOs' Responses

In both schools, SENCOs noted that teachers are well trained in identifying students with learning difficulties, with SENTs following up after teachers raise concerns. Assessments involve observing students' reading, writing, and math skills by sitting with them and monitoring signs of boredom or frustration. However, SENTs found it challenging to assess students in an online environment, as standardized tests are not ideal for virtual use. SENCOs agreed that traditional assessments, even in normal circumstances, are less effective in capturing students' learning, especially during the pandemic (Table 2).

Table 2 Summary of barriers to effective identification during distance learning in school 1 and school 2

Barriers to identification during distance learning	Lack of students' interest and engagement
	Lack of effective instruction
	Lack of data
	Lack of resources and knowledge
	Lack of collaboration with school specialists. Teachers
	Lack of validity and reliability of assessments
	Poor student's evaluation
	Lack of knowledge about SLD
	Lack of awareness among teachers and parents
	Lack of collaboration among stakeholders
	Difficulty in to monitoring and observing the students online during class
	Lack of expertise in this field

3.1.5 Collaboration Among Teachers, Specialists, and Parents During Distance Learning in School 1 and 2

Teachers' Responses

Most teachers reported informing the school psychologist if they suspected a student had learning difficulties, considering them vital for identifying SLD students. Others would first notify the SENCO, who observes the student, recommends an intervention plan, and provides special education support, though applying these interventions during distance learning was challenging. A few teachers sought help from the social worker, believing social issues like bullying or divorce might be affecting the student. Some teachers were unsure what to do if interventions failed. Teachers also highlighted the lack of parental follow-up, stating that students often forgot what they learned and needed extra support at home. Many parents focused solely on grades and resisted acknowledging their child's learning difficulties, hindering the identification process. Additionally, some teachers admitted they were unsure how to help students with learning disorders.

SEN Teachers' and SENCOs' Response

Responses from both schools indicated that parents are not paying enough attention to their children, or they are helping them in inappropriate ways. Therefore, parents need to know when and how to support their children. In addition, some parents leave their children alone while they are being taught remotely without providing them with adequate supervision. Some parents have difficulty understanding what is required. Some of them do not have enough time to supervise their children and are totally dependent on the teacher.

Parents' Response

Parents reported a lack of direction and support during the transition to distance learning. They felt their children received inadequate assistance and were overwhelmed by managing multiple online platforms, finding the content and instruction unsatisfactory. While recognizing the need for additional support and intervention, parents did not link their children's difficulties to a disability. Instead, they were concerned about time management and discipline issues during distance learning.

Discussion of Findings

This study examined the impact of distance learning on identifying primary students with SLD in two schools in Ajman, UAE, during the COVID-19 pandemic. Key issues included teacher and parent awareness of SLD, identification procedures, barriers to effective identification, and collaboration among educators, parents, and special education teachers.

Teachers recognized SLD as learning challenges but lacked understanding of its causes, often citing unrelated factors like poor nutrition or overuse of devices. Both teachers and parents showed limited awareness of SLD, with parents often rejecting support programs due to stigma. Teachers also lacked sufficient training in special education, especially for online instruction, which hindered their ability to identify SLD students during distance learning.

Identification methods like observation, checklists, and pre-tests were used, but many teachers struggled to set appropriate interventions. SENCOs and SEN teachers, however, used multiple data sources and worked with parents to assess students. Distance learning posed significant challenges, including reduced student engagement, difficulty in designing effective instruction, and unreliable assessments. Additionally, teachers cited a lack of collaboration with parents and inadequate resources as barriers to identifying SLD students.

Collaboration between teachers, parents, and special education staff was also insufficient, with teachers needing more support and guidance. While there are clear guidelines on inclusive education in the UAE, challenges remain in identifying and supporting SLD students, exacerbated by distance learning and misconceptions about disabilities.

4 Conclusion and Recommendations

This study examined the impact of distance learning on identifying primary students with Specific Learning Disabilities (SLD) in two schools in Ajman, UAE, during the COVID-19 pandemic. The analysis highlighted key issues: (1) awareness of SLD among teachers and parents, (2) identification procedures, (3) barriers to effective identification, and (4) the need for collaboration between teachers, parents, and special education staff. Teachers and parents showed limited understanding of SLD, often misidentifying its causes. Teachers lacked the training to identify SLD, especially in an online environment, and struggled with insufficient resources and support. Additionally, parents' misconceptions about SLD and concerns about stigma further hindered the identification process.

Teachers primarily relied on classroom observations, checklists, and academic performance to identify SLD, but distance learning posed challenges such as lack of engagement, difficulty in collecting reliable data, and cheating concerns during assessments. Collaboration between teachers, parents, and special educators was limited, further complicating identification efforts. To address these challenges, the study recommends increased in-service training for teachers and special educators, improved data collection systems, enhanced collaboration with parents, and greater use of technology to support SLD identification.

The study also calls for standardized assessment tools and further research to explore effective identification methods, especially in online settings. Although the study focused on only two schools, it emphasizes the need for broader research across UAE schools to develop inclusive education policies and improve SLD identification methods. Effective SLD identification requires increased awareness, better teacher training, and collaboration among stakeholders. Further research is needed to develop practical, inclusive strategies for identifying SLD students in both traditional and online learning environments.

References

Adebisi RO et al (2014a) Barriers to special needs education in Nigeria. Int J Educ Res 2(11):451–462

Adebisi RO, Liman NA, Longpoe PK (2014b) Issues and challenges in the implementation of inclusive education in Nigeria. J Educ Pract 5(7):12–17

Ahmad NA, Hanifah NA (2015) Special education teacher's level of knowledge in dealing with learning disabilities student. Asia Pac Educ 30:73–88

Ajman-The Official Portal of the UAE Government (2020). https://u.ae/en/about-the-uae/the-seven-emirates/ajman

Ayda NK, Bastas M, Altınay F, Altinay Z, Dagli G (2020) Distance education for students with special needs in primary schools in the period of COVID-19 epidemic. Propósitos y Representaciones 8(3):43

Gaad E, Khan L (2007) Primary mainstream teachers' attitudes towards inclusion of students with special educational needs in the private sector: a perspective from Dubai. Int J Spec Educ 22(2):95–109

Gresham FM (2002) Responsiveness to intervention: an alternative approach to the identification of learning disabilities. Identif Learn Disabil: Res Pract 467519

Gunawardena CN, McIsaac MS (2013) Distance education. In: Handbook of research on educational communications and technology. Routledge, pp 361–401

Hudson D (2015).Specific learning difficulties-what teachers need to know. Jessica Kingsley Publishers

Joshi A, Vinay M, Bhaskar P (2020) Impact of coronavirus pandemic on the Indian education sector: perspectives of teachers on online teaching and assessments. Interact Technol Smart Educ

Kohli A, Sharma S, Padhy SK (2018) Specific learning disabilities: issues that remaine unanswered. Indian J Psychol Med 40(5):399–405

Kuzniak A, Rauscher JC (2011) How do teachers' approaches to geometric work relate to geometry students' learning difficulties? Educ Stud Math 77(1):129–147

Lerner JW (2000) Learning disabilities: Theories, diagnosis, and teaching strategies, 8th edn. Houghton Miffin Company, Boston

Lingeswaran A (2013) Assessing knowledge of primary school teachers on specific learning disabilities in two schools in India. J Educ Health Promot 2

Lopes RCF, Crenitte PAP (2013) Analytical study of teachers' knowledge about learning disorders. Revista CEFAC 15:1214–1226

López-Vargas A, Ledezma A, Bott J, Sanchis A (2021) IoT for global development to achieve the united nations sustainable development goals: the new scenario after the COVID-19 pandemic. Ieee Access 9:124711–124726

Moser A, Korstjens I (2018) Series: practical guidance to qualitative research. Part 3: sampling, data collection and analysis. Eur J Gen Pract 24(1):9–18

Özüdoğru G (2021) Problems faced in distance education during COVID-19 pandemic department of educational science, Kırşehir Ahi Evran University. Kırşehir, Turkey, ORCID: 0000–0002–5128–1258

Palvia S, Aeron P, Gupta P, Mahapatra D, Parida R, Rosner R, Sindhi S (2018) Online education: worldwide status, challenges, trends, and implications. J Glob Inf Technol Manag 21(4):233–241

Papanastasiou F (2017) The Psychological implication of learning disabilities. Acta Psychopathol 4:1

Pierangelo R, Giuliani GA (2002) Assessment in special education: a practical approach. Allyn and Bacon, Boston

Pokhrel S, Chhetri R (2021) A literature review on impact of COVID-19 pandemic on teaching and learning. High Educ Futur 8(1):133–141

Regier DA, Narrow WE, Clarke DE, Kraemer HC, Kuramoto SJ, Kuhl EA, Kupfer DJ (2013) DSM-5 field trials in the United States and Canada, Part II: test-retest reliability of selected categorical diagnoses. Am J Psychiatry 170(1):59–70

Richards C, Pavri S, Golez F, Canges R, Murphy J (2007) Response to intervention: building the capacity of teachers to serve students with learning difficulties. Issues Teach Educ 16(2):55–64

Shani M, Hebel O (2016) Educating towards inclusive education: assessing a teacher-training program for working with pupils with special educational needs and disabilities (SEND) enrolled in general education schools. Int J Spec Educ 31(3):n3

Shin M, Lee H, McKenna JW (2016) Special education and general education preservice teachers' co-teaching experiences: a comparative synthesis of qualitative research. Int J Incl Educ 20(1):91–107

Sivin-Kachala J, Bialo ER (1994) Report on the effectiveness of technology in schools, 1990–1994. Interactive Educational Systems Design, Inc. https://eric.ed.gov/?id=ED371726

Sofologi M, Kougioumtzis GA, Efstratopoulou M, Skoura E, Sagia S, Karvela S, Salli PM, Makri E, Bonti E (2022) Specific learning disabilities and psychosocial difficulties in children. In: Advising preservice teachers through narratives from students with disabilities. IGI Global, pp 31–54

Tartavulea CV, Albu CN, Albu N, Dieaconescu RI, Petre S (2020) Online teaching practices and the effectiveness of the educational process in the wake of the COVID-19 pandemic. Amfiteatru Econ 22(55):920–936

Tavares AP, Joia LA, Fornazin M (2021) Digital transformation initiatives in public administration during the covid-19 pandemic in Brazil: unveiling challenges and opportunities. In: International conference on electronic government. Springer, Cham, pp 16–28

Tirivayi N, Richardson D, Gavrilovic M, Groppo V, Kajula L, Valli E, Viola F (2020) A rapid review of economic policy and social protection responses to health and economic crises and their effects on children: lessons for the COVID-19 pandemic response

Tremmel P, Myers R, Brunow DA, Hott BL (2020) Educating students with disabilities during the COVID-19 pandemic: lessons learned from commerce independent school district. Rural Spec Educ Q 39(4):201–210. https://doi.org/10.1177/8756870520958114

The Impact of Flexible Organizational Structure on Universities Employee's Performance and the Moderated Role of Commitment Culture

Esraa Awawdeh

Abstract In today's volatile environment, where effective interaction with organizational structures is essential, flexibility plays a pivotal role in providing organizations with a competitive edge. This research aims to investigate how the flexibility of organizational structures impacts the performance of employees in universities in the UAE, moderated by commitment culture.

This quantitative study tests a hypothesized model using data from 100 university employees. The findings indicate that flexible organizational structure significantly influences employee performance; an increase in organizational flexibility correlates with enhanced employee performance. However, the study reveals that commitment culture does not strengthen the relationship between flexible organizational structure and employee performance.

Keywords Flexibility · Organizational structure · Employee performance · Commitment culture

1 Introduction

Organizations recognize the critical importance of their structures, as a robust structure serves as a foundation for achieving goals effectively and efficiently, defining roles and guiding activities within the entity (Funiminiya 2018). Moreover, organizational structure directly influences employee performance and morale.

In today's volatile business environment, organizations are compelled to seek viable solutions. They understand the necessity of developing flexible structures to navigate market risks and mitigate environmental instability. Some organizations pursue flexible structures to enhance their market position, ensuring alignment with foundational goals and market demands.

E. Awawdeh (✉)
Faculty of Business and Law, The British University in Dubai, Dubai, UAE
e-mail: 20002397@student.buid.ac.ae

K. Al Marri et al. (eds.), *BUiD Doctoral Research Conference 2024*, Lecture Notes in Civil Engineering 587, https://doi.org/10.1007/978-3-031-84371-6_31

Similarly, universities play a pivotal role in societal development and future advancements, equipping societies with essential skills to foster robust economies. To achieve competitive advantage and deliver quality education, university decision-makers must strengthen their organizational structures (Kereto 2021). However, research on the effectiveness of flexible organizational structures in universities remains limited. Scholars argue that flexibility is crucial for adapting to dynamic environments (Yang et al. 2006).

Additionally, understanding commitment culture is essential, as it significantly influences organizational changes and employee performance. Not all employees readily embrace change, and some prefer maintaining established practices despite external pressures (Kereto 2021). Thus, comprehending commitment culture enhances employee performance within flexible organizational structures.

The purpose of this quantitative research is to explore the impact of flexible organizational structures on employee performance in universities, specifically in the UAE, while examining the moderating role of commitment culture. This study aims to investigate how flexible organizational structures influence employee performance in UAE universities, with commitment culture as a moderator.

As a quantitative study, data will be collected via questionnaires and analyzed using WARP-PLS SEM. The paper proceeds with an explanation of the methodologies employed, followed by the presentation and discussion of results. The research concludes with implications, recommendations, and limitations.

2 Literature Review

2.1 Organizational Structure

The framework governing relationships within an organization aims to achieve defined goals by assigning duties and tasks to individuals and teams, while coordinating their interactions during operational activities (Ahmady et al. 2016). Organizational structure encompasses all relationships among components of an organization, delineating responsibilities and coordinating tasks among employees to achieve unified objectives (Monavarian et al. 2007). It influences organizational processes by defining internal relationships, hierarchical reporting structures, and clarifying roles and responsibilities for each employee (Arnold, Feldman, 1986).

Moreover, organizational structure facilitates the flow of information across different levels within the organization (Monavarian et al. 2007), ensures discipline, and manages conflicts among employees. It begins by aligning with the organization's goals and objectives, followed by grouping similar activities and delegating authority based on employee skills within each department. The horizontal structure delineates divisions within the organization, while the vertical structure outlines reporting channels and authority delegations.

3 Employee Performance

Employee performance significantly impacts an organization's productivity and success, particularly within the context of universities (Ezejiofor and Ezekwesili 2021). Defined as the output and productivity resulting from employee development, it encompasses all factors influencing organizational and workforce productivity (Hameed and Waheed 2011). According to Hameed and Waheed (2011), enhanced employee performance directly contributes to achieving organizational goals, although verifying such productivity can be challenging (Yang 2008). Studies suggest that satisfying and motivating employees leads to increased productivity and improved performance (Ahmad et al. 2012).

Employee performance is pivotal for ensuring the smooth operation of organizational activities (Diamantidis and Chatzoglou 2018). Wright and Snell (2009) underscored in their research that optimizing employee performance enhances strategic advantages. Additionally, management support significantly influences employee performance improvement; increased support from management correlates with enhanced employee performance (Armstrong 2012).

4 Commitment Culture

Commitment among employees signifies their adoption of organizational goals and their dedication to achieving them (Miller and Lee 2001). According to Meyer and Allen (1991), commitment involves a sense of belonging to the organization, intentions to remain or depart, and a willingness to contribute effectively to the organization's success.

Numerous studies emphasize that organizational commitment correlates positively and significantly with organizational performance and the ability of employees to perform their tasks effectively (Wong et al. 2001). It is considered a crucial factor contributing to organizational success and performance improvement (Anggraeni 2018). Commitment also significantly impacts human resource outcomes such as loyalty, turnover rates, and absenteeism; employees with high commitment demonstrate positive outcomes in these areas (Kuvaas 2006).

5 Flexibility Concept in the Organizational Context

Flexibility in organizations is essential for navigating uncertainty in highly competitive and rapidly changing environments, enabling adaptation to dynamic conditions (D'Aveni 1994; Brown and Eisenhardt 1997; Volberda 1999). The current fast-paced global changes necessitate organizational structures and operations that are flexible; relying on rigid structures that resist change is increasingly risky (Yang et al. 2006).

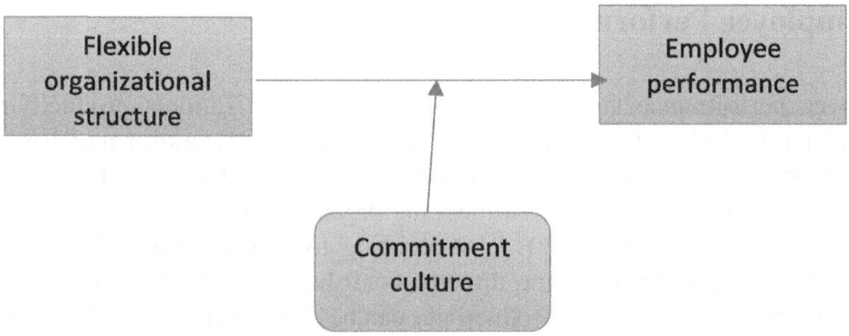

Fig. 1 The conceptual framework

Organizational flexibility is not defined in a singular manner but rather as a blend of managerial capabilities and practices designed to swiftly respond to shifts in the environment (Teece et al. 1997).

The primary objective of flexibility is to facilitate knowledge sharing across various organizational levels within an unstable environment (Tyulkova 2014).

6 Conceptual Framework and Hypothesis Development

Contingency theory underpins the relationships between the variables in this research, providing a foundation to understand the interplay between flexible organizational structure and employee performance, with the moderating impact of commitment culture. Contingency theory posits that the effectiveness of a flexible organizational structure depends on its alignment with contextual factors such as organizational culture. Specifically, a commitment culture ensures that employees are well-positioned to leverage the benefits of flexibility within the organizational structure.

Definitions and Dimensions of Key Concepts: According to the research objectives and a review of the literature, a conceptual framework was developed to determine the nature of the relationships between the variables. The flexible organizational structure is the dependent variable, employee performance is the independent variable, and commitment culture is the moderating variable (Fig. 1).

7 Hypothesis Development

H1: There is a significant relationship between the flexible organizational structure and employee performance.

H2: the commitment culture strengthens the relationship between the flexible organizational structure and employee performance.

8 Research Method

8.1 Variables

The key concepts in this research include one dependent variable, the flexible organizational structure; one independent variable, university employee performance; and one moderating variable, commitment culture. Each variable was measured using well-established scales: items for flexible organizational structure were taken from Camisón (2005), items for employee performance were adopted from Wiedower (2001), and items for commitment culture were sourced from Meyer and Allen (1991).

9 Data Collection

As part of this quantitative research, data will be collected through questionnaires. The questionnaire was created using Microsoft Forms and distributed via email through the relevant departments in various private and public universities in the UAE using a non-random sampling technique, specifically the snowballing technique. The questionnaire consists of 18 items. Questions 1 and 2 collect demographic information to describe the research sample. Questions 3 to 7 assess the employee performance variable, questions 8 to 12 evaluate the flexible organizational structure variable, and questions 13 to 18 measure the commitment culture moderator.

10 Analysis and Results

10.1 Respondents Profile

As shown in Table 1, there were 95 respondents to the online questionnaire, with 75 females and 20 males. The questionnaire was randomly distributed among employees in both private and public sectors on a fair basis, giving any university employee the opportunity to respond. The age range (Table 2) of the employees was categorized into five groups from 18 to 64 years old. Specifically, 42.1% of respondents were from the 35–44 age group, 32% from the 25–34 age group, 14% from the 45–54 age group, 8% from the 18–24 age group, and 3% from the 55–64 age group. No missing data was found, as indicated in Table 1.

Table 1 Demographic table

Descriptives		
	What is your gender?	What is your age group?
N	95	95
Missing	0	0
Mean	1.79	2.73
Median	2	3
Standard deviation	0.410	0.928
Minimum	1	1
Maximum	2	5

Table 2 Age group frequencies

Frequencies of age group?				
What is your age group?		Counts	% of Total (%)	Cumulative (%)
1	18–24	8	8.4	8.4
2	25–34	30	31.6	40.0
3	35–44	40	42.1	82.1
4	45–54	14	14.7	96.8
5	55–64	3	3.2	100

11 Model Assessment Using PLS-SEM

The hypotheses developed in the theoretical framework were tested using Partial Least Squares Structural Equation Modeling (PLS-SEM) for several reasons: its user-friendly software, its suitability for hypotheses derived from a predictive perspective, and its ability to handle small sample sizes (95 respondents) while still yielding respectable results due to the homogeneity of the sample. Additionally, PLS-SEM is preferred when working with latent variables controlled by a moderator (Risher et al. 2018).

After evaluating the measurement model, we identified one reflective latent variable, Employee Performance, and two formative variables, Flexible Organizational Structure and Commitment Culture. We conducted the PLS-SEM analysis using WarpPLS to assess the structural model, as it effectively analyzes both reflective (indicators of varying weights) and formative variables (constructed from multiple indicators of different weights) without causing identification problems (Fernandez et al. 2019).

To assess the data using WarpPLS, we need to evaluate both the measurement model and the structural model (Chin 2010). The measurement model assessment ensures the validity and reliability of the data from the indicators, while the structural model assessment examines the relationships between the constructs (Chin 2010).

12 Reliability and Validity

We begin the analysis by assessing the quality of the data from the measurement model. For reflective variables, such as employee performance, we evaluate reliability and convergent validity, which includes both convergent and discriminant validity. Both types should exceed 0.5. Additionally, we assess collinearity, ensuring that the full VIF is less than 5. Reliability of the construct is examined by checking indicator loadings, which should be above 0.708 to be considered reliable (Risher et al. 2018).

For formative variables (flexible organizational structure and commitment culture), we assess convergent validity and collinearity, ensuring the full VIF is less than 5.

To assess the quality of the data for employee performance as a reflective variable, we verify its reliability and validity by conducting composite reliability and Cronbach's alpha tests, both of which should exceed 0.7. The composite reliability for employee performance is 0.811, and its Cronbach's alpha is 0.708. Next, we check convergent validity, ensuring all contributing items exceed 0.5. Although the average variance extracted (AVE) for the variable is 0.464, which is below the 0.5 threshold, it is still acceptable. The full collinearity VIF for employee performance is 1.411, indicating no collinearity issues. The discriminant validity for employee performance items is higher than for other items.

For the formative variables (flexible organizational structure and commitment culture), we focus on two main aspects. First, we check the weights of the indicators to ensure all items significantly contribute to the constructs. Second, we verify that the items are not collinear. The p-values for all formative indicators are significant, and the individual VIFs are less than 5. Finally, we confirm that the full VIF for all variables is less than 5, ensuring acceptable collinearity levels.

13 Assessment of the Structural Model (Hypothesis Testing)

When assessing the relationship between variables, three key measures should be considered and reported: the p-value, beta coefficient, and R^2. The p-value indicates the significance of the relationship, the beta coefficient tests the strength and direction of the relationship, and R^2 explains the variance in the dependent variable caused by the model or the independent variable.

Upon examining these measures, we found the following:

1. The p-value between the flexible organizational structure and employee performance is less than 0.01, indicating a significant effect of the flexible organizational structure on employee performance. This low p-value suggests the result is not due to chance.

2. The beta coefficient for this relationship is 0.44, indicating a positive relationship between these variables. As the flexibility within the organizational structure increases, employee performance also increases.
3. The R^2 Value is 0.23, Which, Although Relatively Low, is Considered Acceptable in Social Sciences Where Human Behaviors Are Challenging to Predict.

In conclusion, there is a significant relationship between flexible organizational structure and employee performance, supporting H1: there is a significant relationship between flexible organizational structure and employee performance.

For the moderator, commitment culture, the p-value is 0.26, which is greater than 0.05. Thus, commitment culture does not significantly influence employee performance, leading to the rejection of H2: commitment culture strengthens the relationship between flexible organizational structure and employee performance. This indicates that commitment culture does not have a significant moderating influence on the relationship between flexible organizational structure and employee performance.

14 Discussion

The research findings indicate that a flexible organizational structure significantly enhances the performance of university employees. When flexibility within the organizational culture increases, employee performance also improves. However, a commitment culture does not strengthen the relationship between a flexible organizational structure and employee performance.

The first finding aligns with previous studies, such as those by Martin and Healy (2009) and Altindag and Siller (2014), which demonstrated that flexible organizations enhance employee performance and productivity. These studies also noted that employees increasingly prefer flexible organizations due to the better work-life balance they offer. Additionally, scholars like Ramendran et al. (2013) have described flexibility as an investment for companies, as it not only boosts performance and productivity but also provides a competitive advantage by increasing organizational responsiveness and customer satisfaction.

The second finding, which indicates that commitment culture does not enhance the relationship between a flexible organizational structure and employee performance, contradicts some previous research. Studies by Adam et al. (2018) have shown that a strong commitment culture positively impacts employee performance, as committed employees tend to work smarter and harder. This discrepancy can be explained by considering the highly diverse workforce in the UAE, with employees from varied cultural backgrounds and expectations. Many employees view their employment in the UAE as temporary, reducing the effectiveness of a commitment culture. Additionally, bureaucratic tendencies within universities make employees prefer stability and adherence over flexible structures.

15 Conclusion

The purpose of this research is to investigate the effect of flexible organizational structures on employee performance, with commitment culture as a moderating factor, in private and public universities in the UAE. Our aim is to enhance the conceptualization of the relationships between these variables to develop theories that can be utilized by university administrations and academics to improve the academic sector in the UAE. This study is particularly relevant in the post-pandemic era, where flexibility is recognized as a crucial element for enhancing the agility of organizational structures and operations, ensuring the continuity of educational processes under challenging circumstances.

The findings of this research indicate that a flexible organizational structure has a significant positive impact on employee performance, while the commitment culture does not influence this relationship. These results align with our first hypothesis (H1) but contradict our second hypothesis (H2).

Although this research focuses on the university sector in the UAE, the insights gained on managing flexible organizational structures and commitment culture to enhance employee performance are applicable across various sectors. Management should prioritize adopting the appropriate flexible organizational structures, as doing so can increase organizational outputs by 40%. Flexibility has become crucial in today's rapidly changing market.

16 Limitations and Future Recommendations

This research encountered several limitations that constrained its contribution to the field. Bureaucratic obstacles from the universities' administrations limited the sample size. Additionally, the diverse staff within UAE universities may have influenced the results, as the heterogeneous context of the UAE can lead to widely varying employee perceptions.

For future research, it is recommended to develop a more comprehensive scale to equally capture the perceptions of respondents. Paying closer attention to the diversity among staff in UAE universities will yield more practical and applicable insights.

References

Adam A, Yuniarsih T, Ahman E, Kusnendi K (2018) The mediation effect of organizational commitment in the relation of organization culture and employee performance, pp 1–5
Ahmad MB, Wasay E, Malik SU (2012) Impact of employee motivation on customer satisfaction: study of airline industry in Pakistan. Interdiscip J Contemp Res Bus 4(6):531–539
Ahmady G, Mehrpour M, Nikooravesh A (2016) Organizational structure, pp 1–8

Altindag E, Siller F (2014) Effects of flexible working method on employee performance: an empirical study in Turkey, pp 1–7
Anggraeni AI (2018) The effect of psychological contract perceived organizational support, and value congruence on organizational citizenship behavior: social exchange theory perspectives. Calitatea 19(162):67–72
Armstrong M (2012) A Handbook of Human Resource Management Practice. Kogan Page, London
Arnold HJ, Feldman DC (1986) Organizational behavior. NewYork 1
Brown SL, Eisenhardt KM (1997). The art of continous change: linking complexity theory and time- paced evolution in relentlessly shifting organizations. Adm Sci Quarterlv 42:1–34
Camisón C (2005) On how to measure managerial and organizational capabilities: multi-item models for measuring distinctive competences, pp 1–22
Chin WW (2010) How to Write Up and Report PLS Analyses. In: Esposito Vinzi V, Chin WW, Henseler J, Wang H, eds, Handbook of partial least squares: concepts, methods and applications, Springer, Heidelberg, Dordrecht, London, New York, pp 655–690. https://doi.org/10.1007/978-3-540-32827-8_29
D'Aveni RA (1994) Hyper-Competition. The Free Press, New York
Diamantidis AD, Chatzoglou P (2018) Factors affecting employee performance: an empirical approach, pp 1–25
Ezejiofor RA, Ezekwesili TP (2021) Organizational structure and employee performance: evidence from pharmaceutical companies in Anambra state, Nigeria. Innovations, 67
Fernandes AAR, Hutahayan B, Arisoesilaningsih E, Yanti I, Astuti AB, Amaliana L (2019) Comparison of curve estimation of the smoothing spline nonparametric function path based on pls and pwls in various levels of heteroscedasticity.10P Conference Series: Materials science and engineering, Forthcoming issue
Funminiyi AK (2018) Impact of organisational structure on employee engagement: Evidence from North Central, pp 1–11
Hameed A, Waheed A (2011) Employee development and its affect on employee performance a conceptual framework, pp 1–6
Kereto S (2021) Perceptions of university managers on the influence of organizational structures on quality of institutional management in public and private universities in Kenya, pp 70–73
Kuvaas B (2006) Work performance, affective commitment, and work motivation: the roles of pay administration and pay level. J Organ Behav 27(3):365–385. https://doi.org/10.1002/job.377
Martin B, Healey J (2009) Changing work organization and skill requirements. Australia Bulletin of Labour 35(3)
Meyer JP, Allen NJ (1991) A three-component conceptualization of organizational commitment. Hum Resour Manag Rev 1(1):61–89
Miller D, Lee J (2001) The people make the process: commitment to employees, decision- making and performance. J Manag 27:163–189
Monavarian A, Asgari N, Ashena M (2007) Structural and content dimensions of knowledge-based organizations. In: the first national conference of knowledge management
Natalia Tyulkova N (2014) A flexible organizational structure as a way of knowledge management in SMEs, pp 1–9
Overholt MH (1997) Flexible organizations: using organizational design as a competitive advantage. Human Resources Planning, 20, pp 22–32
Ramendran C, Raman G, Mohamed RKMH, Beleya P (2013) Organizational flexibility and its implications on employee's productivity, pp 1–20
Risher J, Ringle CM, Sarstedt M (2018) When to use and how to report the results of PLS-SEM, pp 1–25
Teece DJ, Pisano G, Shuen A (1997) Dynamic capabilities and strategic management. Strat Manag J 18:509–533
Volberda HW (1998) Building the flexible firm-how to remain competitive. Oxford University Press, pp 120–133
Volberda HW (1999) Building the flexible firm: how to remain competitive, pp 1–4

Wang P (2006) Rigid flexibility: the logic of intelligence. Springer
Wiedower KA (2001) A shared vision: the relationship of management communication and contingent reinforcement of the corporate vision with job performance, organizational commitment, and intent to leave. Unpublished Doctoral Dissertation. Alliant International University, California
Wong CS, Wong YT, Hui C, Law KS (2001) The significant role of Chinese employees' organizational commitment: Implications for managing employees in Chinese societies. J World Bus 36(3):326–340
Wright PM, Snell SA (2009) Human resources, organizational resources, and capabilities. In: Storey J, Wright P, Ulrich D (eds) The Routledge Companion to Strategic Human Resource Management. Routledge, London, pp 345–356
Yang G, Huang W, Zhau ZY (2006) Application flexibility strategy to organizational structure design in China's international construction enterprises, pp 1–9
Yang (2008) Individual attitudes and organisational knowledge sharing, pp 345–353

[illegible]

The Impact of Organizational Learning on Business Continuity Through the Moderating Effect of Organizational Culture: A Study of UAE's Public Sector Organizations

Loubna Saad Saoud, Muna Al Suleimani, and Jude Eid

Abstract Despite the UAE's vast growth, its public sector remains essential to its stability, where effective crisis management and fast change are unique challenges. It is crucial to understand the relationship between organizational learning (OL), business continuity (BC), and organizational culture (OC). This paper examines the OL and BC relationship through the moderating role of OC, specifically the hierarchical culture that is applied in the UAE's public sector. The study aims to expand OL and BC literature. Empirical data was collected using the questionnaire method. The study hypothesized that OL has a positive impact on BC and that OC moderates the relationship between OL and BC. The study findings indicate that the OL process has a direct and positive impact on BC in the context of the UAE public sector. However, the study reveal that the OC does not moderate the relationship between OL and BC, which sheds light on the complexity of this relationship and can guide future research. Thus, the study findings have the potential to assist UAE public sector organizations in developing their OL activities, fostering an OC, and enhancing effective strategies for BC.

Keywords OL · BC · Hierarchal OC

Author Note: This paper was prepared for buiness excellence module, taught by Prof. Bader Yousef Bader Obeidat.

L. Saad Saoud (✉) · M. Al Suleimani · J. Eid
Department of Philosophy in Business Management, Faculty of Business and Low, The British University in Dubai, Dubai, UAE
e-mail: 22001908@student.buid.ac.ae

K. Al Marri et al. (eds.), *BUiD Doctoral Research Conference 2024*, Lecture Notes in Civil Engineering 587, https://doi.org/10.1007/978-3-031-84371-6_32

1 Introduction

In today's fast-changing environment, organizations face a variety of challenges that can affect their operations and sustainability. This emphasizes the need for businesses to assure be resiliency. As the UAE becomes an international economical hub, the public sector's capacity to efficiently manage disruptions and maintain continuity becomes a crucial aspect of sustaining national progress (Al Ameri and Musa 2021). In the public sector, government agencies play a fundamental role in delivering key services, which is necessary for effective business continuity. The relationship between OL and BC provides fertile ground for research. However, the complex link between OL and BC within the public sector requires significant examination.

2 Research Purpose and Objectives

The paper aims to examine the relationship between the three constructs (OL, BC, and OC) within the UAE's public sector and understand how public entities may use their internal learning process to enhance their strategies management and recover from crisis.

The papers' objectives are:

1. To examine the relationship between OL and BC within the UAE's public sector entities.
2. To examine the moderating effect of OC on the relationship between OL and BC in the UAE's public sector entities.
3. To provide a set of implications for the UAE's public sector entities that can potentially enhance their BC plans and strategies.

3 Problem Statement

Effective knowledge management requires the development of specialised organizational units and the provision of managerial support for learning processes. Research on OL and its relationship with BC is still in its early stage, requiring further examination (Naimoli and Saxena 2018). A significant gap exists in comprehending the relationship between disasters and their impact on OL processes. Furthermore, limited intention has been directed into the corporation of learning with other multilevel processes throughout time in fostering organizational resilience (Fasey et al. 2021). Empirical research is required to better understand the relationship between OL and BC. This gap highlights the need for a comprehensive study on how organizations can maintain continuity while promoting the learning process. Knowledge sharing and retention are essential for business continuity, and a positive organizational culture facilitates trust and collaboration. Conversely, a lack of communication and transparency has a negative effect on business continuity by resulting in an increase in turnover and the loss of organizational knowledge. Different cultural dimensions, have varying effects on knowledge management (Wanjala 2024). For example, clan cultures, which prioritise collaboration and a familial atmosphere, are more advantageous to knowledge exchange than hierarchical cultures, which may prioritise form structure and control (De Long and Fahey 2000).

4 Research Questions

1. How does OL impact BC in the context of the public sector in the UAE?
2. Does a hierarchal organizational culture moderate the relationship between OL and BC in the context of the public sector in the UAE?

5 Literature Review

5.1 Organizational Learning

OL is the process of the generation, acquisition, and transmission of knowledge within an organization by using information to create knowledge management systems for enhancing organizational business continuity. There are significant reasons to address OL in the public sector. First, the structure of the originations in the public sector is substantial enough to demand attention from management and academics (Ferlie et al. 2003). Second, public sector organizations have experienced significant transformation in recent decades, emphasizing the need to produce OL. Third, the focus on the distinctive features that are related to learning in public organizations may contribute to the expansion of the learning field across all their kinds.

5.2 Business Continuity

BC is defined as a management strategy that fosters the organization's sustainability (Rashid et al. 2014). Public sector organizations have implemented BC to create a smooth flow of operations with minimum interruptions. Al Ameri et al. (2021) conducted a study in the UAE to assess the impact of BC on organizational performance, and they concluded that BC programs have a significant impact on the public sector's organizations survival and interruption recovery, and the factors that led to BC programs' success were top management support, an up-to-date BC plan, and employee training.

5.3 Organizational Culture

OC is a set of shared values that might have a direct or indirect effect on behaviours. culture is fundamental for employees to learn and enhance their performance (Schein 1990). Based on different theories, of organizational environment that encourages knowledge and enables innovation (Al-Hammadi and Adnan 2020). The most relevant OC approach is the Competing Values Framework (CVF) by Cameron and Quinn (2006), which has been applied in several empirical studies (Stock et al. 2007). It describes four types of culture:

- **Clan Culture**: which has a flexible structure and focuses on internal processes.
- **Bureaucracy Culture**: a hierarchical structure where procedures, rules, and policies govern the behavior of employees.
- **Market Culture**: it focuses on increasing market share and achieving higher targets.
- **Adhorarcy Culture**: it promotes a fast reaction to the competing market (Fig. 1).

The UAE public sector is adopting a hierarchal culture that includes governmental rules and regulations. It is a non-commercial aim and lacks adaptability to changing market conditions (Al Dari et al. 2021).

Fig. 1 CVF by Cameron and Quinn (2006)

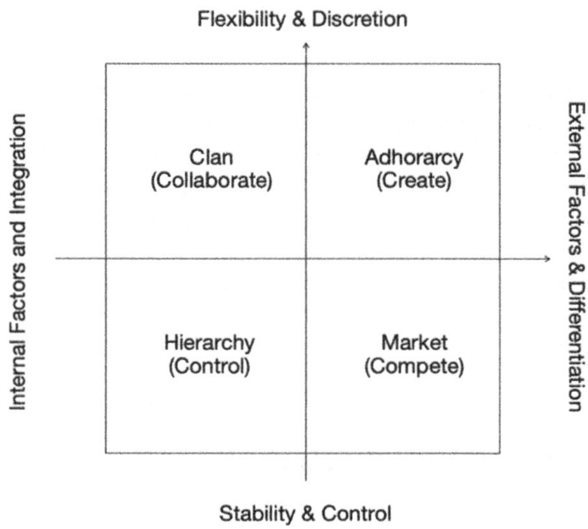

6 Theoretical Framework and Hypothesis Development

6.1 *Resource-Based View (RBV)*

Wernerfelt (1984) established RBV theory on the organization internal capabilities and encouraged the company to concentrate on strategic management.

6.2 *Organizational Development Theory (ODT)*

The ODT by Lewin's (1951) addresses the process of organizational change. It demonstrates the three phases of organizational change: defrosting, transformation, and dicing. ODT is devoted to realigning an organization with its fast-changing environment via knowledge management and OL for sustainability (Cummings 2004).

6.3 *Knowledge-Based View (KBV)*

KBV extends RBV; it emphasizes distinctive knowledge stocks as the foundation of competitive advantage, therefore reinforcing the most important internal resources (Hoskisson et al. 1999).

This study integrates three theoretical frameworks: RBV, KBV, and ODT to reveal the influence of OL on BC through OC. RBV emphasises the significance of internal resources and competencies as determinants of competitive advantages, with organizational learning serving as a crucial resource within this paradigm. KBV enhances RBV by highlighting the significance of knowledge as an essential resource for maintaining organizational capacities. Ultimately, ODT examines how organizations navigate change and respond to new challenges, with OL serving as a fundamental mechanism for attaining resilience and continuity.

6.4 *Conceptual Framework*

The conceptual framework model is developed based on the provided literature review theoretical framework. Table 1 shows the study's measurement table.

The conceptual model (Fig. 2) is hypothesized that OL can positively impact BC through critical knowledge. Also, it is hypothesized that hierarchical OC can serve as a moderator between OL and BC.

- H-1. OL positively impacts BC.

Table 1 Measurement table

Variable	Definition
OL	Is the process of the generation, acquisition, and transmission of knowledge within an organization (Fang and Chen 2016)
BCM	Is a holistic management strategy that offers a framework for creating resilience and the competence to respond effectively to possible risks while protecting the interests of the organization (Business Continuity Management 2005, p. 6)
HC	Is a type of organizational culture that follows official regulations and policies (Cameron and Quinn 2006)

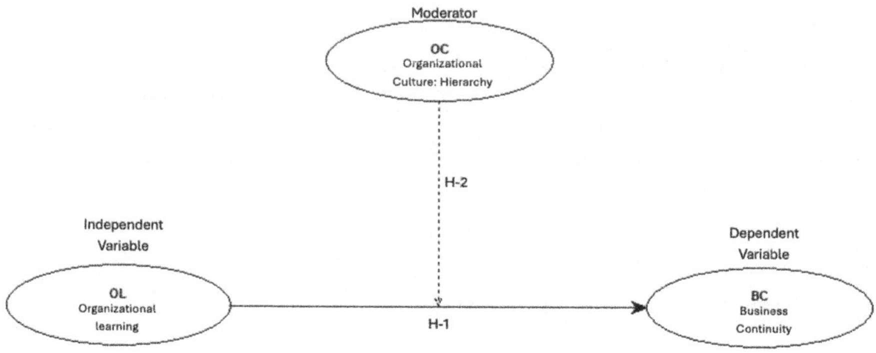

Fig. 2 The research conceptual model

- H-2. OC moderates the relationship between OL and BC.

7 Research Methodology

This paper employs a deductive and non-experimental approach using a questionnaire survey, to collect data from UAE public sector organizations The selected research strategy is a survey-based approach, this method was selected due to the study's measurable concepts, association with a deductive approach for generalizing results, and its effectiveness to gather diverse opinions and identifying cause-and-effect relationships to achieve research objectives (Ghauri et al., 2020). This paper follows quantitative survey methods, the participants are from various levels provide diverse perspectives. The study adopts non-probability sample, forming a simple random sample between 350 and 400 to capture a comprehensive view. A questionnaire was employed as a data collection tool using previous published studies as below:

- Independent variable: OL; five items were adopted from (Watkins and Marsick 1993)

- Dependent variable: BC. Five items were taken from the BC maturity model.
- Moderating variable OC (hierarchy), five items were taken from Cameron and Quinn (2011).

The research questionnaire was developed in English, then translated to Arabic and back translated to English to ensure accuracy and it employs a 5-point Likert scale (1 = strongly disagree, 5 = strongly agree) for assessing responses. Accompanied by a cover letter, the survey aims to establish mutual understanding, clarify objectives and confidentiality, and express appreciation for participation.

7.1 Data Collection

Primary data, collected through questionnaires, and secondary data from literature, were used in this study. Questionnaires targeted 400 public sector employees across different levels in the UAE. Online questionnaires via email and social media were chosen for cost-effectiveness, time efficiency, and participant anonymity to reduce bias and increase response rates.

8 Data Analysis

8.1 Descriptive Analysis

Data were collected from 417 participants via an online questionnaire, exceeding the target. After quality and error-checking, 409 responses remained. Moreover, the descriptive analysis was conducted using Jamovi software (Tables 2, 3, 4 and 5).

The following tables represent participants demographics:

Table 2 Gender

Frequesncies of gender			
Gender	Counts	% of Total	Cumulative %
Female	94	23.3	23.3
Male	310	76.7	100.0

Table 3 Age group

Frequencies of age			
Age	Counts	% of Total	Cumulative %
18–24 years old	171	41.8	41.8
25–34 years old	59	14.4	56.2
35–44 years old	95	23.2	79.5
45–54 years old	63	15.4	94.9
55–64 years old	19	4.6	99.5
65 + years old	2	0.5	100.0

Table 4 Job level

Frequencies of job			
Job	Counts	% of Total	Cumulative %
Entry-level	171	41.8	41.8
Executive-level	32	7.8	49.6
Junior-level	56	13.7	63.3
Mid-level	108	26.4	89.7
Senior-level	42	10.3	100.0

Table 5 Department/ Function

Frequencies of department			
Department	Counts	% of Total	Cumulative %
Admin	140	34.2	34.2
Banking	1	0.2	34.5
Consultation	1	0.2	34.7
Creative	12	2.9	37.7
Customer Service	23	5.6	43.3
Education	14	3.4	46.7
Engineering	16	3.9	50.6
Financial	29	7.1	57.7
Healthcare	20	4.9	62.6
Legal	9	2.2	64.8
Managerial	62	15.2	80.0
Physician	32	7.8	87.8
Research	7	1.7	89.5
Sales and Marketing	32	7.8	97.3
Technical	11	2.7	100.0

Table 6 Reliability test

Reliability test	OL	BC	OC	OC*OL
Composite reliability	0.956	0.967	0.935	1
Cronbach's alpha	0.942	0.958	0.913	1

8.2 Inferential Analysis

8.2.1 Measurement Model

After data cleaning, data was transferred to PLS-SEM Software for measurement model assessment. It was performed through reliability test, convergent validity, and discriminant validity tests Hair et al. (2011).

8.2.2 Reliability

Composite reliability and Cronbach's alpha were used to assess the reliability, with values of 0.6 + accepted in exploratory research, and 0.7–0.9 considered satisfactory as per (Nunnally and Bernstein 1994). Table 6 represents the values of the test results. The values obtained are accepted and considered satisfactory.

8.2.3 Validity

Convergent validity measures correlation between similar constructs, while discriminant validity determines if constructs are distinct based on specific criteria. The table below represents the values for convergent validity which was measured through AVE and items loadings.

The study followed Chin (2010) guidelines where item loadings and AVE values above 0.5 are considered acceptable. Results showed item loadings and AVE values met this threshold. Discriminant validity was assessed using the square root of AVE values in PLS-SEM software, with each variable's square root of AVE correlating highest with itself, meeting Hair et al., (2011) criteria. As per Table 8, the values are considered acceptable.

The next test involved assessing the full collinearity among the constructs, the constructs' collinearity must have values lower than 5 to be accepted. However, Table 7 illustrates that OL and BC have values higher than 5. Therefore, to eliminate collinearity issues, some items had to be dropped and the tests had to be run again. As a result, the following measurement model was re-established:

The values in the table all meet the thresholds and are considered acceptable in the study. The discriminant validity was also assessed using Formell and Larcker criterion (Hair et al. 2014) to ensure its acceptance (Table 9 and 10).

Table 7 Measurement model

Constructs/Items	Items loading	CR	Cronbach's alpha	AVE	Collinearity
Organizational learning		0.956	0.942	0.812	5.198
OL1	0.877				
012	0.894				
0L3	0.917				
OLA	0.911				
0L5	0.904				
Business continuity management		0.967	0.958	0.856	5.888
BC1	0.920				
BC2	0.937				
BC3	0.932				
BC4	0.922				
BC5	0.916				
Organizational culture		0.935	0.913	0.742	2.930
OC1	0.900				
OC2	0.869				
OC3	0.876				
OC4	0.863				
OC5	0.797				

Table 8 HTMT ratio

	OL	BC	OC	OC*OL
OL	0.901			
BC	0.884	0.925		
	0.748	0.800	0.862	
OC*OL	–0.508	–0.412	–0.446	1.000

8.2.4 Structural Model

Hair et al. (2014) outline the structural model's evaluation using significance (p value), path coefficient (β), and R square (R^2). A p value ≤ 0.05 indicates hypothesis acceptance. The test results of the structural model are represented in the Table 11.

As noted above, H-1, which stated that OL positively impacts BC, is considered significant with a p value of less than 0.01. The β coefficient is 0.82 which indicates a strong positive relationship between organizational learning and business continuity.

Table 9 Second measurement model

Constructs/Items	Loading	CR	Cronbach's alpha	AVE collinearity
Organizational learning		0.944	0.912	0.850 3.957
OL2	0.919			
OL3	0.921			
OL5	0.926			
Business continuity management		0.957	0.932	0.881 4.499
BC3	0.939			
BC4	0.934			
BC5	0.942			
Organizational culture		0.935	0.913	0.742 2.930
OC1	0.900			
OC2	0.869			
OC3	0.876			
OC4	0.863			
OC5	0.797			

Table 10 Fornell and Larcker test

	OL	BC	OC	OC*OL
OL	0.922			
BC	0.840	0.939		
OC	0.737	0.792	0.862	
OC*OL	–0.497	–0.378	–0.447	1.000

Table 11 Structural model

	Hypothesis	β Coefficients	p value	Supported
H-1	OL → BC	0.82	p < 0.01	Yes
H-2	OL*OC → BC	0.08	0.06	No
R^2			66%	

On the other hand, H-2, which stated that OC moderates the relationship between OL and BC, is non-significant with a p value of 0.06. The R^2 is 66% which indicates that 66% of business continuity is influenced by organizational learning (Fig. 3).

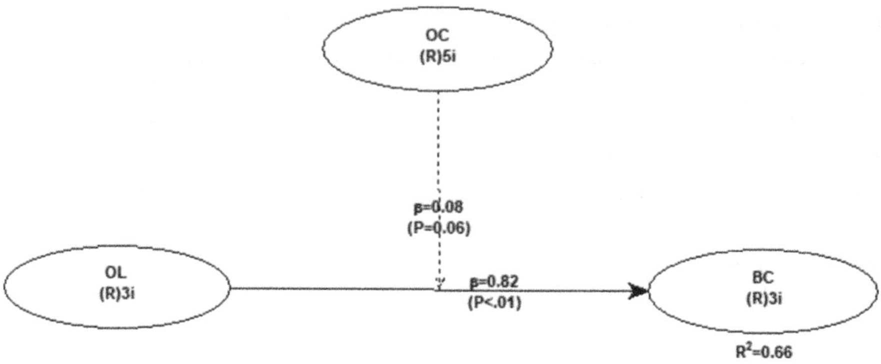

Fig. 3 Structural model

9 Discussion

OL and BC

The study results show a significant correlation between OL and BC. Based on our current literature review, OL refers to the generation, acquisition, and sharing of knowledge within an organization. Our results show a strong and positive relationship between OL and BC, which indicates that there is a direct and strong improvement in the effectiveness of BC techniques as OL rises. The structural model shows a value of 66%, which indicates that OL has a strong explanatory power on BC. The statistical significance of the relationship between OL and BC reinforces its importance. This connection is based on a cycle of continuous improvement, where reflecting on mistakes leads to learning and adaptation. Scholars in KBV theory consider learning and knowledge accumulation as strategic resources for enhancing organizational capabilities (Conner 1991). This indicates that organizations that actively learn from their experiences are better prepared to manage unexpected disruptions. These findings are consistent with previous study results (Abdullah et al. 2014; Abu Bakar et al. 2019).

OC as a Moderator Between OL and BC

The findings of the study have rejected the second hypothesis which claimed that OC (hierarchal) moderates the relationship between OL and BC. The results showed no effect of OC on the relationship between OL and BC in UAE's public sector, even though previous literature review has implied otherwise. Nevertheless, the reasons contributing to this finding can be attributed to UAE's unique strategic approach in the public sector. The UAE has implemented nationwide strategies to enhance citizen and resident experiences while maintaining international competitiveness such as "Vision 2021" and "We the UAE 2031", cascade down to public sector organizations, emphasizing value-adding practices like research, innovation, knowledge-based growth, and business continuity (BC). Despite hierarchical OC, top-down strategic alignment emphasizes the importance of practices like OL and BC in line with national strategies. The UAE's collective focus on national development fosters shared responsibility, transcending hierarchical differences, and underscores the significance of OL and BC for organizational effectiveness and resilience in the public sector (Al-khouri, 2014). This collective approach encourages a culture of innovation and excellence, aligning with the UAE's goals of global sustainably. Additionally, mature organization often possess established standers for knowledge transfer and continuity, making OC less impactful in affecting these processes. In these organizations, standardised procedures override cultural differences, and effective leadership may further reduce the significance of organizational culture since workers choose authoritative orders over cultural norms. Furthermore, the legal framework of the public sector, emphasising compliance and efficiency, may restrict OC's impact on OL and BC. The heterogeneity inside public sector organizations, characterised by many departments and missions, may diminish OC's moderating influence. The rapid advancement of technology and worldwide competitive pressures force organizations to emphasise agility rather than internal cultural dynamics, which may clarify the little influence of OC in the UAE's public sector.

10 Conclusion

This study examines the relationship between OL and BC through the moderating role of OC in the UAE's public sector. The results highlighted a strong positive impact of OL on BC, emphasizing the significance of knowledge and learning. Surprisingly, OC did not moderate this relationship, possibly due to the UAE's distinct leadership and management style. This research contributes novel insights to the field by exploring these constructs together within the UAE's public sector, suggesting cultural context can influence study outcomes. Future studies are encouraged to explore how cultural differences may affect OL-BC relationships in other countries and to identify alternative moderators impacting this relationship. Practitioners in the UAE's public sector is advised to invest in knowledge management and OL strategies to enhance BC effectiveness during disruptions.

References

Abdullah NAS, Noor NLM, Ibrahim ENM (2014) Information technology service management (ITSM): contributing factors to it service disruptions–a case of Malaysia public service agencies. In: Proceedings - Pacific Asia conference on information systems, PACIS

Abu Bakar Z, Yaacob NA, Udin ZM, Hanaysha JR, Loon LK (2019) Business continuity management implementation in the Malaysian public sector. Int J Bus Technol Manag 1(1):18–27

Al Ameri M, Musa M (2021) The impact of business continuity management on the performance of public organizations in Uae. J Leg, Ethical Regul Issues 24(6):1
Al Dari T, Jabeen F, Hussain M, Al Khawaja D (2021) How types of organizational culture and technological capabilities contribute to organizational learning. Manag Res Rev 44(3):2040–8269. https://doi.org/10.1108/MRR-02-2020-0090
Al-khouri AM (2014) Fusing knowledge management into the public sector : a review of the field and the case of the emirates identity authority. J Knowl Manag, Econ, Inf Technol 4(3):1–89
Al-Hammadi FAA, Adnan AAZ (2020) The moderating effect of organisational culture on the relationship between workplace learning and employees' performances in the United Arab Emirates. Eur J Econ Bus Stud 6(2):96–108
Cameron KS, Quinn RE (2006) Diagnosing and changing organizational culture based on the competing values framework the Jossey-Bass business and management series
Cameron KS, Quinn RE (2011) An introduction to changing organisational culture: based on the competing values framework. Diagnosing and chaning organisational culture
Chin, 2010 Chin WW (2010) How to write up and report PLS analyses. In Handbook of Partial Least Squares. https://doi.org/10.1007/978-3-540-32827-8_29
Conner KR (1991) A historical comparison of resource-based theory and five schools of thought within industrial organization economics: do we have a new theory of the firm? J Manag 17(1):121–154. https://doi.org/10.1177/014920639101700109
Cummings JN (2004) Work groups, structural diversity, and knowledge sharing in a global organization. In Management Science, vol 50, no 3, pp 352–364. https://doi.org/10.1287/mnsc.1030.0134
De Long DW, Fahey L (2000) Diagnosing cultural barriers to knowledge management. Acad Manag Exec 14(4):113–127. https://doi.org/10.5465/ame.2000.3979820
Fasey KJ, Sarkar M, Wagstaff CRD, Johnston J (2021) Defining and characterizing organizational resilience in elite sport. Psychol Sport Exerc 52:101834. https://doi.org/10.1016/j.psychsport.2020.101834
Fang SC, Chen HK (2016) Strategic intent, organizational environment, and organizational learning mechanisms: a multiple-case study in the construction industry in Taiwan. Pers Rev 45(5):928–946
Ferlie E, Hartley J, Martin S (2003) Changing public service organizations: current perspectives and future prospects. Br J Manag 14(SPEC. ISS), S1–S14. https://doi.org/10.1111/j.1467-8551.2003.00389.x
Ghauri et al., 2020Ghauri P, Grønhaug K, Strange R (2020) Research methods in business studies. In research methods in business studies. https://doi.org/10.1017/9781108762427
Hair JF, Ringle CM, Sarstedt M (2011) PLS-SEM: Indeed a silver bullet. J Mark Theory Pract 19(2):139–152. https://doi.org/10.2753/MTP1069-6679190202
Hair JF, Sarstedt M, Hopkins L, Kuppelwieser VG (2014) Partial least squares structural equation modeling (PLS-SEM): an emerging tool in business research. Eur Bus Rev 26(2):106–121
Hoskisson RE, Hitt MA, Wan WP, Yiu D (1999) Theory and research in strategic management: swings of a pendulum. J Manag 25(3):417–456. https://doi.org/10.1177/014920639902500307
Lewin K (1951) Field theory in social science: selected theoretical papers (Edited by Dorwin Cartwright.). In Research center for group dynamics
Mohd Rashid AH, Loke SP, Ooi KB (2014) Strengthening supply chain risk management for business continuity: a case study approach. Int J Manag Enterp Dev 13(3–4):278–301. https://doi.org/10.1504/IJMED.2014.069173
Naimoli JF, Saxena S (2018) Realizing their potential to become learning organizations to foster health system resilience: Opportunities and challenges for health ministries in low-and middle-income countries. Health Policy Plan 33(10):1083–1095. https://doi.org/10.1093/heapol/czy100
Nunnally J, Bernstein I (1994) Psychometric Theory, 3rd edn, 1994. McGraw-Hill, New York, vol 3, pp 248–292
Schein EH (1990) Organizational culture

Stock GN, McFadden KL, Gowen CR (2007) Organizational culture, critical success factors, and the reduction of hospital errors. Int J Prod Econ 106(2):368–392. https://doi.org/10.1016/j.ijpe.2006.07.005

Wanjala M (2024) The impact of organizational culture on knowledge management practices. Eur J Inf J Manag ISSN 3(2). https://www.researchgate.net/publication/382229886_The_Impact_of_Organizational_Culture_on_Knowledge_Management_Practices

Watkins KE, Marsick VJ (1993) Sculpting the learning organization: consulting using action technologies. New Dir Adult Contin Educ (58). https://doi.org/10.1002/ace.36719935813

Wernerfelt B (1984) Consumers with differing reaction speeds, scale advantages and industry structure. Eur Econ Rev 24(2):257–270. https://doi.org/10.1016/0014-2921(84)90112-0

[illegible]

The Indirect Impact of Enterprise Architecture on Successful Project Management in UAE Healthcare Organizations

Hadeel Atallah and Rajish Pai

Abstract The purpose of this research was to examine the indirect impact of enterprise architecture on successful project management in the UAE healthcare sector under the mediating role of effective decision making. The current research adopted a quantitative methodological framework to rigorously examine the subject matter above and study the phenomenon of the proposed relationships in the case of UAE healthcare entities. A survey was distributed to 500 high level healthcare practitioners and decision makers across the UAE (mainly, Abu Dhabi, Dubai, and Sharjah). The findings of the study revealed that, the variable of effective decision making positively mediated the relationship between enterprise architecture and successful project management in the UAE healthcare industry. The combined path regression value for the independent variable of enterprise architecture, through the mediating variable of effective decision making, to the dependent variable of successful project management, was $R = 0.789$, at a p value of $< .0001$. These findings revealed that when the enterprise architecture is synchronized through effective decision making, then implementing any change project and managing it to a successful end would not be a challenge. In agreement with evidence from literature, the findings of the current study further illuminated why it is critical to have synergy across organizational functions through a fully integrated enterprise architecture. The challenge of lack of integration and the bottlenecks it can cause to successful project implementation were implied by virtue of the findings confirming the importance of effective decision making across the enterprise as an enabler of success.

Keywords Enterprise architecture · Effective decision making · Project management

H. Atallah (✉) · R. Pai
Swiss Business School, Abu Dhabi, UAE

K. Al Marri et al. (eds.), *BUiD Doctoral Research Conference 2024*, Lecture Notes in Civil Engineering 587, https://doi.org/10.1007/978-3-031-84371-6_33

1 Introduction

Enterprise architecture (EA) is identified by multiple evidences of research as a significant element in the establishment of and integration of multiple operational frameworks of operations for organizations (Zachman 1996; Shanks et al. 2018; Takeuchi and Yamamoto 2019). Simply put, according to the cited researchers, EA enables the advisory of organizational plans especially through the integration of technological projections to foster the understanding of the dynamics embroiled with certain decisions that an organization is looking to take whether the decisions involve change or continued project management. To this extent and having clear knowledge of the fact that enterprise architecture is behind the assessment of the organization as is and the development of the future road maps that inform organizational strategies, considering its application in an actual organizational setting cannot be done without considering the debate surrounding decision making processes in organizations.

1.1 Problem Statement

The problem that the research has identified based on preliminary review of literature is the fact that, enterprise architecture is ambiguous and abstract in nature. This notion, is not denied by scholars and proponents of enterprise architecture who in simplest of terms identify enterprise architecture as a medium of understanding processes and states in an operation in a way that one can be able to understand the present, anticipate the future and clearly detail out the plans that lead to the anticipated future (Shanks et al. 2018; Alwadain et al. 2016; Gong and Janssen 2021). The afore-cited scholars among a host of others that have been intermittently cited in the other areas of the text indeed point out that enterprise architecture is just the medium through which an organization is able to effectively plan its operations. Shank and colleagues actually claim that the benefits of enterprise architecture are intangible owing to the fact that enterprise architecture as a tool, does not in itself proffer any benefits that can be deemed to stem directly from it, rather, the benefits enjoyed are as a result of its application to understand the operations of other functions (Anisforov et al. 2019). This position provides the current study with the impetus of studying the concept of enterprise architecture in detail in an attempt to establish the tangibility of its utility in organizations.

1.2 Research Aim and Objectives

The aim of the current study was to investigate the role and application of enterprise architecture in project management cycles in contemporary organizations and the mediating role of effective decision making.

In pursuing the above aim, the research will seek to fulfil the following objectives:-

- Conduct an in-depth investigation seeking to make connections between enterprise architecture, project management and effective decision making.
- To develop a model for the effective integration of enterprise architecture, effective decision making and project management that would help organizations in appreciating the benefits of their relationships.

1.3 Research Contributions

Theoretically, the current research will have an impact on the body of knowledge of around successful project management by bringing together and highlighting the tripartite relationship between the architecture of the organization, the efficacy of the decision-making process and the outcomes of project management. This is a connection that is yet to be established in literature and therefore the research will contribute in filling up the knowledge gap.

Practically, the current research will contribute to the integration of the currently isolated functions that contribute to success in project management in healthcare organizations in the UAE. Success in said projects has been the offshoot of the unconscientious application and deployment of functions. However, through this research practical understanding of the role of enterprise architecture in effective decision-making and how these impacts successful project management will lead to increased success rates of projects.

2 Literature Review

Perez-Castillio et al. (2020) mentions that enterprise architecture has in the recent years been identified as an effective tool for efficient and effective decision making in organizations. The current discourse into enterprise architecture and its impact on decision making has gained traction as organizations continually appreciate the fact that through enterprise architecture, they can be able to have a systematic and logical way of decision-making, and implementation of ideas as opposed to previously where decision making could at many times be crooked at different points therefore affecting proper implementation. Basu (2017) in agreement with Perez-Castillio et al. (2020) acknowledges that enterprise architecture, by means of being able to define the current state of an organization and by use of clear steps and metrics to project the desired future steps of an organization, is able to reinforce the decision making model of organizations and ensure better ethical decisions.

The agreeably cordial relationship between enterprise architecture and effective decision making and decision making as a whole in the organization is not without blemish. According to opponents of the relationships, enterprise architecture does not effectively support effective decision making processes especially owing to the opinion that, it is founded on the view that the IT department, despite being respponsible for designing the future of the organization in the actual sense, has found its position usurped and sweeped under the enterprise architecture ambigous cover (Hsing and Souza 2012; Cristobal et al. 2018; Dang and Pekkola 2017). Hsing and Souza (2012) further inform that this instance of the IT department loosing its recognition has been considered by opponents of enterprise architecture as an effective overstepping and has been responsible for breeding cause for opposing any changes brought about through enterprise architecture. In a review of the factors affecting the adoption of enterprise architecture, research reveals that, the organizational structure is among the factors and makes it particular that in many organizations, reorganization of the IT department under enterprise architecture without due recognition of the fact that the IT department is the backbone of enterprise architecture and therefore should be considered with sufficient respect makes it impossible for enterprise architecture to lead decision making (Dang et al. 2019; Gampfer et al. 2018). This sentiments of the lack of ethics that informs the usurping of ITs efforts by the creation of enterprise architecture which research evidence identifies to be a descriptive and advisory role/tool and not in itself the driver of performance is shared by many IT professionals and they are keen to highlight how it borders effectively on infringement. According to Graves (2019), the reference of enterprise architecture as higher than IT function in the organization works to create a bottleneck in decision making in many organizations as IT professionals, being the ones handling the actual design of the organization and modeling enterprise architecture consider the role of enterprise architecture inferior.

Together with the above challenge to the ability of enterprise architecture to support effective decision making in the organization, there is a challenge on the technological approach of the entire modeling of enterprise projections. According to evidence of research, aspects such as ethics and motivation in the organization cannot be modelled and therefore, the strictly technology-centered enterprise architure models are not able to support effective decision making owing to the fact that they are digital and out of touch with non-digitized aspects such as effective decision making (Zappa et al. 2019; Gellweiler 2020). Zappa et al. (2019) conduct an in-depth empirical study on the application of enterpise architecture to support change implementation in the tourism industry. The scholars collect qualitative data and utilize sentiment analysis to extract meaning from the data collected. According to the scholars, despite the fact that enterprise architecture is effectuve in structuring the information and streamlining the analysis process as well as mappping different information to certain goals and targets of the project as a whole, it is not able to make the actual meaning of the data that can then be used by the executors of the information. In a simple term, the researchers find out that despite the fact that many enterprise architecture models are able to effectively model scenarios that organizations should anticipate in the process of implementing changes or running projects, they are not able to provide the actual decision and meaning of the scenarios without the human input. This understanding relates with the previous paragraphs sentiments on enterprise architecture and its inability to lead to sound decision making processes

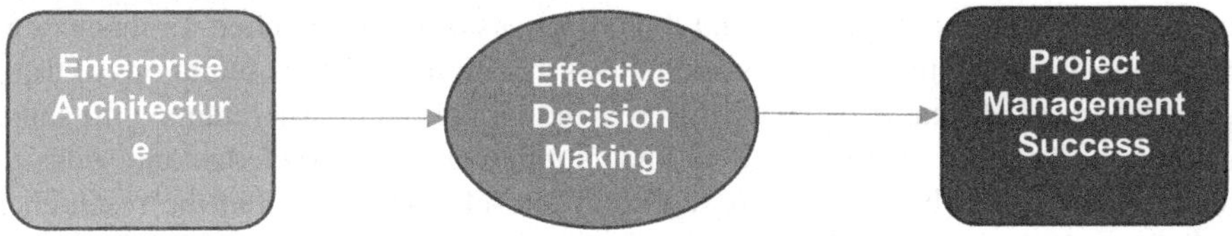

Fig. 1 Research conceptual model

let alone effective ones. Based on Haki and Legner (2020) research, enterprise architecture is useful but the fact that it in the long run relies on human judgement to make the decision of how important its scenario models are or which one is best to go with, makes it such that it is not entirely able to facilitate effective decision making. Logically, this can be attributed to the fact that, the end decision maker being human can choose to make an effective or an uneffective decision even when enterpise architecture had successfully modeled all potential scenarios and made them clear.

As far as enterprise architecture and project management is concerned, there is sufficient evidence that highlight how enterpirse architeture can support the project management efforts in the organization. Shanks et al. (2018) noted that there is sufficient anectodal evidence of how enterprise architecture benefits organizations through making their initiatives streamlined and successful. In support of the above assertion, concurrent evidence of research from Bhattacharya (2018) highlights that through effective enterprise architecture and the alignment of the objectives of a project inside the architecture thereof, the business can be able to benefit from sound visibility of the different project stages and thus be able to predict the project managemenmt requirements at different stages of the project management cycle. Literature is replete with evidences of research that support this notion further and brings different other dimensions. For instance, in an earlier research, scholars Trad and Kalpic (2016), the application of enterprise architecture in the healthcare and social works environment has played a significant role in ensuring that the course of projects is stayed based on the simple fact that the information of the different requirements and changes of the project from stage to stage had been almost accurately modeled. The scholars in agreement with other scholars cited are of the notion that enterprise architecture is critical for the future of project management especially as far as efficiency in the process is concerned (Fig. 1).

The below is the conceptual model of the research:-

3 Methodology

Subject to the purpose of this research, which is to investigate an area of knowledge that has minimally been investigated, this research adopts the positivist philosophy of research. Then selection of the research philosophy was guided by the conviction of the researcher about the truth and the objectives of the researcher. To this extent therefore, the conviction of the researcher was that numerical representations of truth are valid and can be passed down to different individuals while still retaining the same meaning. The study ventures into an understudied area of knowledge and seeks to test a hypothesis. This hypothesis represent the relationship(s) that the researcher believes exist between the three constructs of this research.

A quantitative method of data collection based on a 5-point Likert scale questionnaire was used in the collection of data for the research. The methods of analysis will be quantitative based on several statistical tests including structural equation modelling to test the strengths of the relationships between the variables. A survey was developed and distributed to 500 high level healthcare practitioners and decision makers across the UAE (mainly, Abu Dhabi, Dubai, and Sharjah). In doing the quantitative analysis, the research first assessed the validity and reliability of the measurement instrument and the table with the results of the constructs validity and reliability is attached in the appendix. The research then conducted structural equation modelling (SEM) to test the relationships as expressed in the conceptual model of the research shared previously.

4 Findings and Discussion

The findings of the study revealed that, the variable of effective decision making positively mediated the relationship between enterprise architecture and successful project management in the UAE healthcare industry. The combined path regression value for the independent variable of enterprise architecture, through the assessment sub-sonstruct of the mediating variable of effective decision making, to the dependent variable of successful project management, was $R = 0.789$, at a p value of < 0.0001. These findings revealed that when the enterprise architecture is synchronized through assessment sub-construct of the construct of effective decision making, then implementing any change project and managing it to a successful end would not be a challenge. Based on evidence earlier shared from literature, the findings of the current study further illuminate why it is critical to have synergy across organizational functions through a fully integrated enterprise architecture. The challenge of lack of integration and the bottlenecks it can cause to successful project implementation were implied by virtue of the findings of the current study confirming the importance of effective decision making across the enterprise as an enabler of success.

According to concurrent research evidence by Tamm et al. (2011), EA relies heavily on the multi-level, vertical and horizontal interaction between key players in the organization so that to transform an organization from a current state to a future desired state. Communication, governing and implementation form part of the required ingredients that help enterprise architecture to thrive, without which, then enterprise architecture remains in the realm of propositions of paths to desired states, with little to no impact tangible impact on the same (The Open Organization,

2011). The foregoing statement heralds the need for effective decision-making which enhances communication and effectively coordinates the different functions under enterprise architecture in order to realize the ideal state. According to evidence of research earlier gathered, the end state when organizational decisions are swiftly translated into action based on the action plans and with everyone on board, is the successful implementation of change, policy or any other organizational plan (PMI, 2021). This end state is supported by a number of steps in the decision-making process and as PMI (2021) research reports, it is essential that organizational leaders ensure an efficient decision-making process because this can either be an enabler or a bottleneck to the smooth running of organizations.

5 Conclusions

When the above is considered in the context of the UAE healthcare organizations, it is noted that, when the organization is properly aligned, implying that all actors responsible for a particular project in the organization are on board, and effective decision-making with regards to which of the actors will do what and the points of convergence and divergence, then the process of project management remains smooth and success is imminent. This has been confirmed by statistical results which objectively studied the matter at hand and inquired from professionals in the field of project management in UAE healthcare organizations. To this extent therefore, the current research notes that it is critical that healthcare organizations in the UAE, in the process of conduct of their projects, strive to ensure the alignment among actors involved in the processes of the project, and from this point, effectively decide on how to embark and move on with a plan. This way, the project will likely succeed as there are minimal ambiguities and variances.

Appendix

Measurement instrument and reliability/validity assessment results

Construct/item of measurement	Loading of items	Composite Reliability	Cronbach alpha	Average variance extracted (AVE)	Full Collinearity VIFs
Enterprise Architecture (EA) – Organizational alignment		0.955	0.941	0.81	4.842
Enterprise architecture promotes alignment of organizational goals	0.908				

(continued)

(continued)

Construct/item of measurement	Loading of items	Composite Reliability	Cronbach alpha	Average variance extracted (AVE)	Full Collinearity VIFs
Enterprise architecture is used as the underlying support for organizational strategy	0.919				
Enterprise architecture leads to the effective prioritization of objectives	0.901				
Enterprise architecture is used to integrate different organization functions	0.886				
Enterprise architecture leads to smooth operations in the organization	0.884				
Enterprise Architecture (EA) – Information availability		0.931	0.853	0.872	3.923
Enterprise architecture through technology leads to better information	0.934				
Enterprise architecture helps in the analysis and understanding of data	0.834				
Enterprise architecture promotes access to information	0.944				
Enterprise architecture supports clarity of information on organization strategy	0.894				
Enterprise architecture supports easy sharing of information across organization departments	0.902				
Enterprise Architecture (EA) – Resource Portfolio Optimization		0.946	0.929	0.779	7.658
Enterprise architecture supports better resource management	0.822				
Enterprise architecture informs the organization on the performance of projects	0.926				

(continued)

(continued)

Construct/item of measurement	Loading of items	Composite Reliability	Cronbach alpha	Average variance extracted (AVE)	Full Collinearity VIFs
Enterprise architecture helps in making decisions regarding organization projects' resources	0.915				
Enterprise architecture enhances the identification of projects resources	0.922				
Enterprise architecture reduces wastage of resources in projects	0.868				
Enterprise Architecture (EA) – Resource Complimentarity		0.954	0.94	0.806	5.288
Enterprise architecture enhances the integration of projects resources	0.902				
Enterprise architecture informs on the joint use of resources	0.931				
Enterprise architecture supports decision making on resource allocation to projects	0.902				
Enterprise architecture enhances the management of multiple projects	0.857				
Enterprise architecture supports the planning of resource allocation to complement projects	0.894				
Effective Decision Making - Assessment		0.784	0.856	0.778	4.786
Assessment of project decisions is critical to effective decision making	0.691				
The assessment process is taken seriously by project managers	0.756				
Project managers effectively assess the legal implications of decisions	0.812				

(continued)

(continued)

Construct/item of measurement	Loading of items	Composite Reliability	Cronbach alpha	Average variance extracted (AVE)	Full Collinearity VIFs
Project managers critically reflect on the ethical implications of decisions	0.578				
Assessment is based on the code of conduct of the organization and field industry	0.631				
Successful Project Management		0.756	0.874	0.689	3.678
Enterprise architecture and effective decision making lead to proper identification of projects	0.779				
Enterprise architecture and effective decision making promote good use of resources	0.856				
Enterprise architecture and effective decision making reduces project redundancies	0.789				
Enterprise architecture and effective decision making reduce unproductive projects	0.593				
Enterprise architecture and effective decision making help in improving costs savings in projects	0.813				
Enterprise architecture and effective decision making leads to project and strategy alignment	0.972				
Enterprise architecture and effective decision making lead to the identification of project gaps	0.791				
Enterprise architecture and effective decision making promote time savings for product delivery	0.691				

(continued)

(continued)

Construct/item of measurement	Loading of items	Composite Reliability	Cronbach alpha	Average variance extracted (AVE)	Full Collinearity VIFs
Enterprise architecture and effective decision making enhance innovation	0.745				

References

Akhelmann F, Legner C, Lux J (2021) A resource-based perspective of value generration through enterprise architecture management. Inf Manag, 58(1)

Alaedini M, Salekfard S (2013) Investigating the role of enterprise architecture project in business-IT alignment in Iran. Inf Syst Front 15(1):67–88

Alwadain A, Fielt E, Korthaus A, Rosemann M (2016) Empirical insights into the development of a service-oriented enterprise architecture. Data Knowl Eng 105:39–52

Anisiforov A, Dubgorn A, Lepekhin A (2019) Organizational and economic changes in the development of enterprise architecture. In: International science conference

Basu R (2017) Quality management tools and techniques in major infrastructure projects. In: 6th International conference on reliability, infocom technologies and oprimization, Noida

Bhattacharya P (2018) Aligning enterprise systems capabilities with business strategy: an extension of strategi alignment model using enterprise architecture. In: International conference on project management

Borkowski S, Knop K, Szklarzyk P (2015) Meaning of quality inspection and control methods during manufacturing process of metal sheet stampings. Manuf Technol 15(3):266–274

Cristobal SRJ et al. (2018) Complexity and project management: a general overview. Complexity

Dang DD, Pekkola S (2017) Systematic literature review on enterprise architecture in the public sector. Electron J E-GovMent 15(2):132–154

Dang T, Vatiainen T, Pekkola S (2019) Patterns of enterprise architecture adoption in the public sector: a resource-based perspective. In: 27th European conference of infrastructural systems

Gampfer, F, Andreas J, Markus M, Ruediger B (2018) Past current and future trends in enterprise architecture—a view beyond the horizon. Comput Ind 100:70–84. https://doi.org/10.1016/j.compind.2018.03.006

Gellweiler C (2020) Connecting enterprise architecture and project portfolio management: a reveiw and a model for IT project alignment. Int J Inf Technol Manag 11(1):1–16

Gong Y, Janssen M (2021) Roles and capabilities of enterprise architecture in big data analytics technology adoption and implementation. J Theroretical Appl Electron Commer Res 16(1)

Haki K, Legner C (2020) The mechanics of enterprise architecture principles. J Assoc Inf Syst

Hsing WC, Souza DA (2012) Management practices and influence on IT architecture decisions. J Inf Syst Technol Manag 9(3)

Perez-Castillio R, Ruiz F, Piattini M (2020) A decision making support system for enterprise architecture modelling. Decis Support Syst 131

Shanks G, Gloet M, Someh IS, Frampton K, Tamm T (2018) Achieving benefits with enterprise architecture. J Strat Inf Syst 27:139–156

Takeuchi H, Yamamoto S (2019) AI service system development using enterprise architecture modelling. In: 23rd International conference on knowledge-based and intelligent information and engineering systems. Science Direct

Tamm T, Seddon PB, Shanks G, Reynolds P (2011) How does enterprise architecture add value to organisations? Commun Assoc Inf Syst 28. https://doi.org/10.17705/1CAIS.02810

Trad A, Kalpic D (2016) A transformational framework proposal for managers in business innovation and business transformation projects - enterprise pattern based architecture. In: Conference on health and social care information systems and technologies

Zachman AJ (1996) Enterprise architecture: the issue of the century. Database Program Des 1–13

Zappa G, Murga J, Raymundo C, Mateos DF (2019) Business information architecture for successful project implementation based on sentiment analysis in the tourists sector. J Intell Inf Syst 53(1)

The Influence of Green Innovation and Digital Transformation Strategy on the Relationship Between Sustainability and Service Sector Project Success in the Context of UAE

Abdulazeez Othman

Abstract The aim of the study is to shed light on how green innovation and digital transformation strategies influence the connection underlying to sustainability practices and project success. Through an analysis of the moderating impacts of green innovation and digital transformation, the study will offer important insights into how these elements may boost project success in a sustainable way. The study will be using a quantitative methodology, distributing surveys to directors or executives and project managers in UAE service industries in public sector. Employing structural equation modelling, the gathered data will be analysed to look at both the direct and indirect implications of the suggested factors. The study is expected to determine the extent to which digital transformation and green innovation impact the correlation between sustainability and project success. The presentation will emphasize specific advancements in digitalization and environmentally-friendly innovations that, when integrated with sustainable practices, enhance the results of projects. This study project is distinctive since it concentrates primarily on analyzing the service industry in the public sector of the United Arab Emirates. The market has undergone considerable expansion and an increasing focus on sustainability. This study investigates the importance of digital transformation and green innovation as mediators in the association between project success and sustainability, aiming to fill the existing gap in knowledge.

The findings will have a significant impact on organizational decisions and contribute to the development of optimal strategies for sustainable project management in the public sector of the UAE's service industry. Overall, this research has the capacity to stimulate positive transformation and ingenuity in the UAE service industry-public sector, creating opportunities for enhanced sustainability and prosperous initiatives in the future.

This paper was prepared for Research Proposal, taught by BUID

A. Othman (✉)
British University in Dubai, Dubai, UAE
e-mail: 21000986@student.buid.ac.ae

K. Al Marri et al. (eds.), *BUiD Doctoral Research Conference 2024*, Lecture Notes in Civil Engineering 587, https://doi.org/10.1007/978-3-031-84371-6_34

Keywords Sustainability · Project success · Service sector · Green innovation · Digital transformation strategy

1 Introduction

The aim of the UAE's Vision 2021 is to extend the economy and lessening dependency on oil. The National Strategy for Green Growth (NGGS) promote for the improvement of sustainable development. The UAE is in the procedure of shifting in the direction of renewable energy sources and is keen to enhancing the well-being of its residents while also diminishing carbon emissions (Krzymowski 2020). The service industry plays a vital role in the economy of the UAE. Nevertheless, these enterprises must strive to enhance their sustainability. Green innovation and digital transformation are regarded as viable answers (Vial 2019). The success of a project is closely connected to its ability to be sustained over time. Businesses are forced to embrace sustainable methods as a consequence of competition, laws, and public awareness (Shaukat et al. 2021). Employing green innovation strategies can enhance a company's environmental performance and strengthen its competitiveness. Executing digital transformation may elevate productivity, upsurge profitability, and improve complete operations (Tsou and Chen 2023).

2 Background and Rationale of the Study

As cited by Kayani (2021), the service sector forms a vital proportion of the UAE economy, but it has various obstacles related to sustainability. Mensah (2019) has stated that the government is making business commitment to sustainable development, distinguishing green innovation and digital transformation as feasible alternatives.

Green innovation encompasses strategies aimed at minimizing environmental harm and optimizing resource utilization (Vial 2019). Digital transformation entails leveraging technology to enhance operational efficiency and optimize decision-making processes (Feroz et al. 2021).

Prior studies have not systematically investigated the effect of sustainability practices on the success of projects, mostly in emerging nations such as the UAE. As well as investigation is essential to grasp the process of including sustainability into project management so as to achieve superior results (Shaukat et al. 2021).

3 Problem Statement of the Research

The aim of the work is to fill a literature gap with an empirical analysis of the relationship that has developed independently among the research domains, using digital technology, organizational innovation, and firm performance (Vial 2019), have so far proposed a DT framework for investigating the links between these ideas.

The research is set out to investigate the relationship between organizational innovation, digital transformation strategy, use of digital technology, and success service project. The mediational role of such factors, arising outcomes from the utilization of digital technology, and their implications on corporate performance are researched. This research furthers existing literature by showing how this could be done within an appropriate framework in which GS and DTS serve as strategic resources in attaining sustainable performance. This study by and Ja-Shen Chen investigates the mediating role of organizational innovation and digital transformation strategy in the relationship linking the digital supply chain platform, firm performance, and digital technology use.

The relationship of organizational innovation, digital transformation strategy, and digital technology adoption is not well-analyzed in existing research, nor the implications of these relationships on company performance. This paper tries to extend the research by focusing on the strategies concerning digital transformation and organizational innovation as mediators (Tsou and Chen 2023).

4 Primary Research Question

To what extent do digital transformation strategies and green innovation mediate the relationship between sustainability practices and project success in the security service sector of the UAE?

5 Research Objectives

- Analyze the current state of green innovation and digital transformation strategies in service sector organizations in the UAE
- Investigate the relationship between green innovation, digital transformation strategy, and sustainability in service sector projects
- Examine the influence of green innovation and digital transformation strategy on service sector project success in the UAE context

6 Research Hypothesis

H1: Digital Transformation Strategy mediate the relationship between sustainability social dimension & service sector project success

H2: Digital Transformation Strategy mediate the relationship between sustainability Environmental dimension & service sector project success

H3: Digital Transformation Strategy mediate the relationship between sustainability Economical dimension & service sector project success

H4: Green Innovation mediate the relationship between sustainability Economical dimension & service sector project success

H5: Green Innovation mediate the relationship between sustainability Economical dimension & service sector project success

H6: Green Innovation mediate the relationship between sustainability Economical dimension & service sector project success

7 Scope of the Study

The research investigates how green innovation and digital transformation impact public service initiatives in the UAE. The survey will be designed to capture perceptions from different stakeholders at large, including project managers and citizens, in order to clearly understand the influences these practices have on the success and sustainability of projects. It will also examine how environmentally sustainable technology and digital tools help in better resource allocation and efficiency in the execution of projects. The study aims to eventually come up with a framework that will guide and aid in the delivery of prosperous and sustainable public service projects in UAE.

8 Literature Review

8.1 Overview of the UAE (PESTLE)

The UAE has emphasized a lot on green innovation and sustainable development through various means, including investment in renewable energy, especially solar power, green buildings, and waste management. It also encourages and finances the research related to renewable energy and sustainable technologies. The adopted government programs, such as UAE Vision 2021, also tend to promote sustainable behaviour and reduce CO_2 emissions into the atmosphere. The UAE actively participates in all international agreements associated with addressing climate change problems and achieving sustainable development goals (Alnaqbi et al. 2023).

Political Sustainability: With seven emirates, UAE is a monarchy-based federation known for political sustainability, rapid economic expansion, and solid diplomatic ties. Its aggressive foreign policy involves trade, alliances, and humanitarian aid (Modarress et al. 2020).

Economic Overview: The UAE, historically reliant on oil and gas, is diversifying its economy by focusing on technology, real estate, banking, and tourism, and establishing free trade zones for global trade and business (such as UAE Vision 2021 and UAE Vision 2030; Alnaqbi et al. 2023).

Sociocultural Factors: The United Arab Emirates is home to a large expatriate community who coexist with a diversified section of the Emirati population. While Islam is the recognized religion and Islamic customs are the bedrock of Emirati culture, the country does accept and accommodate differences in religion and culture.

There have been major initiatives to improve gender equality, healthcare, education, and social welfare (Elkady et al. 2019)

Technological Landscape: UAE is committed to supporting technological innovation and has made significant investments in fields such as space exploration, renewable energy, and artificial intelligence. Embedding technology in effective public services, transport, and government is turning Dubai and similar cities into "smart cities" (Krzymowski 2020).

Legal Environment: Almost all aspects of life, such as business, family, criminal, etc. are regulated through federal and municipal laws. The country has legislated reform in order to relax the company procedures and invite foreign investment for the country to develop a friendly business environment (Alsuawaidi et al. 2020).

Environmental Aspects: The UAE has acknowledged the need to be sustainable by promoting renewable sources of energy and decreasing carbon emissions. The different ecosystems conserved include the mangrove and desert environments. Water scarcity is tackled through innovative means (Rodriguez-Ubinas et al. (2020).

9 The UAE Service Sector Overview

The service industry is the major contributor to the economy of the UAE, wherein most of the sub-sectors like tourism, financial, real estate, health, education, and professional services show sturdy growth). In the real estate sector, marvellous infrastructural projects and high-rise buildings are developing). The UAE places a high importance on healthcare and education, which helps to attract skilled individuals due to its exceptional amenities (UAE Ministry of Finance 2022). The implementation of smart cities and digital transformation is enhancing the efficiency and effectiveness of service delivery (Alzaabi 2023). Policies, free zones, and infrastructure investment are some of the areas where there is active assistance extended by the government to the service sector. (Ansari et al. 2021). Owing to its favourable geographical location and advanced transportation framework, the United Arab Emirates has emerged as a significant global hub for commerce and trade (Krzymowski 2020).

10 Sustainability and Dimensions (Social, Environmental, Economical)

Mensah et al. (2019) highlighted in the study the phrase "sustainable development" (SD) a widely accepted development paradigm, has gained significant attention but is flawed because it has been diluted and stolen by other organizations, and because it lacks specific actions and quantifiable results. The foundation for long-term strategies for businesses transitioning to sustainability was established by this model, which was grounded on the three fundamental principles of sustainable development: social equality, economic gains, and environmental quality (Mensah et al. 2019).

According to experts, the concept of Correia (2019) The People social component deals with an organization's impact on the welfare of its employees and the broader community. It includes ethical hiring practices, community involvement, funding toward education, and other philanthropic activities.

Conduct social impact investigations to identify areas of development; implement policies on diversity, ethical procurement, and welfare of workers; conduct community and non-governmental organization partnership activities; and measure and communicate the social impacts of transparency and trust-building activities.

It is appreciated that environmental sustainability is an integral factor of project performance within the UAE service industry including banking, healthcare, and tourism. Eco-friendly approaches can help lead to a decrease in costs, brand reputation, compliance, risk management, and innovativeness (Salim Karimi Takalo et al. 2021).

Energy-efficient equipment and smart building technology will help in saving money while saving the environment. Reduced emissions will thus improve regulatory compliance of businesses and avoid fines. Green innovation can thus attract environmentally sensitive investors and spur novel ideas.

It is in this environment that UAE service organizations strategically invest in economic sustainability in order to leverage resources, reduce operational costs, increase income, and raise brand recognition for creating a responsible service industry and environmental sustainability.

11 Establishing Connection between Sustainability and Project Success

Sustainability is a concept that many academics and business experts believe has several benefits when it comes to managing portfolios, programs, and projects (Silvius and Schipper 2015). In order to assure the success of their projects, firms continue to develop them using the finest project management strategies. Accordingly, for a project to be successful, sustainable development and project management procedures must be properly integrated (Watfa et al. 2021). More to that, many businesses prioritize sustainability, and when they successfully include it into their company strategy, net profit obviously increased from 23 to 37% (Watfa et al. 2021). According to Khalifeh et al. (2019) and many other authors like (Watfa et al. 2021) have stated that a fairly recent systematic literature study, "more research is needed in this topic area". The analysis shows that there aren't many empirical contributions that integrate sustainability and project management.

12 Role of Digital Transformation

Digital transformation is described as using the latest technologies—such as AI, Big Data, and mobile gadgets—that allow enterprises to enhance performance and improve efficiency in business performance. It is essential for organizations to maintain their competitive edge, providing advantages associated with enhanced decision-making processes and heightened operational efficiency (Utegenova et al. 2020).

Digital transformation encompasses modifications to business models, processes, and frameworks; however, it does not possess a unified definition. To successfully manage this transition, organizations necessitate a clearly articulated strategy and framework (Ziyadin et al. 2020).

One of the primary effects of digital transformation is its impact on sustainability. Companies are engaging in programs for reducing wastes, pollution, and the general ecological footprint through artificial intelligence and big data analytics (Feroz et al. 2021).

Setting objectives is very critical in the initiation of digital transformation, coupled with an analysis of the organization's prevailing conditions. This would create the realization of the inherent strengths, challenges, and opportunities for improvement. There will be a need to understand the prevailing trends and the competitor's action thus conducting market research (Albukhitan 2020).

Digital transformation is dramatically influencing different industries, including manufacturing, government, and healthcare. The technologies are utilized by enterprises to maximize productivity, deliver better services, and manage customer relationships (Mathew 2021).

13 Importance of Green Innovation (GI)

GI has been an increasingly important and key strategy for firms to develop green products and processes, improving the economic performance in their operations, according to Borsatto and Bazani (2020). Through its focus on waste, pollution, and resource reduction, GI provides a great many benefits, which Wang (2019) has stated include improved efficiency, cost savings, and enhanced brand reputation. Successful implementation of Green Infrastructure (GI) by companies may be accomplished through a comprehensive strategy encompassing three key elements: organizational readiness with a robust environmentally conscious culture, technology preparedness with the adoption of green practices, and a deep understanding of the applicable environmental concerns and legislation (Zhang et al., 2020a, b). It has been confirmed through research that the mere presence of a concept of GI translates to several financial advantages. On the bottom line, GI entitles enterprises and the environment to a "win–win" situation wherein enterprises continue to thrive with reduced harmful impacts on the environment.

14 Establishing Connection between Green Innovation and Digital Transformational Strategy

This study intends to close this research gap by examining the linkages between organizational innovation, digital transformation strategy, and digital technology utilization and how these factors effect service sector project success.

The collation of research by various authors, such as that of Thanh Tiep Le on social responsibility and green innovation, and that of Hung-Tai Tsou on digital technology, was used to perform a statistical test of the two factors. Other independently developed themes of the literature review are organizational innovation, use of digital technology, and strategy of digital transformation.

The only authors that provide a DT framework to explore the relationships between these concepts are Vital (2019) and Hess and Benlian (2015). Furthermore, the application of digital technology in businesses continues to be a significant but little-researched field (Tsou and Chen 2023).

Green innovations, such energy-efficient technology and sustainable practices, have the potential to drastically lower utility and resource consumption costs. These innovations can boost business performance, productivity, process improvements, inventory management, product quality, value generation, and flexibility. Service innovation competence has a favourable correlation with improved business performance. Companies that invest in green innovation skills are more likely to generate long-term business performance results (Yudi Fernando et al. 2019).

15 Modern Thinking and Emerging Theories

Several theories can provide the basis for a project with digital technology strategy, sustainability, green innovation, and success in the service sector as its focus. Some of the few examples are given below:

1. Stakeholder Theory

According to Freeman, stakeholder theory of firms basically incorporates various interests of the stakeholders in order to achieve better results on a project. Stakeholder management, therefore, implies effective determination of their requirements and how they participate in decision-making (Le 2022).

Another study shows the relationship between green supply chains and the long-term sustainable performance of firms. The green supply chain aids an organization to achieve a competitive benefit and stakeholders also stand with the organization. According to the stakeholder theory, green supply chains are seen as a method for managing environmental issues; this step will be useful in enhancing the ecological proficiency and attaining the sustainable success (Le 2022).

2. *Resource-Based View*

According to RBV theory, the long-term continuing success of a company would depend upon whether its valuable, rare, and hard-to-imitate resources are available. Resources may be tangible and include factories; it also refers to the intangible type, such as brand reputation (Le 2022).

RBV posits that including environmental and social considerations in addition to financial issues can result in a durable competitive advantage. Therefore, it is imperative for enterprises to employ eco-friendly innovation in order to achieve a harmonious equilibrium between resource consumption and conservation.

RBV treats knowledge and skills as major intangible resources that are rather difficult to imitate and appreciate with time. This approach helps to integrate social and environmental goals within commercial goals that enhance the chances for long-term success of firms (Al-Aloula 2019).

16 Conceptual Framework

16.1 Resource-Based View Theory and Sustainability

The Resource-Based View presents the idea that the strategic use of Strategic Performance Management can improve the performance of a firm (Shaukat et al. 2021). Strategic Project Management ensures the completion of projects within the desired time and cost limits. The paragraph also emphasizes that NRBV promotes environmentally sustainable methods (Alhemairy 2022).

17 Stakeholder Theory and Sustainability

The stakeholder theory prioritizes the inclusion of all relevant parties in order to achieve project success and long-term viability. Ensuring the participation of stakeholders at every stage of the project's duration is crucial for attaining project success. Stakeholder involvement fosters elevated levels of dedication, collaboration, and congruence among stakeholders. All of these characteristics are crucial for achieving efficient project management (Shaukat et al. 2021).

18 Alignment of Research Strategy with the Conceptual Framework

The stakeholder theory underscores the necessity for firms to take into account several stakeholders in order to achieve success, a crucial aspect for the viability of initiatives in the long term. Developing nations encounter difficulties in adopting sustainable practices, however, the emergence of innovative technologies and increasing public consciousness are fostering their efforts. The research emphasizes that sustainability is a strategic methodology for achieving project success, taking into account environmental and social concerns. It helps to achieve higher productivity with less operation waste, which encourages innovation. The research mainly focuses on establishing a connected relationship between sustainability and project success for service-based projects (Fig. 1).

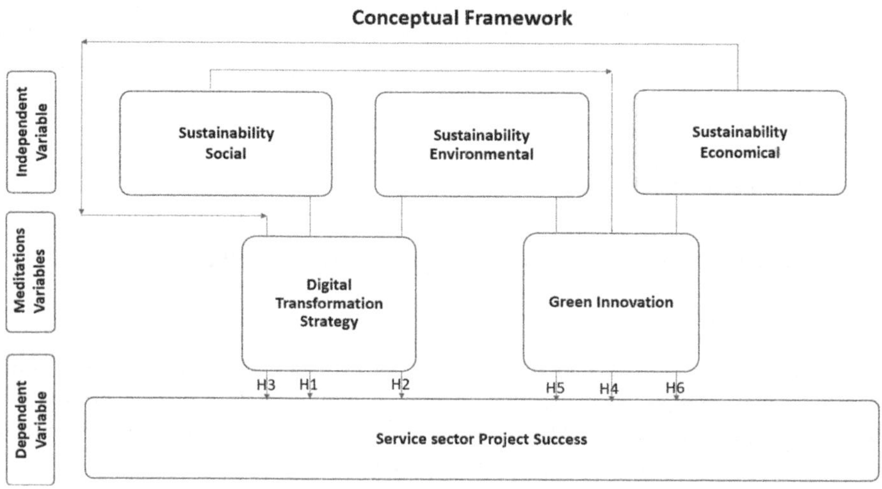

Fig. 1 Conceptual framework model. *Note* Othman 2023

19 Methodology

19.1 Research Design

The study investigates how green initiatives and digital transformation influence the project's success in UAE's service industry. The research design for this study is deductive, utilizing questionnaires in order to gauge the degree of impact these factors have. Considering the goal is those who have relevant experience, it is therefore stratified. The sampling will be representative of the population. Questionnaires are to be used with numerical data, while one section will have written comments. The project itself will serve as the unit of analysis on how these practices impact performance, sustainability, and team dynamics. Ultimately, the UAE service sector will be accurately represented since it will take into purposeful selection and random sampling procedures to establish an adequate sample size.

20 Data Collection Procedure

This information will be collected based on a stratified survey across different service professional categories in the UAE. The questionnaire will be divided into three parts: the first will validate competence, the second will address environmentally friendly practices, digital methods, and project achievements, and, finally, an open comment section. It will be based on a Likert scale. The questionnaire design will consider well-established research principles and shall be pre-tested to achieve accuracy and reliability. Base success on the project: stakeholder satisfaction, budgetary compliance, and deadline observance. This research recognizes the functions of digital transformation and organizational innovation in impacting environmentally friendly practices and project results.

21 Data Analysis

Data is collected from professionals working in various service fields through a pre-tested questionnaire comprising three sections. Data collected is analysed with the help of software (SPSS) for assessing the impact of green digital initiatives on project performance.

Validity and reliability tests were necessary to ascertain whether the data was indeed accurate. Validity refers to how correct the instrument of measurement really actually measures what it is supposed to measure, while reliability denotes the degree to which responses from a measurement are consistent. Statistical methods include descriptive analysis, correlation analysis, and regression analysis applied in the study of relationships between variables with a view to arriving at an understanding of the influence which green digital initiatives have on project success.

References

Ahmad N, Ghazali N, Abdullah MF, Nordin R, Najihah I, Nasir M, Farid AM (2020) Green marketing and its effect on consumers' purchase behaviour: an empirical analysis. J Int Bus, Econ Entrep 5(2)

Ai YH, Peng DY, Xiong HH (2021) Impact of environmental regulation intensity on green technology innovation: from the perspective of political and business connections. Sustainability (Switzerland) 13(9). https://doi.org/10.3390/su13094862

Al-Aloula WAS (2022, September) Reconfiguring strategic resources to enhance the performance of a firm during economic uncertainty using the framework of the resource-based theory

Al-Dabbagh R (2021) Waste management strategy and development in Ajman, UAE. Renew Energy Environ Sustain 6:14. https://doi.org/10.1051/rees/2021005

Al Ali HM, Ajmal MM, Gunasekaran A, Helo PT (2019) Managing change in constant evolving sustainable energy market: an exploratory study of motivators, enablers, and barriers. Sustain Prod Consum 20:84–97. https://doi.org/10.1016/j.spc.2019.05.005

Albukhitan S (2020) Developing digital transformation strategy for manufacturing. Procedia Comput Sci 170:664–671. https://doi.org/10.1016/j.procs.2020.03.173

Alhemairy ZHS (2022) The Influence of innovation practices on human capital development: the mediation role of innovation management [Thesis]. Universiti Tun Hussein Onn Malaysia

Alkhaldi M, Moonesar IA, Issa ST, Ghach W, Okasha A, Albada M, Chelli S, Takshe AA (2023) Analysis of the United Arab Emirates' contribution to the sustainable development goals with a focus on global health and climate change. Int J Health Govance 28(4):357–367. https://doi.org/10.1108/IJHG-04-2023-0040

Alnaqbi SA, Alami AH (2023) Sustainability and renewable energy in the UAE: a case study of Sharjah. Energies 16(20). https://doi.org/10.3390/en16207034

Alosani MS, Al-Dhaafri HS (2023) Service innovation in government: evidence from the UAE. Manag and Sustain: Arab Rev 2(3):239–254. https://doi.org/10.1108/msar-11-2022-0057

Alsuwaidi AA, Rosman ASB, Shehab AAA (2020) Evaluation of the case management system in the United Arab Emirates judicial system. J Southwest Jiaotong Uni 55(3). https://doi.org/10.35741/issn.0258-2724.55.3.10

Alturki R (2021) Research onion for smart IoT-enabled mobile applications. Scien Program. https://doi.org/10.1155/2021/4270998

Alzaabi D (2023) The impact of technology and digital transformation on innovation in the UAE. Int J Technol Syst 8(2):82–99. https://doi.org/10.47604/ijts.2170

Ashtankar OM, Kale S, Kakade K (2019) Role of service sector in the economic growth of India. https://www.researchgate.net/publication/46435409_The_Service_Sector_as_India's_Road_to_Economic_Growth

Aslanova IV, Kulichkina AI (2020) Digital maturity: definition and Model. https://sloanreview.mit.edu

Bataineh O (2019) Sustainable development goal in UAE. https://www.researchgate.net/publication/337428594

Beavers AS, John W, Lounsbury JW, Richards JK, Huck Gary SW, Skolits GJ (2013) Practical considerations for using exploratory factor analysis in educational research. Pract Assess Res Eval 18(6). https://doi.org/10.7275/qv2q-rk76

Borsatto JMLS, Bazani CL (2020) Green innovation and environmental regulations: a systematic review of international academic works. Environ Sci Pollut Res 28(45):63751–63768. https://doi.org/10.1007/s11356-020-11379-7

Browning MHEM, Rigolon A (2019) School green space and its impact on academic performance: a systematic literature review. Int J Environ Res Public Health 16(3). https://doi.org/10.3390/ijerph16030429

Carthey J (2020) User group consultation: design quality and project success. Health Environ Res des J 13(2):143–169. https://doi.org/10.1177/1937586719873143

Chien F, Ngo QT, Hsu CC, Chau KY, Iram R (2021) Assessing the mechanism of barriers towards green finance and public spending in small and medium enterprises from developed countries. Environ Sci Pollut Res 28(43):60495–60510. https://doi.org/10.1007/s11356-021-14907-1

Chukhrova N, Johannssen A (2019) Fuzzy regression analysis: systematic review and bibliography. Appl Soft Comput J 84. https://doi.org/10.1016/j.asoc.2019.105708

Ciric D, Delic M, Lalic B, Gracanin D, Lolic T (2021) Exploring the link between project management approach and project success dimensions: a structural model approach. Adv Prod Eng Manag 16(1) 99–111. https://doi.org/10.14743/apem2021.1.387

Correia MS (2019) Sustainability: an overview of the triple bottom line and sustainability implementation. https://doi.org/10.4018/IJoSE.2019010103

Crossley RM, Elmagrhi MH, Ntim CG (2021) Sustainability and legitimacy theory: the case of sustainable social and environmental practices of small and medium-sized enterprises. Bus Strateg Environ 30(8):3740–3762. https://doi.org/10.1002/bse.2837

Elkady L, Khaimah A, Al Khaimah R, Sediadi E (2019) The role of architecture in the sustainable development of the United Arab Emirates (UAE)

Eltom A, Yatiban A, Rusdi O (2023) Sustainability reporting practices in UAE business firms, the society, and stakeholders' insights. Emir J Bus, Econ Soc Stud 2(1). https://doi.org/10.54878/ejbess.291

Farhan Nasir M, Abdullah Al Mumin Professor M (n d) E-government in United Arab Emirates (UAE)

Fabiani C, de Albuquerque F, Landi F, Cabeza LF, Pisello AL (2024) Sustainability of international research: evidence from an H2020 European project. Sustain Sci 19(2):423–448. https://doi.org/10.1007/s11625-023-01421-1

Fernando Y, Chiappetta Jabbour CJ, Wah WX (2019) Pursuing green growth in technology firms through the connections between environmental innovation and sustainable business performance: does service capability matter? Resour Conserv Recycl 141:8–20. https://doi.org/10.1016/j.resconrec.2018.09.031

Feroz AK, Zo H, Chiravuri A (2021) Digital transformation and environmental sustainability: a review and research agenda. Sustainability (Switzerland) 13(3):1–20. https://doi.org/10.3390/su13031530

George G, Schillebeeckx SJD (2022) Digital transformation, sustainability, and purpose in the multinational enterprise. J World Bus 57(3), Elsevier Inc. https://doi.org/10.1016/j.jwb.2022.101326

Graves L, Dalgarno N, Van Hoorn R, Hastings-Truelove A, Mulder J, Kolomitro K, Kirby F, Van Wylick R (2023) Creating change: Kotter's change management model in action. Can Med Educ J. https://doi.org/10.36834/cmej.76680

Grubb M, Mclachlan K (n d) Screening approaches for detecting fetal alcohol spectrum disorder in children, adolescents, and adults: a systematic review screening approaches for detecting fetal alcohol spectrum disorder in children, adolescents, and adults

Gulseven O, Al Harmoodi F, Al Falasi M, ALshomali I (2020) Three pillars of sustainability: In search of conceptual origins. https://www.researchgate.net/publication/341099486

Gulseven O, Al Harmoodi F, Al Falasi M, ALshomali I (2020) How the COVID-19 Pandemic will affect the UN sustainable development goals? SSRN Elect J. https://doi.org/10.2139/ssrn.3592933

Gumilyov LN (n d) Ethical implications of large language models a multidimensional exploration of societal, economic, and technical concerns Kassym-Jomart Tokayev. Int J Soc Anal

Gün Ö, Kumru PY, Aladağ Z (2020) Developing a model for measuring project performance with software life cycle process metrics and calculating project success score. Sakar Univ J Science 24(3):536–554. https://doi.org/10.16984/saufenbilder.649588
Hanif S, Ahmed A, Younas N (2023) Examining the impact of environmental management accounting practices and green transformational leadership on corporate environmental performance; the mediating role of green process innovation. J Clean Prod 414. https://doi.org/10.1016/j.jclepro.2023.137584
Hao X, Chen F, Chen Z (2022) Does green innovation increase enterprise value? Bus Strateg Environ 31(3):1232–1247. https://doi.org/10.1002/bse.2952
Hasan A, Khan MZ, Rehman UU (n d) Moderating role of organizational culture between adoption of agile project management methodology and project success. https://doi.org/10.13135/2704-9906/5158
He Q, Wang T, Chan APC, Li H, Chen Y (2019) Identifying the gaps in project success research: a mixed bibliographic and bibliometric analysis. In Eng, Constr Arch Manag 26(8):1553–1573. https://doi.org/10.1108/ECAM-04-2018-0181
Hermundsdottir F, Aspelund A (2021) Sustainability innovations and firm competitiveness: a review. J Clean Prod 280. https://doi.org/10.1016/j.jclepro.2020.124715
Hindarto D, Wedha BY (2023) Maximizing ERP benefits with enterprise architecture: a holistic approach. J Comput Netw, Arch High-Perform Comput 5(2):703–713. https://doi.org/10.47709/cnahpc.v5i2.2790
Hosken DJ, Buss DL, Hodgson DJ (2018) Beware the F test (or, how to compare variances). Anim Behav. https://doi.org/10.1016/j.anbehav.2017.12.014
Hughes DL, Rana NP, Dwivedi YK (2020) Elucidation of IS project success factors: an interpretive structural modelling approach. Ann Oper Res 285(1–2):35–66. https://doi.org/10.1007/s10479-019-03146-w
Iqbal MDA (2023) Sustainable technology readiness of apparel professionals in Bangladesh ProQuest Central. https://search.proquest.com/openview/b0105974124aa622a05cadef0de06488/1?pq-origsite=gscholar&cbl=18750&diss=y
Jerónimo HM, Henriques PL, Carvalho SI (2022) Being inclusive boosts impact of diversity practices on employee engagement. Management Research 20(2):129–147. https://doi.org/10.1108/MRJIAM-05-2021-1175
Jilcha Sileyew K (2020) Research design and methodology. In Cyberspace. IntechOpen. https://doi.org/10.5772/intechopen.85731
Karimi Takalo S, Sayyadi Tooranloo H, Shahabaldini parizi Z (2021) Green innovation: a systematic literature review. J Clean Prod 279. https://doi.org/10.1016/j.jclepro.2020.122474
Kayani FN (2021) Renewable energy and economic growth nexus: a case of United Arab Emirates. Int J Energy Econ Policy 11(5):504–509. https://doi.org/10.32479/ijeep.11559
Khan S, Saher N, Sohail Yunis M (2019) Project planning, project success and project risk. Glob Soc Sci Rev IV(I):219–225. https://doi.org/10.31703/gssr.2019(IV-I).29
Khan AJ, Ul Hameed W, Iqbal J, Shah AA, Tariq MAUR, Ahmed S (2022) Adoption of sustainability innovations and environmental opinion leadership: a way to foster environmental sustainability through diffusion of innovation theory. Sustainability (Switzerland) 14(21). https://doi.org/10.3390/su142114547
Khanra S, Kaur P, Joseph RP, Malik A, Dhir A (2022) A resource-based view of green innovation as a strategic firm resource: present status and future directions. Bus Strateg Environ 31(4):1395–1413. https://doi.org/10.1002/bse.2961
Kocaoglu B, Demir E (2019) The use of McKinsey s 7S framework as a strategic planning and economic assestment tool in the process of digital transformation. Pressacademia 9(9). https://doi.org/10.17261/pressacademia.2019.1078
Korachi Z, Bounabat B (2019) Integrated methodological framework for digital transformation strategy building (IMFDS). In IJACSA Int J Adv Comput Sci Appl 10(12). www.ijacsa.thesai.org

Krzymowski A (2020) Sustainable development goals in arab region – United Arab Emirates case study. Problemy Ekorozwoju 15(1): 211–220. https://doi.org/10.35784/pe.2020.1.22
Kumar V, Pandey A, Singh R (2023) Critical success factor models for project success. J Eng Proj Prod Manag 13(2):148–158. https://doi.org/10.32738/JEPPM-2023-0015
Lafortune G, Fuller G, Moreno J, Schmidt-Traub G, Kroll C (2019) Sustainable development report. In Bertelsmann Stiftung (ed), sustainable development solutions network report 2019. Bertelsmann Stiftung website. https://www.bertelsmann-stiftung.de/en/publications/publication/did/sustainable-development-report-2019/
Le TT (2022) How do corporate social responsibility and green innovation transform corporate green strategy into sustainable firm performance? J Clean Prod 362:132228. https://doi.org/10.1016/j.jclepro.2022.132228
Lee JH, Wood J, Kim J (2021) Tracing the trends in sustainability and social media research using topic modeling. Sustainability (Switzerland) 13(3):1–23. https://doi.org/10.3390/su13031269
Li L (2022) Digital transformation and sustainable performance: the moderating role of market turbulence. Ind Mark Manage 104:28–37. https://doi.org/10.1016/j.indmarman.2022.04.007
Luo X, Zhang W (2021) Green innovation efficiency: a threshold effect of research and development. Clean Technol Environ Policy 23(1):285–298. https://doi.org/10.1007/s10098-020-01977-x
Martínez-Caro E, Cegarra-Navarro JG, Alfonso-Ruiz FJ (2020) Digital technologies and firm performance: the role of digital organisational culture. Technol Forecast Soc Chang 154. https://doi.org/10.1016/j.techfore.2020.119962
Martínez-Peláez R, Ochoa-Brust A, Rivera S, Félix VG, Ostos R, Brito H, Félix RA, Mena LJ (2023) Role of digital transformation for achieving sustainability: mediated role of stakeholders, key capabilities, and technology. Sustainability (Switzerland) 15(14). https://doi.org/10.3390/su151411221
Mathew S (n d) The need for digital transformation in business. In Int J Adv Res Ideas Innov Technol
McDougall N, Wagner B, MacBryde J (2019) An empirical explanation of the natural-resource-based view of the firm. Prod Plan Control 30(16):1366–1382. https://doi.org/10.1080/09537287.2019.1620361
Melnikovas A (2018) Towards an explicit research methodology: adapting research onion model for futures studies. J Futur Stud 23(2):29–44. https://doi.org/10.6531/JFS.201812_23(2).0003
Mensah J (2019) Sustainable development: meaning, history, principles, pillars, and implications for human action: literature review. Cogent Social Sci 5(1). https://doi.org/10.1080/23311886.2019.1653531
Mishra BK, Kumar P, Saraswat C, Chakraborty S, Gautam A (2021) Water security in a changing environment: concept, challenges and solutions. Water (Switzerland) 13(4). https://doi.org/10.3390/w13040490
Modarress B, Ansari A, Ansari A (2020) Sustainable development and ecological deficit in the United Arab Emirates. Sustainability (Switzerland) 12(15). https://doi.org/10.3390/su12156180
Mukherjee C (2023) Multi-dimensional leader capabilities for organizational agility in the future of work. Indian school of business. https://eprints.exchange.isb.edu/id/eprint/2016/1/Multi-Dimensional%20Leader%20Capabilities%20for%20Organizational%20Agility%20in%20the%20Future%20of%20Work.pdf
Nauman S, Musawir AUl, Zahra Malik S (n d) Servant leadership and project success: parallel mediation by work engagement and project work withdrawal servant leadership and project success: parallel mediation by work engagement and project work withdrawal BAM Track 13: leadership and leadership development
Onyeaka H, Tamasiga P, Nwauzoma UM, Miri T, Juliet UC, Nwaiwu O, Akinsemolu AA (2023) Using artificial intelligence to tackle food waste and enhance the circular economy: maximising resource efficiency and minimising environmental impact: a review. Sustainability (Switzerland) 15(13). https://doi.org/10.3390/su151310482
Ostrom AL, Field JM, Fotheringham D, Subramony M, Gustafsson A, Lemon KN, Huang MH, McColl-Kennedy JR (2021) Service research priorities: managing and delivering service in turbulent times. J Serv Res 24(3):329–353. https://doi.org/10.1177/10946705211021915

Pace M (2019) A correlational study on project management methodology and project success. J Eng Proj Prod Manag 9(2):56–65. https://doi.org/10.2478/jeppm-2019-0007
Pensia R, Singh R (2022) Relationship between sustainability and heritage sites: a case study of heritage sites in India. Int J Res in IT and Management (IJRIM) 12. https://www.euroasiapub.org
Pirotti A, Keshavarzsaleh A, Rahim FAM, Zakaria N (2020) Effective factors on project success in Malaysian construction industry. J Eng Proj Prod Manag 10(1):1–10. https://doi.org/10.2478/jeppm-2020-0001
Prabhakar MM, Ashok A, Kamal Ch, R Dr (2021) a study of industrial green initiatives management. https://www.researchgate.net/publication/353792736_A_study_of_Industrial_Green_initiatives_Management
Principles of correlation and regression analysis (n d)
Puri-Mirza A (2022) Contribution to the gross domestic product of the emirate of Dubai in the United Arab Emirates in the first quarter of 2021, by industry Statista Report. https://www.statista.com/statistics/297605/uae-gross-domestic-product/
Purvis B, Mao Y, Robinson D (2019) Three pillars of sustainability: in search of conceptual origins. Sustain Sci 14(3):681–695. https://doi.org/10.1007/s11625-018-0627-5
Ravazzoli E, Valero DE (2020) Social innovation: an instrument to achieve the sustainable development of communities, pp 1–10. https://doi.org/10.1007/978-3-319-71061-7_108-1
Rehman SU, Kraus S, Shah SA., Khanin D, Mahto RV (2020) Analyzing the relationship between green innovation and environmental performance in large manufacturing firms. Technol Forecast Soc Change 163:120481. https://doi.org/10.1016/j.techfore.2020.120481
Rodriguez-Ubinas E, Alzarouni S, Alhammadi N, Alantali M, Behzad F, Sgouridis S (2020) Sustainability through energy conservation building codes: comparative analysis of green building regulations in the middle east. WIT Trans Ecol Environ 249:85–100. https://doi.org/10.2495/SC200081
Salvioni DM, Almici A (2020) Transitioning toward a circular economy: the impact of stakeholder engagement on sustainability culture. Sustainability (Switzerland) 12(20):1–30. https://doi.org/10.3390/su12208641
Schöggl JP, Stumpf L, Baumgartner RJ (2020) The narrative of sustainability and circular economy - a longitudinal review of two decades of research. Resour Conserv Recycl 163. https://doi.org/10.1016/j.resconrec.2020.105073
Shahzad M, Qu Y, Javed SA, Zafar AU, Rehman SU (2020) Relation of environment sustainability to CSR and green innovation: a case of Pakistani manufacturing industry. J Clean Prod 253. https://doi.org/10.1016/j.jclepro.2019.119938
Shareefdeen Z, Youssef N, Taha A, Masoud C (2020) Comments on waste to energy technologies in the United Arab Emirates (UAE). Environ Eng Res 25(1):129–134. https://doi.org/10.4491/eer.2018.387
Sharma V, Ramachandran M, Chinnasamy S, Saravanan V (n d) A review on structural equation modeling and its classification. Copyright@ REST Publisher 135 REST J Emerg Trends Model Manuf 7(4):2021. www.restpublisher.com/journals/jemm
Shaukat MB, Latif KF, Sajjad A, Eweje G (2021) Revisiting the relationship between sustainable project management and project success: the moderating role of stakeholder engagement and team building. Sustain Dev 30(1):58–75. https://doi.org/10.1002/sd.2228
Sheikh Hamdo S (n d) Change management models: a comparative review. https://doi.org/10.13140/RG.2.2.24741.01764
Shawish AA, Nabhan T, Almadidy A (2019) Potable water in UAE: an overview of water characteristics and sources of contamination. J Environ Toxicol Stud 3(2). https://doi.org/10.16966/2576-6430.120
Shibin KT, Dubey R, Gunasekaran A, Hazen B, Roubaud D, Gupta S, Foropon C (2020) Examining sustainable supply chain management of SMEs using resource-based view and institutional theory. Ann Oper Res 290(1–2):301–326. https://doi.org/10.1007/s10479-017-2706-x
Siegfried P (2013) The importance of the service sector for the industry. https://www.researchgate.net/publication/348311067

Silvius AG, Schipper R (2015) A conceptual model for exploring the relationship between sustainability and project success. Procedia Comput Sci 64:334–342. https://doi.org/10.1016/j.procs.2015.08.497

Soewarno N, Tjahjadi B, Fithrianti F (2019) Green innovation strategy and green innovation: the roles of green organizational identity and environmental organizational legitimacy. Manag Decis 57(11):3061–3078. https://doi.org/10.1108/MD-05-2018-0563

Song M, Fisher R, Kwoh Y (2019) Technological challenges of green innovation and sustainable resource management with large scale data. Technol Forecast Soc Chang 144:361–368. https://doi.org/10.1016/j.techfore.2018.07.055

Soto Setzke D, Riasanow T, Böhm M, Krcmar H (n d) Pathways to digital service innovation: the role of digital transformation strategies in established organizations. https://doi.org/10.1007/s10796-021-10112-0/Published

Sovacool BK, Griffiths S (2020) Culture and low-carbon energy transitions. Nat Sustain 3(9): 685–693. https://doi.org/10.1038/s41893-020-0519-4

Sovey S, Osman,K, Mohd-Matore MEE (2022) Exploratory and confirmatory factor analysis for disposition levels of computational thinking instrument among secondary school students. Eur J Educ Res 11(2):639–652. https://doi.org/10.12973/eu-jer.11.2.639

Sparx Systems (2021) Enterprise architect user guide: enterprise architecture guidebook. Sparx Systems website. https://sparxsystems.com/resources/user-guides/16.0/guidebooks/enterprise-architecture.pdf?ref=eaRes

Tasleem M, Khan N, Nisar A (2019) Impact of technology management on corporate sustainability performance: the mediating role of TQM. Int J Qual Reliab Manag 36(9):1574–1599. https://doi.org/10.1108/IJQRM-01-2018-0017

Tham KW, Dastane O, Johari Z, Ismail, NB (2019) Perceived risk factors affecting consumers' online shopping behaviour. J Asian Financ, Econ Bus 6(4):246–260. https://doi.org/10.13106/jafeb.2019.vol6.no4.249

The Relationship between customer satisfaction, corporate social responsibility activities towards customer loyalty in TNB Kuantan Pahang (n d)

Tiwari V, Thakur S (2021) Environment sustainability through sustainability innovations. Environ Dev Sustain 23(5):6941–6965. https://doi.org/10.1007/s10668-020-00899-4

Tokayev KJ (2023) Ethical implications of large language models: a multidimensional exploration of societal, economic, and technical concerns. Int J Soc Attitudes (IJSA) 08:17–33. https://norislab.com/index.php/ijsa/article/download/42/34

Trischler J, Johnson M, Kristensson P (2020) A service ecosystem perspective on the diffusion of sustainability-oriented user innovations. J Bus Res 116:552–560. https://doi.org/10.1016/j.jbusres.2020.01.011

Tsou H, Chen J (2023) How does digital technology usage benefit firm performance? Digital transformation strategy and organisational innovation as mediators. Technol Anal Strateg Manag 35(9):1114–1127. https://doi.org/10.1080/09537325.2021.1991575

UAE Cabinet (2023) The UAE centennial plan 2071. UAE government website. https://u.ae/en/about-the-uae/strategies-initiatives-and-awards/strategies-plans-and-visions/innovation-and-future-shaping/uae-centennial-2071

UAE Government, 2019, Vision 2021, United in ambition and determination (n d)

United Arab Emirates Ministry of Finance (2022) The Annual Report 2022: remarkable achievements shaping promising financial prospects. UAE Ministry of Finance website. Accessed 1 Nov 2023. https://mof.gov.ae/wp-content/uploads/2023/06/MoF-Interactive-Annual-Report-2022-English-v11.pdf

United Arab Emirates and THE 2030 AGENDA for sustainable development excellence in implementation: executive summary voluntary national review un high level political forum 2018 national committee on sustainable development goals (n d)

Vargo SL, Akaka MA, Wieland H (2020) Rethinking the process of diffusion in innovation: a service-ecosystems and institutional perspective. J Bus Res 116:526–534. https://doi.org/10.1016/j.jbusres.2020.01.038

Verina N, Titko J (2019). Digit Transform: Concept Fram. https://doi.org/10.3846/cibmee.2019.073
Vial G (2019) Understanding digital transformation: a review and a research agenda. J Strat Inf Syst 28(2): 118–144. https://doi.org/10.1016/j.jsis.2019.01.003
Virtanen PK, Siragusa L, Guttorm H (2020) Introduction: toward more inclusive definitions of sustainability. Curr Opin Environ Sustain 43:77–82. Elsevier BV. https://doi.org/10.1016/j.cosust.2020.04.003
Wang CH (2019) How organizational green culture influences green performance and competitive advantage: the mediating role of green innovation. J Manuf Technol Manag 30(4):666–683. https://doi.org/10.1108/JMTM-09-2018-0314
Wang X, Cheng Z (2020) Cross-sectional studies: strengths, weaknesses, and recommendations. Chest 158(1):S65–S71. https://doi.org/10.1016/j.chest.2020.03.012
Watfa MK, Abdelrehim A, Shahin N, Jaafar K (2023) A structural equation model to assess the impact of sustainability management on the success of construction projects. Int J Constr Manag 23(10):1653–1664. https://doi.org/10.1080/15623599.2021.1998302
Waxin MF, Knuteson SL, Bartholomew A (2020) Outcomes and key factors of success for ISO 14001 certification: evidence from an emerging Arab gulf country. Sustainability (Switzerland) 12(1). https://doi.org/10.3390/su12010258
Webber KL, Zheng H, Zheng HY (n d) Data analytics and the imperatives for data-informed decision-making in higher education Chapter 1 data analytics and the imperatives for data-informed decision making in higher education. https://ihe.uga.edu/rps/2019_004
Weiland S, Hickmann T, Lederer M, Marquardt J, Schwindenhammer S (2021) The 2030 agenda for sustainable development: transformative change through the sustainable development goals?. Polit Govan 9(1):90–95 https://doi.org/10.17645/PAG.V9I1.4191
Xue L, Zhang Q, Zhang X, Li C (2022) Can digital transformation promote green technology innovation? Sustainability (Switzerland) 14(12). https://doi.org/10.3390/su14127497
Yimer N (2020) School of graduate studies the effect of political marketing mix elements on voters' loyalty (The case of marketing in prosperity party)
Yin S, Yu Y (2022) An adoption-implementation framework of digital green knowledge to improve the performance of digital green innovation practices for industry 5.0. J Clean Prod 363. https://doi.org/10.1016/j.jclepro.2022.132608
Zaoui F, Souissi N (2020) Roadmap for digital transformation: a literature review. Procedia Computer Science 175:621–628. https://doi.org/10.1016/j.procs.2020.07.090
Zhanbayev RA, Irfan M, Shutaleva AV, Maksimov DG., Abdykadyrkyzy R, Filiz Ş (2023) Demoethical model of sustainable development of society: a roadmap towards digital transformation. Sustainability (Switzerland) 15(16). https://doi.org/10.3390/su151612478
Zhang J, Kang L, Li H, Ballesteros-Pérez P, Skitmore M, Zuo J (2020) The impact of environmental regulations on urban green innovation efficiency: the case of Xi'an. Sustain Cities Soc 57. https://doi.org/10.1016/j.scs.2020.102123
Zhang Y, Sun J, Yang Z, Wang Y (2020) Critical success factors of green innovation: technology, organization and environment readiness. J Clean Prod 264. https://doi.org/10.1016/j.jclepro.2020.121701
Ziyadin S, Suieubayeva S, Utegenova A (2020) Digital Transformation in Business. In lecture notes in networks and systems, vol 84, pp 408–415, Springer. https://doi.org/10.1007/978-3-030-27015-5_49
Zoppelletto A, Bullini Orlandi L, Rossignoli C (2020) Adopting a digital transformation strategy to enhance business network commons regeneration: an explorative case study. TQM J 32(4):561–585. https://doi.org/10.1108/TQM-04-2020-0077

The Relationship Between Bilingualism and Achievement

Marwa Elkhouly and Tendai Charles

Abstract This paper addresses the cognitive and academic impacts of bilingualism, particularly focusing on students in the UAE, where bilingual education programs have been significantly promoted. Bilingual students are hypothesized to have enhanced executive control and cognitive flexibility, potentially leading to superior academic performance compared to monolingual peers. Empirical evidence and theoretical frameworks such as Krashen's Second Language Acquisition Theory and various cognitive development theories are explored to understand how bilingualism may affect academic achievement. The study utilizes a mixed-method approach, employing interviews and questionnaires among educators in UAE schools to gauge perceptions and real-world impacts of bilingual education on student achievement. Findings are expected to contribute to the ongoing debate regarding the efficacy of bilingual education programs and provide insights into optimal instructional strategies in bilingual settings.

Keywords Bilingual education · Academic achievement · Cognitive development

1 The Relationship between Bilingualism and Achievement

Bilingualism and bilingual learners have become a matter of concern for educationalists and sociolinguists since the 1900s century. Over the past few years, many studies have been conducted to investigate the impact of bilingualism on children's academic performance. Some studies have tried proving the assumption that bilingual children might outperform monolingual children academically and cognitively. Costa (2009) has argued that bilingual children are thought to develop and control more language function competency than monolingual learners. He believed that bilingual students can moderate more communicative interaction skills than children who speak one language; and thus, bilingual children manage to control the surpassing interference of both spoken languages. According to Colzato et al. (2008), they explained

M. Elkhouly (✉) · T. Charles
Faculty of Education, The British University in Dubai, Dubai, United Arab Emirates
e-mail: 20205122@student.buid.ac.ae

K. Al Marri et al. (eds.), *BUiD Doctoral Research Conference 2024*, Lecture Notes in Civil Engineering 587, https://doi.org/10.1007/978-3-031-84371-6_35

that the bilingual children must appropriately shift between the two languages and choose the specific lexicalization. This kind of cognitive ability illustrates a dominant control over the conflict-monitoring and goal-orienting capacity that bilingual children outperform monolingual children.

Additionally, the United Arab Emirates (UAE) has recently adopted a bilingual system based on the Abu Dhabi Education Council (ADEC) tactical plan for 2009–2018. The policy framework's purpose was to promote the bilingual approach in the educational system from K to 12 in public and private schools. ADEC has directed the schools and educational institutions to implement the model of the New School policy within the bilingual program that seeks to show its remarkable results on the standardized and international exams. Based on the above considerations, this paper seeks to investigate the phenomena focussed on empirical studies and UAE educational vision and implementation of public and private schools. The paper will also explore the teachers' understanding of the bilingual program and instructions in the UAE context. This paper looks at the cognitive impact of bilingualism on children's academic achievement; and explores teachers' attitudes towards bilingual education instructions. Thus, this paper will try exploring the following questions: (1) Do bilingual students academically outperform monolingual learners? (2) How does the teacher's attitude towards bilingualism and Bilingual instructions impact the learners' achievement when learning the second language (L2)? This paper has used a qualitative method to address the achieved progress that bilingual students may indicate in the bilingual educational system. It assumed that bilingual learners academically outperform monolingual learners based on the recent empirical studies conducted by researchers such as (Chin 2015; Harris 2012) on the bilingual academic progress in a dual-educational program.

2 Literature Review

This paper is based on the constructivist's perspective, human beings build up their perception based on the way they perceive the information from the surrounding culture. To this end, some cultural aspects may interrupt the children's cognitive ability to appropriately acquire the required competency skills of two or multiple languages. It will also be based on the Krashen's theory framework in second language acquisition. The framework theory has investigated the strategical and pedagogical support needed for how the second language in a bilingual environment is acquired and achieved. Krashen's theory illustrated the fact that any second language is acquired in a natural setting and communication. He has set five main components in his theory assumption: learning the second language acquisition, natural setting, observation, input, and filtering the information. The researcher has argued that formal instruction happened because of natural communication between the learners in a natural setting. However, Krashen's theory has been argued among researchers and linguists due to the shortage of the empirical evidence and content. Terrell and Brown (1981) have supported Krashen's theory by explaining that second language acquisition is both a conscious and unconscious process that needs a natural setting to cognitively develop. Terrell argued that in order for the children to acquire a second language, they should be exposed to a formal bilingual instructional programme and communicate the language in a natural setting.

There has been a controversy among researchers about the impact of bilingualism on the children's educational progress. According to Gordon and Hoxby, they have investigated the influence of bilingualism education programmes on the early grades. The researchers have found that there is a positive effect on the early grade' academic performance of those who have been taught in a bilingual education system. By the same token, Green (1998) and August and Shanahan have concluded in their studies that children in the bilingual programme have accomplished better results than those who were settled in a monolingual educational programme. Additionally, Robinson has concluded in his study that learners who have been included in bilingual or multiple language systems have developed more cognitive abilities and mastery of concept information than those who have learnt in a monolingual system. The researcher has illustrated that the children have shown a positive performance in the standardised tests on the core subject skills of English, maths, and social studies.

On the other hand, Chin (2015) has argued that the bilingual educational system and English only educational system have a similar impact on the children's performance who have achieved the same results on the US standardised test. Chin has concluded that the quality of instructions should be an effective element when teaching students with limited English proficiency. The researcher has suggested that the given instructions should be a matter of adequate intensity and focus on the quality of the instruction to achieve a better result. Furthermore, Chin (2015) has demonstrated some benefits of a bilingual system on children's academic performance. He confirmed that a bilingual system might assist low-efficient English learners when learning other subjects, such as maths and science. The researcher's findings show that students might learn other subjects in their native language while they continue to improve their English proficiency skills.

According to Chin (2015), utilising the native language instructions might help improve certain language skills when learning a new language. He suggested that the same native language reading strategies can be appropriately used to teach reading in English. Furthermore, Chin has specified the enormous advantage of the bilingual educational system in supporting cultural diversity and people's involvement in the community. Chin has seen that bilingual settings may be a helpful tool for parents with English deficiency skills; teachers and parents may communicate using bilingual languages that facilitate the means of understanding and communication. However, some limitations and drawbacks of Bilingual programmes have been discussed by Chin (2015) in his study. Chin has argued that bilingual education programmes may disrupt the second language acquisition due to the less exposure of appropriate instructions or using the native language instructions when teaching bilingual children. According to Chin (2015), the shortage of certified bilingual teachers is considered one of the basic drawbacks of the bilingual educational programme. Finally, he has found that the bilingual teaching resources might not be available in many native grades and languages. Similarly, Rodriguez has confirmed that teachers should be provided with strategic and instructional bilingual resources, appropriate tools, and professional development to academically support their students' and improve their linguistic competences.

Another recent study conducted by Harris (2012) has discussed the significant benefit of dual-language programmes. Harris has specified the importance of the proper implementations of both languages: native language and the second language. For example, he has suggested that during the reading class, teachers should have two different versions of the same reading article in the English and the Spanish languages. The researcher has also confirmed that teachers and educators should not only deliver standard content knowledge but also include the everyday communicative language in the bilingual programme.

Suwanarak has confirmed and explored in his study the impact of bilingual learning instructions on the progress of the second language communicative competence. The study has been conducted on the early grades (grade 3 students) in the Thai context. The study followed the interpretive approach, where the theory is based on social reality and undertaken in a natural context. The researcher study data were collected based on classroom observations and interviews with teachers. Suwanarak's study was sought to inspect the effect of a Thai-English bilingual programme on children's communicative competence development in Thailand. The study has also focused on teaching strategies and pedagogies in a bilingual Thai context. The researcher has concluded that students' communicative competences were showing significant progress. He has observed the children and noticed the implementation of the code-switching strategy when children did not have the appropriate vocabulary to express their ideas and thoughts. Teacher interviews have confirmed the assumption that children lack lexical components in their second language, which has helped them communicate more effectively with their teachers in their native language.

3 Bilingualism and Cognitive Ability

Recent research continually emphasizes the beneficial association between bilingualism and cognitive development. Bilingual people, especially those who regularly utilize both languages, exhibit improved executive functioning, including problem-solving, multitasking, and cognitive flexibility. The cognitive benefits and its development are mainly ascribed to the continuous requirement of managing two linguistic systems, which improves the brain's capacity to eliminate distractions and concentrate on pertinent information. As an illustration, Bialystok (2017) discovered that those who can speak two languages scored better than those who can only speak one language in activities that require cognitive control. This implies that the ability to switch coding languages enhances cognitive processes.

Furthermore, Costa et al. (2018) conducted a study that demonstrated how bilingualism can potentially postpone the beginning of dementia, hence highlighting its beneficial influence on cognitive well-being. These findings emphasize the wider cognitive advantages of being bilingual, which go beyond just being proficient in many languages and include cognitive development of mental agility and resilience.

4 Bilingualism in UAE Context

Although the United Arab Emirates (UAE) has witnessed a significant development in the educational system and policymaking, there has been a noticeable shortage of research studies on bilingual education in the UAE and the Gulf countries in general. ADEC has established a wide range of educational reforms on the ongoing pedagogies, circular and school leadership and management (Gallagher, 2011). Since the UAE government established the Ministry of Education (MoE) in 1972, English has been considered a foreign language in the school curriculum until 1985, when the MOE decided to design its own curriculum and adopt the English language as a second language. By 2018, English has become a dominant and official medium language of communication and learning in the UAE. One of the essential graduation requirements of any university student in the UAE is to obtain a medium level of English competency. This urges the authorities in the educational sector to officially issue reforms and policies to schools to prepare the students for the study in English particularly in higher education (Al Hammeli and Underwood 2014).

A study that has been recently conducted to examine the bilingual education system in the UAE was performed by Gallagher (2011). The study aimed to investigate and analyze the current obligatory bilingual educational program in the Abu Dhabi state of the UAE. The study aimed to comprehensively investigate the socio-cultural, linguistic, and pedagogical aspects of the UAE reform's implementation in bilingual education schools.

According to (Gallagher 2011), the change from a monolingual education (Arabic) system to a bilingual educational system (Arabic-English) has been described as "monumental". The researcher has explained the terminology used, stating that all subjects in the UAE were previously taught in Arabic, with English being taught as a second language. All learners_ regardless of their parental choice or agreement_ will be required to participate in a bilingual educational program where English predominates over Arabic. The study has highlighted the benefits of implementing bilingual educational programs in the UAE context. According to the 2006 International English Language Testing System conducted in the UAE, half of the diploma-level learners have failed to meet the minimum requirements for English language proficiency.

Thus, there was a remarkable drawback in the students' attainment in college due to the lack of English language competencies obtained at schools. Consequently, there was an urgent need to resolve this problem. Loughrey et al. has seen that to solve that radical and unachieved performance, the educational sector had to design new reforms that witnessed the English Immersion Program, which is a new school-based curricular. The reform has based its priors on the appropriate bilingual programme that supports and improves the English language competence. In conclusion, the researcher has seen that the bilingual education system in the UAE is very promising. The two most powerful and dominant languages (English and Arabic) are being taught in a bilingual system from the kindergarten to high school stage in well-provided resources and applicable circular and pedagogic conditions. Finally, the researcher has suggested more studies based on the appropriate implementation of the English-Arabic bilingual system in the UAE.

5 Methodology

This paper is based on a mixed method of research and has adopted the question-based research in the form of a questionnaire and an interview with a curriculum developer to explore the impact of bilingual instructions on students' achievement. It also seeks to investigate the teachers' perceptions of the bilingual education program in the UAE. Considering the limited studies of bilingualism and bilingual programs in the UAE, this paper aims to shed light on the relationship between bilingualism and academic achievement. In addition to examining the impact of bilingualism on students' academic achievement, the study looked into the correlation between bilingualism and cognitive ability.

The paper utilized a qualitative method questionnaire to analyze the effective implementation of the bilingual instructions program in a private school in the UAE. According to Kabel, he argued that the Arabic and English languages have shared prevailing and predominant attributes. He described both languages as a means that was used to colonize and globalize the world. He argued that the bilingual program in Abu Dhabi schools could be described as "a discourse of opportunity" that is based more on economic and political beliefs and norms.

The sample participants who participated in the questionnaire were selected as teachers with at least 5–10 years of teaching experience in the UAE. The teachers who teach different subjects (Math, English, Science) in English as a second language were supposed to respond to the questions to provide their viewpoint on the implementation of the bilingual program in a non-governmental school. The sample participants were 28 teachers who are currently teaching in a bilingual system in a private school in the UAE. Additionally, the collected data was based on an interview with an expert curriculum developer who had more than 20 years of experience in curriculum development and design. The interview has been virtually conducted using Zoom meetings. It lasted for 30 min discussing the viewpoint of a global bilingual program in comparison to the UAE context.

6 Ethical Consideration

Prior to the participants' engagement in the questionnaire and interviews, they were formally notified about the research aims and the intended utilization of the data. That process was taken to guarantee ethical consideration and protect the participants' identifications. In addition, consent forms have been given to the participants and the curriculum developer via email. Furthermore, the participants' privacy and confidentiality were carefully protected to ensure the security of their identities and maintain the credibility of the research. The participants' identities were safely stored in an encrypted folder, and the collected data was labelled with the researcher's unique code.

7 Findings and Discussion

First, the sample participants were divided into three main categories: teachers, coordinators, and heads of the department. The highest percentage of participation was recorded by teachers 67%, then 21% head of the department, and 10% coordinators. Thus, teachers felt that they were the core of the study, and the high rate of participation has proved the importance of the subject matter. According to Johnson and Swain (1997), teachers are considered a significant factor in the multilingual and the Immersion programs. Bilingual teachers obtain the necessary language proficiency in L2 to communicate with the students and develop their language competencies. However, the researcher has explained that the UAE scenario is different. He argued that the implementation of the Bilingual program will be affected by the teachers who might communicate with the students in their native language, Arabic. As a result, the learners' achievement in the second language in the bilingual program will be worthless. The quality of the teachers is a predominant element that determines the overall teaching outcomes in any program. To this end, there should be more focus on the teacher's professional development, particularly in the bilingual teaching program (Spolsky 2004).

Second, the participants showed 82% when they were asked if bilingualism positively impacts the student's learning capacity.

Based on the (Yeganeh and Malekzadeh 2015) study, they have confirmed that bilingual students obtained a high cognitive ability to judge the grammatical sentence and make comparisons between the two languages more than the monolingual learners. Thus, the bilingual students are assumed to acquire metalinguistic abilities that outperform the monolingual child. Third, when the teachers were asked about the bilingual program implementation and its impact on the students' behaviour using the two languages, the participants showed close results between agreement and disagreement on the subject matter. The results have shown that 42% of the teachers have agreed, while 39% disagree.

Fourth, some studies on bilinguals have suggested that teachers should attend professional development when they decide to teach in a bilingual setting. In this regard, the impact of the teachers' instructions and pedagogies will develop the learners' skills. The analysis has shown that 64% of the participants have agreed, while 10% disagree. On the other hand, 25% were not sure about the topic. Additionally, the participants have shown a positive attitude towards the statement that assumes learners who obtained good competency skills in the first language reading and writing would naturally improve in the second language competencies. The analysis data showed that 60% agree, while 35% disagree (Fig. 1).

Ultimately, the participants expressed dissent towards the notion that early exposure to a second language would have an impact on their proficiency in their original language. The participants showed 46% disapproval, while 28% agreed with the statement because the second language might affect the first language acquisition (Fig. 2).

In the domain of teaching and learning, the researcher interviewed a curriculum developer expert. The interview was conducted in the second phase of the data collection process. The participant has accumulated almost two decades of experience as an American curriculum developer. The conversation commenced by contrasting bilingualism and monolingualism within the context of 21st-century abilities and

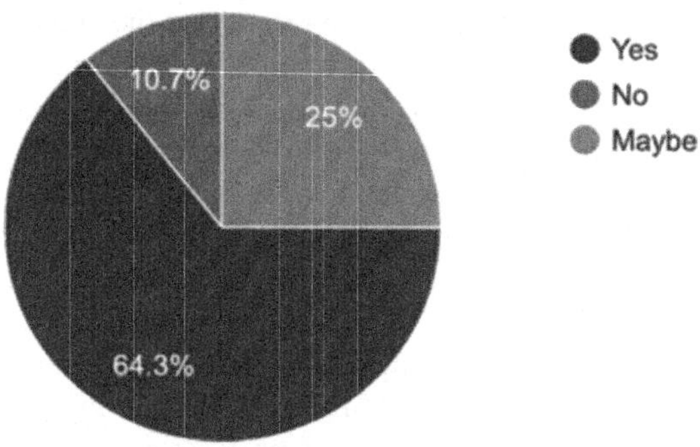

Fig. 1 Do you think providing bilingual instructions would impact the bilingual learners' academic achievement?

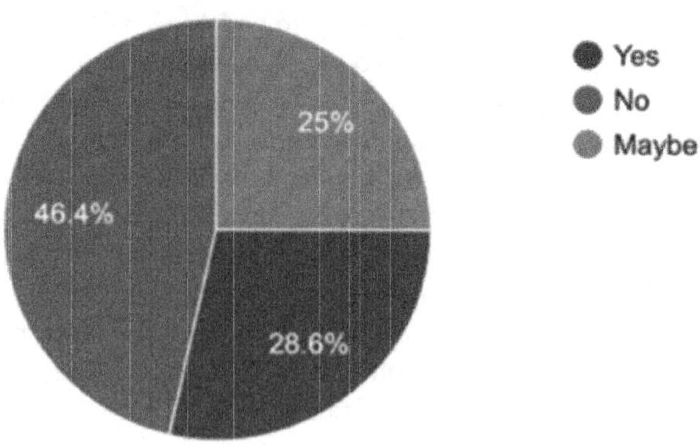

Fig. 2 Do you believe students must learn the second language in early age even if it means the loss of the native language

its relationship to the learners' cognitive abilities. The participant has affirmed that the sociocultural aspect offers significant potential for children to acquire a second or even a third language in an authentic environment with suitable communication and interaction. The interviewee has also verified that bilingualism may have both advantageous and detrimental effects on the academic achievement of young learners.

The interviewee confirmed that the negative effect of learning a second language on young individuals can hinder their competency in their primary or secondary language. In addition, she has explained that the bilingual student doesn't need to outperform the monolingual counterpart. She sees that every language has its own skills and strategies. The evident example from her experience was enacted through the two core subjects English and Maths; she illustrated the example by clarifying that some children obtained fewer reading competencies that might affect the students' skills in maths when they have a problem-solving equation. Ultimately, the interviewee has proposed that schools should offer enhanced opportunities for students to engage in language practice to strengthen their language competencies. As she stated in the interview, "The more they practice, the more proficient they will become."

Eventually, the interviewee affirmed that teacher professional development training is a means to enhance the teacher's practices, which in turn have an implicit impact on the quality of instruction for the pupils.

8 Conclusion

The results of the questionnaire have given a holistic vision of the teacher's positive attitude towards the bilingual program. Additionally, the interview has been pointed out and dug deep into further considerations; such as the curriculum components in a bilingual program and the applicable resources and materials that should be provided and embedded in the dual program. In Conclusion, this paper and the descriptive analysis have specified the impact of bilingualism on the students' academic performance; however, further studies and analysis should be conducted if the UAE government will analyze the bilingual program content and scope to achieve the ultimate benefits of the program. The educational sector should also question and investigate the program implementation across UAE schools to access a better quality of education.

References

Al Hameli A, Underwood M (2014) Challenges of abolishing foundation year in the UAE. The National Newspaper

Al-Khatib MA (2006) Aspects of bilingualism in the Arab world: an introduction. Int J Biling Educ Biling. https://doi.org/10.1080/13670050608668626

Bialystok E (2017) The bilingual adaptation: how minds accommodate experience. Psychol Bull 143(3):233–262. https://doi.org/10.1037/bul0000099

Colzato LS, Bajo MT, van den Wildenberg W, Paolieri D, Nieuwenhuis S, Heij WL, Hommel B (2008) How does bilingualism improve executive control? A comparison of active and reactive inhibition mechanisms. J Exp Psychol Learn Mem Cogn 34(2):302–312

Chin A (2015) Impact of bilingual education on student achievement. IZA World of Labor, pp 10

Costa A, Hernández M, Costa-Faidella J, Sebastián-Gallés N (2009) On the bilingual advantage in conflict processing: now you see it, now you don't. Cognition 113(2):135–149. https://doi.org/10.1016/j.cognition.2009.08.001

Costa A, Hernández M, Sebastián-Gallés N (2018) Bilingualism aids conflict monitoring and slows age-related cognitive decline. Psychol Sci 29(9):1371–1376

Gallagher K (2011) Bilingual education in the UAE: factors, variables and critical questions. Educ Bus Soc: Contemp Middle East Issues.https://doi.org/10.1108/17537981111111274

Green DW (1998) Mental control of the bilingual lexico-semantic system. Biling: Lang Cogn 1(2):67–81. https://doi.org/10.1017/S1366728998000133

Harris R (2012) Caught between two languages. Educ Dig 78(2):57

Johnson R, Swain M (1997) Immersion education: international perspectives. Cambridge University Press, Cambridge

Robinson DW (1998) The cognitive, academic, and attitudinal benefits of early language learning. Critical issues in early second language learning: building for our children's future, pp 37–43

Spolsky B (2004) Language policy. Cambridge University Press, Cambridge

Suwanarak K (2014) Bilingual learning and its effects on students Commun Competence 3(1):60–70

Tamiya S, Okada Y, Koterazawa K, Miyata H, Sanz M (2014) What is it like to grow up to be bilingual?: a survey report on bilingual high school students. REMIE: Multidiscip J Educ Res 4(3):317–338. https://dialnet.unirioja.es/servlet/articulo?codigo=4842681

Terrell TD, Brown HD (1981) Principles of language learning and teaching

Yeganeh MT, Malekzadeh P (2015) The effect of bilingualism on the developing of English reading skill. Procedia–Social Behav Sci 192:803–810. Elsevier B.V. https://doi.org/10.1016/j.sbspro.2015.06.099

The Role of Digital Transformation in Improving Intellectual Capital: The Context of Integrated Reporting

Dania M. Kurdy and Farzana Asad Mir

Abstract Business operations and individual behavior have been transformed by digitalization. Businesses are expected to sustainably create value over time, while also adapting their business models to rapidly changing business environments. In order to demonstrate how knowledge management (KM) systems can direct corporate practices toward digital transformation and the adoption of digital technologies, this study explores whether integrated reporting (IR) supports the dissemination of Intellectual Capital (IC). A systematic review of the literature identified 42 peer-reviewed articles in English-language publications between 2013 and 2023. The results show that researchers have mainly concentrated on value creation, IR and IC; along with how these elements interact, and how digital transformation can potentially act as a tool to improve these existing relationships. The study highlights the importance of digital transformation in implementing IC dissemination models through KM systems, ultimately leading to improved access and utilization of intellectual capital.

Keywords Digital technologies · Intellectual capital · KM systems · Integrated reporting · Value creation · Disclosure systems

1 Introduction

Digital transformation involves utilizing cutting-edge technologies to facilitate significant commercial benefits. It has a big impact on value creation in the modern era of affordable high performance computers and communication networks (Tomita 2022). The growth of an organization requires effective KM encompassing use of digital transformation (DT) to link communication, strategy, and operational processes, along with development of IC —which is defined as a combination of knowledge and experience used to create value (Di Vaio et al. 2020) DT has made

This Paper was prepared for the knowledge management Module, taught by Dr. Farzana Asad Mir

D. M. Kurdy (✉) · F. A. Mir
Faculty of Business and Law, The British University in Dubai, Dubai, United Arab Emirates
e-mail: kurdy.dania@gmail.com

K. Al Marri et al. (eds.), *BUiD Doctoral Research Conference 2024*, Lecture Notes in Civil Engineering 587, https://doi.org/10.1007/978-3-031-84371-6_36

access to large amounts of data which has led to the need for measurement and reporting tools that can meet the needs of a wider range of stakeholders (Di Vaio et al. 2020). The International Integrated Reporting Framework (IIRF) was created to better connect investment decisions, company behavior and reporting and thereby to support sustainable development. Its major goal is to show potential investors in financial capital how a business creates value over time (IIRC 2013). Companies need strong KM and IC to develop new ideas and gain an edge. A company's ability to innovate is mostly reliant on its IC, which includes its human, organizational, and social resources (Dabić et al. 2018). Sustainable Development Goals (SDGs) push businesses to adopt new business models that integrate sustainability. This includes using modern non-financial reporting to meet stakeholder demands for information on how sustainability creates value (Di Vaio et al. 2020). IR helps businesses understand value creation and improve sustainability performance. This framework is crucial for businesses to adapt and achieve sustainable value creation in today's changing world (Tomita 2022). The study is guided by the following research question:

What role does digital transformation play in increasing IC in the context of integrated reporting?

This study contributes to existing knowledge by investigating the under-explored link between DT and IC within the framework of IR. Moreover, our study will offer valuable guidance for companies seeking to leverage DT to strengthen their IC and improve the effectiveness of their IR. This study addresses several key gaps in the existing literature. First, while prior reviews have examined IR, IC, and DT separately, there is a lack of research synthesizing how these concepts intersect and influence each other. It provides a holistic view of their interrelationships. Second, the role of DT in enhancing IC disclosure through IR has not been comprehensively explored. Our work sheds light on how DT can improve IC reporting practices.

Also, there is limited understanding of how companies can leverage digital tools to create value from their IC in the context of IR. This study offers insights into effective strategies for value creation. Previous reviews have not fully captured recent developments in this rapidly evolving field. By including literature up to 2023, this study provides an up-to-date analysis of the latest trends and findings.

2 Methodology

This research reviews 42 articles published between 2013 and 2023 on IR, IC, DT, and value creation. It used Scopus and Web of Science databases along with a manual search to find relevant articles. The initial set of 437 papers was refined to a final sample of 42 through a two-step process of filtering and analysis. Figure 1 illustrates the detailed data collection approach.

Overview of Reviewed Literature

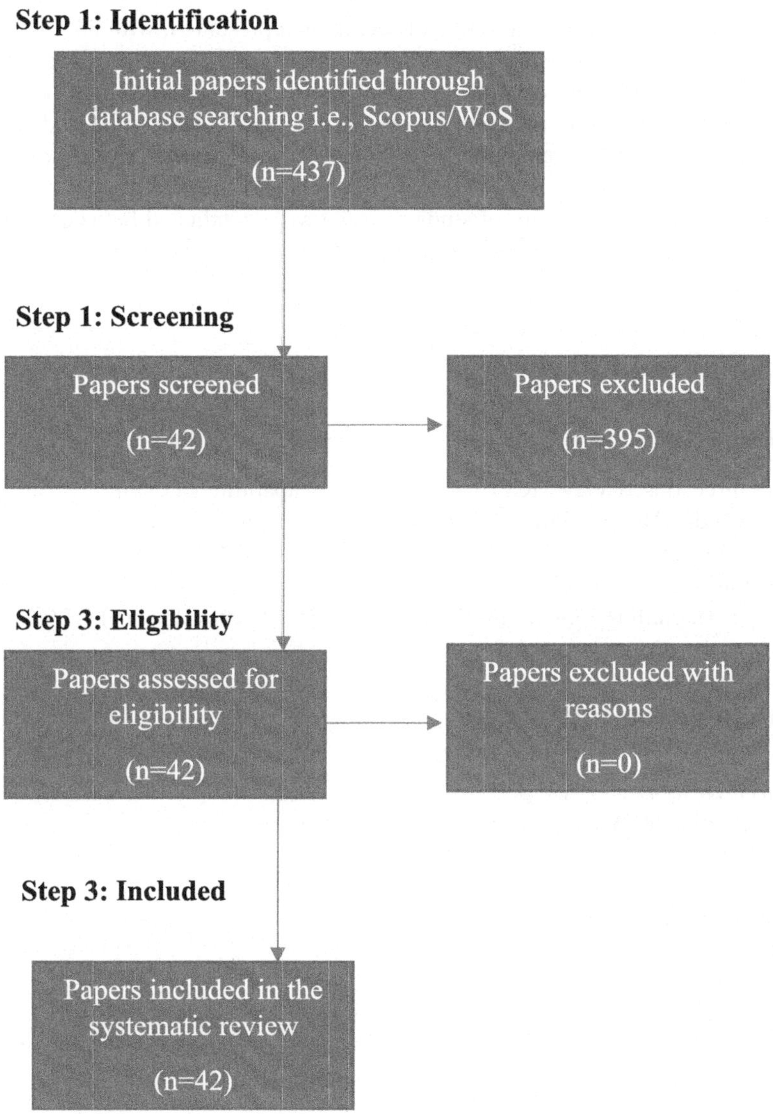

Fig. 1 A comprehensive approach for data collection

3 Fieldwork Characteristics

Geographic Distribution: The reviewed studies span multiple regions, with a predominance of research conducted in Europe (38%) and North America (29%). Asia–Pacific accounts for 19% of the studies, while 14% are global in scope.

Industry Sectors: A diverse range of sectors is represented, with 33% of studies focusing on multiple industries. The technology sector is most prominent (24%), followed by financial services (19%), manufacturing (14%), and others (10%).

Organization Types: Large corporations are the focus of 48% of the studies, while 29% examine SMEs. The remaining 23% include a mix of organization types or focus on public sector entities.

Time Frame: The majority of studies (76%) were conducted between 2018 and 2023, reflecting the recent nature of digital transformation research.

4 Theoretical Frameworks

The most prevalent theoretical approaches in the reviewed literature are:

1. Resource-Based View (RBV) and Dynamic Capabilities (33%)
2. Knowledge-Based View (KBV) (24%)
3. Stakeholder Theory (19%)
4. Institutional Theory (14%)
5. Others (including Innovation Diffusion Theory, Agency Theory) (10%).

5 Methodological Aspects

Research Design:

Quantitative: 48%
Qualitative: 33%
Mixed Methods: 19%

Data Collection Methods:

Surveys: 38%
Interviews: 29%
Case Studies: 19%
Archival Data Analysis: 14%

Sample Sizes:

Large (> 500): 24%
Medium (100–500): 43%
Small (< 100): 33%

Analytical Methods:

Regression Analysis: 33%
Content Analysis: 24%
Structural Equation Modeling: 19%
Thematic Analysis: 14%
Others (including Grounded Theory, Network Analysis): 10%

6 Literature Review

A. Digital Transformation, Intellectual Capital, and Knowledge Management

IC is the collection of knowledge and expertise held by an organization's staff that, when combined with information archives, accurately reveals the potential for long-term business benefit (Joshi et al. 2013). This activity directs companies towards competitiveness (Kianto et al. 2014). However, the function of IC in the execution of organizational processes for value creation was investigated from a dynamic standpoint to determine the interactions between KM and IC to develop and optimize the benefits of IC. IC is being researched largely as a quantitative resource for value generation (Kianto et al. 2014). To enhance an organization's potential to create competitive advantages, researchers are looking for strategic solutions that prioritize people and incorporate "biophysical, economic, and social sustainability components" which are founded on knowledge and ethical principles. This can't be done, by disregarding the role of digital technologies, which are engrained in people's knowledge, make information more accessible, and shorten organizational learning curves for greater profitability (Natalicchio et al. 2019).

The link between IC components and company performance is debated. While some studies suggest all IC aspects contribute, others argue only specific components matter. KM practices are seen to improve IC, but research hasn't explored how well-matched IC and KM strategies impact performance (Wang et al. 2016). DT strengthens a company's competitive edge by improving how they manage knowledge. This focus on applying knowledge is even more important in today's changing environment (Buntak et al. 2020).

B. Relevance of Knowledge Management in the Digital Age

Companies should implement KM practices to ensure progress, improvement, and sustainability. A key focus is structural knowledge, which can provide a competitive edge. This requires a secure information system within the company to facilitate knowledge sharing. However, integrating new technologies for KM presents challenges, including data classification, performance activity tracking, and initial data handling (Mohajan 2017). This shift requires moving from outdated knowledge systems to embracing DT. However, even innovative companies need strong structural knowledge, which can be improved through digital advancements.

C. Intellectual Capital, Integrated Reporting

The International Integrated Reporting Council (IIRC), founded in 2010, created Integrated Reporting (IR) as a more comprehensive reporting approach. According to its definition, IR provides a picture of "the relationships and resources used and influenced by an organization" (IIRC 2013). A study found that there will be a significant and strong link between IR and IC since IR will have the ability to clarify the potential of IC in corporate management processes, hence strengthening IC's function in governance procedures (Badia et al. 2019). Few IR studies are relevant to IC because researchers tend to concentrate on environmental and social aspects of IR rather than the elements associated to IC (de Villiers and Hsiao 2017). Addressing ethical and relational considerations is viewed as vital for long-term value creation, although it can be difficult for managers to balance between the needs of stakeholders and the financial, environmental, and social issues. As relational capital in the context of IC, this study emphasized the significance of stakeholder engagement and management (de Villiers and Hsiao 2017).

Many of the traits of the non-financial reporting methodologies, comprising IC, business social responsibility, and strategic management, are included into IR. By examining existing academic research and publications, we can identify how DT are being used to improve the collection, analysis, and presentation of data for IR.

D. Results and Discussion

Results from a bibliometric analysis of 42 studies that emerged during the course of this research are presented below. Since information is the engine for economic growth and wealth creation, acquiring and using it have become crucial goals for companies looking to build long-lasting competitive advantage and achieve commercial success. Consequently, in today's knowledge economy, the capacity to manage knowledge has become essential (Dalkir 2017). In the transition from the industrial to the digital era, the value of labor, natural resources, and machinery has diminished, while the value of intangible resources such as information and IC has increased. In a knowledge-based economy, a company's most asset is its information and expertise. IC and digital transformation, which are the foundational elements of the knowledge economy, are both connected to and derived from knowledge. They are the main factors that mediate between knowledge and economic development (Xi 2021).

Digital transformation necessitates a fresh look at business reporting practices. The IIRC recognizes this and sees potential in DT for IR. IR can leverage these technologies to create connections between information and provide easier access for stakeholders, improving their ability to assess a company's performance (Lombardi and Secundo 2021). By promoting Integrated Thinking, which promotes the alignment of strategy and resources, the IR structure fosters a comprehensive view of the organizational dynamics of the company and enables the incorporation of IC in the management and decision-making process. This enables the business to respond to the shifting demands of a market that has been completely transformed by DT (Di Vaio et al. 2020). Dynamic capabilities, a framework for integrating digital technologies and intellectual capital (IC), is key for companies to adapt and create value in today's fast-paced environment. It involves sensing opportunities, seizing them through skill development, and transforming internal processes to leverage a stronger IC (Izzo et al. 2022) (Fig. 2).

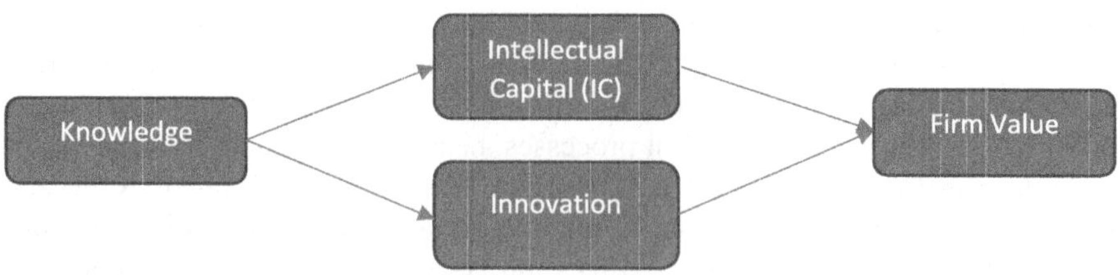

Fig. 2 Conceptual framework

Academic research has shown that organizational knowledge and DT are based on IC (Dinu 2022). By fostering knowledge creation and sharing, companies can fuel transformation and develop new technologies, ultimately driving value creation (Dinu 2022). Researchers have found that while IC affects novelty in small businesses; larger businesses are better equipped to use information and communication technology systems to connect structural capital and boost digital transformation (Verbano and Crema 2016). IC is crucial in promoting technological absorptive capacity (TAC) and transformation. New technologies are crucial for creating value because they provide the tools for better performance and KM (Andersson et al. 2021). However, the benefits of technology are not beyond debate. A study found that while digital technologies boost business productivity, they also have the potential to harm relational and human capital by reducing interactions and standardizing education. However, there is a limit to how useful digital transformations (Usai et al. 2021). KM need to be capable of comprehending and utilizing the potential of these technologies in order to create a new kind of value (Matt et al. 2015). Unfortunately, the shortcomings of organizational processes and people's resistance to change frequently prevent new technologies from producing the expected benefits. Businesses need to manage knowledge in order to aid digital transformation and derive benefit from cutting-edge technologies (Sarka et al. 2019). A study claims that digitalization is the application of digital technology that is changing markets and enterprises. In this regard, the introduction and utilization of digital technologies affect a variety of business tasks and alter the ways in which they create value (Vial 2019). Organizational role, degree of digitization, and technology of choice all have an impact on the value creation processes. The use of technology to generate certain types of value and alter business practices is strongly related to digital transformation. The IIRC's essential idea of "connectivity," which is a phenomenon in IR, is presented in great detail in a study that sees social media and digital tools as bridges to transparency and stakeholder engagement in IR, aligning with Legitimacy, Reputation Risk, and Stakeholder theories (Rivera-Arrubla and Zorio-Grima 2016). The foundation of legitimacy theory is the idea of a social contract between a firm and society, together with the presumption that the company will implement disclosure techniques to demonstrate conformity with societal expectations. In this regard, the stakeholder theory proves that these disclosures are intended for certain stakeholders who are giving the business resources (Rivera-Arrubla and Zorio-Grima 2016). Therefore, IR may be used to respond to and modify that social contract and potentially have an impact on global change. Digital business strategies require more dynamic ecosystems and cross-organizational collaboration to create and capture value across the entire product, process, and service chain (Pagani 2013). The link between digital business models and how they capture value is also increasingly unclear (Kazan et al. 2015).

Following the review of the literature and investigation of the above-mentioned findings, it is crucial to highlight the role of the conceptual framework designed to demonstrate the role of dynamic capabilities in regulating the relationship between IC, transformation, and value creation of a firm. Based on the above and our literature review, we have developed the following conceptual framework:

Future Research Agenda

Based on our comprehensive review of the literature on DT, IC, and IR, we have identified several promising avenues for future research:

1. Cross-cultural studies: Given the geographic concentration of existing research, there's a need for more studies in underrepresented regions, particularly in developing economies. This could provide insights into how cultural and economic contexts influence the relationships between digital transformation, IC, and IR.
2. Longitudinal research: Most existing studies are cross-sectional. Longitudinal studies could help understand the long-term impacts of digital transformation on IC and IR practices.
3. SME focus: While large corporations dominate current research, more studies on SMEs are needed to understand how size and resource constraints affect digital transformation and IC management.
4. Emerging technologies: Future research should explore the impact of specific emerging technologies (e.g., AI, blockchain, IoT) on IC and IR practices.
5. Stakeholder perspectives: Studies incorporating multiple stakeholder perspectives (e.g., employees, customers, investors) on the value of digitally-enhanced IC reporting could provide valuable insights.
6. Measurement and metrics: There's a need for research developing and validating new metrics for measuring digitally-transformed IC and its impact on organizational performance.
7. Integration of theoretical perspectives: Future studies could benefit from integrating multiple theoretical lenses (e.g., combining RBV with institutional theory) to provide a more holistic understanding of the phenomena.
8. Mixed-method approaches: Given the predominance of quantitative studies, more mixed-method research could provide both breadth and depth in understanding these complex relationships.
9. Sustainability and ESG integration: Research exploring how digital transformation of IC and IR practices can support broader sustainability and ESG reporting efforts is needed.
10. Regulatory and policy implications: Studies examining how regulatory environments influence the digital transformation of IC and IR practices could inform policy development.

7 Conclusions and Recommendations

This study emphasizes the importance of KM for IC dissemination in digital transformation companies. It finds a strong link between IR and IC in value creation. The research identifies gaps in the current understanding of how digital technologies impact IR and IC disclosure, particularly for disruptive transformations. It highlights the need for further research on identifying and leveraging valuable IC components. Finally, the study acknowledges limitations in existing theoretical frameworks for dealing with intangible resources like IC.

References

Andersson M, Kusetogullari A, Wernberg J (2021) Software development and innovation: exploring the software shift in innovation in Swedish firms. Technol Forecast Soc Chang 167:120695

Badia F, Dicuonzo G, Petruzzelli S, Dell'Atti V (2019) Integrated reporting in action: mobilizing intellectual capital to improve management and governance practices. J Manag Governance 23(2):299–320

Buntak K, Kovačić M, Martinčević I (2020) Impact of digital transformation on knowledge management in organization. Adv Bus-Relat Sci Res J 11(1):36–47

Dabić M, Lažnjak J, Smallbone D, Švarc J (2018) Intellectual capital, organisational climate, innovation culture, and SME performance: evidence from Croatia. J Small Bus Enterp Dev 26(4):522–544

Dalkir K (2017) Knowledge management in theory and practice. MIT press

de Villiers C, Hsiao P-CK (2017) Integrated reporting and the connections between integrated reporting and intellectual capital. In The routledge companion to intellectual capital. Routledge, pp 483–491

De Villiers C, Maroun W (2017) Sustainability accounting and integrated reporting. Routledge

Di Vaio A, Hassan R, Palladino R (2020) Digital innovation and disruptive technologies in the intellectual capital (IC) and knowledge management systems (KMS) disclosure: a bibliometric analysis. In: Paper presented at the 2020 IEEE international conference on technology management, operations and decisions (ICTMOD)

Di Vaio A, Syriopoulos T, Alvino F, Palladino R (2021) Integrated thinking and reporting towards sustainable business models: a concise bibliometric analysis. Meditari Account Res 29(4):691–719

Dinu E (2022) A systematic review of the literature on intellectual capital management, technology and innovation. Ekonomicko-Manazerske Spektrum 16(1):58–75

IIRC (2013) The international integrated reporting council: the international<IR> framework. https://www.integratedreporting.org/wp-content/uploads/2013/12/13-12-08-THE-INTERNATIONAL-IR-FRAMEWORK-2-1.pdf

Izzo MF, Fasan M, Tiscini R (2022) The role of digital transformation in enabling continuous accounting and the effects on intellectual capital: the case of oracle. Meditari Account Res 30(4):1007–1026

Joshi M, Cahill D, Sidhu J, Kansal M (2013) Intellectual capital and financial performance: an evaluation of the Australian financial sector. J Intellect Cap

Kazan E, Tan C-W, Lim ET (2015) Value creation in cryptocurrency networks: towards a taxonomy of digital business models for bitcoin companies

Kianto A, Ritala P, Spender J-C, Vanhala M (2014) The interaction of intellectual capital assets and knowledge management practices in organizational value creation. J Intellect Cap

Lombardi R, Secundo G (2021) The digital transformation of corporate reporting–a systematic literature review and avenues for future research. Meditari Account Res 29(5):1179–1208

Matt C, Hess T, Benlian A (2015) Digital transformation strategies. Bus Inf Syst Eng 57:339–343

Mohajan H (2017) The impact of knowledge management models for the development of organizations

Natalicchio A, Ardito L, Messeni Petruzzelli A, Del Giudice M (2019) The origins of external knowledge inflows and the impact of university technologies. R D Manag 49(4):639–651

Pagani M (2013) Digital business strategy and value creation: framing the dynamic cycle of control points. Manag Inf Syst Q 617–632

Rivera-Arrubla YA, Zorio-Grima A (2016) Integrated reporting, connectivity, and social media. Psychol Mark 33(12):1159–1165

Sarka P, Heisig P, Caldwell NH, Maier AM, Ipsen C (2019) Future research on information technology in knowledge management. Knowl Process Manag 26(3):277–296

Tomita A (2022) Aligning digital transformations with value creation based on international integrated reporting. In: Paper presented at the 2022 Portland international conference on management of engineering and technology (PICMET)
Usai A, Fiano F, Petruzzelli AM, Paoloni P, Briamonte MF, Orlando B (2021) Unveiling the impact of the adoption of digital technologies on firms innovation performance. J Bus Res 133:327–336
Verbano C, Crema M (2016) Linking technology innovation strategy, intellectual capital and technology innovation performance in manufacturing SMEs. Technol Anal Strat Manag 28(5):524–540
Vial G (2019) Understanding digital transformation: a review and a research agenda. J Strateg Inf Syst 28(2):118–144
Wang Z, Wang N, Cao J, Ye X (2016) The impact of intellectual capital–knowledge management strategy fit on firm performance. Manag Decis
Xi R (2021) Intellectual capital, knowledge, and knowledge management: exploring the value creation link

The Role of Judicial Precedent as a Source of International Investment Law

Julius Galisonga and Abba Ali Kolo

Abstract This paper examines the position of judicial precedent in international investment law in the context of international investment arbitration, interrogating its rationale and implications under the provisions of Article 38 (1) (d) of the ICJ Statute. The paper seeks to explain what appears to be an 'over reliance' on previous decisions by investment treaty arbitral tribunals in their task of resolving investor-state disputes. It argues that although in theory, previous decisions of courts and tribunals are non-binding on subsequent courts and tribunals as they constitute only a 'subsidiary' source of law, nevertheless in practice, investment treaty tribunals rely heavily on previous decisions in arriving at their own decisions arguably to promote a harmonious development of investment law and to meet the legitimate expectations of the community of States and investors towards the certainty of the rule of law. The paper demonstrates that investment tribunals have, through interpretation of treaties, created and defined remedies, rights, and obligations under international investment law much more than formal sources set out those rights thereby influencing content of treaty standards of the international investment law and the phrasing of the terms of the treaties. The paper proposes that to support the role of precedent in investment dispute resolutions, investment treaties should explicitly provide for application of precedent.

Keywords Judicial precedent · International investment law · Investment arbitration · Investment tribunals

---

J. Galisonga (✉)
Faculty of Business and Law, The British University in Dubai, Dubai, United Arab Emirates
e-mail: 22001730@student.buid.ac.ae

A. A. Kolo
Department of Law, The British University in Dubai, Dubai, United Arab Emirates

K. Al Marri et al. (eds.), *BUiD Doctoral Research Conference 2024*, Lecture Notes in Civil Engineering 587, https://doi.org/10.1007/978-3-031-84371-6_37

1 Introduction

Article 38 of the ICJ Statute (hereinafter the statute) defines the law and hierarchy thereof, applicable by a court deciding by International Law, disputes submitted to it. Specifically, under clause 38 (1) (d), the statute provides for the application of judicial decisions (precedent) as subsidiary means for the determination of rules of law, subject to Article 59, which provides that decisions of a court in any given case, only bind the parties to that case.

Various authorities have explained the import of Article 38 (1) (d) to mean that where a judicial decision or legal scholarly writing states a rule of international law, it will be attributable to a treaty, custom or general principle of law. Such decision or writing will not proclaim it as law because the presiding judge or author declared it but because the judge or writer deems it as emanating from one of the three principal sources indicated in Article 38 (Thirlway 2019).

The paper, discusses the use of "judicial precedent" in international law generally and under international investment law in particular, its rationale, and implications, in light of the above provisions. Section 2, gives the contextual background of precedent in settling disputes generally and international investment disputes in particular, as well as tracing the evolution of the doctrine of "precedent" in resolving international disputes. Section 3, examines the present application and degree of reliance on precedence by arbitral tribunals and implications arising from such use. Section 4 discusses the rationale for relying on decided cases under the above provisions and the shortcomings of doing so. Section 5 concludes.

2 Evolution of Precedent in International Law

The doctrine of "precedent" is rooted in the eighteenth-century English common law, when William Blackstone defined it as establishing a solid presumption that judges, to promote stability in the law, would "abide by former precedents, where the same points come again in litigation" unless such precedents were "flatly absurd or unjust." (Blackstone, 1765) Also known as "*stare decisis*," in the English Common Law, the doctrine of "precedent" provides that a court follows earlier judicial decisions when the same points arise again in litigation, such that a decided case furnishes a basis for determining later cases involving similar facts or issues (Black's Law Dictionary 2014).

Within the national legal setting, whether under the English Common Law, or under Civil Law, where judges do not make law, the position of precedent is quite clear, Judges, when determining disputes, generally rely on solutions provided in previous decisions, hence the principles of ***stare decisis, jurisprudence constante*** or ***standige Rechtsprechung***, (Guillaume 2011).

At the international level, the position on precedent remains incomplete, and the principles governing its use in international law (Jennings 1967), in all dispute resolution fora, including international commercial arbitrations, are not clearly defined.

At the inception of the Permanent Court of Arbitration, while some expected that when several arbitral tribunals solved a contentious legal issue in one way, that solution would become part of the body of international law, (Lapradelle, 1798–1855), the drafters of the statute of the Permanent Court of Justice never intended that the court would create rules, and thus specified that the court cannot legislate and that decided cases would only state but not create law, hence Article 38, which excludes judicial decisions as a source of law save for a supplemental role for determination of the rule of law.

The *Fisheries case, 1974* cemented this view when court held that it could not render judgment "***sub specie legis ferendae***" or anticipate the law before the legislator has laid it down. Similarly, in the *Advisory Opinion on the Threat or Use of Nuclear Weapons, 1996* the ICJ held that under the state of international law it is only limited to recording the state of the law without being able to substitute its assessment for the will of sovereign States.

The above history and provisions of the statute now beg the question, why the ICJ and other international dispute resolution fora rely on precedent as if they are a primary source of law under Article 38 (1) (a–c)?

The ICJ when resolving Nigeria's objections premised on Article 59 in *Land and Maritime Boundary case, Cameroon and Nigeria (Preliminary Objections), 1998* explained that while the court's decisions only bind the parties to and in respect of a particular case, "the real question is whether, in this case, there is cause not to follow the reasoning and conclusions of earlier cases."

The court's reasoning in *Land and Maritime,* which renders historical justification for reliance on precedent, in spite of Article 59, supports the argument of Thirlway 2019, that Article 59 does not oust earlier decided cases per se because, "...when the 'decisions' of a court are being referred to as a source of law, it is evident that it is primarily the reasoning that is being contemplated...here the 'decision' is the reasoning, which may cause the case to become a precedent, faire jurisprudence".

Notably, while jurisprudence of the ICJ is weighty, decisions cited are not simply automatically applied to another case, even when they are in pari-materia, as the ICJ has clarified that even for it, previous decisions are not binding precedents, (Thirlway 2019).

3 The Role of Judicial Precedent in International Investment Law and its Implications

Thus far it is has been demontrated that in the hierarchy of sources of international law, as a matter of formality, judicial decisions are a subsidiary source of law but practically international tribunals treat them much more than that and regularly rely on precedent to determine cases that are in *pari-materia*, even when the rules of those various fora do not permit it, (Bjorklund 2008).

Primarily this is because Article 38 is crafted merely as a directive to consider precedent for proof of the existence of rights and obligations and not setting up a hierarchy and that case law plays a cardinal function rather than just pointing to other sources of law, (Bjorklund 2008), and whereas the ICJ statute refers to "judicial" decisions, in practice, those "decisions" are now coopted and used as jurisprudential sources, (Zimmermann, Tomuschat and Koellers-Frahm 2006).

Therefore, while arbitral tribunals generally advise that precedent is not binding, they rely on it where, applying a particular legal principle precedent is definitive as an argument and not as a source of law and contributes to the harmonious development of investment law thereby meeting legitimate expectations to certainty of rule of law. The tribunal in Saipem V. Bangladesh 2007 held that;

The tribunal considers that it is not bound by previous decisions. At the same time, it is of the opinion that it must pay due consideration to earlier decisions of international tribunals. It believes that, subject to compelling contrary grounds, it has a duty to adopt solutions established in a series of consistent cases. It also believes that subject to the specifics of a given treaty and of the circumstances of the actual case, it has the duty to seek to contribute to the harmonious development of investment law and thereby meet the legitimate expectations of the community of States and investors towards certainty of rule of law.

Consequently through preceedent, tribunals have created and defined rights and obligations under international investment law through wide interpretation of treaties much more than formal sources set out those rights, (Schill 2017) and redefined the ambit of protection provided by the treaties.

For example, under Article 47 of the ICSID convention tribunals lack jurisdiction to create law on provisional measures, and any measures under the article are merely recommendations, but tribunals have exploited the provision to create rights to provisional measures, (*Quiborax v Bolivia, 2002*). Tribunals have also created rights like compound interest instead of simple interest, (*Compania De Desarrollo Da Santa ELena, S.A. v Costa Rica, 1996*), defined principles and rights on indirect expropriation, (Metalclad v. Mexico ICSID, 1997), National Treatment, (Myers v Canada 2000) complete protection and security, (*Ampal-American v. Egypt, 2011*), etc.

In *Waste Management v Mexico, 2000* for example, the tribunal's definition of Fair and Equitable Treatment depended on the tribunal's understanding of the standards as developed in arbitral jurisprudence rather than from an independent application of methods of treaty interpretation. Ultimately these declarations inform the content of future BITs and MITs (Schill 2017).

Resultantly, precedent influences the content of treaty standards, as well as the drafting of treaty terms, not through the methods of interpretation under the Vienna Convention or reference to customary international law, (Schill 2017). Indeed, several countries have revised their model BITs and investment chapters in PTAs and updated their wording, content, and structure to incorporate the lessons learned from investment-related litigation, (Yannaca-Small, 2018).

The conclusion to draw is that while precedent is a subsidiary source, arbitral tribunals rely on them as a principal reference point for understanding the substance of those rights and responsibilities (Yannaca-Small, 2018).

So important has precedent become that citing past continues to rise as the number of cases grows, with one study on ICSID citation practice showing that before 1994, the number of ICSID precedents cited per ICSID decision was about two; from 1994 and 2002, the average number was between two and four; and between 2002 and 2006, the average number was ten, (Chen, 2019).

Evidently, tribunals have morphed from settlers of individual investment disputes to makers of international investment law, effectively determining how investment treaties are interpreted, with several implications, including;

(a) Efforts to cultivate greater precision in the scope of the definition of investment. Interpretations of what "investment" meant by tribunals were deemed too broad and beyond what the contracting parties conceived as 'investment' when negotiating the IIA, (Yannaca-Small, 2018). For instance, in the case of *Pope & Talbot v Canada, 2000*, the tribunal found that a market share through trade could be regarded as part of the assets of an investment,while in *S.D. Myers v Canada, 2000* the arbitral tribunal held that the establishment of a sales office and commitment of marketing time formed a sufficient investment. These definitions by tribunals encompassed matters that were not anticipated as falling under investment at the time of negotiations of the IIA.

States reacted by incorporing definitions of "investment", adopting concepts developed in ICSID practice including the requirement of commitment of capital or other resources, the expectation of gain or profit or the assumption of risk. They included exclusions of named kinds of assets, as not to falling within the ambit of covered investments under the agreement, (Yannaca-Small, 2018).

(b) Renegotiation and BITs to clarify the wording of treaties to better define obligations and rights, such as international minimum standard of treatment, expropriation, clarification that investment protection should not be pursued at the expense of other public policy objectives and to achieve greater transparency between the contracting parties and in the process of domestic rule-making, (Yannaca-Small, 2018).
(c) Innovations in ISDS procedures, to not only provide guidance generally but also provide greater control of the contracting parties over arbitration procedures, reduce costs involved in dispute resolution, etc. (Yannaca-Small, 2018).

The commentaries to the *SADC Model Bilateral Investment Treaty, 2017* reflect the foregoing consequences, thus; "The Second Edition also sees the narrowing of several of the options provided in the first edition, *reflecting the growing desire to limit the breadth of certain types of clauses in order to avoid unintended consequences, and a greater determination to put forward "made in Africa" solutions on issues such as the definition of investment and investor, pre-establishment rights, compensation levels for expropriation, and dispute settlement. (SADC Model BIT with Commentary, 2nd Edn, June 2017)*

4 Rationale for Relying on Precedent, the Draw Backs Thereto and How to Harness the Use of Precedent to Better Serve International Investment Law

Because treaties contain vague provisions that need grounded interpretation, yet the Vienna Convention on the Law of Treaties only provides a framework for treaty interpretation, but each treaty and dispute require special consideration, reference to prior decided cases becomes inevitable, as they provide guidance to the tribunals. The following have been cited as justification for reliance on precedent;

I. Predictability of the law, rights, obligations, etc. Justice Brandeis' observation that in most matters, it is more important that a question be settled than that it be settled right, (Burner v Colorado, 1932) (Brandeis, J dissenting,), underscores the importance of precedent. Whether an answer is correct or not, if settled, it allows for proper planning with better understanding of legal consequences. This predictability facilitates legitimate expectation so that an investor will chose a particular host state because of an expected stable legal framework, (Chen, 2019).
II. Provisions of bilateral investment treaties (BITs) are ambiguously and vaguely crafted and susceptible to many constructions. Interpretations avail rules which can then be relied on in subsequent cases and inform parties while negotiating, agreements. Therefore complying with precedent benefits from the collective wisdom of earlier decision makers, (Chen, 2019).
A. Legitimacy of the tribunals and the decisions of the tribunals. Viability of arbitration tribunals depends on the perception that users have of them as legitimate. Legitimacy is cultivated by being predictable and accurate in their decisions, resulting in decisions gaining wide acceptance and consideration as justified, (Cheng, 2006).

There are several arguments against precedent in resolving International investment disputes, including that treaties do not provide for the doctrine of *stare-decisis*, so there is lack of continuity and hierarchy. There is no way of managing predictability because the doctrine of stare-decisis, the bedrock of common law is not embedded in international investment law, (Chen, 2019). Therefore, shaping international investment law is fragmented as it arises from so many different investment treaties.

Moreover, predictability can only be if tribunals dealing with an issue agree on a particular interpretation. However if tribunals give incompatible construction of comparable clauses, confusion arises, on the position that should be applied, resulting in endless debate with no hope of settling the same.

Divergent interpretation and construction of analogous clauses in bilateral investment treaties (BITs) by two different ICSID tribunals, for example has resulted in uncertainty of contractual forum selection clauses and umbrella clauses (Chaise, 2023). In *SGS v Pakistan, 2013* the tribunal found that an umbrella clause does not automatically elevate a contract breach to a treaty violation when a contract contained a valid forum-selection clause, in *SGS v Philippines, 2006* the tribunal decided that exclusive forum-selection clauses implied that contractual claims are inadmissible before the tribunal, even when the tribunal is still vested with jurisdiction.

The above two decisions have resulted in a dualist approach on the use of umbrella clauses. On the one hand, an expansive application of umbrella clauses, to accommodate investment treaty based arbitration arising from any breach of an investment-related promise by the host State, see *CMS v. Argentine, 2008.* On the other hand a limited function to umbrella clauses which restricts it only to a host "states sovereign conduct" is advanced, and invoking an umbrella clause is conditioned on violating an investor-state contract resulting from a sovereign act of the host State, *Sempra v. Argentina, 2016*. By these decisions, application of forum selection clauses as well as the degree of protection of umbrella clauses was rendered uncertain.

The confusions supports the argument that precedent without a higher court or tribunal to settle the debate, the dualist debate will not be resolved, thereby undermining predictability supposed to be rendered by precedent.

Tribunals can however ameliorate the effects of the confusion created by conflicting decisions through distinguishing previous cases, and through in-depth analysis and evaluation of previous decided cases, as well as evidence before them as was the case in *Masdar Solar v. Spain, 2014* in distinguishing the decision of *Charanne B.V. v. Spain, 2012.*

When tribunals give explanations and rationales for reaching the decisions they reach, leaving no doubt as to the fundamental principles of international law developed in previous decisions, or that are being enunciated, then a degree of predictability will be cultivated.

Recommendations

To harness the benefits of precedent in international investment law, it is proposed that;

Foremost, states adopt the responsive approaches and actions by states in reaction to precedent highlighted in Section 3 (a) to (c) above.

Further, future treaties should define the application of precedent as a source of the applicable substantive laws. Article 38 (1) (d) of the ICJ Statute allows its application, as subsidiary means for the determination of rules of law.

The premise of this suggestion is that when identifying sources of foreign investment law, many authors use the categories set out under Article 38 (1) of the Statute of the International Court of Justice (ICJ), which include judicial decisions. Treaties such as the Argentina-Netherlands (under Article 10) and the ICSID Convention under Article 42(1) of the ICSID Convention, have defined laws to be applied Article 26 (6) Energy Charter Treaty, Article 10 of the Argentina-Netherlands BIT of 1992 to mention just a few.

A framework must be devised within the international investment disputes resolution fora, to provide for an appellate process to review decisions and to clarify if the principles developed by tribunals reflect the true position of the law, intentions of the parties or the changes of the time. For example what amounts to an investment may be percieved differently as times change. This appellate process will help guide he development of precedent as a source of law.

5 Conclusion

Conclusively, it is apparent that judicial precedent has taken center stage in international investment law. Treaty arbitration tribunals rely on arbitral precedent as a principal source for rules of international law, much more than just as a subsidiary source of law for determining the rules of law.

Through the application of precedent, the practical requirements of international dispute resolution and the dynamic nature of needs of all players in international investment have been served, as judicial interpretation quickly plugs any loopholes while providing better understanding of the law. The reason being that when a court decides that a particular consequence results from a rule deriving from a recognized source 'added value' is obtained for that particular rule.

The practical role played by the various dispute resolution fora under IIL, implies that a judicial decision, of whatever kind, enriches the law on the subject of the dispute rendering the principle that a judicial body does not create law fiction, because courts do have, in effect, a legislative role and competence.

In conclusion, judicial precedent no longer merely plays the role of handmaidens for determining rules of law, as anticipated under Article 38 (1) (d) of the ICJ statute. In some cases, more often than not, it has become the case of the tail wagging the dog, whereby judicial precedent now determines the content of the law.

References

Ampal-American and others v. Egypt ICSID case no. ARB/12/11 (2004) In occidental exploration and production company v Ecuador, Award

Bjorklund AK (2008) Investment treaty arbitral decision as jurisprudence constant. UC Davis legal studies research paper series. Research paper no. 158

Black's law dictionary (2014) defining stare decisis and id. at 1366 defining precedent. 10th ed

Blackstone W. Commentaries on the laws of England 69–70 (describing precedent as a permanent rule, which it is not in the breast of any subsequent judge to alter or vary from, according to his private sentiments)

Brabandere, Tarcision G (2012) International investment law: sources of rights and obligations. Chapter 1 and 9

Burner v Colorado Oil and Gas Co. (1932) 285 US 393, 406 (Brandeis, J dissenting,) as cited by Richard C

Chaisse J (2023) Rethinking umbrella clauses in international agreements. Columbia centre for sustainable investment. No. 368

Chen RC (2019) Precedent and dialogue in investment treaty arbitration. Harv Int Law J 60:47–53

Cheng T (2006) Precedent and control of investment treaty arbitration. Fordham Int Law J 30:1026

CMS Gas v. Argentina (2007) ICSID (W. Bank) case no. ARB/01/8, Decision on annulment, paras 89–95

Compania de Desarrollo da Santa ELena, S.A. and The Republic Costa Rica case no. ARB/96/1 para 97

Constante. UC Davis. Legal studies research paper series. Research paper no. 158

Florian G (2014) The sources of foreign investment law. In: Douglas Z et al (ed) The foundations of international investment law: bringing theory into practice. Oxford University Press. Chap 7, pp 213–33

Guillaume G (2011) The use of precedent by international judges and arbitrators. J Int Disput Settl 2(1)

Jennings R (1967) General course on principles of international law. Acad Int Law Recueil des Cours 121(2)

Kohler GK (2007) Arbitral precedent: dream, necessity or excuse?—The 2006, Freshfields lecture. Arbitr Int 23(3):357

Lapradelle A, Politis N (2011) Recueil des arbitrages internationaux (1798–1855) preface by Renault L vii, as cited by Guillaume G in: the use of precedent by international judges and arbitrators. J Int Disput Settl 2(1)

Legality of threat or use of nuclear weapons (advisory opinion) (1996) ICJ Rep 226

Metalclad Corporation v. The United States of MexicoICSID case no. ARB (AF)/97/1, Award of 30 August 2000

Myers SD (2000) Inc. v. Canada, UNCITRAL Arbitration, First Partial Award of 2000

Paulsson J (2006) International arbitration and the generation of legal norms: treaty arbitration and international law. Transnatl Disput Manag

Pope and Talbot, Inc. v. Government of Canada. UNCITRAL, Interim Award on Merits (June 26, 2000); Award on merits (Apr. 10, 2001); Award on damages (May 31, 2002); Award on costs (Nov. 26, 2002)

Quiborax SA and others v Plurinational state of Bolivia. ICSID case no. ARB/06/2, Award, para 582

Roberts A (2010) Power and persuasion in investment treaty interpretation: the dual role of states. Arab J Int Law 104:179

Small KY (2018) Arbitration under international investment agreements: a guide to the key issues 2nd edn. Oxford University Press

SADC model BIT template with commentary, 2nd edn, p 3

Saipem Spa v Bangladesh. ICSID case no. ARB/05/07, decision on jurisdiction and provisional measures, para 67

Schill SW (2011) System-building in investment treaty arbitration and law making. Ger Law J 12:1083

Schill SW. Sources of international investment law: multilateralisation, arbitral precedent, comparativism, soft law, in The Oxford handbook of the sources of international law (Edited by Samantha Besson and Jean d'Aspremont) p 1122 (1096 intext)

Sempra v. Argentine Republic (2007) ICSID (W. Bank) case no. ARB/02/16, Award, paras 305–14

SGS v Pakistan. ICSID case no. ARB/01/13

SGS v. Philippines, ICSID (W. Bank) case no. ARB/02/6, Decision on objections to jurisdiction, paras 113–28

Small KY Arbitration under international investment agreements, a guide to the key issues 2nd edn. Oxford University Press

Thirlway H (2019) The sources of international law

Thirlway H (2019) The sources of international law, 2nd edn. Oxford University Press

United Kingdom v Iceland (1974) ICJ Rep 23–24 Fisheries jurisdiction (Merits) (Judgement)

Waste Management, Inc v Mexico, ICSID case no Arb (AF)/00/3 (NAFTA), Award (2004), para 98

Zimmermann CT, Koellers-Frahm (2006) The statute of the international court of justice: a commentary. Oxford University Press, pp 785–788

Theoretical Estimation of the Dispersion Curves in an Orthotropic Composite Plate

Abdellah Nouala, Faeez Masurkar, and Alaa A. Ameer

Abstract In this work, root-finding using the bisection algorithm is proposed to solve the transcendental dispersion relations and evaluate the dispersion curves of ultrasonic-guided waves propagating in an orthotropic composite plate. The implementation of the dispersion relations are accomplished via a coding scheme conducted in the MATLAB software. The obtained phase and group velocity dispersion curves from MATLAB using the in-house developed code are validated with a freely available Dispersion software and found to be in good agreement. Further, qualitative, and quantitative analysis is conducted considering different layers and fiber direction of the composite plate to verify the of accuracy of the obtained results.

Keywords Composite material · Dispersion curves · Root-finding method · Numerical method · Bisection algorithm

1 Introduction

Engineering structures such as those used in mechanical and electrical industries, civil infrastructure, nuclear reactors, aerospace industries are often susceptible to damages while in service due to random loads acting on them. It is therefore inevitable to keep a track and monitor their health on a time basis to ensure that can be used safely. Further, this monitoring followed by any maintenance also helps to improve

---

This paper was prepared as a part of the research work undertaken by Abdellah Nouala for his undergraduate project work supervised by Dr. Faeez Masurkar and co-supervised by Prof. Alaa A Ameer and for his bachelor's degree in the Electromechanical engineering programme at the British University in Dubai, Dubai, United Arab Emirates.

A. Nouala (✉) · F. Masurkar · A. A. Ameer
Faculty of Engineering and IT, The British University in Dubai, Dubai International Academic City, Dubai, United Arab Emirates
e-mail: 20001071@ug.buid.ac.ae

F. Masurkar
e-mail: faeez.masurkar@buid.ac.ae

K. Al Marri et al. (eds.), *BUiD Doctoral Research Conference 2024*, Lecture Notes in Civil Engineering 587, https://doi.org/10.1007/978-3-031-84371-6_38

the performance and reliability. Thus, it is clear that early-stage monitoring and diagnostics is crucial so that repairs can be accomplished with an affordable budget and less downtime (Rose 2011; Masurkar and Yelve 2015; Wu et al. 2021).

The early-stage monitoring could be achieved by non-destructive testing (NDT) methods based on ultrasonic waves. This approach can detect any underlying changes in the strength of the material without damaging the part or the structure. However, employing the ultrasonic waves in practice is challenging since they are highly multimodal and dispersive in nature. Thus, it is paramount to study the dispersion curves of all the engineering structures before applying the ultrasonic wave-based NDT method on them. The calculations of the dispersion curves can then help to estimate the excitation frequency and corresponding ultrasonic mode to be employed for testing purposes (Masurkar et al. 2014; James and Giurgiutiu 2020; Masurkar and Yelve 2017).

This research therefore delves deeper into the calculation of dispersion curves in orthotropic composite plates, with the aim of developing an efficient and precise method to characterize their dispersive behaviour and with and in-house developed code that can be customized and used for other orthotropic or isotropic materials readily. The proposed method uses bisection algorithm to efficiently calculate the transcendental dispersion relations, thereby facilitating the accurate evaluation of guided wave dispersion curves. The coding for obtaining the dispersion curves is accomplished using MATLAB software. Further, validation of the derived relationships will be performed against freely available dispersion software, ensuring the robustness and accuracy of proposed approach. Additionally, comprehensive qualitative and quantitative analyses is carried out, considering the different layer arrangements and number of plies in the composite plate. By meticulously examining the influence of layer configurations on the characteristics of dispersion curves, it is strived to clearly understand how material composition affects wave propagation.

Ultimately, this research aims to contribute to the advancement of ultrasonic NDT methods for composite structures. By equipping researchers and engineers with a powerful tool to analyse wave behaviour and identify the damage, we are paving the way to improve structural integrity monitoring and ultimately safeguard critical infrastructure in an increasingly technological landscape.

2 Theory of Lamb Waves

Lamb waves are multimodal elastic waves that propagate in a solid plate with traction-free boundaries on both of their surfaces and it was firstly discovered and presented by Horace Lamb in 1917. Lamb waves are guided waves that offer benefits for NDT because they can propagate large distances with less attenuation. These waves can be either symmetric (S) or antisymmetric (A) and their selection depends on the application. In the case of composite plates, the dispersion equations are given for the symmetric case (S) (Masurkar et al. 2014):

$$\frac{\sin(kA_1h)\cos(kA_2h)}{\sin(kA_2h)\cos(kA_1h)} = \frac{A_1\left[A_2^2 - \frac{C_{11}}{C_{13}}\left(\frac{C_P^2}{c_{1L}^2} - 1\right)\right]}{A_2\left[A_1^2 - \frac{C_{11}}{C_{13}}\left(\frac{C_P^2}{c_{1L}^2} - 1\right)\right]} \times \frac{\left[\frac{C_{13}}{C_{33}}\left(\frac{C_{13}}{C_{44}} + 1\right) + A_1^2 + \frac{C_{11}}{C_{44}}\left(\frac{C_P^2}{c_{1L}^2} - 1\right)\right]}{\left[\frac{C_{13}}{C_{33}}\left(\frac{C_{13}}{C_{44}} + 1\right) + A_2^2 + \frac{C_{11}}{C_{44}}\left(\frac{C_P^2}{c_{1L}^2} - 1\right)\right]} \quad (1)$$

and for antisymmetric case (A):

$$\frac{\sin(kA_2h)\cos(kA_1h)}{\sin(kA_1h)\cos(kA_2h)} = \frac{A_1\left[A_2^2 - \frac{C_{11}}{C_{13}}\left(\frac{C_P^2}{c_{1L}^2} - 1\right)\right]}{A_2\left[A_1^2 - \frac{C_{11}}{C_{13}}\left(\frac{C_P^2}{c_{1L}^2} - 1\right)\right]} \times \frac{\left[\frac{C_{13}}{C_{33}}\left(\frac{C_{13}}{C_{44}} + 1\right) + A_1^2 + \frac{C_{11}}{C_{44}}\left(\frac{C_P^2}{c_{1L}^2} - 1\right)\right]}{\left[\frac{C_{13}}{C_{33}}\left(\frac{C_{13}}{C_{44}} + 1\right) + A_2^2 + \frac{C_{11}}{C_{44}}\left(\frac{C_P^2}{c_{1L}^2} - 1\right)\right]} \quad (2)$$

where:

$$A_1 = \sqrt{\frac{-A + \sqrt{A^2 - 4AB}}{2}};\ A_2 = \sqrt{\frac{-A - \sqrt{A^2 - 4AB}}{2}};$$

$$A = \frac{C_P^2}{c_{3L}^2} + \frac{C_P^2}{c_{3T}^2} - \frac{(C_{11}C_{33} - C_{13}^2 - C_{13c_{44}})}{C_{33}C_{44}};$$

$$k = \frac{\omega}{C_\rho};\ c_{1L}^2 = \frac{C_{11}}{\rho};\ c_{3L}^2 = \frac{C_{33}}{\rho};\ c_{3T}^2 = \frac{C_{44}}{\rho};$$

$$C_{11} = \frac{E_1(1 - \nu_{32}\nu_{23})}{(1 + \nu_{12})(1 - \nu_{12}2\nu_{32}\nu_{23})};\ C_{12}\frac{E_2(\nu_{12} + \nu_{32}\nu_{23})}{(1 + \nu_{12})(1 - \nu_{12}2\nu_{32}\nu_{23})};$$

$$C_{33} = \frac{E_3(1 - \nu_{12})}{(1 - \nu_{12}2\nu_{32}\nu_{23})};\ C_{13} = \frac{\nu_{23}E_3}{(1 - \nu_{12}2\nu_{32}\nu_{23})};\ C_{55} = G_{13}$$

k is wavenumber, ω is angular frequency, C_p is phase velocity, C_{ij} are the elements of stiffness tensor, E_i are the young's modulus in different directions, and ν_{ij} are the Poisson ratios, ρ is the material density, G_{ij} is the shear modulus, A's and B's are the substitution factors, h is the plate thickness, and L and T stands for longitudinal and transverse directions.

In this research work, the group velocity (C_g) is calculated from the phase velocity (C_p) using the relation:

$$C_g = C_p\left[1 - \frac{\omega}{C_p}\frac{\partial C_p}{\partial \omega}\right]^{-1} \quad (3)$$

Table 1 Material properties of the orthotropic composite

Elements (C_{ij}) of stiffness matrix for single lamina '0' degree fiber (GPa)									
Density (ρ)	C_{11}	C_{12}	C_{13}	C_{22}	C_{23}	C_{33}	C_{44}	C_{55}	C_{66}
1590	165.24	6.63	6.63	14.34	6.40	14.34	3.98	5.17	5.17

Equations (1) to (3) are implemented in MATLAB and solved using the bisection algorithm described in following section to get the dispersion curves. The mechanical properties of the specimen are presented in following section (Table 1).

3 Mechanical Properties of the Specimen

In this study, two types of stacking sequence of the composite material are chosen. The first one is $[0, +45, -45, 90]_s$. (Composite fibres are oriented at 0°, ±45°, 90° and 's' specifies the symmetry of the individual layers/laminas). Each layer will have a thickness of 0.25 mm. The single lamina material here is IM7/8552. The lamina thereby has 16 plies making the overall laminate thickness 2 mm. The second one for a qualitative and quantitative comparison of results is chosen with the following stacking sequence $[0, +30, -30, +60, -60, +90, -90]_s$.

4 Bisection Algorithm

The Bisection Method is a numerical root-finding technique used to solve equations that lack analytical solutions and to find the roots of polynomial equations. It works by dividing an interval into smaller intervals and narrowing down the range of possible solutions iteratively. By evaluating the function at the endpoints of each interval and determining the new interval based on the signs of these function values, the Bisection Method guarantees convergence to at least one root within the initial interval. It is a simple and reliable approach that can handle various types of equations, making it a valuable tool in computational mathematics. The steps for the employed bisection algorithm are shown below in four steps:

Step 1: Find points a and b such that $a<b$ and $f(a)\cdot f(b)<0$.

Step 2: Take the interval $[a,b]$ and find next value $x_0 = \frac{a+b}{2}$

Step 3: If $f(x_0)=0$ then x_0 is an exact root,

else if $f(a)\cdot f(x_0)<0$ then $b=x_0$,

else if $f(x0)\cdot f(x_0)<0$ then $a=x_0$.

Step 4: Repeat steps 2 & 3 until $f(x_i)=0$ or $|f(x_i)|\leq$Accuracy

5 Estimate the Effective Stiffness Matrix

The composite material supposed to be used for getting the dispersion curves has several plies according to the stacking sequence mentioned earlier. Thus, for testing purpose, homogenization of the composite material is an important step. This can be done by calculating the effective stiffness matrix. Once the effective stiffness matrix is obtained, it is used in Eqs. (1) and (2) to find the dispersion curves from MATLAB using root-finding methods. To estimate the effective stiffness matrix, we used Dispersion software and by adding the material properties of each lamina's layer-by-layer and it is shown in Figs. 1 and 2 along with the stacking sequence.

6 Results and Discussions

The obtained results from the developed in-house MATLAB code and from the dispersion software are shown in Figs. 2 and 3. The phase and group velocity dispersion curves for the stacking sequence $[0, +45, -45, 90]_s$ are presented side-by-side to have a comparison of the results from a qualitative and quantitative perspective. It can be seen that there is a very good agreement between the results. In order to further validate, the another stacking sequence was considered as $[0, +30, -30, +60, -60, +90, -90]_s$ and the results are compared and shown in Figs. 4 and 5. It can be seen that for both the stacking sequences, the results are in good agreement. Furthermore, we did a quantitative and qualitative comparison of the results and presented in a graphical and tabular format and is discussed in following subsection.

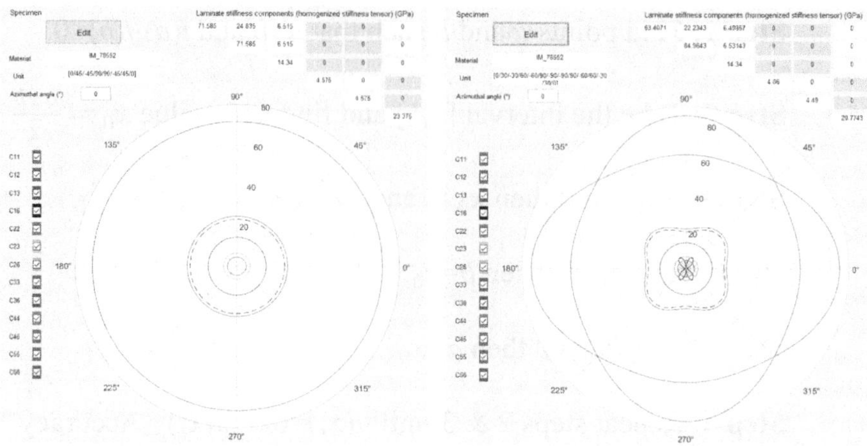

(a) (b)

Fig. 1 Homogenized stiffness tensor. **a** $[0, +45, -45, 90]_s$ **b** $[0, +30, -30, +60, -60, +90, -90]_s$

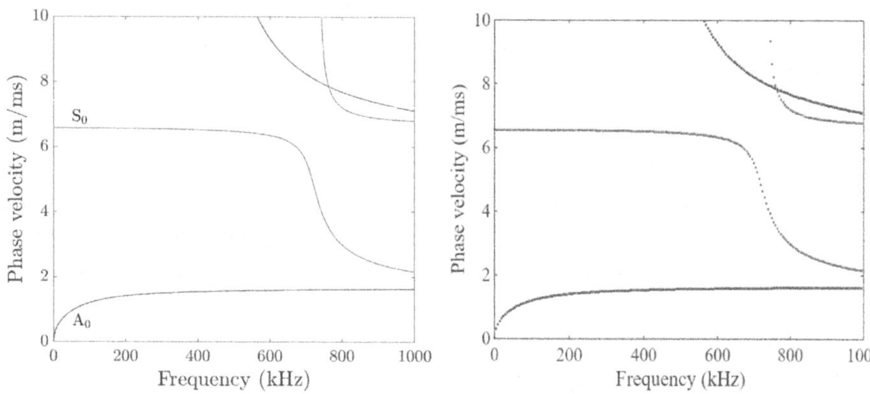

(a) (b)

Fig. 2 Phase velocity dispersion curve. $[0, +45, -45, 90]_s$ **a** Dispersion software **b** MATLAB

6.1 Quantitative and Qualitative Analysis of the Results

In this subsection, the phase and group velocities are compared from the results of developed in-house MATLAB code with that of the dispersion software. Since, overlapping of the results could potentially be confusing, it is wise to compare the results qualitatively and quantitatively at a range of different frequencies as shown in Figs. 6 and 7 and also Table 2 of Appendix. The figures show that for both the phase and group velocity dispersion curves and at different frequencies, there is a good

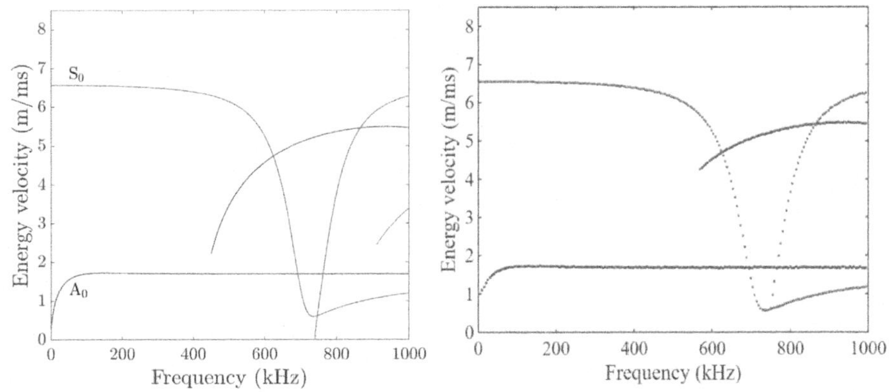

(a) (b)

Fig. 3 Group velocity dispersion curve. [0, +45, −45,90]$_s$ **a** Dispersion software **b** MATLAB

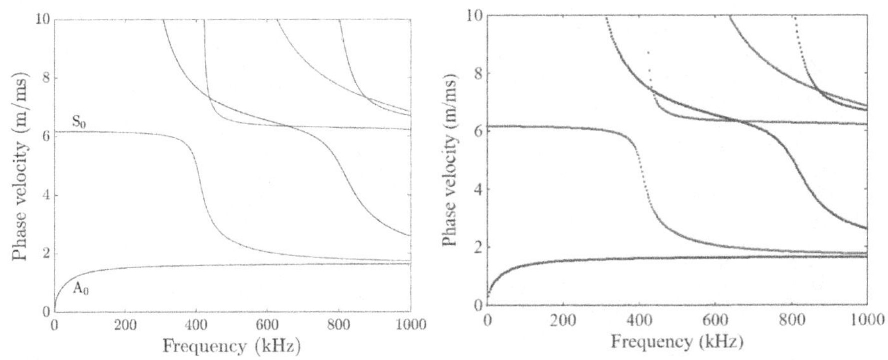

(a) (b)

Fig. 4 Phase velocity dispersion curve. [0, +30, −30, +60, −60, +90, −90]$_s$ **a** Dispersion software **b** MATLAB

agreement between the results. The agreement confirms that there is a quantitative and qualitative harmony between these results which confirms the accuracy of the developed in house code in MATLAB software. The concluding remarks from the presnet study are presented in following section.

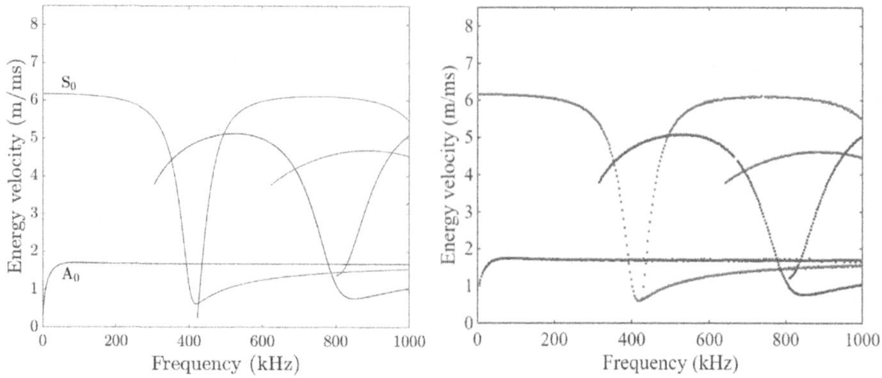

(a) (b)

Fig. 5 Group velocity dispersion curve. $[0, +30, -30, +60, -60, +90, -90]_s$ **a** Dispersion software **b** MATLAB

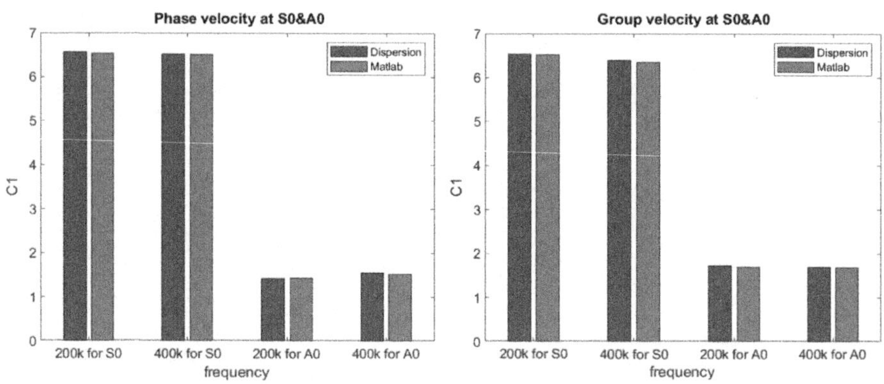

(a) (b)

Fig. 6 Qualitative comparison of results between MATLAB and dispersion software for stacking sequence of $[0, +45, -45, 90]_s \rightarrow$ **C1**

7 Conclusions

This work presents a comparison between results of phase and group velocity dispersion curves obtained from the in-house developed MATLB code and that from the

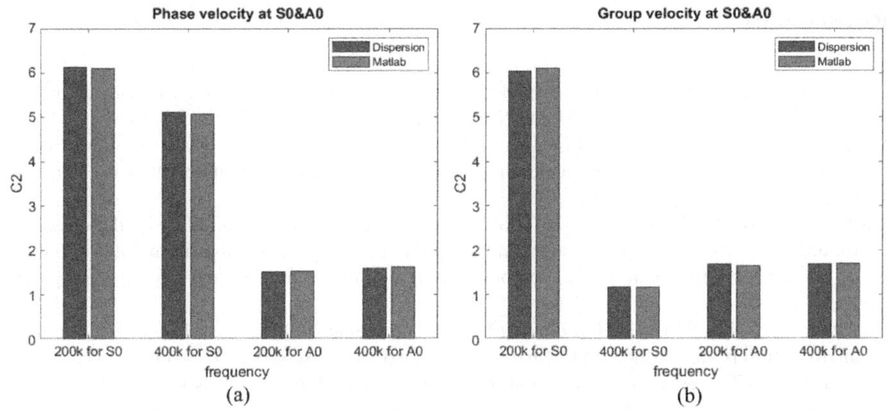

Fig. 7 Qualitative Comparison of results between MATLAB and dispersion software for stacking sequence of $[0, +30, -30, +60, -60, +90, -90]_s \rightarrow$ **C2**

dispersion curves software for an orthotropic composite plate. The comparison is done for two different stacking sequences that are totally different from each other in terms of the fiber direction, stacking sequences, and the plate thickness. The results show that there is a good agreement between the results of in-house developed code and that of the dispersion software.

Appendix

See Table 2.

Table 2 Quantitative comparison of results between MATLAB and dispersion software for stacking sequence of $[0, +45, -45, 90]_s \rightarrow$ **C1**

Mode	Frequency (kHz)	Phase velocity		Group velocity	
		MATLAB	Dispersion	MATLAB	Dispersion
S_0	200	6.531	6.65	6.52	6.5415
A_0	200	1.43	1.4231	1.7	1.7149
S_0	400	6.51	6.52286	6.35	6.39281
A_0	400	1.52	1.5511	1.68	1.69727

References

James R, Giurgiutiu V (2020) Towards the generation of controlled one-inch impact damage in thick CFRP composites for SHM and NDE validation. Compos B Eng 203:108463

Masurkar FA, Yelve NP (2015) Locating a damage in an aluminium plate using lamb waves. In: Proceedings of the Indian national seminar & exhibition on non-destructive evaluation, Hyderabad, India, pp 26–28

Masurkar FA, Yelve NP, Mitra M (2014) Lamb wave based experimental and FE simulation studies for damage detection in metallic and composite plates using geodesic algorithm. In: Proceedings of 7th ISSS international conference on smart materials, structures, and systems, Bangalore, India

Masurkar FA, Yelve NP (2017) Optimizing location of damage within an enclosed area defined by an algorithm based on the Lamb wave response data. Appl Acoust 120:98–110

Rose JL (2011) The upcoming revolution in ultrasonic guided waves. In: Nondestructive characterization for composite materials, aerospace engineering, civil infrastructure, and homeland security 2011. SPIE, vol 7983, pp 21–50

Wu Y, Shen X, Li D (2021) Numerical and experimental research on damage shape recognition of aluminum alloy plates based on lamb wave. J Intell Mater Syst Struct 32(18–19):2273–2288

Towards Inclusive Education: A Critical Analysis of Inclusive Education Practices in Pakistan

Kamal Nosheen and Eman Gaad

Abstract Many countries have adopted inclusive education policies addressing diversity, but implementation challenges remain, especially in Pakistan. Here, inclusive education reflects the country's willingness to allow every student to study at quality institutions while recognizing cultural and intellectual differences. Major obstacles include public attitudes, infrastructure shortages, and unprepared teachers. This paper reviews sociocultural factors within policy frameworks and suggests strategies like awareness programs, professional training, and policy reform. Prioritizing inclusivity, equity, and social justice can help Pakistan create a fair system that fosters a tolerant community, allowing every individual to reach their full potential.

Keywords Inclusion · Educational equity · Social and cultural context · Pakistan

1 Introduction

The context and the situation of inclusive education in Pakistan are culturally complicated, standardized, and governed by policies and administration. The principle of inclusive education is about giving equal learning opportunities for every student from diverse backgrounds or circumstances to achieve quality education. However, the situation in the country does not correspond with the fact that the main obstacle to achieving the goals of inclusive education is what is perceived. In 2005, the Islamabad was established. According to UNESCO (2021), Pakistan reportedly adopted a "Declaration on Inclusive Education" to provide not only girls but also others, such as disabled ones, with equal dignity, respect, education, healthcare, employment, and other aspects of life, as well as fair opportunities to unleash their academic skills,

This Paper Was Prepared for Inclusive and Special Education Module, Taught By Prof. Eman Gaad

K. Nosheen (✉) · E. Gaad
Faculty of Education, The British University in Dubai, Dubai, United Arab Emirates
e-mail: 23000039@student.buid.ac.ae

K. Al Marri et al. (eds.), *BUiD Doctoral Research Conference 2024*, Lecture Notes in Civil Engineering 587, https://doi.org/10.1007/978-3-031-84371-6_39

physical fitness, emotional growth, and social recognition, and finally the capability that allows the development of skills, talents and future prospects.

1.1 Rationale/Aim/Gaps

1.1.1 Rationale

In Pakistan, a society that promotes inclusive education is considered an essential element of educational reform and social justice (Malik et al. 2014). This evaluative report will explore the nature of inclusivity in Pakistan's inclusive education to offer informed recommendations on the practical implementation of such policies. Although the legislation has been passed, text doubles still face substantial problems that require a thorough consideration of the current situation. The rationale behind this enterprise revolves around the ethical obligation to help learners of all backgrounds and capabilities to access quality education. Therefore, the main objective is to develop an environment where everyone is treated with respect and can fully utilize their abilities to maximize their contribution to society.

1.1.2 Aim

This report aims to conduct an extensive study on the state of inclusive education in Pakistan and spot the main barriers and windows of opportunity in its implementation. Based on the current status of existing policies, practices, and societal mindsets, there is a need to understand the factors hindering the effective implementation of inclusive policies in the education sector.

1.1.3 Gap

In Pakistan, although inclusive education has been the focus for a while, some gaps still hinder the attainment of equal educational opportunities for all learners (Khan et al. 2017). The first disparity among these subordinate aspects is the inefficient utilization of resources and infrastructure to provide inclusive practices. However, besides the findings by Khan et al. (2017), it cannot be assured that all schools possess sufficient facilities and imminent features to implement inclusive policies that cover diverse students, including those with special education conditions. Furthermore, as Chow et al. (2023) suggest, there is a challenge among most teachers to adopt the inclusive approach of educating diverse students because they lack the skills, knowledge, and information about involving all students in the education process. Moreover, the deep-rooted societal prejudice and stigma of disability or special needs add to the discriminatory outlook and behavior, and this only widens the gap between individuals and their families.

2 Literature Review

A literature review remains the critical aspect of the report about the inclusion condition in Pakistan. This section offers a comprehensive analysis of inclusivity in Pakistan's education sector based on existing literature.

Implementing an education system that recognizes diversity requires policy-makers, educators, and other stakeholders to consider several factors. Malik et al. (2014) asserted that inclusive practices incorporate diversity, empowerment, and collaboration. In an inclusive system, learning is for all, ensuring equal educational rights. Every learner should feel valued and have their needs met. However, Ehsan (2018) stresses that implementing inclusive education requires collective efforts from all stakeholders and supportive policies that ensure diversity, equality, and good governance within educational institutions.

Another vital aspect that needs to be emphasized in the support for students and families with disabilities is the need for constant and reliable support to those students and families. Grimes et al. (2021) suggest supplying aids, therapy services, and counseling can be life-changing in fostering the success of inclusive education. Therefore, inclusive education programs must also be evaluated along with other factors mentioned above. Grimes et al. (2021) suggested that self-evaluation tools can be used to examine the outcomes and overall impact of inclusive education visits.

To comply with UNESCO's idea of inclusive education, defined as a process of identifying and addressing the diversity of needs of all learners through increasing their participation in education, culture, and community and reducing exclusion from and within education, Pakistan is trying to achieve an inclusive education as indicated by UNESCO (2021). In the 2005 Islamabad Declaration on Inclusive Education, all of these provincial and federal governments signed a working definition of equitable education that could ensure every child's opportunity, dignity, and respect, regardless of their gender, disabilities, abilities, socioeconomic, cultural, and ethnic backgrounds (UNESCO 2021). At this point, UNESCO's inclusive education criteria were embraced in a new dimension.

The government of Pakistan has taken various legislative and policy steps to empower inclusive education. The National Policy for Persons with Disabilities, enacted in 2002, has the objective of ensuring that mainstream education is inclusive of children who are disabled alongside the 2017 National Education Policy that supports the UNESCO concept of inclusive education (UNESCO 2021). Also, Grimes et al. (2021) highlighted that all provinces have passed laws on disability motivating equal education and demanding non-discrimination policy guidelines to achieve this. Nevertheless, even though these efforts have been made, the realization of inclusive education is still a problem because the education sector plans of some provinces still focus on special education rather than inclusive education.

Behlol (2011) expresses that Pakistan confronts some obstacles in realizing inclusion in education, including the social attitude toward disability, inadequate teacher training, and unsteady policies and legislation. According to the authors, Pakistan can gain from adopting a more coordinated strategy for enhancing inclusive education policies through the joint participation of stakeholders, teachers, and parents. Similarly, it is equally crucial to understand that establishing inclusive education cannot be simple and demands continuous effort in the long run (Chow et al. 2023).

Although a lot of effort is being made to revive inclusive education, it is still complicated and challenging to introduce inclusive education techniques. As Trani et al. (2019) put it, one of the main difficulties is developing transparent policies, guidelines, and regulations that make inclusive education as full-fledged as possible. Policymakers have to come up with and implement policies that protect disabled students from discrimination. As per Singal (2016), other facets, such as social perceptions about disabilities and inadequate knowledge about the inclusive education practices of teachers and families, must also be tackled.

The establishment of inclusive education policies also necessitates significant support and building of capability for teachers and school heads. Khan et al. (2017) raised a point that teachers need to be given the appropriate training to help them develop pedagogies that meet the different needs of students, both with and without disabilities. Behlol (2011) argued that school heads are vital for creating an inclusive environment whose policies promote the respect of diversity and equity within the education system.

Nevertheless, Ahmad et al. (2014) observed that bringing about inclusive education in institutions is a complex process, as the adoption of inclusive education needs a paradigm shift in educational institutions' pedagogical approaches, structures, and practices. Teachers and school administrators need to be supported; initiatives to strengthen and enhance the teaching workforce are being implemented in countries such as Pakistan (Thakur and Abbas, 2017).

Mendoza and Heymann (2022) underscored the need to adopt an inclusive education theory based on research-based pedagogical approaches that can respond to disabled students. The literature suggests that inclusive education can be enhanced through assistive technologies, actively involving learners with disabilities. Policymakers should create systems tailored to these contexts. In India, inclusive education focuses more on integrating children with special needs than ensuring quality education for all learners (Singal 2016).

Researchers argue for a critical and contextual look at inclusive education that considers the multiple contextual factors that affect the implementation of inclusive education (Ireland 2007). Researchers need to look at issues that are context-specific, like the attitudes of society towards disability, cultural beliefs, politics, and the laws. These issues can either enhance or inhibit the practices of inclusive education. Also, researchers need to look into the challenges and opportunities the pandemic posed to the education system (Ahmad et al. 2014). HEC (2021) recommends that policies and practices be adjusted to meet particular conditions and ensure collaboration among all stakeholders to promote inclusivity and equality.

3 Relevant Studies

3.1 Exploration Framework

To illuminate the landscape of inclusive education in Pakistan, a research framework emphasizing depth and context was employed. This choice was informed by the complexity of social issues like inclusive education, where a nuanced understanding of its dynamics is essential (Busetto et al. 2020). This study delves into the operational dynamics of inclusive education in Pakistan, focusing on key issues facing the nation.

3.2 Data Collection

The study used secondary data from an extensive literature review and document analysis. The study process included an in-depth analysis of the literature feed, which involved using various resources, including previous studies, articles, statistics, government data, UNESCO, and other secondary relevant resources. 'The National Policy for Persons with Disabilities' and 'The National Education Policy' and relevant provincial regulations were significant sources for reviewing the inclusive education policy documents as far as they are related to the comprehension and current uptake of the policy.

3.3 Analytical Exploration

Data analysis was conducted employing a theme-based approach. Policy documents and literature reviews were organized according to the theme and categorizations, resulting in prominent themes and categories. This induction process involved finding relationships, similarities, or differences in the data (Naeem et al. 2023), which was used to conclude and result in a comprehensive picture of inclusive education in Pakistan.

3.4 Limitations

This research intended to present a generalist picture; however, some limitations should be mentioned. The study relied heavily on secondary data, which limits the researcher from manipulating it because it forces the researcher to rely on predetermined themes and conclusions, limiting the scope of the data. Along with time and resource limitations, the range of data collection was limited to a particular time frame and a specific geographic area of Pakistan only. Moreover, the inclusiveness, ongoing modification, and development of inclusive education policy and practice might make some findings obsolete.

4 Findings

The research reveals the current state of inclusive education in Pakistan, identifying key challenges and potential solutions. The findings highlight that societal attitudes toward disability and diversity remain significant barriers, with deep-rooted stigmas and stereotypes persisting (Malik et al. 2022). The need for awareness campaigns and community involvement is emphasized to challenge these biases (Chow et al. 2023). Inadequate resources and infrastructure, such as mobility aids and specialized learning tools, also hinder progress. Despite commitments, insufficient funding continues to be a major obstacle (Thakur and Abbas 2017).

Teacher readiness is another critical issue, with many educators lacking the necessary training and skills to support inclusive classrooms (Chow et al. 2023). Professional development and ongoing training are essential to equip teachers with effective strategies. The study also points out gaps in policy implementation, with inconsistent monitoring and coordination among stakeholders (Trani et al. 2019).

Regional inequalities exacerbate these challenges, particularly in rural areas with limited facilities and awareness of inclusive education (HEC 2021). Despite these obstacles, promising practices like the Inclusive School Network of Baluchistan and the Inclusive Education Resource Centers of Sindh demonstrate that progress is possible when government, NGOs, and communities collaborate effectively (Grimes et al. 2021). Strengthening these partnerships and supporting targeted interventions can help advance inclusive education across Pakistan.

5 Discussion

The discussion explores critical themes and problems as indicated by the literature in the context of inclusive schooling, stressing the multifaceted nature of the issue and the need for systemic reform.

5.1 Societal Attitudes and Cultural Perceptions

Sociocultural constructions of disability and special educational needs in Pakistan reinforce the pattern of divisive behavior. Those with stigma and misconceptions typically go to the side of individuals and families who are perceived as different and thus not welcome into mainstream institutions (HEC 2021). Combating the deeply rooted bias requires a combination of comprehensive awareness raising, community engagement, and intervention at the level of target groups and eliminating stereotypes that lead to the formation of a culture of inclusivity and acceptance (Malik et al. 2022).

5.2 *Infrastructure and Resource Allocation*

Pakistan faces infrastructural and resource-allocation barriers for inclusive education. Most schools lack suitable facilities and accommodations to support special needs students (Thakur and Abbas 2017). This resource constraint compromises the quality of education and stimulates the cycles of exclusion and inequality. Targeted investments can close infrastructural deficits, especially in physical infrastructure, assistive devices, and accessible resources.

5.3 *Continuous Development of Teachers and Teacher Preparedness*

The fact that educators are prepared is crucial for effectively expanding inclusive education in Pakistan. However, as opined by Chow et al. (2023), most teachers cannot render specialized support to pupils with diverse learning needs. This problem is not addressed because there are insufficient technical training courses and professional development opportunities for instructors to improve their skills. Teacher preparedness development assumes complex procedures that include course curriculum, teaching methods, and access to the best inclusive education practices (Chow et al. 2023).

5.4 *Policy Frameworks and Implementation Methods*

While inclusive education policies have been introduced in Pakistan, there is a need for bottleneck assessments to help make these a reality. Due to fragmentation, the adoption of different approaches to policy-systems management and monitoring is demonstrated by inadequate implementations and accountability among the various regions/provinces (Ehsan 2018). A multi-stakeholder approach supports a policy framework and practices; it's a synergy of government agencies, non-governmental organizations, and educational organizations. Quality guidelines, appropriate monitoring systems, mobilization of resources, and continuous implementation underline the process of effective policy formulation and implementation of inclusive education (Behlol 2011).

5.5 Considerations of Regional Imbalance and Equity Issues

The regionally-tiered access to inclusive educational institutions in Pakistan intensifies the difficulties in promoting the balance of educational opportunities (Malik et al. 2022). Although some provinces have made good progress, others are slow, as there are differences in political will, administrative capacity, and resource allocation. Tackling regional differences is only achievable with special measures sensitive to equity and distributing financial resources based on the level of demand.

6 Recommendations and Conclusion

6.1 Recommendations

The education landscape of Pakistan is a complex aspect associated with multiple factors that need to be addressed to enhance inclusive education systems. In the wake of high demand for education, various elements are crucial for improving inclusivity in the entire system. The following recommendations are essential for this move:

6.1.1 Align with UNESCO's Guidelines

To further align with UNESCO's guidelines and achieve inclusive education, Pakistan must address the following recommendations: safeguard against the denial of education rights irrespective of background, establish the non-discriminatory principle in education, remove the discriminatory hate materials against religious minorities in textbooks and classroom, make the necessary infrastructure and learning materials accessible and up to date, as well as improve training of teachers on inclusive education practices (UNESCO 2021). Though the government, non-governmental organizations, and international community partners are present, their collaboration is vital for this engagement. Using the realization of these projects, Pakistan can attain the point where education is an inclusive concept. As a result, learners from different backgrounds and abilities can access learning and skills enhancement.

6.1.2 Development of Teacher Training Programs

It will be up to Pakistan to realize the concept of inclusive education by running high-quality teacher training sessions. This program should equip the teachers with the required skills, know-how, and techniques to identify and help the learners who do not learn in the traditional style. Attention should be given to training practices that are practical and applicable to the teachers to enhance inclusivity best practices.

6.1.3 Development of Inclusive Curriculum and Materials

Curriculum frameworks and instructional materials development must be inclusive and meet the needs of learners with diverse styles. Curriculum developers must, therefore, work jointly with inclusive education experts to guarantee that students acquire instructional materials that are accessible, culturally relevant, and tend to cater to all student needs (Malik et al. 2022). Moreover, there should be an attempt to stimulate the representation of different points of view and experiences throughout the curriculum content.

6.1.4 Promotion of Community Engagement and Collaboration

The main focus of Pakistan should be to enhance community engagement and collaboration while designing and implementing inclusive education programs. This includes building partnerships between schools, families, communities, and relevant partners to create supportive contexts that facilitate inclusive practices (Mendoza and Heymann 2022). Community-based programs, like parent support groups and community outreach programs, can be instrumental in mobilizing support, countering stigma, and promoting the rights of those with different abilities.

6.2 Conclusions

Both difficulties and possibilities characterize the path to inclusive education in Pakistan. This can be concluded by analyzing the social viewpoints, infrastructure limitations, staff development requirements, policy frameworks, and regional gaps that define the current state of inclusive education. As a result, further action is needed to enhance inclusivity in education. Nevertheless, these obstacles can be overcome by following the suggestions in this study and by establishing a collaborative, evidence-informed system that considers diversity, promotes equity, and empowers each learner to achieve their fullest potential.

References

Ahmad I, Rehman K, Ali A, Khan I, Kha FA (2014) Critical analysis of the problems of education in pakistan: possible solutions. Int J Eval Res Educ (IJERE) 3(2):79–84. ISSN: 2252–8822

Behlol MG (2011) Inclusive education: preparation of teachers, challenges in the classroom and prospects. Pak J Educ 10108(2)

Busetto L, Wick W, Gumbinger C (2020) How to use and assess qualitative research methods. Neurol Res Pract 2(1):14. https://doi.org/10.1186/s42466-020-00059-z

Chow WSE, de Bruin K, Sharma U (2023) A scoping review of perceived support needs of teachers for implementing inclusive education. Int J Incl Educ 28(3):3321–3340

Ehsan M (2018) Inclusive education in primary and secondary schools of Pakistan: role of teachers. Am Sci Res J Eng Technol Sci (ASRJETS) 40(1):40–61
Grimes P et al (2021) Disability-inclusive education practices in Pakistan. United nations children's fund regional office for South Asia, Kathmandu
Ireland (Ed.) (2007) Inclusion of students with special educational needs: post-primary guidelines. Stationery Office
Kamran M, Siddiqui S, Adil MS (2023) Breaking barriers: the influence of teachers' attitudes on inclusive education for students with mild learning disabilities (Mlds). Educ Sci 13(6):606
Khan IK, Hashmi SH, Khanum N (2017) Inclusive education in government primary schools: teacher perceptions. J Educ Educ Dev 14(1)
Malik R, Raza F, Rose R, Singal N (2022) Are children with disabilities in school and learning? Evidence from a household survey in rural Punjab, Pakistan, compare. A J Comp Int Educ 52(2):211–231
Malik AA, Shah D, Saqlain AM, Hanif M, Hussain S, Bano S (2014) Inclusive education and relevance. AEPAM, Islamabad
Mendoza M, Heymann J (2022) Implementation of inclusive education: a systematic review of studies of inclusive education interventions in low- and lower-middle-income countries. Int J Disabil Dev Educ 1–18
Naeem M, Ozuem W, Howell K, Ranfagni S (2023) A step-by-step process of thematic analysis to develop a conceptual model in qualitative research. Int J Qual Methods 22:16094069231205788. https://doi.org/10.1177/16094069231205789
Higher Education Commission (HEC) (2021) Policy for students with disabilities at higher education institutions in Pakistan (Amended)
Singal N (2016) Education of children with disabilities in India and Pakistan: critical analysis of developments in the last 15 years. Prospects 46(1):171–183
Thakur I, Abbas F (2017) Inclusive education in Punjab: challenges and ways forward. J Incl Educ 1:15–26
Tomaszewski LE, Zarestky J, Gonzalez E (2020) Planning qualitative research: design and decision making for new researchers. Int J Qual Methods 19:160940692096717. https://doi.org/10.1177/1609406920967174
Trani J-F, Bakhshi P, Mozaffari A, Sohail M, Rawab H, Kaplan I, Ballard E, Hovmand P (2019) Strengthening child inclusion in the classroom in rural schools of Pakistan and Afghanistan: what did we learn by testing the system dynamics protocol for community engagement? Res Comp Int Educ 14(1):158–181
UNESCO (2021) Pakistan: inclusion. https://education-profiles.org/central-and-southern-asia/pakistan/~inclusion

Using XAI Techniques to Detect Targeted Data Poisoning Attacks on Healthcare Applications of Machine Learning Systems

Eyad Dhaher Megdadi and Usman Javed Butt

Abstract This research study explores the application of Explainable Artificial Intelligence (XAI) methods for detecting targeted data poisoning attacks in healthcare machine learning systems. As machine learning becomes increasingly integrated into critical fields like healthcare, the integrity and security of training data have become paramount concerns. Data poisoning attacks, which manipulate training datasets to influence model behaviour, pose a significant threat to the reliability and effectiveness of these systems. Our study presents a novel approach that leverages XAI techniques, particularly focusing on global explanations of selected features, to identify signs of data manipulation. We propose a method of monitoring the impact level of carefully chosen features as an indicator of potential poisoning, using predetermined thresholds to trigger warnings when unusual patterns are detected. The research methodology involves applying global explanation method, to measure and monitor features impact in healthcare datasets, then explore the effectiveness of this approach using a case study on hypothyroid diagnosis, where data poisoning could lead to delayed treatment with potentially life-threatening consequences. Research findings suggest that XAI techniques can provide valuable insights into the behaviour of machine learning models, enabling more effective detection of subtle, targeted poisoning attacks. However, we also acknowledge limitations, including the need for some prior knowledge of potential attack goals and the risk of false positives or negatives.

Keywords Artificial intelligence · Data poisoning · Explanation artificial intelligence · Intrusion detection · Machine learning

---

Author Note This paper was prepared for INF613 Computer Network and Data Security, taught by Usman Javed Butt.

E. D. Megdadi (✉)
Student of Engineering and IT, The British University in Dubai, Dubai, United Arab Emirates
e-mail: 23000035@student.buid.ac.ae

U. J. Butt
Faculty of Engineering and IT, The British University in Dubai, Dubai, United Arab Emirates
e-mail: usman.butt@buid.ac.ae

K. Al Marri et al. (eds.), *BUiD Doctoral Research Conference 2024*, Lecture Notes in Civil Engineering 587, https://doi.org/10.1007/978-3-031-84371-6_40

As there is a lack of research in the field of using XAI for cyber security and almost no research yet for using XAI in data poisoning detection, this research aims to explore this potential application and a research gap in using XAI in data poisoning detection. As a limitation this research effective against targeted attacks, it might not work for all poisoning types and requires some prior knowledge about the attacker's goals. Additionally, there's a risk of false positives or negatives, which is critical in sensitive fields like healthcare. Also, real-world deployment faces challenges like data privacy concerns and integration complexity.

1 Literature Review

1.1 A. Cyber Security Attacks

Cyber security attacks as malicious attempts to gain unauthorized access, disrupt, or steal data can cause financial losses and serious damage to operations, most common cyber security attacks can be categorized as per working principle and target type as Malware (malicious software in general), Viruses which is Self-replicating so it can be considered the most dangerous as it can replicate themselves and spread from one computer or device to another, affect machine processes and operations. This self-replication ability is what makes them particularly dangerous, as they can quickly spread and cause widespread damage.

While Worms can spread without requiring user action, which enables them to infect entire networks and cause huge disruption by looking for other computers on the network and exploiting network protocols to enable the attacker to steal sensitive data and disrupt normal operations. One of the most famous worms is the WannaCry worm which was discovered in 2017 (Akbanov et al. 2019) was utilizing vulnerable gaps in Windows systems to spread and encrypt users' and entity's data, demanding payment to enable the user to recover and get back their data (Akbanov et al. 2019).

1.2 B. Data Poisoning

Data poisoning which this research explores is considered to have the most negative impact on AI systems as it intends to manipulate the datasets leading to manipulation of the behaviour of AI models and machine learning systems. Attackers will have in some cases a target of their attack by making a change in the AI model output to achieve another desired output. So, attacks can be classified as Targeted attacks in which the attacker manipulates the model for a pacific purpose, and Nontargeted attacks which target the model's overall performance, causing errors in functionality (Dunn et al. 2020).

An attacker can achieve his target through many techniques such as label modification or flipping and injecting manipulated data (Akbanov et al. 2019) in this research a focus will be on adding malicious data by adding new data points to influence the AI system's behaviour and modifying existing data by altering original dataset points.

Figure 1 clarifies that by label flipping the labels are modified but other feature data points in the training set remain the same, while injecting new data points involves producing entirely new data points into the training set or the same effect can be obtained but manipulating the existing data points features for a specific predetermined repurpose.

The effect of data poisoning is very dangerous as it is difficult to detect and can lead to biased decisions and incorrect output causing a security risk for, critical rule applications. Dat poisoning effect on machine learning modules can be in many forms such as reduced accuracy and performance as poisoned data can lead the model

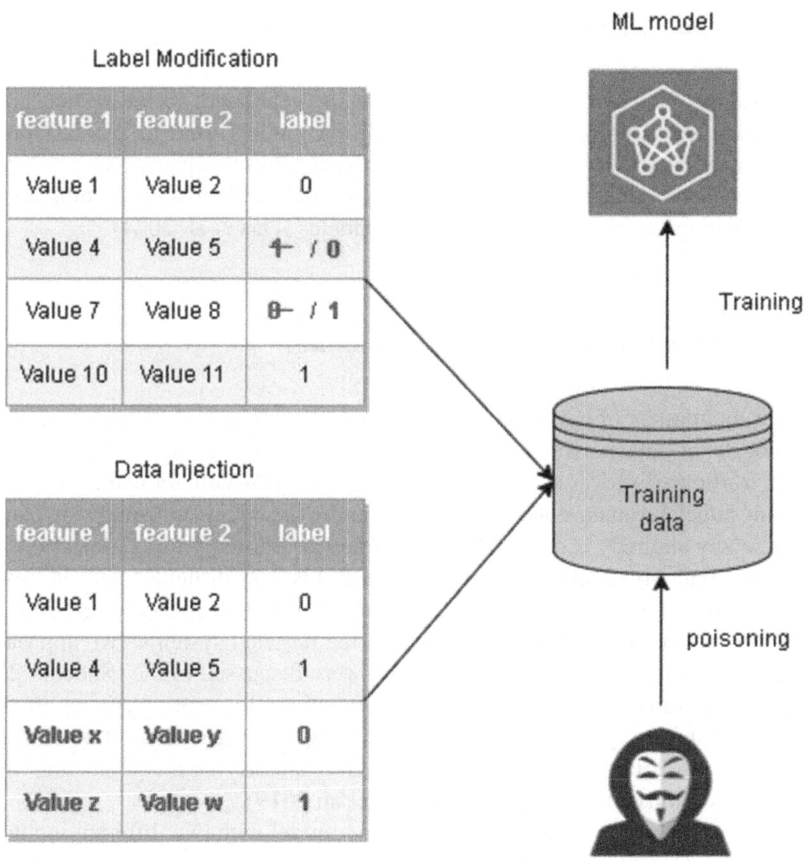

Fig. 1 Data poisoning and its effects on machine learning models

to produce incorrect patterns and relationships. Data poisoning can lead to biased outputs by injecting data to obtain specific outcomes. Also, Data poisoning can create model vulnerability and create "backdoors" that allow attackers to manipulate the model's behaviour. In some critical areas that rely nowadays on machine learning and AI systems like healthcare, finance, or autonomous vehicles, affected Moldes by data poisoning can lead to economic losses, safety issues, and even threaten human lives.

The emergence of Distributed machine learning (DML) systems based on massive databases involves machine learning models across multiple computers by using more than one machine working together by dividing the training data among these machines to solve complex models using different techniques (Chen et al. 2020). This causes more challenges in managing the security of the database as many machines share and contribute to the same massive database.

1.3 C. Impact of Data Poisoning on Health Care Systems

Data poisoning can have significant impacts on the healthcare sector, compromising the accuracy and reliability of machine learning models used in critical applications. Here are some research articles that discuss these impacts. As it is impact the performance of different machine learning models under data poisoning attacks, specifically in predicting the probability of diabetes (Das et al. 2024).

1.4 D. Defence and Detecting Techniques

Different techniques of cyber security intrusion detection systems can be used for protecting from data poisoning, this can be achieved by continuously monitoring network traffic to discover any suspicious behaviour, this will allow the early detection of any attack by analyse the attacker's signature like identifying malware patterns or analyse any anomaly activity which is defined as Anomaly-based detection.

Once the data poisoning attack is successful, it is very difficult to detect in some cases, as some attackers can insert some data points with the same original data retheme in case he has enough knowledge of the targeted systems, two approaches to detect the data poisoning in deep learning were discussed in the (Subedar 2019) which are: a) using the probability density functions by using statistics methods to measure dataset input uncertainty, b) using Bayesian deep neural networks By analysing the uncertainty estimates on test data by classifying the points into higher or lower uncertainty than expected (Subedar et al. 2019).

The Fig. 2 shows the same machine learning model with two different inputs one has original data and another poisoned data, with deferent out for each input.

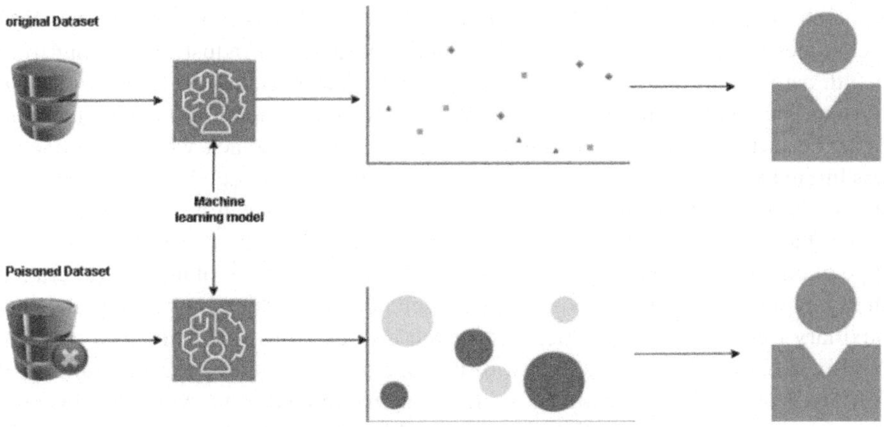

Fig. 2 Explaining the system fooling model by data poisoning

1.5 E. Explore Explainable Artificial Intelligence (XAI) Methods

XAI methods aims to simplify the complexity of AI systems decisions, making AI decisions and predictions understandable to humans. This gives trust, transparency, and ethical property to AI and machine learning models. In general Local explanation is used, which relates to individual data points, clearing the influenced feature for a specific output. Techniques like LIME (Local Interpretable Model-Agnostic Explanations) or SHAP (Shapley Additive explanations) highlight the features' contribution to the outcome for a single specific data point but there is a need for an explanation to explain feature's impact in general without relying on specific points as Global explanation which give a general view of the Machine learning model's and overall behaviour of the system prediction, identify the general patterns that affect the model's predictions. Methods like decision trees or rule-based models intrinsically provide global explainability for many applications.

1.6 F. Using of (XAI) Methods in the Cyber Security Field

Many researchers are exploring XAI methods in the cyber security field in both detecting and protecting techniques, one of the sate of the art research in this field (Zhang et al. 2022) is discussing the (AXI) Applications in Cyber Security, the research focusing on cyber security fields such as intrusion and malware detection systems, and spam filtering. However, the research is not exploring the XAI techniques for the data poisoning matter, but it considers as an example of using the XAI in the cyber security systems, which current work exploring and studying, in the research of (Zhang et al. 2022) the aim of the researcher is to implement the XAI techniques in cyber security as XAI is applied in deferent industries starting from health care, smart cities, Smart Farming, Smart Transportation to Smart Financial System, the research highlighted a point that XAI models them self can be target to cyber security attacks with a technique called black box attack where the attacker has limited knowledge about the working principle of the model, by interacting with the model through its input and output, black box attacks can be classified as two types: the first for a single point of the XAI explanation results, while the other type of attacks are targeting to compromise the integrity the explainer in general, at the same time the research indicate that a defence methods can be through the sing auxiliary tool (Zhang et al. 2022). Research also explores the challenges of using XAI in cyber security like: the lack of big size of available data for the training of XAI, indicating future directions for increasing the reliance on XAI models in cyber security by using High-Quality Datasets and Multimodal XAI, as understanding AI decisions with diverse data sources (Zhang et al. 2022).

Another research Walia (2022) (Walia et al. 2022) explores the use of XAI in detecting Image manipulation, by conducting error analysis on images that are affected by manipulation by detecting the binary pattern and then convert into feature maps. The method error level analysis (ELA) method will take place based on that local binary patterns are discriminative in nature, the research utilized the JPEG format analysis based on textural features and then explained it using the XAI techniques.

Another technique has been discussed in the same concept of using XAI in detecting Anomalies Using SHAP (Antwarg et al. 2019), by creating an Explanatory Feature Set as part of the explanation method. further research can be conducted to use the same technique in detecting data poisoning.

2 Methodology

It is very difficult to discover manipulating features by a targeted attack by looking into a single point, as the attacker injects a false set of points that change the feature impact at the end, while for big data the single point will have almost no effect on the feature impact. This research focuses on using the global XAI explanation for monitoring the feature impact.

An example of data poisoning in the healthcare sector, as machine learning in this field has an important impact on the sector operations and development by utilizing huge amounts of datasets to improve diagnoses, treatment planning, drug discovery, and more. According to Mozaffari (2015) Huang et al. (2020) which exploring data poisoning attacks in the healthcare field as a dangerous threat, because manipulation of existing data can lead to delayed treatment of hypothyroid diagnosis due to not discovering the early stage of the disease, this highlights the importance of detecting data poisoning as in similar cases if poisoning attacks detected then specialists can ignore the results of machine learning until they success to clean the dataset from malicious data. The research paper Mozaffari (2015) work based on considering the attacker has information about the training dataset and uses this knowledge to insert malicious data, this knowledge is because the data is available publicly or because of some techniques by attacker eavesdropping on network transactions or reach the database storage area. Targeted datasets can be classified as fixed types in case it does not include a mechanism to add data input by users during the attack or evolving datasets in which new inputs are to be added by users to the dataset during the time of machine learning model training. To gain the required knowledge about the dataset by an attacker in case accessing the original dataset is difficult, the Attacker needs to generate a "proxy" dataset by exploring the statistics of the original training dataset test the system inputs to the ML module, and observing its output (Akbanov et al. 2019).

The Mozaffari (2015) research will be used in this research for using of XAI method in detecting data poisoning. As Mozaffari et al. (2015) provides an example of a targeted attack by specifying the attacker's exact goal, a research paper details an attacker's aim to delay the diagnosis of hypothyroid disease by changing the feature weights, which can have a life-threatening impact at the same time a false positive prediction can lead to unnecessary concern.

For the same collected data (Mozaffari 2015), another approach using XAI techniques can be applied such as the Tree SHAP method approach as a global explanation method to monitor the predetermined feature based on the attacker goal by setting up a bandwidth or technique to distinguish between the normal feather impact and the unnormal feature impact generated based on poisoned injected data. In the same research for countermeasures against manipulating data to maintain accuracy, the validation set must be updated periodically to match updates in the training data.

Figure 3 illustrates a scenario where attackers target a specific machine learning model with the goal of manipulating its output. They achieve this through careful planning and research, focusing on the target's AI models, datasets, and security weaknesses. As attackers employ various techniques to gain access to the system, such as exploiting software vulnerabilities or compromising user accounts. Once inside, they may establish persistent access to manipulate the system or steal data for an extended period. The attackers' primary strategy involves manipulating the data used to train or operate the AI model. They inject modified data specifically designed to influence the model's output. This data manipulation disrupts the AI model's normal functioning, causing it to generate inaccurate or misleading predictions. Understanding these targeted attack methods allows network security engineers to implement security measures and protect their models from manipulation. One of the suggested methods is by monitoring the feature impact in XAI output to discover any abnormal or model output to achieve the attacker target.

Figure 4 shows a systematic approach for poisoning attacks across machine learning algorithms and Thyroid disease datasets. It shows that these attacks successful, demonstrating the vulnerability of healthcare ML systems to such attacks, the figure shows the impact on accuracy Mozaffari-Kermani (2015), in the same time output is a combination of mutable feature impact which mean lower accuracy with manipulated feature importance.

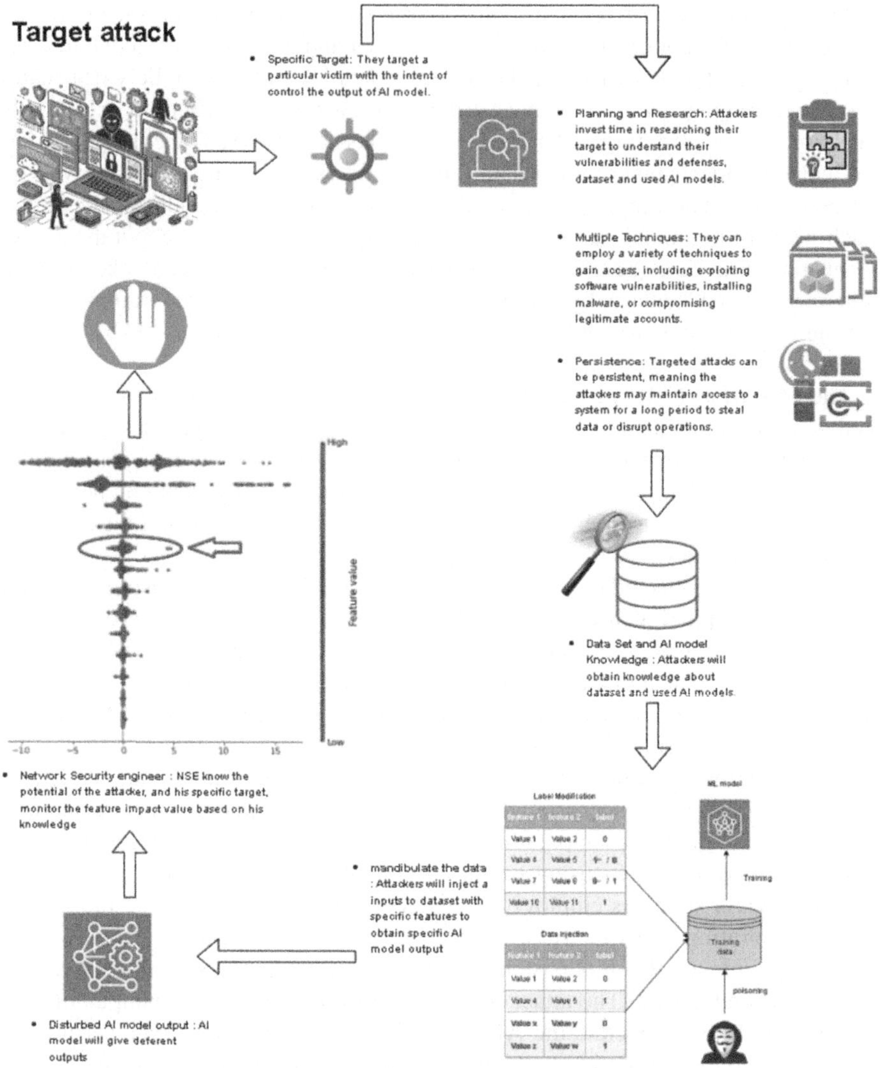

Fig. 3 Explaining diagram of the research method to detect data poisoning

3 Limitations

Detecting data poisoning attacks in healthcare applications presents several unique challenges due to the sensitive and complex nature of healthcare data. Firstly, the heterogeneity of healthcare data, which includes structured data (e.g., patient records, lab results) and unstructured data (e.g., doctor's notes, medical images), complicates the identification of anomalies. Data poisoning can subtly alter critical features

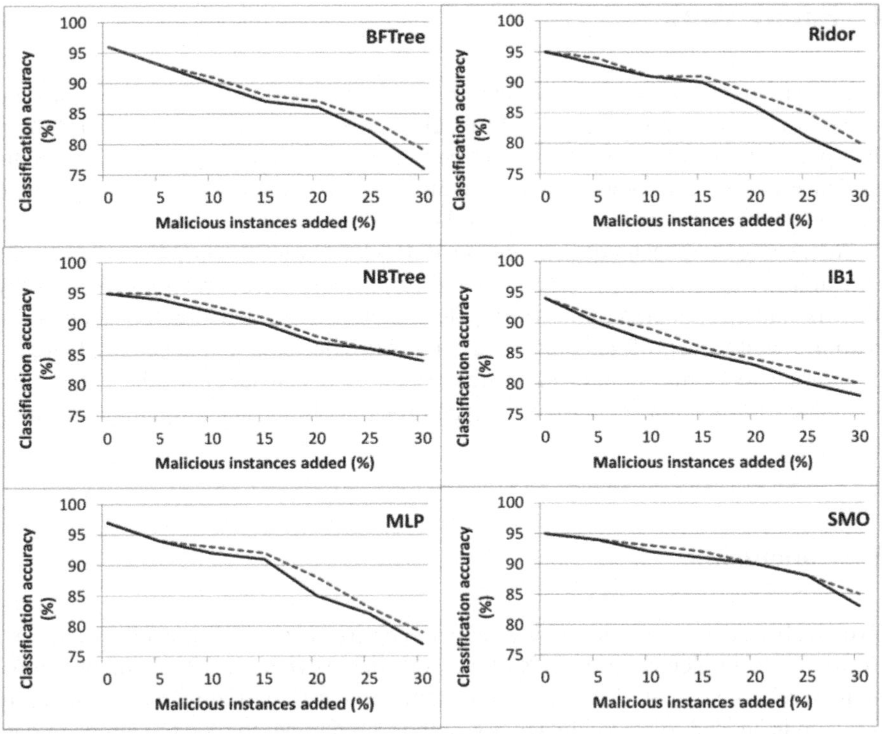

Fig. 4 Health care machine learning algorithm output and data poising impact on accuracy

in these datasets, making it difficult to distinguish between normal variability and malicious alterations. This research has penial and exploring a new method of data poisoning detection, but at the same time some limitations and barriers exist like the need to address the attacker's aim and to have previous knowledge about the poisoning attack aim, this means that specific types of poisoning the explored method can detect like a targeted attack.

Also, the chance of generating false alarms (false positives) or missing actual poisoning attempts (false negatives), will have a huge effect, especially in some critical fields like health care. And for the Real-world deployment challenges, there are practical limitations to deploying the method in real-world healthcare systems, such as data privacy or integration complexity. Also, the high stakes involved in healthcare decisions mean that even minor data integrity issues can lead to significant consequences, such as misdiagnoses or incorrect treatment plans, amplifying the need for precise and reliable detection mechanisms.

4 Future Work

The future work to enhance the level of explainability which can be conducted in the context of the subjective XAI results, as XAI can be in different levels in data poisoning identification, XAI can be with more details to generate actionable guidance to help managers develop quality defence techniques by identifying the aim of attackers and goal and detecting the data poisoning.

Advancing XAI techniques to enhance the detection of data poisoning attacks in healthcare applications involves several promising future directions by not only identify unusual patterns but also provide context-rich explanations for anomalies, enabling healthcare professionals to make informed decisions about potential threats. Integrating real-time monitoring systems with XAI can facilitate the continuous assessment of incoming data streams, promptly identifying and mitigating poisoning attempts.

5 Conclusion

Explainable AI is very important in cyber security, as several research studies are exploring the implementation of XAI methods in the cyber security field, but almost no research in the field of data poisoning detection or trial of implementation, more research is needed to be conducted not only to implement the XAI methods in data poisoning detection filed but also for improving the defence techniques by identifying the aim of attackers and goal.

References

Akbanov M, Vassilakis VG, Logothetis MD (2019) WannaCry ransomware: analysis of infection, persistence, recovery prevention and propagation mechanisms. J Telecommun Inf Technol 1(1):113–124

Antwarg L, Miller RM, Shapira B, Rokach L (2019) Explaining anomalies detected by autoencoders using SHAP. arXiv:1903.02407

Baniecki H, Kretowicz W, Biecek P (2022) Fooling partial dependence via data poisoning. In: joint European conference on machine learning and knowledge discovery in databases (ECML PKDD), Springer, pp 702–720

Butt UJ, Hussien O, Hasanaj K, Shaalan K, Hassan B, Al-Khateeb H (2023) Predicting the impact of data poisoning attacks in blockchain-enabled supply chain networks. Algorithms 16(12). https://doi.org/10.3390/a16120549

Chen Y, Mao Y, Liang H, Yu S, Wei Y, Leng S (2020) Data poison detection schemes for distributed machine learning. IEEE Access 8:7442–7454. https://doi.org/10.1109/ACCESS.2019.2962525

Das D, Roy S, Sahoo B (2024) Impact of data poisoning attack on the performance of machine learning models. In: Namasudra S, Trivedi MC, Crespo RG, Lorenz P (eds) Data science and network engineering. ICDSNE 2023. Lecture notes in networks and systems, vol 791. Springer, Singapore. https://doi.org/10.1007/978-981-99-6755-1_32

Dunn C, Moustafa N, Turnbull B (2020) Robustness evaluations of sustainable machine learning models against data poisoning attacks in the internet of things. Sustainability (Switzerland) 12(17). https://doi.org/10.3390/SU12166434

Hariharan S, Robinson RRR, Prasad RR, Thomas C, Balakrishnan N (2023) XAI for intrusion detection system: comparing explanations based on global and local scope. J Comput Virol Hacking Tech 19(2):217–239

Huang T, Chen Y, Yao B, Yang B, Wang X, Li Y (2020) Adversarial attacks on deep-learning-based radar range profile target recognition. Inf Sci 531:159–176. https://doi.org/10.1016/j.ins.2020.03.066

Meng Y, Yang N, Qian Z, Zhang G (2021) What makes an online review more helpful: an interpretation framework using xgboost and shap values. J Theor Appl Electron Commer Res 16(3):466–490. https://doi.org/10.3390/jtaer16030029

Mozaffari-Kermani M, Sur-Kolay S, Raghunathan A, Jha NK (2015) Systematic poisoning attacks on and defenses for machine learning in healthcare. IEEE J Biomed Health Inform 19(6):1893–1905. https://doi.org/10.1109/JBHI.2014.2344095

Subedar M, Ahuja N, Krishnan R, Ndiour IJ, Tickoo O (2019) Deep probabilistic models to detect data poisoning attacks. http://arxiv.org/abs/1912.01206

Walia S, Kumar K, Agarwal S, Kim H (2022) Using XAI for deep learning-based image manipulation detection with shapley additive explanation. Symmetry 14(8):1611

Yuan X, He P, Zhu Q, Li X (2019) Adversarial examples: attacks and defenses for deep learning. IEEE Trans Neural Netw Learn Syst 30(9):2805–2824

Zhang Z, Hamadi HAl, Damiani E, Yeun CY, Taher F (2022) Explainable artificial intelligence applications in cyber security: state-of-the-art in research. IEEE Access 10:93104–93139. https://doi.org/10.1109/ACCESS.2022.3204051

Utilization of a Grid Connected PV System to Power a Water Purifying Plant with Economic and Environmental Feasibility Analysis

Ayman Rayees Mulla and Bashar Aldbaiat

Abstract Grid connected Photo Voltaic (PV) system is designed to power a water purification plant mounted on the roof of a commercial building. The system is scaled to the water plant's requirements, and the tilt angle of the panels is calibrated to produce the most energy possible throughout the year. The return on investment must be attained in less than six years from commissioning, which is a lofty goal. Net metering feature which allows exporting excess energy to the grid is discussed and emphasized. The viability of this technique is highlighted, and simulations of the designed system are shown to back up its theoretical promises. Throughout the course of the project, the building's carbon footprint is reduced as well by over 200 tons of CO_2 emissions. The study provides an environmental analysis to showcase this improvement.

Keywords PV system · Grid connected solar system · Powered desalination plant · Net metering

1 Introduction

Assuring the provision of high-quality, purified water is crucial for maintaining optimal infection control. One of the effective methods employed to achieve this is the utilization of reverse osmosis (RO) systems and while RO systems are highly effective in removing impurities and producing purified water, it is important to consider their energy consumption and the associated environmental impact of such consumption. In this discussion a solar-powered RO system is considered to create system's energy independence. By harnessing the power of the sun, these systems can operate off-grid and on grid eliminating the reliance on traditional electricity

A. R. Mulla · B. Aldbaiat (✉)
Faculty of Engineering and IT, The British University in Dubai, Dubai, United Arab Emirates
e-mail: bashar.aldbaiat@buid.ac.ae

A. R. Mulla
e-mail: 20194702@ug.buid.ac.ae

K. Al Marri et al. (eds.), *BUiD Doctoral Research Conference 2024*, Lecture Notes in Civil Engineering 587, https://doi.org/10.1007/978-3-031-84371-6_41

sources partially or fully (Al-Hinti et al. 2013; Chaturvedi and Jha 2018). This is particularly beneficial in remote or off-grid locations where access to a stable grid power supply is limited or costly and even within urban areas where mitigating carbon footprint and promoting renewables are highly encouraged and appreciated. Solar-powered RO systems have a significantly lower environmental impact compared to on-grid systems. By utilizing clean, renewable solar energy, these systems contribute to reducing greenhouse gas emissions and dependence on fossil fuels. Solar power is a clean and sustainable energy source, and when coupled with the water purification efficiency of reverse osmosis, it provides an eco-friendly alternative for water treatment needs. Some of the papers in the relevant literature aim to critically analyze the economic and environmental aspects of water purification systems in commercial buildings. This study seeks to identify key gaps and weaknesses in the current knowledge regarding the cost-effectiveness and environmental impact of such systems. By addressing these gaps, the research aims to provide valuable insights into the optimal use of resources and contribute to the current understanding of sustainable water purification practices in commercial settings.

The growing demand for fresh water and the limited availability of water sources have led to an increased requirement for desalinated seawater in recent years. Among various methods of desalination, Reverse Osmosis (RO) has emerged as a cost-effective and leading process but desalination in general is an energy-intensive operation that necessitates a significant amount of energy (Al-Hinti et al. 2013). In a theoretical study conducted by Abuzaid and Samara (2022) a Photovoltaic (PV) system was employed to convert solar energy into electricity, which was subsequently utilized to operate the RO plant. The study estimated the cost of water desalination from an RO plant powered by a PV system (Alhammami and An 2020). This cost comparison highlights the potential cost savings associated with employing solar power for RO desalination. A recent study by Almulali and Sab (2018) concluded that as the costs of PV installations decrease, while grid electricity prices rise, and the industry gains more expertise.

PV will progressively become a financially advantageous electricity source across broader geographic areas. Furthermore, the United Arab Emirates (UAE), known for its oil and gas reserves, has emerged as a leader in deploying solar energy sources in the Middle East since 2006 (Ghafoor 2020). UAE has been a host to one of the largest Solar sites in Al Dharfa region (Apostoleris and Chiesa 2023) making UAE a motivated country towards sustainable green future and a best place for solar powered RO water purification system. Motivated by population growth, economic development, and environmental responsibility, the UAE has prioritized solar energy due to its high sunlight exposure (IEA 2020) While currently heavily reliant on natural gas power plants, the UAE aims to generate 50% of its energy from clean sources by 2050, demonstrating its commitment to sustainable energy transition (MOCCAE, 2017). An economic analysis by Kannan and Vakeesan (2016) of Khalifa City and Zayed City in Abu Dhabi, UAE exhibited significant energy potential from residential rooftop PV systems, with estimated annual kWh ranging from 116,302,894 to 211,693,508 and 248,985,515 to 448,174,049, respectively. Additionally, Khalifa

City's chosen building's rooftop has the capacity to provide 11–20% of the city's annual electricity needs (Kazim 2007).

Future studies might examine the economics of all the buildings in the study area, analyse the best solar panel types with varying costs and efficiency, and evaluate the impact of RO systems. PV panels, which turn sunshine into power, are also excellent for lowering CO_2 emissions (Subedi 2021). PV solar power is being evaluated as a potential grid electricity substitute to reduce additional CO_2 emissions because the rising demand for RO distillation systems needs higher electricity consumption (Subedi 2021). Both the environmental and economic assessments are done in this study. The net metering feature is used to give the project a profitability component. Research made at MIT proposes as desalination plants are on the rise and the world produces around 27 billion gallons of water in a single day through these plants. The fact energy required to produce can't be ignored. Whereas it is also mentioned that almost the same amount of brine water is produced (Chandler 2019). Brine water is the highly salt concentrated water which is the byproduct of the desalinated water (Root 2019). With the use of PV system, we will be also discussing how the disposal or the usage of this produced by product can be handled in an efficient sustainable manner.

This article is divided into four sections: an introduction, system model and optimization, an economic and environmental evaluation, and finally a conclusion.

2 System Model and Optimization

A reverse osmosis pump, a UV filter, and a softener are all powered by the system's PV panels and inverter. A business building's water requirements are reportedly going to be met by the water plant in the picture above. The solar system is well designed to supply enough power for the purification machinery and to send the extra energy collected to the grid. The building's water demand will be first evaluated, and the necessary electrical energy to power the system appropriately will be defined, to avoid oversizing the system since these solar panels will be put on the building's roof where there is a limited amount of open space. The efficiencies of the system components are considered in the PV system design, as explained within the subsection demand data (Fig. 1).

2.1 Demand Data

Demand dataset shows the weakly needs for water supply, The rate of water flow required to fully serve the needs of this building could peak during daytime due to, a regular load profile indicate a slow growth in the number of admissions throughout the day while max load profile which will be considered as the reference profile assumes a constant flowrate of 60 gal/h for 13 hrs as depicted below in Figure 4: Figure 4

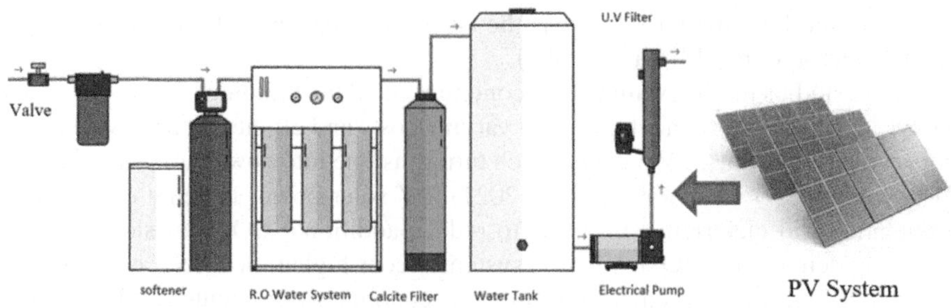

Fig. 1 Schematic of PV system powering water purification system

water demand profile (regular vs. max)Therefore, our daily water consumption can be estimated as the following:13 hours * 60 gal/hr = 780 gallons/day. Total weekly water consumption will be: 5 * 780 gallons = 3900 gallons/week. Figure 2 illustrates the variable flow rate needed to sufficiently meet the building demands.

Demand in terms of electricity consumption can be seen as an equivalent power rating of our water station components multiplied by their electrical efficiencies and then by the duration to conclude overall energy. The whole power rating of the system is calculated as 1.1 kW to fulfil the water requirements. With taking in consideration the different efficiencies of the system's components. And hence the estimated daily Energy would be: 1.1 * 13 = 14.3 kWh /day as shown in Table 1.

Both the UV filter and the softener have sufficient diameter to permit such capacity to pass through while the nominal power of the reverse osmosis unit is selected based on the maximum flow rate of our demand data, as depicted in Fig. 2.

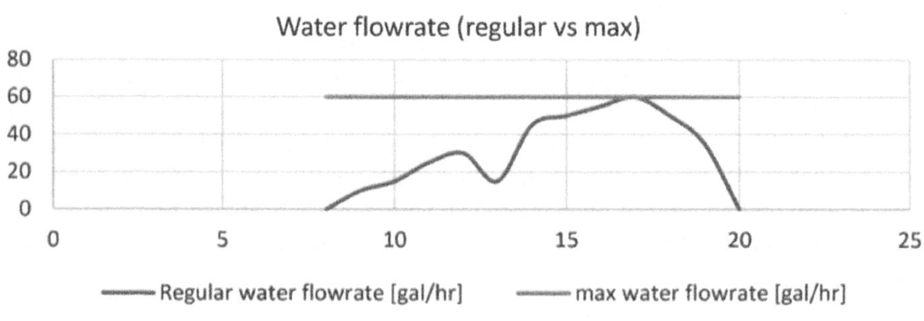

Fig. 2 Water flowrate requirements at the studied case

Table 1 Reverse osmosis system components versus nominal power and efficiencies

Device	Nominal power [W]	(η) Efficiency
R.O unit	790 W	0.75
U.V filter	40	0.85
Softener	20 W	0.9

2.2 PV Productivity Optimization

Utilizing rooftops to install solar panels is becoming increasingly popular as more people recognize the benefits of clean renewable energy. Rooftop solar PV systems typically consist of a set of solar panels, an inverter to convert DC electricity to AC electricity, and a maximum power point tracking (MPPT) to export the maximum active power out of the system. Installing a solar PV system on a rooftop can help to reduce carbon emissions and lower electricity bills. Considering the advantage of net metering feature provided by many grid operators where customers have the choice to export the excess of their generated energy back to the grid in return of financial gains. It is also a good choice for those who have limited space and cannot install solar panels on the ground. However, before installing a solar PV system, it is important to ensure the system is well designed and not oversized as we are looking at limited vacant area and each square meter counts.

The available area in our case was confined to 40 m^2, this can fit approximately 14 solar panels. Mono crystalline solar panels were selected as they feature higher efficiency than their polycrystalline counterparts (Chaturvedi and Jha 2018). To optimize the benefits of these solar panels, a fixed inclination and azimuth angles were defined across the year as shown in Fig. 3. Location-based simulation in Fig. 4 illustrates the enhancement of this optimization over none optimized solar system. Location has clearly the potential of utilizing solar power, and this was not a surprise by any means considering the metrological characteristics of Dubai in terms of solar irradiance. The tilt angle was suited for a better yearly productivity, and it was defined as 29.8 deg.

The intermittent nature of such renewable energy resource is easily solved thanks to the net metering feature which allows the system to withdraw energy from the grid whenever sudden shading occurs or during night hours and export the excess of generated energy to the grid as well.

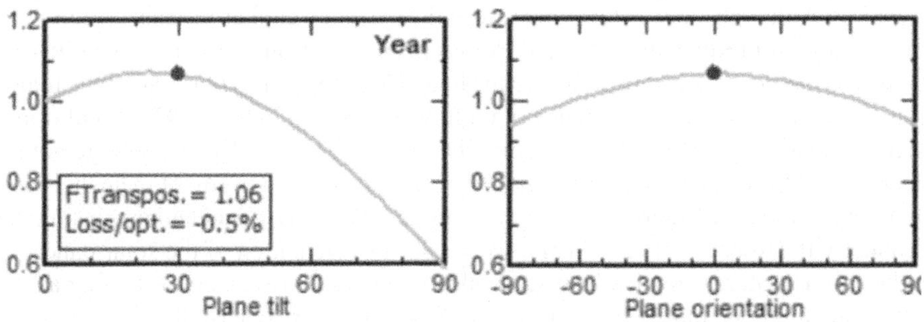

Fig. 3 Inclination angle optimized on a yearly basis

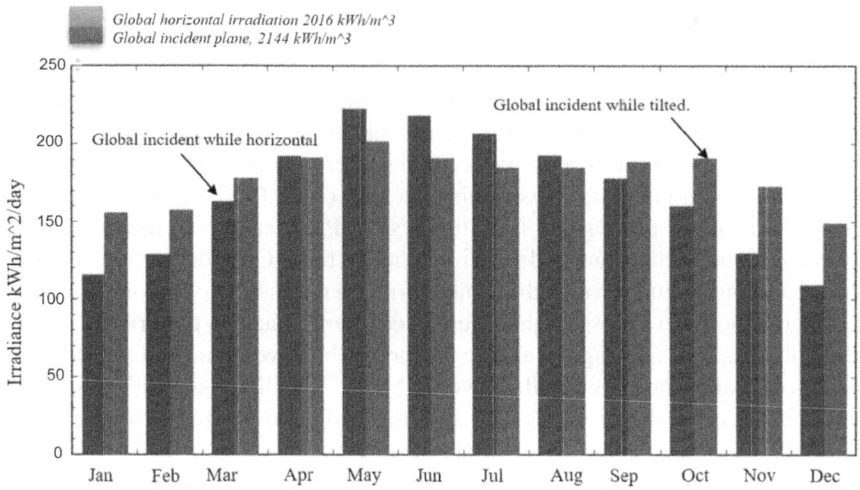

Fig. 4 Global horizontal irradiance versus global incident in cell plane

2.3 PV System Losses

PV system losses refer to the various factors that can reduce the amount of energy a photovoltaic (PV) system produces compared to its theoretical maximum output. Figure 5 shows the various losses in the system. One of the biggest losses is the temperature loss, which accounts for 11.6% in this case.

The optimization carried out throughout the design phase managed to reduce the losses. For instance, as demonstrated above, choosing proper inclination angle managed to save nearly 6.4% of total energy. There are many ways to reduce PV system losses even further, including reducing shadowing, choosing suitable components with high efficiency ratings, and performing routine system maintenance and cleaning. Although such a step would raise the system's capital expense. According to an article from Kennedy (2023), the losses occur in various form, listing out few of them and their causes and how these can be avoided. Temperature is one of the major causes of energy loss, the reference point however is said to be 25 °C, an additional increase of 1 °C after the reference point increases the losses. The percentage output decreases. This can be avoided by optimistic orientation and tilting of the panels and avoiding flatbed orientation. Thus, increasing the life of the panel and increasing the productivity (Kennedy 2023). While we set up the orientation of the solar panels to reduce temperature effective losses, we also reduce the losses due to shading as the optimistic angle is found (Kennedy 2023). Wire losses are one of the other losses to be considered but they can be neglected when compared to other losses and may be brought to 1% that of the whole system if thicker wires are used at shorter distances.

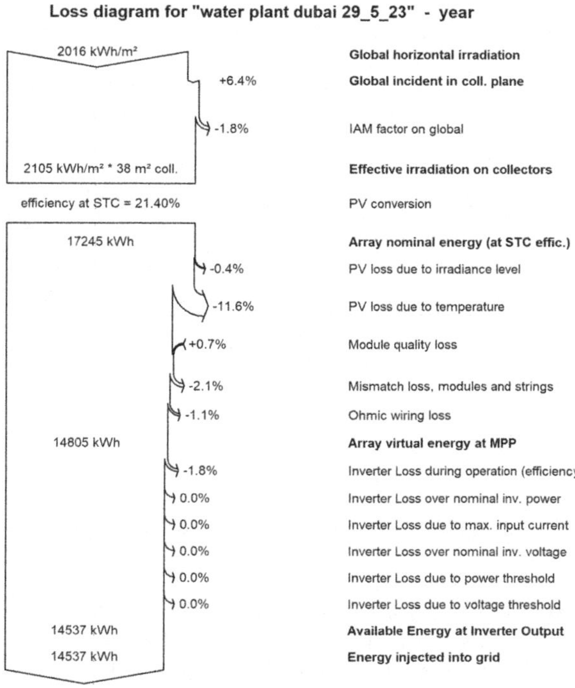

Fig. 5 PV system losses due to multiple factors

3 Economic and Environmental Evaluation

To ensure the profitability and sustainability of solar-powered projects, an assessment of their environmental and economic aspects is necessary. In terms of the economy, it includes evaluating startup costs, return on investment. In terms of the environment, it evaluates resource conservation, the utilization of renewable energy, and the decrease of carbon emissions. These assessments not only help with decision-making but also emphasize the long-term advantages of solar energy, such as reduced costs, energy independence and the reduction of climate change.

3.1 Economic Evaluation

The system design objective to meet demand and achieve a return on investmnet in less than 6 years is achieved as shown below through Fig. 6. Net metering proved it's value as an attractive factor to encourage more renewable installations. Exporting electricity to the grid enabled the system to become profitable and cover the total cost of comissioning in relatively short period. Attaining a return on investment point is made possible by the fact that the government of the United Arab Emirates values

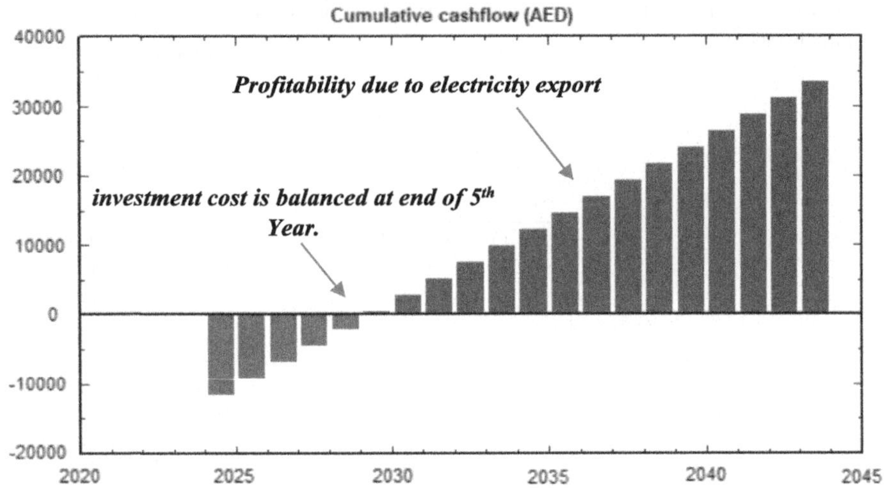

Fig. 6 Cumulative cashflow of the PV system across the project's lifespan

exported energy at the same tariff rate as consumption, which in our case averages about 4 fils per kwh per month (IEA 2020).

An additional factor could be considered to increase the profit as well as making the earning of the profit quicker. The brine water which is produced during the systems running. The brain water is highly concentrated with salt. Major companies pump this water back to the ocean. But this increases the pumping cost, more electricity consumption as well as leaves behind a carbon footprint. According to the article (Chandler 2019) this brine water with the help of simple chemical reaction setup on site can be treated and the salts can be separated. Mainly NaOH sodium hydroxide can be retrieved, and HCL hydrochloric acid can be retrieved. This is also a sustainable way as carbon emission is avoided and the marine life endangering is reduced as the highly concentrated brine water is not let into the ocean water body (Chandler 2019).

Figure 7 depicts the amount of energy expenditure to the grid. A peak of export is noticeable during sunny months as the system was capable of injecting relatively high power to the grid.

3.2 *Environmental Evaluation*

Cutting off dependency on fossil fuel electricity and replace it with renewable energy resource is going to help mitigating carbon footprint of the building and might open the door for further similar investments in the future, grid electricity in UAE is mainly generated from natural gas turbines and CCGTs (Chaturvedi and Jha 2018), carbon emissions accompanied with each kWh generated by fossil fuel power plants, fossil fuels has the lion's share of electricity generation in UAE. Accompanied

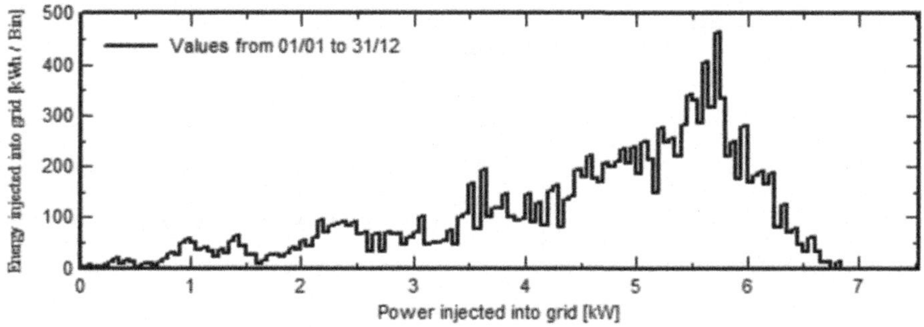

Fig. 7 Exported energy to the grid by the PV system

carbon emissions with these conventional electricity generations are depicted below in Table 2.

From the table above and considering a total consumption of 3450 kWh every year we can calculate carbon amount accompanied with such generation: 3450 * 0.91 = 1417 kg. This is the same carbon sequestered by 40.3 seedings grown for the next 10 years Abuzaid and Samara (2022). Many design decisions were based on their environmental benefits. Buffering tank was considered to reduce the number of batteries needed, although the proposal didn't achieve its goal, having a physical product storage will surely mitigate losses and decrease charging/discharging cycles of Li-ion batteries which will help expanding their lifespan. Having said that, solar panels have a lifecycle expectancy of 25–30 years and after that decommissioning creates a new environmental challenge, economic viability of recycling PV panels is still unfavourable. Alhammami and An (2020) therefore, disposing PV panels and batteries should be done properly to achieve the best environmental enhancement results, Fig. 8 shows cumulative carbon emission savings over project lifetime:

Carbon emission balance can be calculated using the following Eq. (1) (UAE Ministry of Climate Change and Environment 2017):

$$\textit{Carbon balance} = \textit{E gird} * \textit{Project lifetime} * \textit{life cycle emission (LCE)(System)} \quad (1)$$

Table 2 Accompanied carbon emissions with different conventional electricity generations

	Electricity generation million kWh	CO_2 emissions million metric tons	Million short tons	Pounds per kWh
Coal	947,891	952	1,049	2.21
Natural gas	1,358,047	560	617	0.91
Petroleum	15,471	15	17	2.13

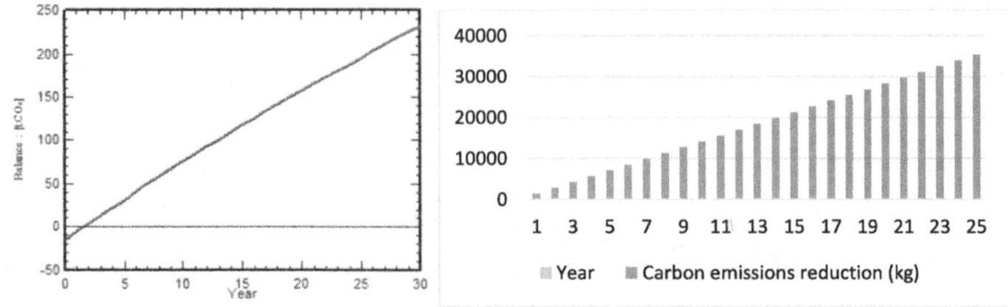

Fig. 8 Cumulative carbon emissions savings with consideration of commissioning carbon emission (**left**) and without commissioning emissions (**right**)

Based on the international energy agency estimations, the life cycle of carbon emissions in UAE is around 653 gCO_2/kWh (Woodard 2022). The energy exported to the grid across the life cycle is estimated through simulations as 14.5 MWh, the life cycle of emission by the system itself is 15. 1 ton of CO_2. Accordingly, we can use the previously mentioned equation to calculate the yearly carbon balance which is 0.994 ton of CO_2. Consequently, the total saving of CO_2 emissions across the whole life cycle of the project is 232.023 tons of CO_2 which is very optimistic figure considering the scale of this project.

4 Conclusion

The feasibility of a grid connected system in powering desalination plant was studied here both economic and environmental evaluations proved its effectiveness. Moreover, given the project's high potential for profitability, it would be difficult not to consider PV grid connectivity as it is true win–win situation. Dubai's implementation of the Net metering featured a few years ago has made PV systems more economically feasible and, in some cases, profitable, like in this case. Even when compared to residential roof investments, a PV system set up to power a 1.1 kW water station for 13 h is a much too small installation. However, on the bright side, the environmental benefit is significant throughout the life cycle of this project. 7 tonnes of carbon emissions could be avoided annually. This is the same amount of carbon sequestered by around 160 seedings grown for ten years (UAE Ministry of Climate Change and Environment 2017). This system also avoids the waste brine water from getting into oceans and rather make it producing and further stoppage of carbon footprint emission. From a business point of view such system help the owner for the stoppage of outflow cash and increase the inflow by the net metering availability and by the selling of the salts. Overall using RO system reduces the usage of packaged water which in turn again reduces the production of plastic and the carbon emission caused while producing them. Maintenance of such systems can be tricky to handle as UAE is a sand borne country and the solar radiation is high, but maintenance is of no harm as it gives good output and increases the life of the PV panels. Overall, the system is a fantastic form of using renewable energy to its best. This system can only not be a future to developing countries but can be a great asset to poorer countries who are facing crisis for the water as this system will provide them pure and clean water and as the system is powered through solar energy the cost of generating such high electricity is not a burden on such countries.

References

Abuzaid H, Samara F (2022) Environmental and economic impact assessments of photovoltaic rooftop system in the United Arab Emirates. Energies 15(22):8765

Alhammami H, An H (2020) Techno-economic analysis and policy implications for promoting residential rooftop solar photovoltaics in Abu Dhabi, UAE, Renewable Energy. https://www.sciencedirect.com/science/article/abs/pii/S0960148120318413. Accessed 28 May 2023

Al-Hinti I et al (2013) A comparative analysis of energy indicators and CO_2 emissions in 15. Int J Environ Waste Manag. https://www.researchgate.net/publication/264816843_A_comparative_analysis_of_energy_indicators_and_CO2_emissions_in_15_Arab_countries Accessed: 27 May 2023

Almulali U, Sab C (2018) Energy consumption, CO_2 emissions, and development in the UAE, energy sources, part b: economics, planning, and policy. https://www.tandfonline.com/doi/full/https://doi.org/10.1080/15567249.2012.689796. Accessed 29 May 2023

Apostoleris H, Chiesa M (2023) Solar energy in the United Arab Emirates. In: the handbook of environmental chemistry, Springer, Berlin, Heidelberg

Branker K, Pathak M, Pearce J (2011) A review of solar photovoltaic levelized cost of electricity, renewable and sustainable energy reviews. https://www.sciencedirect.com/science/article/abs/pii/S1364032111003492. Accessed 29 May 2023

Chandler DL (2019) Turning desalination waste into a useful resource, MIT news office. Turning desalination waste into a useful resource | MIT News | Massachusetts Institute of Technology. Accessed 15 May 2024

Chaturvedi P, Jha KK (2018) Performance and efficiency comparison between single crystalline, poly-crystalline and amorphous silicon photovoltaic cells. Renew Sustain Energy Rev 87:223–231

Ghafoor A et al (2020) Techno-economic feasibility of solar based desalination through reverse osmosis. Desalination. https://www.sciencedirect.com/science/article/abs/pii/S0011916419316820. Accessed 28 May 2023

IEA (2020) United Arab Emirates—countries & regions, IEA. https://www.iea.org/countries/united-arab-emirates. Accessed 28 May 2023

Kannan N, Vakeesan D (2016) Solar energy for future world: a review, renewable and sustainable energy reviews. https://www.sciencedirect.com/science/article/abs/pii/S1364032116301320. Accessed 28 May 2023

Kazim A (2007) Assessments of primary energy consumption and its environmental consequences in the United Arab Emirates, renewable and sustainable energy reviews. https://www.sciencedirect.com/science/article/abs/pii/S1364032105000390. Accessed 27 May 2023

Kennedy R (2023) Guide to understanding solar production losses, pv magazine International. Guide to understanding solar production losses—pv magazine International (pv-magazine.com). Accessed 15 May 2024

Ministry of Climate Change and Environment (2017) The UAE's response to climate change, UAE net zero 2050—the official portal of the UAE Government. https://u.ae/en/information-and-services/environment-and-energy/climate-change/theuaesresponsetoclimatechange/uae-net-zero-2050. Accessed 25 May 2023

Mokri A, Ali M, Emziane M (2013) Solar energy in the United Arab Emirates: a review, renewable and sustainable energy reviews. https://www.sciencedirect.com/science/article/abs/pii/S1364032113004838. Accessed 25 May 2023

Muhammad B (2019) Energy consumption, CO_2 emissions and economic growth in developed, emerging and middle east and North Africa countries. Energy. https://www.sciencedirect.com/science/article/abs/pii/S0360544219305420. Accessed 28 May 2023

Root T (2019) Desalination plants produce more waste brine than thought, national geographic. Desalination plants produce more waste brine than thought (nationalgeographic.com). Accessed 15 May 2024

Singh G (2013) Solar power generation by PV (photovoltaic) technology: a review, energy. https://www.sciencedirect.com/science/article/abs/pii/S0360544213001758. Accessed 28 May 2023

Subedi S (2021) Evaluation and sizing of solar powered reverse osmosis water desalination system. https://egrove.olemiss.edu/. https://core.ac.uk/download/pdf/427483915.pdf Accessed: 25 May 2023

UAE Ministry of Climate Change and Environment (2017) The UAE's response to climate change, The UAE's response to climate change—the official portal of the UAE Government. https://u.ae/en/information-and-services/environment-and-energy/climate-change/theuaesresponsetoclimatechange. Accessed 28 May 2023

Woodard J (2022) What is a reverse osmosis system and how does it work? Fresh water systems. https://www.freshwatersystems.com/blogs/blog/what-is-reverse-osmosis. Accessed 26 May 2023

Waste Management in the Construction Industry in Dubai—UAE

Byan Aldeeri and Basim Touqan

Abstract The United Arab Emirates (UAE) focuses on achieving sustainable development. One of the main factors that has to be taken into consideration is the construction waste management techniques. This study focused on exploring the causes behind waste generation in construction sites and the effective techniques implemented to manage the construction waste at construction sites in the UAE. Also, it addresses the challenges that may face implementing effective strategies. This research focused on qualitative research methods to reach the findings. Initial site visits to Four construction projects were followed by interviews targeting contractors' viewpoints on waste causes, waste reduction strategies, and their associated challenges. The data have been analysed using NVIVO software as a tool to conduct a thematic analysis approach. The findings highlighted that material waste stems from factors such as limited awareness, excess off-cuts, improper planning and coordination and rework due to changes in the design. Commonly used waste management practices include proper material storage, landfilling, reuse of some materials and just-in-time material delivery. Areas like waste coordinating and planning and segregation require enhanced efficiency to meet waste management and reduction objectives. Additionally, the study revealed that contractors mainly view waste reduction as a financial concern aimed at cost reduction, often overlooking the environmental impacts associated with waste management.

Keywords Construction · Material waste · Waste management · Sustainability · UAE

This Paper Was Prepared for Research Methods Course, Taught By Dr. Basim Touqan

B. Aldeeri (✉) · B. Touqan
Faculty of Engineering and IT, The British University in Dubai, Dubai, United Arab Emirates
e-mail: 23001504@student.buid.ac.ae

K. Al Marri et al. (eds.), *BUiD Doctoral Research Conference 2024*, Lecture Notes in Civil Engineering 587, https://doi.org/10.1007/978-3-031-84371-6_42

1 Introduction

1.1 Overview

Sustainable development has become one of the most important issues that many countries are working on. To achieve more sustainable environments, countries are dedicated to maintaining a balance between protecting the natural environments and developing the built environments. The key to attaining this balance lies in transitioning from conventional production methods to a cyclic approach focused on preventing, reusing, and recycling material waste (Al-Hajj and Hamani 2011). Considering these efforts towards sustainability, it's important to understand the significant role of the construction sector in national development. The construction sector plays a vital role in the advancement of any country. Despite its positive impacts, it is responsible about generating substantial quantities of waste (Sweis et al. 2021). Annually, the global construction and demolition (C&D) sector generates more than 10 billion tons of waste from building projects, infrastructure development, demolitions, and land excavations (Wu et al. 2023).

1.2 Sustainable Construction Waste Management Techniques

Implementing sustainable construction practices is considered as an effective strategy to manage construction waste efficiently. The concept of sustainability in construction waste revolves around implementing optimal processes corresponding to the specific conditions and patterns of waste generation. Fig. 1 shows the hierarchy of waste management.

Prevention can be considered as the best waste management approach necessary for achieving the highest sustainability aiming to minimize avoidance (Zamri et al. 2020). Nevertheless, reducing waste on construction sites is important for promoting environmental sustainability and enhancing cost-efficiency. Precise planning and accurate measurement of materials can minimize overordering and excess waste (Alhawamdeh and Lee 2021). On the other hand, the reuse of construction waste

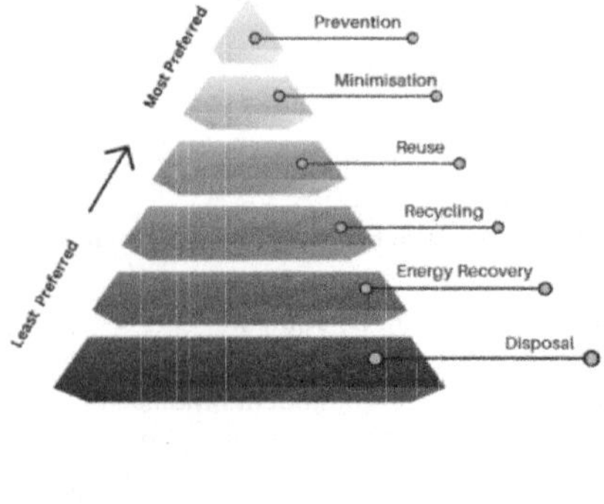

Fig. 1 Sustainable waste management hierarchy

offers an important chance to reduce environmental effects, preserve resources, and enhance sustainability in the construction sector (Ntitanguranwa et al. 2018).

Conversely, Recycling construction waste serves as a valuable approach to prevent the disposal of such waste in landfills (Janani and Kaveri 2021). Regarding materials recovery, this method can address the issue of reducing waste occupying landfill space (Kazerooni Sadi et al. 2012). However, Disposal is typically considered the final option and is regarded as the least desirable method in the hierarchy of sustainable waste management practices.

1.3 Waste Management in Dubai

Construction waste management plays a crucial role within the construction sector of the United Arab Emirates (UAE). With the rapid growth and development in the region, there has been a significant increase in construction activities, leading to a rise in construction waste. Construction and demolition (C&D) wastes make up a substantial 70% of the solid waste produced in the UAE. In UAE, Common sustainability rating systems in UAE are LEED, the Pearl Rating System for Estidama, and EHS Trakhees' internal certification.

In Dubai, according to the regulations set by the Dubai Municipality, Al Sa'fat rating system for sustainable construction stated that third level of certification (Silver Sa'fa) requires that 50% of the waste material produced during the construction or demolition of new buildings needs to be redirected away from landfills, and instead, recycled or repurposed. This rule applies to all new buildings except for those located in the CBD area. (Dubai Municipality & Government of Dubai 2016).

1.4 Research Aims & Objectives

This research study aims to examine the factors contributing to material wastage in construction sites in the UAE and evaluate the strategies currently employed by contractors to reduce such wastage with the focus on the following points.

1. Identify the major causes of waste generation in construction projects within the UAE context.
2. Evaluate the effectiveness of existing waste management strategies and policies in mitigating waste in construction sites.
3. Propose recommendations for improving waste management practices in the UAE construction industry.

2 Literature Review

The construction industry can be considered as the primary contributor to waste accumulation in landfills and a significant consumer of resources. A study conducted by (Tafesse 2021a) indicated that construction activities utilize 40% of the world's energy, make up 5–15% of the GDP, and support 5–10% of the workforce. Moreover, it consumes 40% of the world's raw materials.

2.1 Causes of Construction Waste

The quantity of waste produced at construction sites can vary across different sectors and may be influenced by the type and source of materials used. In the same context, a study Conducted by (Sweis et al. 2021) examined the primary reasons for material wastage, which include design changes made during the construction phase, rework caused by mistakes done by the labors, procurement of items not meeting specifications, incorrect material cutting, and inadequate site organization. Another study by (Fadiya Georgakis and Chinyio 2014) showed that the primary, secondary, and tertiary sources of construction waste are cut off materials, changes in design, and material management, respectively. In the meanwhile, the study investigated by (Soharu Naveen and Sil 2022) revealed that the primary causes of construction waste have been identified during material ordering, transit damage, negligent frame of mind of the working team, wrong work methodologies and improper work planning.

2.2 Construction Waste Management Techniques and Challenges

Several studies have been perused with the aim of identifying effective waste management strategies. A study conducted by (Yahya 2015) investigated solutions for reducing the waste in UAE by studying a case study in Abu Dhabi and conducting interviews. The results showed that most of the waste was demolished by landfilling due to the lack of management in the construction site. Another study conducted by (Hittini n.d.) investigated thirty construction sites in Abu Dhabi by conducting unstructured open interviews. The results showed that the implementation of targeted strategies during employee onboarding and skill development programs has resulted in a significant decrease of 10.2% in overall waste production and a notable 51.4% increase in waste diversion from landfills. Another study conducted by (Spišáková et al. 2021) investigated the waste management techniques used during the renovation of shopping center building in the town of Snina in Northeastern Slovakia. The results showed that most of the materials waste were concrete, bricks, tiles, and ceramics. The preferred approach involves recycling the waste although a significant portion of construction waste was typically disposed of in landfills. Another study conducted by (Umar et al. 2021) studied the waste management techniques for 4 construction sites in Malaysia, and the results showed that Recycling efforts differ, with 80% of the waste from high-rise buildings being sent for recycling compared to only 20% from housing sites.

Moreover, a study conducted by (Ouda et al. 2018) aimed to analyze 81 construction sites located in Saudi Arabia to identify the key factors that significantly impact the effective management of Construction and Demolition waste in the region. It was observed that merely 13.6% of C&D waste undergoes recycling and reusability annually, with the remaining 86.4% being disposed of in landfills. Another study conducted by (Tafesse 2021b) studied the minimization waste management technique in a construction site in Ethiopia. The findings highlighted that utilizing prefabricated or off-site manufacturing of elements, efficient on-site waste management, and implementing a strategy for minimizing material wastage are essential strategies to reduce construction material wastage.

While there is existing research on construction waste management in the UAE, much of it focuses on general trends, regulatory frameworks, and broad strategies rather than providing detailed insights into the effectiveness of current practices at the operational level. Additionally, there is limited empirical research that systematically evaluates the implementation of advanced waste recycling initiatives, and regulatory compliance measures in the context of the UAE's construction sector. Addressing this research gap is essential for informing evidence-based policies and interventions that can drive continuous improvement in construction waste management practices, leading to more sustainable outcomes and mitigating environmental impact in the UAE.

2.3 *Research Questions*

This research investigates the waste management practices in the construction sector in Dubai with the focus of answering these questions.

1. What are the causes behind the waste generation in the construction sites?
2. What are the effective waste management techniques implemented in the construction sites?
3. What are the challenges behind implementing sustainable waste management practices?

3 Research Methodology

For the purpose of this research, a qualitative research method will be conducted by the case study method and interviews, each selected for its particular strengths despite some minor shortcomings for each method. All the data will be analyzed by thematic analysis approach using NVIVO software.

3.1 Selected Case Studies

To explore the reasons behind material wastage in construction sites and evaluate the material wastage management techniques employed by contractors in the Dubai, site visits were made to four projects varying in location, size, type, and construction phase to inspect material waste handling practices and to gather insights from project managers regarding waste minimization strategies. Table 1 shows descriptions and details for the selected case studies.

During site visits, semi structured observations were conducted to systematically document waste management practices and procedures. The observations will focus on Types and quantities of waste generated, segregation methods, storage conditions, recycling initiatives, disposal practices, and compliance with waste management regulations.

Table 1 Selected case studies details and descriptions

	Project 1	Project 2	Project 3	Project 4
Site location	Al Barsha south—Dubai	International financial center—Dubai	Jumeirah village circle—Dubai	Muhaisnah—Dubai
Description of the project	G + 4P + 14 typical + Roof residential tower	High rise residential tower	G + M + 4P + 15 commercial and residential tower	Renovation for G + 1 residential villa
Contractor	Unlimited contracting company	Unlimited contracting company	Unlimited contracting company	Small size company
Stage of the project	Foundation	Structure	Finishings	Renovation
Methodology used	Interviews and site observation	Interviews and site observation	Interviews and site observation	Interviews and site observation

3.2 Interviews

In order to gather in-depth insights and perspectives from participants, semi structured interviews were conducted to allow for adaptability during the interview process. Table 2 summarizes the interviewees' general information.

4 Results and Findings

A thematic results analysis approach was done using NVIVO Software to systemically organize and analyze the qualitative data obtained from the interviews and observations to develop patterns and themes that answer the research questions. Under each research question different codes from the responses of interviews and observations were clustered together based on their similarities and conceptual connection. Then, six themes were developed after grouping and organizing the codes. Figure 2 shows the hierarchy of the analyzed data including generated themes and codes for each research question.

Theme 1: Waste Materials

The theme focuses on the type of waste generated in the construction sites. It represents various waste materials arises in depending on the stage of the construction in the site. The materials includes different structural materials like steel and concrete, some finishing materials, and some waste due to electrical and mechanical work. The highest waste weight contributed to wood, steel, and concrete and this can be related to the major use of these materials during different construction stages (Figs. 3 and 4).

Theme 2: Waste Source

This theme investigates the causes behind the waste generated in the construction sites. Waste sources can be contributed to many reasons including the changes in the design and some human errors especially during concrete pouring and fabricating

Table 2 Interviewees general information

	Position	Background	Experience	Company/authority
1	Project manager	Civil engineering	18 years	Contracting company (Project 1)
2	Quality assurance engineer	Civil engineering	11 years	Contracting company (Project 2)
3	Construction manager	Civil engineering	12 years	Contracting company (Project 3)
4	Site engineer	Civil engineering	4 years	Contracting company (Project 4)
5	Projects coordinating engineer	Civil engineering	10 years	Dubai municipality

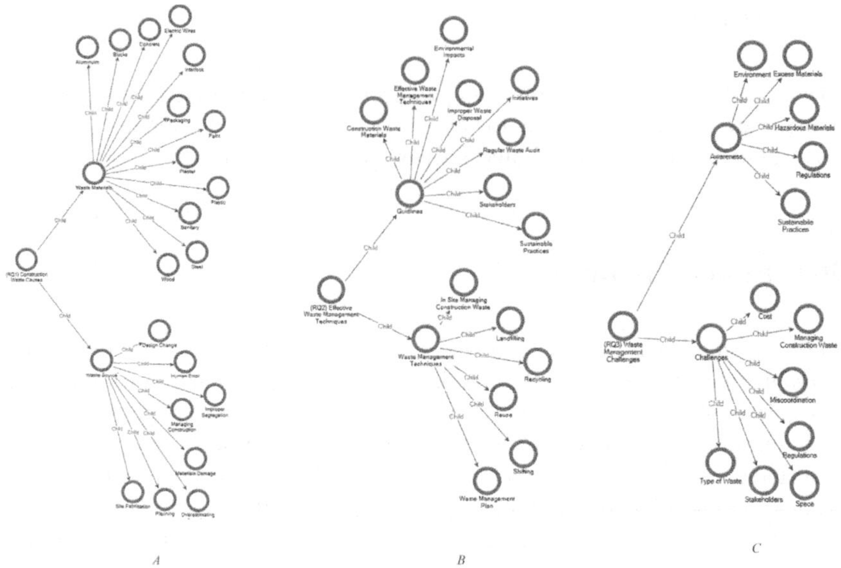

Fig. 2 The hierarchy of the analyzed data **a** developed themes and codes for RQ1 **b** developed themes and codes for RQ2 **c** developed themes and codes for RQ3

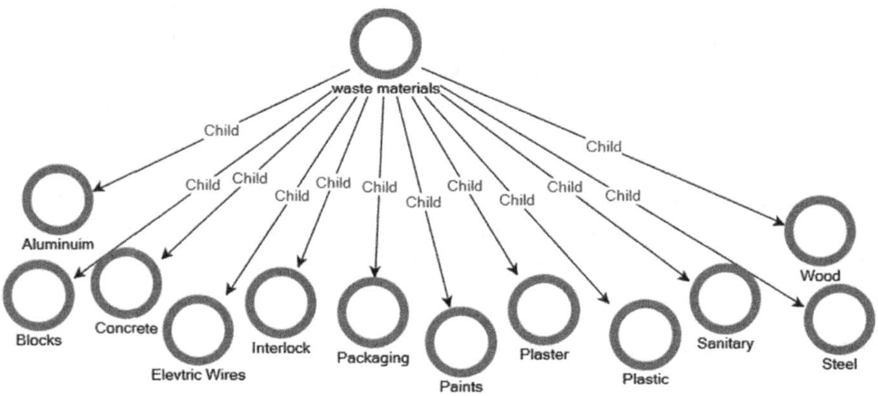

Fig. 3 Theme 1 codes

materials in the site. Also, waste source can be due to improper planning for the project and miscoordination between responsible project parties as shown in Figures 5 and 6.

Participant 1 stated that causes can be contributed to human errors and fabricating steel in the site. Participant 2 focused on design changes as the main cause for construction waste. Participant 3 agreed with participants 1 and 2 that it can be

Fig. 4 Theme 1 codes weights

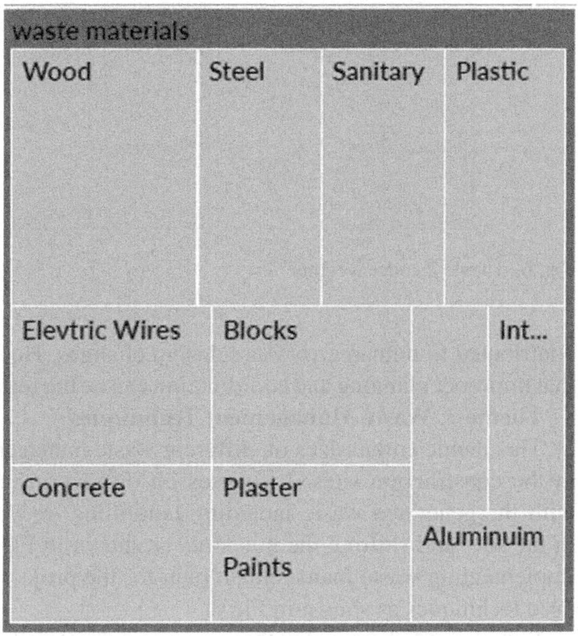

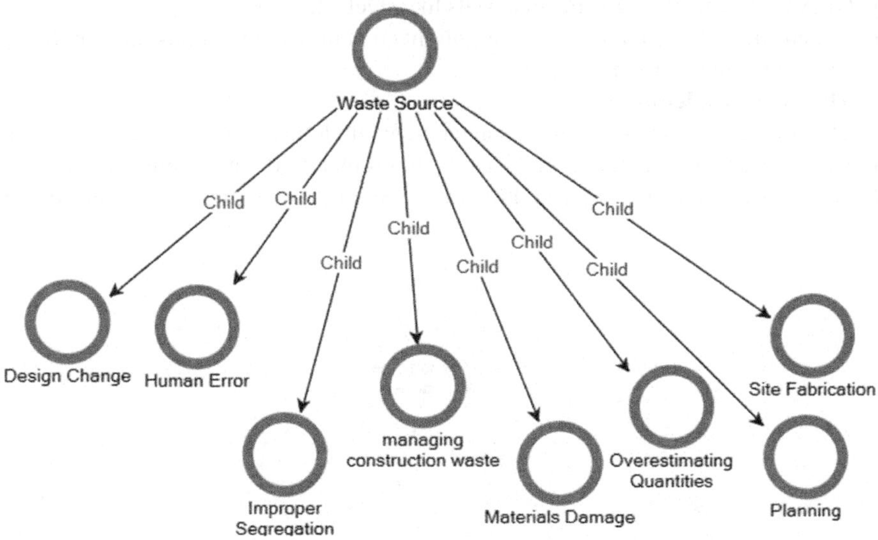

Fig. 5 Theme 2 codes

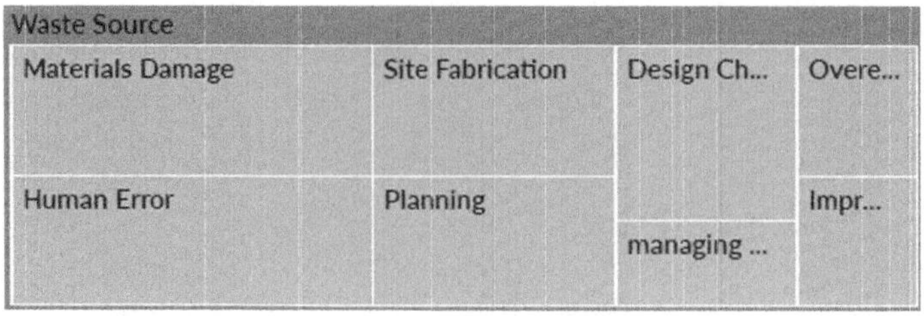

Fig. 6 Theme 2 codes weights

contributed to human errors and design changes. However, participant 4 highlights that improper planning and coordination can be the reason behind construction waste.

Theme 3: **Waste Management Techniques**

This theme emphasizes on different waste management techniques implemented in the construction sites. It focuses on discovering the effective methods to deal with the generated waste including landfilling, recycling, reuse, storing materials in the site, and shifting the materials as shown in Fig 7. The responds showed that implementing waste management plan for the project and landfilling were the most used techniques as shown in Fig 8.

Participant 1 showed that they reuse some materials. However, participant 2 stated that they rely on recycling some materials like steel while the other materials are sent to be landfilled. Participants 3 and 4 highlights that they use methods like landfilling and reusing some materials (Fig. 9).

Theme 4: **Guidelines**

This theme focuses on waste management guidelines and its effectiveness in managing waste management practices. It shows that guidelines include regulations about the environmental impacts, how to manage the waste, initiatives and regular

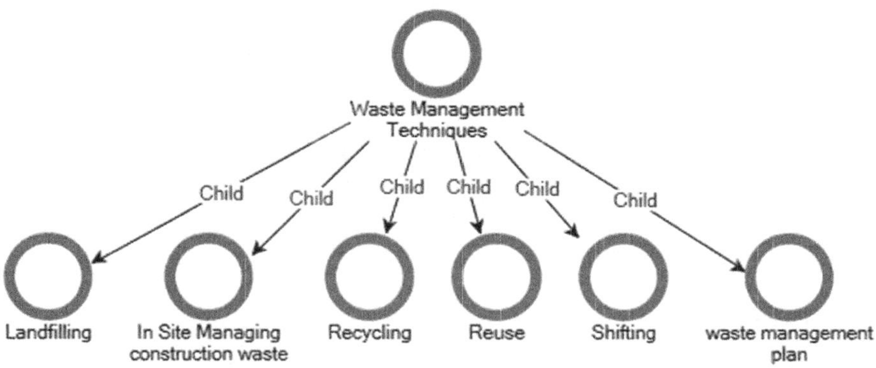

Fig. 7 Theme 3 codes

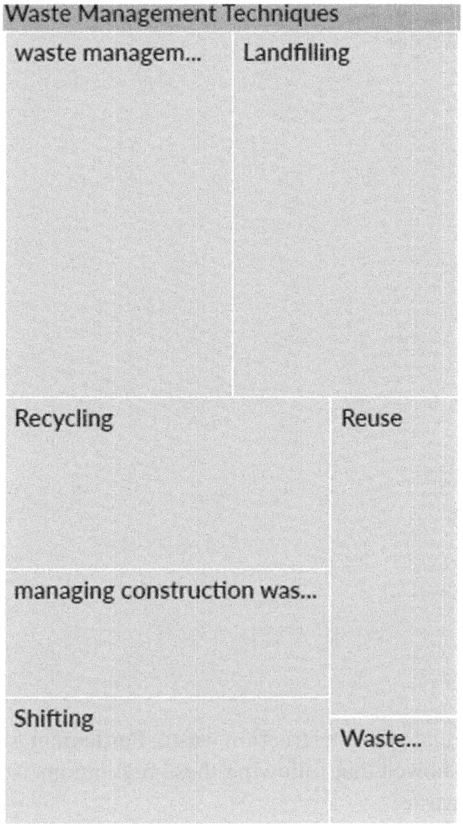

Fig. 8 Theme 3 codes weights

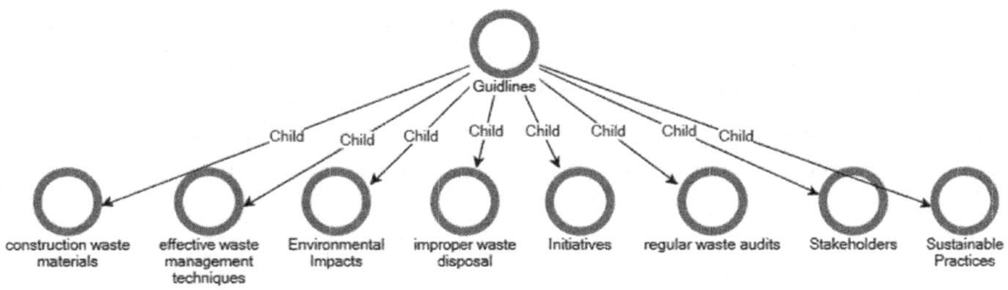

Fig. 9 Theme 4 Codes

audits, the sustainable practices needed. The results revealed that most used guidelines related to regular audits and effective waste management techniques as shown in Fig 10.

Participants 1, 2, and 3 agreed that they follow regulations related to regular waste audits, sustainable practices, environmental impacts, and effective waste management techniques. However, participant 4 mentioned that they focus on the initiatives

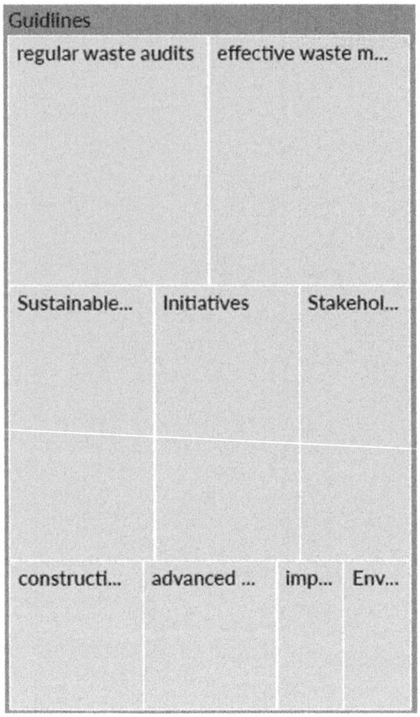

Fig. 10 Theme 4 Codes Weights

related to construction waste. Participant 5 includes all of the needed regulations and showed that following these regulations is very important in managing construction waste.

Theme 5: **Challenges**

This theme captures the challenges that construction site faces regarding managing the waste. This theme showed that most of the challenges faced were related to the cost, miscoordination, regulations, space in the site, stakeholders, the type of the waste and how it can be managed, and the weather conditions as shown in Fig. 11. Also, Fig. 12 shows that the highest contribution from the results was related to challenges in managing the construction waste and the high cost.

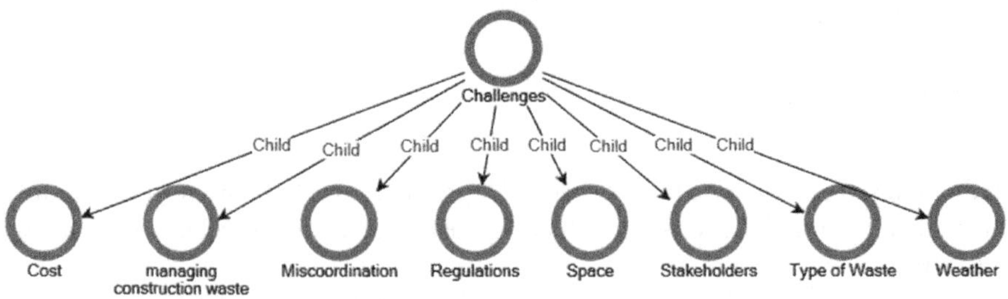

Fig. 11 Theme 5 codes

Fig. 12 Theme 5 codes weights

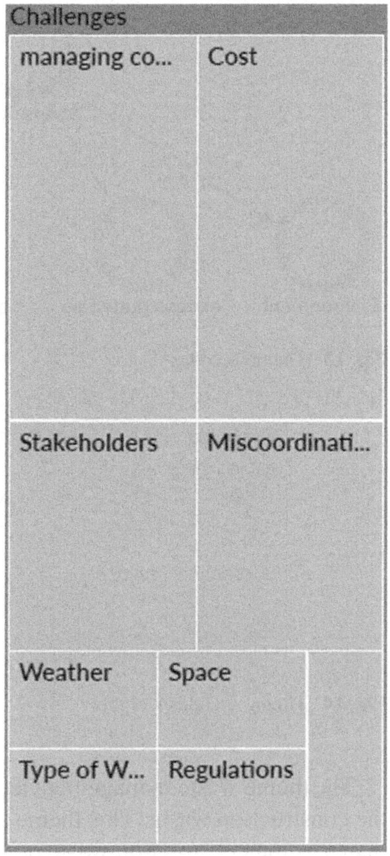

All the participants agreed that the cost is the main challenge they face. However, participant 3 highlights that the type of waste can be challenge as some materials cannot be reused or recycled and participant 4 added the weather as one of the challenges. Participant 5 mentioned that miscoordination between stakeholders, regulations, and following the regulations are the main challenges.

Theme 6: **Awareness**

The sixth theme focuses on the awareness of the waste management techniques as shown in Fig 13. It can be realized from Fig 14 that the biggest problem was the awareness about the dealing with excess materials and sustainable practices.

All the participants mentioned the need of the awareness about the sustainable practices and how to deal with excess materials. Participant 2 revealed the need of the awareness about hazardous materials also. However, participant 5 focused on the importance of all the project parts to be aware about the environment and waste regulations.

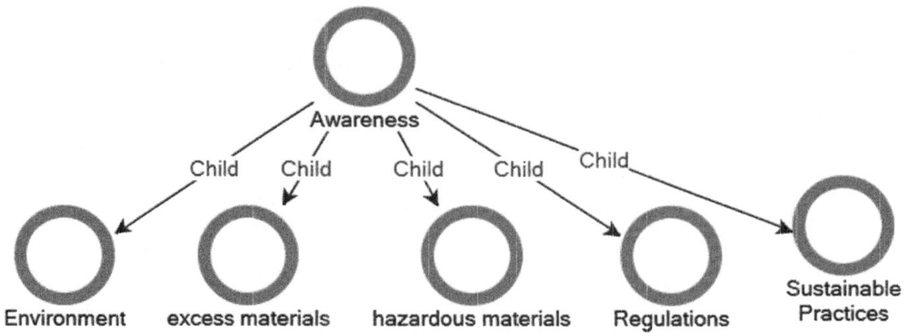

Fig. 13 Theme 6 codes

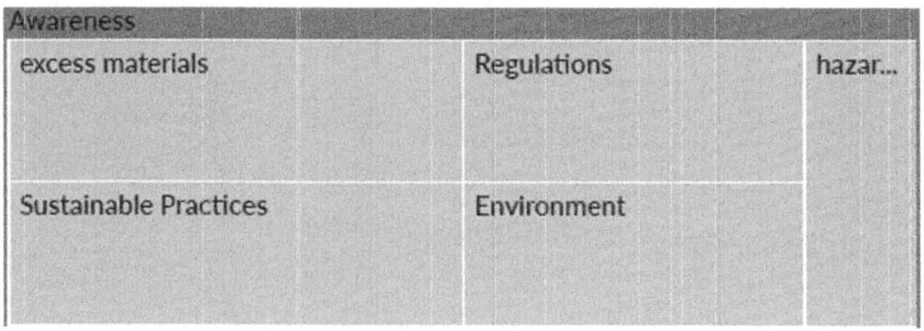

Fig. 14 Theme 6 codes weights

The theme waste management techniques explored the methods used to deal with the construction waste. This theme intersects with the guidelines themes in which it manages the construction waste practices and provide effective techniques.

Moreover, the theme challenges investigated the key factors that act like barriers in the construction sites. This theme aligns with the awareness theme as it shows some solutions to overcome these challenges.

In addition, the theme waste source which listed the causes behind materials waste intersects with the waste materials theme, waste management techniques theme, and the guidelines theme as it depend on the type of the material and how to manage the waste and follow the needed guidelines to ensure effective waste management methods.

This thematic analysis effectively answered the research questions. The first question about the causes of construction waste has been answered by the codes in waste sources theme and waste materials theme. The second question about effective waste management techniques has been answered through the waste management techniques theme and the guidelines theme that can be follow to help in managing construction waste effectively. Moreover, the third research question asking about the waste management challenges has been answered by the codes of the challenges and awareness themes which showed the main challenges faced during managing the waste and the lack of awareness of the effective practices.

5 Discussion

The research problem of this study focused on the waste management techniques implemented in the construction site in Dubai—UAE. The study investigated the causes behind materials waste, the waste management techniques implemented, and the challenges behind implementing effective waste management techniques by exploring the construction engineers experiences and perspectives.

This study reveals six key themes related to waste management techniques. The results obtained from the observations and interviews showed that main causes of waste can be contributed to changes in the design after construction which leads to excess materials in the site. Also, poor planning and coordination between the team can be another major reason. Moreover, it can be shown that although some materials are reused and recycled, most of the sites depends on the landfilling as waste management technique which is not favourable method due to its environmental impacts. The results revealed some challenges such as the high cost and lack of awareness of sustainable waste management practices between stakeholders.

The results obtained shows the importance of increasing the awareness between stakeholders about the effective and sustainable waste management techniques. Also, decreasing the cost of recycling and reusing materials could be a solution to encourage sustainable waste management practices.

The results aligned with the literature reviewed. The causes of construction waste and challenges obtained intersects with what was mentioned in the literature. This research provides deeper understanding about the construction waste techniques and its challenges.

Researching the causes and challenges associated with waste generation can help construction companies and communities stay compliant with these regulations. Also, by understanding the causes of waste generation and implementing effective waste management techniques, communities can significantly reduce their environmental footprint. This reduction in environmental impact can lead to cleaner air, water, and soil, thereby promoting overall environmental health.

In summary, the results from the discussion showed the comprehensive analysis of the research findings and how it can benefit the community and the construction stakeholders in shaping better waste management experience in construction sites.

6 Conclusions and Recommendations

6.1 Conclusions

This study evaluates the factors behind material wastage and the strategies for waste management applied within construction sites in the Dubai through a qualitative research method. It can be concluded from the results that the primary contributors to material wastage on construction selected sites are primarily attributed to workers' insufficient knowledge, inadequate design leading to excessive off-cuts, and the need for rework and modifications. Conversely, issues related to inefficient logistics management and procurement challenges were deemed less impactful.

The current practices implemented in the sites to manage the construction waste are landfilling, reusing some materials like wood and steel, and waste segregation in the site. However, recycling materials was not implemented in all of the sites. Moreover, it can be concluded that the cost and the changes in the project design can be the main challenges affecting construction waste minimization.

It is very important to increase the awareness of all the stakeholders involved in the construction management process and develop a clear and accurate plan from the beginning of the project which can minimize the construction waste. Moreover, incorporating advanced BIM techniques can be very effective in predicting the conflicts which can affect the amount of the waste.

6.2 Recommendations

A more detailed examination of the efficiency of the waste reduction strategies that have been put in place as indicated by the findings of this study is recommended to investigate the effectiveness in applying waste management strategies.

References

Al-Hajj A, Hamani K (2011) Material waste in the UAE construction industry: main causes and minimization practices. Arch Eng des Manag 7(4):221–235. https://doi.org/10.1080/17452007.2011.594576

Alhawamdeh M, Lee A(2021) Construction waste minimization: a narrative review. Int J Environ Sustain 18(1):1–33. https://doi.org/10.18848/2325-1077/CGP/V18I01/1-33

Hittini BY (n.d.) Construction waste management practices in Abu Dhabi. Investigation and evaluation. https://scholarworks.uaeu.ac.ae/architectural_theses/4

Janani R, Kaveri V 2021() A critical literature review on reuse and recycling of construction waste in construction industry. Mater Today Proc 37(Part 2): 3077–3081. https://doi.org/10.1016/J.MATPR.2020.09.015

Kazerooni SMA, Abdullah A, Sajoudi MN, Bin Mustaffa Kamal MF, Torshizi F, Taherkhani R (2012) Reduce, reuse, recycle and recovery in sustainable construction waste management. Adv Mater Res 446–449, 937–944. https://doi.org/10.4028/WWW.SCIENTIFIC.NET/AMR.446-449.937

Ntitanguranwa Paul J, Kabano J, Gasingwa N (2018) Reuse of construction and demolished concrete waste by producing affordable high strength concrete block. Rwanda J Eng Sci Technol Environ 1(1). https://doi.org/10.4314/rjeste.v1i1.14s

Ouda OKM, Peterson HP, Rehan M, Sadef Y, Alghazo JM, Nizami AS (2018) A case study of sustainable construction waste management in Saudi Arabia. Waste Biomass Valorization 9(12):2541–2555. https://doi.org/10.1007/S12649-017-0174-9/METRICS

Spišáková M, Mésároš P, Mandičák T (2021) Construction waste audit in the framework of sustainable waste management in construction projects—case study. Buildings 11(2):61. https://doi.org/10.3390/BUILDINGS11020061

Sweis G, Thneibat M, Hiyassat M, Abu-Khader W (2021) Understanding the causes of material wastage in the construction industry. https://www.researchgate.net/publication/350605048

Tafesse S (2021) Material waste minimization techniques in building construction projects. Ethiop. J. Sci. Technol. 14(1):1–19. https://doi.org/10.4314/EJST.V14I1.1

Umar UA, Shafiq N, Ahmad FA (2021) A case study on the effective implementation of the reuse and recycling of construction & demolition waste management practices in Malaysia. Ain Shams Eng. J. 12(1):283–291. https://doi.org/10.1016/J.ASEJ.2020.07.005

Wu H, Zuo J, Yuan H, Zillante G, Wang J (2023) Investigation of the social and economic impacts of cross-regional mobility of construction and demolition waste in Australia. Resour Conserv Recycl 190:106814. https://doi.org/10.1016/J.RESCONREC.2022.106814

Yahya MM (2015) Sustainable solutions for construction waste reduction in UAE. IGAD about studying sustainable solutions to reduce construction waste in the United Arab Emirates

Zamri GB, Azizal NKA, Nakamura S, Okada K, Nordin NH, Othman N, Akhir FNM, Sobian A, Kaida N, Hara H (2020) Delivery, impact and approach of household food waste reduction campaigns. J Clean Prod 246:118969. https://doi.org/10.1016/J.JCLEPRO.2019.118969

The manufacturer's authorised representative in the EU is Springer
Nature Customer Service Centre GmbH, Europaplatz 3, 69115 Heidelberg,
Germany. If you have any concerns regarding our products, please
contact ProductSafety@springernature.com

Printed and bound by CPI Group (UK) Ltd, Croydon, CR0 4YY
30/09/2026
02225558-0020